Oscillations and Waves

Springer
Berlin
Heidelberg
New York
Barcelona
Budapest
Hong Kong
London
Milan
Paris
Santa Clara
Singapore
Tokyo

Fritz K. Kneubühl

Oscillations and Waves

With 146 Figures and 16 Tables

Springer

Professor Dr. Fritz K. Kneubühl
Institute of Quantum Electronics
Physics Department
Swiss Federal Institute of Technology (ETH)
CH-8093 Zürich
Switzerland

Library of Congress Cataloging-in-Publication Data applied for.

Kneubühl, Fritz Kurt:
Oscillations and waves: with 16 tables / Fritz K. Kneubühl. – Berlin; Heidelberg; New York; Barcelona; Budapest; Hong Kong; London; Milan; Paris; Santa Clara; Singapore; Tokyo: Springer, 1997
ISBN 3-540-62001-X

ISBN 3-540-62001-x Springer-Verlag Berlin Heidelberg New York

Printed in Germany

Typesetting: Typesetting: Camera-ready copy from the author
Cover design: *design & production* GmbH, Heidelberg

SPIN 10534409 57/3144 – 5 4 3 2 1 0 – Printed on acid-free paper

To my wife

Waltraud Dorothea

Preface

In the course of over thirty years of research in various fields of physics and teaching experimental physics to undergraduate and graduate students of physics, mathematics, electrical engineering, chemistry and natural sciences I missed an introductory comprehensive book on the mathematics of linear and nonlinear oscillations and waves from the point of view of physicists and engineers. Oscillations and waves are the playground for all kinds of scientists in spite of the fact that they represent essentially mathematical concepts. In this field, however, the interests of experimentalists and engineers, on one side, and mathematicians, on the other side, usually differ. The latter are most interested and engaged in proofs of general fundamental laws on the existence and behavior of the solutions of basic differential equations and on the convergence of their approximations, whereas the former need explicit analytical and numerical solutions or approximations, computer programs and graphic displays. In the past decades a gap opened between these two groups with the result that they have difficulties in communicating with each other. This comprehensive book represents a novel attempt to bridge this gap.

This book is based on my lecture notes and its predecessor "Lineare und nichtlineare Schwingungen und Wellen" published by B.G. Teubner, Stuttgart, FRG, in 1995. The contents of the previous book have been considerably extended, revised and improved thanks to advice from colleagues and co-workers. Additions to be mentioned are the first classification of two-dimensional autonomous, i.e. time-invariant quadratic systems, the complex representation of ideal flows, the waves on telegraph lines, the polarization of electromagnetic waves, the response theory of electromagnetic waves in linear causal media, standing waves on membranes and in acoustic Helmholtz resonators and, last but not least, an appendix with tables of Fourier series and transforms, convolutions and Laplace and Hilbert transforms. Furthermore, effort has been made to use the same terminology throughout all chapters and topics in the book.
According to my experience, university teaching on oscillations and waves at undergraduate and graduate levels is often focused on linear phenomena with some rudimentary comments on nonlinearity, although nonlinear phenomena dominate in nature and in engineering. Therefore I hope that this book gives some impetus to improve and extend university teaching on this relevant topic.

Here I take pleasure in acknowledging the valued advice and help from many friends with whom I have discussed various parts of this volume; in particular Prof. L. Jansen, University of Amsterdam, NL; Prof. G.D. Boreman, University of Central

Florida, Orlando, USA; Profs. J. Bilgram, J. Fröhlich, B. Keller, H. Melchior, G. Mislin, H.J. Schötzau, H.Ch. Siegmann, M.W. Sigrist, Drs. F. Kärtner, E. Magyari and R. Monnier, ETH Zürich. I am also very indebted to my assistants Dr. D.P. Scherrer and M. Baumgartner for special calculations, computer design of figures, proof reading and valuable comments.

The completion of this work was greatly facilitated by continuous support from Springer-Verlag, Heidelberg, Germany, and B.G. Teubner, Stuttgart, Germany, who gave permission to include in this book many figures from its predecessor.

Furthermore, I wish to record my special gratitude to Ms. I.-M. Hausner for her expert assistance and perseverance in typing the manuscript. I am also grateful to Ms. I. Wiederkehr and J.P. Stucki, Physics Department ETH Zürich, for preparing figures and to Dr. U. Helg and Ms. M. Papadellis, Physics Library ETH Zürich, for their support.

Finally, I owe very special thanks to my wife Waltraud and my children for their cheeriness and their forbearance during the writing process.

Zurich, June 20, 1997 Fritz K. Kneubühl

Contents

1. Introduction

This book addresses students after two years of elementary college mathematics and physics courses, graduate students, university teachers, scientists active in research and engineers involved in advanced development. It constitutes a comprehensive introduction to the various mathematical aspects of oscillations and waves from the point of view of physicists and engineers. Oscillations and waves are basic phenomena in nature as well as in engineering. In order to comply with the requests of physicists and engineers the emphasis of this book is on physical characterization, basic equations, analytical solutions, graphic displays and tables as well as on general laws that govern the wealth of phenomena. As a consequence it also serves as a basis of digital computations dealing with oscillations and waves. Theorems and proofs either lengthy or not particularly illuminating the subject at hand have been omitted in favor of references to the literature. This is compensated by the presentation of many new results and examples.

1.1 Topics

The book covers the relevant topics related to oscillations and waves. It starts in Chapter 2 with the description of *free linear and nonlinear oscillators* defined by basic differential equations. Discussed are the linear harmonic, parametric, continuously modulated, pulse and step modulated oscillators on one hand, and the simple pendulum, the nonlinear Liénard, Duffing and van der Pol oscillators on the other hand. Also considered are linear oscillators with switches because they represent an intermediate between the linear and the nonlinear oscillators.

Linear oscillators can oscillate with arbitrary amplitudes on the contrary to most nonlinear oscillators whose amplitudes of oscillation are determined by the nonlinearity. The amplitudes of a linear oscillation can be influenced by damping, amplification and feedback. The motion of a linear oscillator is usually a solution of a linear differential equation of second order. In the case of a modulated linear oscillator the coefficients of this equation vary with time. Then the solution and its approximations are governed by occurrence and type of the singularities of these coefficients. Periodically modulated oscillators, which are usually called parametric oscillators, exhibit allowed frequency ranges with stationary oscillations, and forbidden frequency ranges or gaps with parametric damping and amplification.

Free *nonlinear oscillations* are characterized by well-defined amplitudes as well as by the occurrence of higher harmonics. The description of these motions as solutions of nonlinear differential equations often requires sophisticated approximations. A standard example are the motions of the van der Pol oscillator which are used to test new methods of approximation.

Chapter 3 is dedicated to *forced oscillations*. These represent the response of oscillators to external excitations. Taken into account are pulse and step excitations as well as harmonic, periodic and broad-band excitations. They require the application of the Fourier and the Laplace transformation. Typical phenomena related to forced oscillations are resonances and transients. Nonlinear forced oscillators exhibit also bistability, instabilities and even chaos.

Oscillations of systems of differential equations are presented in Chapter 4. These systems are relevant in many fields of science and engineering, e.g. in classical mechanics, fluid dynamics and biology. Two- and three-dimensional systems of explicit differential equations of first order describe velocity fields of flows in fluids. Therefore, these systems are characterized in terms of *fluid dynamics*. They are supplemented by concepts of classical mechanics, i.e. the potential, the Hamiltonian and the Lyapunov function. Discussed in detail are the two-dimensional autonomous, i.e. time-invariant systems which correspond to the two-dimensional stationary flows of fluid dynamics, e.g. irrotational and gradient flows, flows without sources or sinks, Hamiltonian flows, and the ideal flows with their complex representations. Special attention is paid to the stable and instable singular, i.e. stationary points of linear autonomous systems and their close environments in the form of nodes, saddles, vortices and spirals. In addition, this book offers the first classification of the singular points of the quadratic autonomous systems according to their various environments. These systems result from the second order approximations of two-dimensional autonomous systems at singular points. Further topics of Chapter 4 are the limit cycles of nonlinear systems, the stability criteria of Lyapunov, the Lotka-Volterra system, conservative linear mechanical systems and time-dependent systems. This chapter ends with an extensive description of the three-dimensional systems and flows.

Chapter 5 is dedicated to the response theory of *linear transfer systems* and *delay systems* in resonant oscillating circuits and feedback-control systems. These systems are of importance in electrical engineering and cybernetics. The response of a linear transfer system corresponds to the convolution of the input with the characteristic impulse response function of the system. The responses to periodic, transient and random inputs are relevant. Random inputs and responses are described with *statistical measures*, e.g. average, square mean, crosscorrelation, autocorrelation and power spectrum. Autocorrelation and power spectrum are linked by the Wiener-Khintchine theorem.

Even small variations of one or more characteristic parameters of a system can cause instabilities, structural changes or catastrophes, and even *chaos*. These scenarios are discussed quantitatively in Chapter 6. It is demonstrated that structural changes occur in various forms of *bifurcations*, e.g. saddle-node, transcritical, pitchfork or Hopf, and that chaos can be generated even by purely deterministic systems, e.g. by the *Lorenz model* and the *logistic map*. Of general interest are the various *routes to chaos* starting from stationary oscillations or motions, e.g. via intermittency, a finite or infinite number of bifurcations.

Chapters 7 to 9 are devoted to the discussion of *waves*. The principle objective of these chapters is to develop an understanding of the basic wave concepts and their relations with one another. Therefore, they are organized in terms of these concepts rather than in terms of observable phenomena such as sound and light.

The *linear waves* are topic of Chapter 7. It starts with the general classification of waves and continues with the description of harmonic waves, wave velocities and dispersion, plane waves in homogeneous isotropic media and electromagnetic waves in linear causal media which are characterized by the Kramers-Kronig relation as well as waves in periodic media and structures. In this context it elucidates the concepts of dispersion relation, phase and group velocity, carrier wave and envelope. It covers scalar, longitudinal and transversal waves, polarization, allowed and forbidden frequency ranges of periodic media and structures, Bloch waves etc., and explains the solutions of a variety of linear wave equations, e.g. Hertz, reduced Hertz, linear Klein-Gordon, telegraph, diffusion, linearized Korteweg-de Vries, Schrödinger and Hill.

Chapter 8 presents the basic equations of *nonlinear waves* and their solutions in the form of shock waves, *solitons* and cnoidal waves which involve Jacobian integrals and functions. It begins with the demonstration of the effect of nonlinearity and then illustrates the wealth of phenomena occurring in waves governed by the nonlinear wave equations of diffusion, Korteweg-de Vries, Klein-Gordon, Schrödinger Maxwell-Bloch. Finally, it commentates the waves of the Toda chain as an example of waves in a nonlinear periodic structure.

Standing waves are described in the final Chapter 9. A standing wave represents an oscillation of an extended medium with amplitude and phase depending on the location. It is determined by the wave equation as well as by the boundary conditions. These constitute an eigenvalue problem whose solutions yield the possible frequencies of oscillation as eigenvalues. The topics of Chapter 9 include free and forced standing waves on a string, free standing waves on membranes and acoustic standing waves in cavities, Sturm-Liouville wave patterns, forced standing waves on telegraph lines and, finally, free and forced nonlinear standing waves rarely discussed in the literature.

Wave mechanics is not a main topic of this book. Nevertheless it presents its basic concepts, its differential operators (Section 7.4.7) and the linear (Section 7.4.7) as well as the nonlinear (Section 8.6) Schrödinger equation. In addition it includes a discussion of the wave solutions of the Schrödinger equations of the harmonic oscillator (Sections 9.5.1&2) and of a particle in a periodic potential (Section 7.6.4). The harmonic oscillator, which is fundamental in quantum mechanics, represents a simple model of molecular vibrations, while the motion of a particle in a periodic potential corresponds to that of an electron in a semiconductor or metal.

1.2 Helpful Notes

For *mechanical and electromagnetic units* this book uses the MKSC system [Alonso & Finn 1967 B] that has been officially approved for scientific work by the US National Bureau of Standards and IUPAP. This system, which is convenient in experimental physics and engineering, is also called rationalized MKS system [Crawford 1968 B], "Système international" or SI [Alonso & Finn 1970 B, Kneubühl 1994 B].

The comprehensive description of oscillations and waves involves a variety of mathematical functions and transformations. The most relevant are listed in the tables of the *Appendix*. They include e.g. Fourier series, the Dirac-δ and the Heavyside step function, convolutions, and the transformations of Fourier, Laplace and Hilbert.

References are quoted by author(s) and year of publication. Furthermore a distinction is made between books (B) and publications in journals (J) throughout the text as well as in the reference list.

2. Free Oscillations

This chapter starts with a general survey of oscillations and proceeds with the detailed description of free well-known linear and nonlinear oscillators. Linear and nonlinear oscillators are discussed separately because of their different behavior. Also considered are relay oscillators since they constitute an intermediate between these two types of oscillators. They act as linear oscillators except for momentary switching. Furthermore, special emphasis is put on linear parametric and classical nonlinear oscillators such as the van der Pol oscillator.

2.1 Survey

An *oscillation* designates an exactly or almost periodic motion or change of state of a physical, chemical, or biological system. A system with the capability to oscillate is called *oscillator* [Andronov et al. 1966 B, Magnus & Popp 1997 B, Pain 1993 B]. *Free* oscillations of a system are those which occur without an external force or other type of excitation, while *forced* oscillations are excited by an external force. Oscillations may be *damped* or *amplified.* Damping and amplification can be varied by *feedback.*

Mathematically, an oscillator is defined by one or several basic characteristic equations that originate in physics, chemistry, biology, economics, sociology, etc. In most cases they represent *ordinary differential equations.* However, other types of equations can be involved, such as *delay equations*. The oscillator and its oscillations are *linear* if the basic equation or the basic system of equations is linear. A *nonlinear* oscillator is characterized either by a nonlinear basic equation or by a system of coupled equations, where at least one is nonlinear. In general, the oscillations of these nonlinear oscillators are considerably more complicated than those of the linear oscillators.

The different behavior of linear and nonlinear oscillators is demonstrated by comparison of the various oscillations of linear harmonic, aperiodically, periodically modulated and *parametrically excited* oscillators with those of nonlinear *relay* oscillators as well as classical nonlinear *Liénard* oscillators such as the simple *gravitational pendulum*, the *Duffing* oscillator, and the *van der Pol* oscillator.

The free oscillations of oscillators are described in this chapter while forced oscillations are discussed in Chapter 3. Oscillations of systems defined by two or more coupled ordinary differential equations are the topic of Chapter 4. Chapter 5 is devoted to oscillations of *transfer systems.*

2.2 Harmonic Oscillators

Harmonic oscillators constitute the simplest type of oscillator because they are linear as well as time-invariant. They can be separated into three classes: undamped, damped and amplified oscillators.

2.2.1 Undamped Harmonic Oscillators

A free harmonic oscillator without damping or amplification is characterized by the basic second-order linear differential equation

$$\ddot{x}+\Omega^2 x=0 \quad \text{or} \quad \frac{d^2}{dt^2}x+\Omega^2 x=0 \quad . \tag{2.2 - 1}$$

The parameter Ω [s^{-1}] > 0 indicates the *characteristic circular frequency* of the harmonic oscillator.

The undamped harmonic oscillator (2.2 - 1) represents the simplest basic type of oscillator. In electronics it is realized by a *LC* circuit without resistance, while in mechanics it corresponds to the undamped linear spring-mass system.

a) Electric *LC* circuit

Linear electric circuits constitute basic elements of electrical engineering. The simplest passive linear electric circuit is the *LC* circuit without resistance shown in Fig. 2.2 - 1.

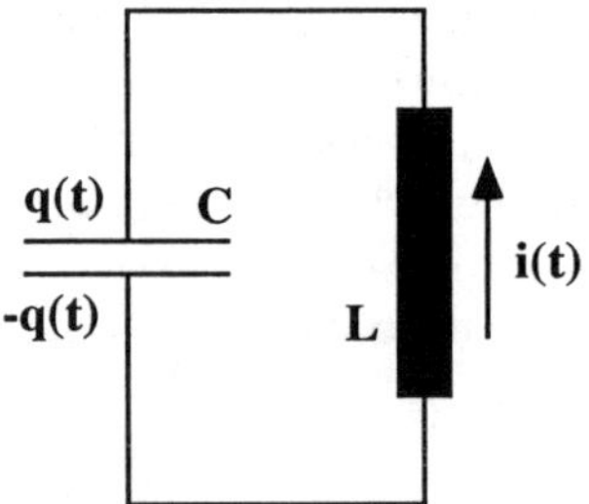

Fig. 2.2 - 1. *LC* circuit without resistance in equilibrium for $q = 0$

This circuit consists of a *capacitor* with capacitance C [farad = F = A s V^{-1}] and a *coil* with self-inductance L [henry = H = V s A^{-1}]. Because it represents a closed circuit, the sum of the potential differences on the two elements vanishes according to *Kirchhoff's law.* This yields the relation

$$L\ddot{q} + C^{-1}q = 0 \quad , \tag{2.2 - 2a}$$

that can be written in the form of (2.2 - 1)

$$\ddot{x} + \Omega^2 x = 0 \quad \text{with} \quad x = q \quad \text{and} \quad \Omega^2 = 1 / LC \quad . \tag{2.2 - 2b}$$

Accordingly, the *LC* circuit represents an undamped harmonic oscillator.

b) Undamped spring-mass system

The linear spring-mass system illustrated in Fig. 2.2 - 2 consists of a constant mass μ [kg] which is attached to a spring suspended at the point *A* of the support. This spring exerts a linear restoring force F_r [N = kg m s^{-2}] on the mass μ which is proportional to the vertical displacement x of the mass μ from its equilibrium position as follows

$$F_r = -f\,x \quad \text{with} \quad f > 0 \quad . \tag{2.2 - 3}$$

The parameter f [Nm^{-1}] represents the force constant. The application of *Newton's second law* of mechanics yields the following equation of motion

$$\mu\,\ddot{x} = F_r = -fx \quad . \tag{2.2 - 4a}$$

It is equivalent to the basic equation (2.2 - 1) of the undamped harmonic oscillator because it can be transformed into

$$\ddot{x} + \Omega^2 x = 0 \quad \text{with} \quad \Omega^2 = f / \mu \quad . \tag{2.2 - 4b}$$

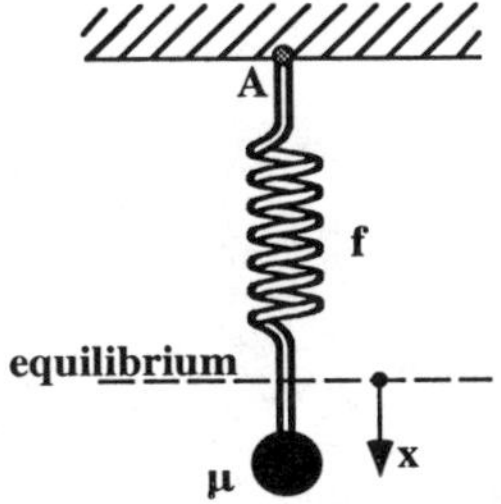

Fig. 2.2 - 2. Spring-mass system in equilibrium for $x = 0$

c) Simple undamped pendulum

For many nonharmonic and nonlinear oscillators the basic equation (2.2 - 1) of the undamped harmonic oscillator represents *a first approximation* of the equation of motion for weak oscillations with small amplitudes. This is well elucidated by the simple undamped pendulum or *mathematical pendulum* shown in Fig. 2.2 - 3. It consists of a point mass μ [kg] suspended at a fixed point *A* of a rigid support by a stiff bar of length a and negligible mass. In stable equilibrium the pendulum is oriented vertically with the point mass μ at its lower end because of gravity. When the pendulum swings, the point mass μ moves on the circle of radius a with the center in A. The momentary deviation of the orientation of the pendulum versus the vertical is described by the angle α. This angle represents the time-dependent variable of the pendulum.

The equation of motion of the simple undamped pendulum can be derived from the law of mechanics which demands that the time derivate of the angular momentum L equals the applied torque T. With respect to the point A of suspension the angular momentum L and the torque T of the pendulum are

$$L = \mu a^2 \dot{\alpha} \quad \text{and} \quad T = -ga \sin \alpha \tag{2.2 - 5}$$

where $g \approx 9.81$ m s^{-2} indicates the acceleration of free fall due to gravity. As a consequence one finds the following equation of motion

$$\dot{L} = \mu a^2 \ddot{\alpha} = T = -ga \sin \alpha \tag{2.2 - 6a}$$

or, after transformation,

$$\ddot{x} + \Omega^2 \sin x = 0 \quad \text{with} \quad x = \alpha \quad \text{and} \quad \Omega^2 = g / a \quad . \tag{2.2 - 6b}$$

This equation of motion is *nonlinear*. It will be discussed in detail in Section 2.5.3.

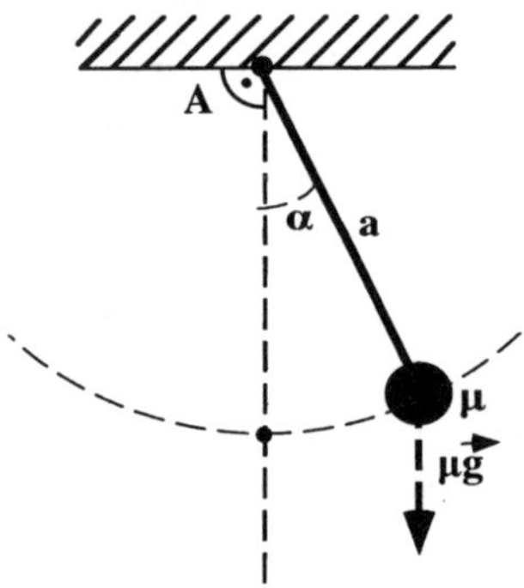

Fig. 2.2 - 3. Simple pendulum

For weak oscillations with small angles $\alpha = x$ one can assume

$$|x| << 1 \quad \text{and} \quad \sin x \approx x \tag{2.2 - 7a}$$

where $|x|$ indicates the modulus of x. Therefore, the *first approximation* of the equation of motion of the simple undamped pendulum for weak oscillations corresponds to the basic equation (2.2 - 1) of the undamped harmonic oscillator:

$$\ddot{x} + \Omega^2 x \approx 0 \quad \text{with} \quad \Omega^2 = g / a \quad . \tag{2.2 - 7b}$$

2.2.2 Harmonic Oscillations

There exists a variety of representations of the periodic motion of an undamped harmonic oscillator. These include real and complex harmonic functions, propagator matrices and phase diagrams. All of these are significant in the general theory on oscillators.

a) General laws on linear oscillations

The basic equation (2.2 - 1) of undamped harmonic oscillators represents a homogeneous linear differential equation of second order. The harmonic oscillations $x(t)$ as its solutions obey therefore the following three general laws of linear oscillations.

α) *Linear superposition*

The superposition $x_{1+2}(t)$ of two solutions $x_1(t)$ and $x_2(t)$ equals their sum

$$x_{1+2}(t) = x_1(t) + x_2(t) \quad . \tag{2.2 - 8}$$

β) *Linear independence*

There exist two linearly independent solutions $x_a(t)$ and $x_b(t)$. In this statement, linearly independent means that $x_b(t)$ is not a multiple of $x_a(t)$

$$x_b(t) \neq C\, x_a(t) \quad \text{with} \quad C = const \quad . \tag{2.2 - 9}$$

γ) *Form of the general solution*

The general solution $x(t)$ of (2.2 - 1) constitutes a linear combination of two linearly independent solutions $x_a(t)$ and $x_b(t)$

$$x(t) = a\, x_a(t) + b\, x_b(t) \quad . \tag{2.2 - 10}$$

The parameters a and b can take arbitrary values. The two linearly independent solutions $x_a(t)$ and $x_b(t)$ form a *basis* of the general solution.

b) The period

The solutions $x(t)$ of (2.2 - 1) represent *periodic* harmonic oscillations. Their periodicity requires

$$x(t+T) = x(t) \quad . \tag{2.2 - 11a}$$

The *period* T [s] in this equation is determined by the characteristic circular frequency Ω of (2.2 - 1)

$$T = 2\pi / \Omega \quad . \tag{2.2 - 11b}$$

The inverse of the period T equals the *frequency* ν [hertz = Hz = s^{-1}]

$$\nu = 1/T \quad . \tag{2.2 - 11c}$$

c) Real representations

Harmonic oscillations $x(t)$ as real solutions of (2.2 - 1) can be represented either as

α) *linear combinations of real circular functions*

$$x(t) = C_1 \cos \Omega t + C_2 \sin \Omega t \tag{2.2 - 12}$$

where C_1 and C_2 are arbitrary real constants, or as

β) *real circular functions with amplitude and phase*

$$x(t) = A \cos(\Omega t - \varphi) \quad \text{with} \quad A > 0 \quad . \tag{2.2 - 13}$$

The parameter A indicates an arbitrary constant positive amplitude and φ an arbitrary constant real phase. This type of oscillation is illustrated in Fig. 2.2 - 4.

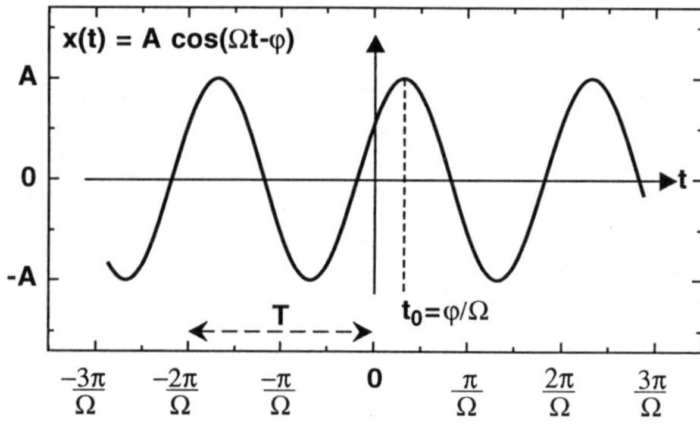

Fig. 2.2 - 4. Harmonic oscillation

The parameters of (2.2 - 12) and (2.2 - 13) are related by the equations

$$A^2 = C_1^2 + C_2^2 \quad \text{and} \quad \tan \varphi = C_2 / C_1 \quad . \tag{2.2 - 14}$$

d) Complex representations

In physics and engineering oscillations are often represented in the complex plane. Consequently, the harmonic oscillations as complex solutions of (2.2 - 1) of the undamped harmonic oscillator are of special interest.

The general complex representation of a harmonic oscillation has the form

$$x(t) = C_1 \exp(-i\Omega t) + C_2 \exp(+i\Omega t) \tag{2.2 - 15a}$$

with

$$C_1 = A_1 \, exp(+i\varphi_1) \quad \text{with} \quad \text{real } A_1 \geq 0 \quad \text{and} \quad \varphi_1 \quad ,$$
$$C_2 = A_2 \, exp(-i\varphi_2) \quad \text{with} \quad \text{real } A_2 \geq 0 \quad \text{and} \quad \varphi_2 \quad . \qquad (2.2 - 15b)$$

In the complex representation (2.2 - 15a) a *real harmonic oscillation* $x(t)$ fulfills the relations

$$x(t) = x^*(t) \quad , \qquad (2.2 - 16a)$$

$$C_1 = A \, exp(+i\varphi) \quad \text{and} \quad C_2 = A \, exp(-i\varphi) = C_1^* \qquad (2.2 - 16b)$$

where the asterisk * indicates the complex conjugate. These relations permit different complex representations of real harmonic oscillations $x(t)$:

In *physics and engineering* real harmonic oscillations are usually represented by

$$x(t) = x^*(t) = Re\left\{C_1 \, exp(-i\Omega t)\right\} = A \, cos(\Omega t - \varphi) \qquad (2.2 - 17)$$

where *Re* indicates the real part.

In modern *optics and quantum electronics*, however, real harmonic oscillations are written as

$$x(t) = x^*(t) = C_1 \, exp(-i\Omega t) + c.c. = 2A \, cos(\Omega t - \varphi) \qquad (2.2 - 18)$$

where *c.c.* is the abbreviation for complex conjugate.

e) Propagator matrices

The state of the simple harmonic oscillator at any time t is completely defined by the momentary values of the variable $x(t)$ and its time derivative $\dot{x}(t)$ because (2.2 - 1) is a linear second-order differential equation. The states at an initial time t_0 and at a later time t are related by a propagator matrix $\boldsymbol{P}(t_0, t)$ as follows

$$\begin{pmatrix} x(t) \\ \dot{x}(t) \end{pmatrix} = \boldsymbol{P}(t_0, t) \begin{pmatrix} x(t_0) \\ \dot{x}(t_0) \end{pmatrix} \quad . \qquad (2.2 - 19a)$$

In general propagator matrices fulfill the following law of multiplication

$$\boldsymbol{P}(t_0, t_2) = \boldsymbol{P}(t_1, t_2) \, \boldsymbol{P}(t_0, t_1) \quad . \qquad (2.2 - 19b)$$

The propagator matrix of undamped real or complex harmonic oscillations has the form

$$\boldsymbol{P}(t_0, t) = \boldsymbol{P}(t - t_0) = \boldsymbol{P}(\Delta t) = \begin{pmatrix} cos \, \Omega \Delta t & \frac{1}{\Omega} sin \, \Omega \Delta t \\ -\Omega sin \, \Omega \Delta t & cos \, \Omega \Delta t \end{pmatrix} \qquad (2.2 - 20a)$$

$$\text{with} \quad \Delta t = t - t_0 \quad .$$

This matrix is *unimodular* because its determinant equals unity

$$det\, \boldsymbol{P}(t_0, t) = 1 \quad . \tag{2.2 - 20b}$$

Its two *eigenvalues* are

$$\gamma_{1,2} = \Lambda\{\boldsymbol{P}(t_0, t)\} = exp\,(\pm\, i\Omega t) \quad . \tag{2.2 - 20c}$$

In addition the propagator matrix permits a shift of the time scale. This yields the following law of multiplication:

$$\boldsymbol{P}(t_0, t_2) = \boldsymbol{P}(t_2 - t_1)\, \boldsymbol{P}(t_1 - t_0) \quad . \tag{2.2 - 20d}$$

2.2.3 Phase Diagrams

In physics the motion of dynamical systems such as oscillators is often represented in the *phase space*. This representation includes a *phase diagram*. The phase space is spanned by generalized space coordinates x_k and momentum coordinates p_k of the dynamical system in consideration. The index k takes the values $k = 1,2 \ldots, f$; where f indicates the degree of freedom of the dynamical system. The dimension of the phase space is therefore $2f$. In the phase space the motion of the dynamical system corresponds to a trajectory with the time t as running parameter. This implies that *periodic motions yield closed trajectories.*

a) Phase diagram of the harmonic oscillator

The harmonic oscillator described by (2.2 - 1) has one degree of freedom, i.e. $f = 1$. Therefore, its phase space is a *plane*. As generalized space and momentum coordinates one can choose

$$x_1 = x \quad \text{and} \quad p_1 = \dot{x} \quad . \tag{2.2 - 21}$$

In order to determine the trajectory in the phase space that corresponds to the undamped harmonic oscillator equation (2.2 - 1) has to be transformed into a nonlinear equation by multiplication with $\dot{x}$

$$\dot{x} \cdot \ddot{x} + \Omega^2\, x \cdot \dot{x} = \frac{1}{2}\frac{d}{dt}\left\{(\dot{x})^2 + \Omega^2\, x^2\right\} = 0 \quad . \tag{2.2 - 22}$$

The subsequent integration over time t results in the first-order nonlinear differential equation

$$\frac{1}{2}(\dot{x})^2 + \frac{1}{2}\Omega^2\, x^2 = \frac{1}{2}\Omega^2\, {x_0}^2 = const \quad \text{with} \quad x_0 = x(\dot{x}=0) > 0 \quad . \tag{2.2 - 23}$$

This equation can also be derived by introducing $d\dot{x}\,/\,dx$ in (2.2 - 1)

$$\ddot{x}=\frac{d}{dt}\dot{x}=\frac{d\dot{x}}{dx}\dot{x}=\frac{1}{2}\frac{d}{dx}(\dot{x})^2=-\Omega^2 x \quad . \tag{2.2 - 24}$$

The integration over x yields again (2.2 - 23).

Equation (2.2 - 23) describes the trajectories in the phase space which correspond to the different motions of the harmonic oscillator determined by the initial conditions. These trajectories are illustrated in Fig. 2.2 - 5. They form concentric ellipses. Each trajectory can be labeled by the corresponding $x_0 = x(\dot{x} = 0)$. As mentioned before the time t acts as running parameter. This yields the following parameter representation of these trajectories

$$\begin{aligned} x &= x(t) \quad \text{with} \quad x_0 = x(t_0) > 0 \quad , \\ \dot{x} &= \dot{x}(t) \quad \text{with} \quad 0 = \dot{x}(t_0) \end{aligned} \tag{2.2 - 25}$$

where t_0 fixes the time scale.

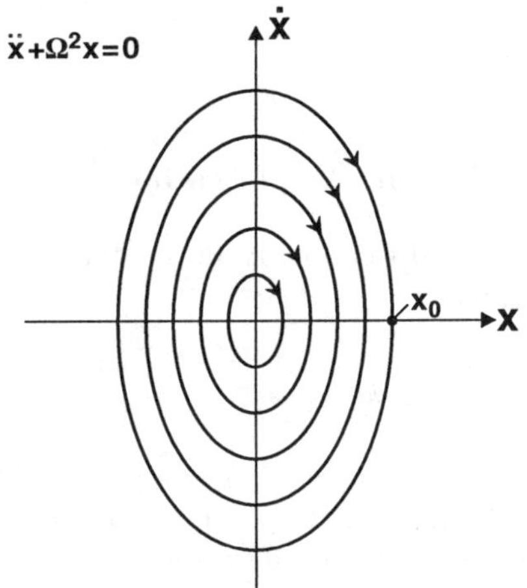

Fig. 2.2 - 5. Phase diagram of the simple harmonic oscillator

b) Clockwise motion of trajectories

As demonstrated above the motions of a harmonic oscillator can be described by trajectories in the two-dimensional phase space $(x, \dot{x})$. In this phase space the trajectories of all dynamical systems with one degree of freedom run clockwise around the center. This is demonstrated by the relation

$$d\,x = \dot{x}\,dt \quad . \tag{2.2 - 26}$$

For $dt > 0$ this relation requires $dx > 0$ for $\dot{x} > 0$ on one hand and $dx < 0$ for $\dot{x} < 0$ on the other hand. As a consequence the trajectories run clockwise around the center (0,0) of the phase plane $(x, \dot{x})$.

c) Periodicity

The relation (2.2 - 11b) between the characteristic circular frequency Ω of a harmonic oscillator and the period T of its motions can be verified with the aid of (2.2 - 23) and (2.2 - 26) by integration of the time t over a closed elliptic trajectory in the phase diagram illustrated in Fig. 2.2 - 5. For this purpose one introduces the parameters u and α defined by

$$u = x / x_0 = sin\,\alpha \quad . \qquad (2.2 - 27a)$$

With the aid of these parameters the time t can be integrated as follows

$$\begin{aligned} T &= \oint dt = \oint (\dot{x})^{-1} dx = \Omega^{-1} \oint \left[x_0^2 - x^2\right]^{-1/2} dx \\ &= \Omega^{-1} \oint \left[1 - u^2\right]^{-1/2} du = \Omega^{-1} \int_0^{2\pi} d\alpha = 2\pi / \Omega \quad . \end{aligned} \qquad (2.2 - 27b)$$

This result agrees with (2.2 - 11b).

2.2.4 Harmonic Oscillators with Damping or Amplification

A free harmonic oscillator with damping or amplification is defined by the homogeneous linear second-order differential equation

$$\ddot{x} + (2/\tau)\dot{x} + \Omega^2 x = 0 \quad \text{or} \quad \frac{d^2}{dt^2} x(t) + \frac{2}{\tau}\frac{d}{dt} x(t) + \Omega^2 x(t) = 0 \quad . \qquad (2.2 - 28)$$

In this equation it is assumed that damping and amplification are proportional to dx/dt. The parameter Ω [s^{-1}] > 0 is the *characteristic circular frequency*. If τ [s] is positive, it represents the *characteristic damping time.* If it is negative, its magnitude $|\tau|$ corresponds to the *characteristic amplification time.*

The oscillation types of a harmonic oscillator with damping or amplification are determined by the *quality factor* or *Q - factor* defined by the product

$$\Omega \tau = 2Q \quad . \qquad (2.2 - 29)$$

These oscillations are listed in Table 2.2 - 1 and discussed in the following.

Table 2.2 - 1 Types of oscillation

$\Omega\,\tau = 2\,Q$	Type of oscillation
$\Omega\,\tau = \pm\,\infty$	undamped
$1 < \Omega\,\tau < +\,\infty$	subcritically damped
$\Omega\,\tau = 1$	critically damped
$0 < \Omega\,\tau < 1$	supercritically damped
$-1 < \Omega\,\tau < 0$	supercritically amplified
$\Omega\,\tau = -1$	critically amplified
$-\infty < \Omega\,\tau < -1$	subcritically amplified

a) Subcritical damping and amplification

Subcritical damping and amplification are characterized by the inequality

$$\Omega\,|\tau| > 1 \quad \text{or} \quad Q^2 > 1/4 \quad . \tag{2.2 - 30}$$

Under these circumstances the real solutions $x(t)$ of (2.2 - 28) represent damped or amplified oscillations with the *quasi-circular frequency* ω_0 and the quasi-period T_0 defined by the relations

$$T_0 = 2\pi/\omega_0 > 0 \quad \text{and} \quad \omega_0^2 = \Omega^2 - (1/\tau)^2 > 0 \quad . \tag{2.2 - 31}$$

These damped or amplified oscillations $x(t)$ are not periodic. Consequently, the *quasi-period* T_0 determines only the period of the zeros of such an oscillation $x(t)$ according to

$$0 = x(t_0) = x(t_0 + \frac{n}{2}T_0) \quad \text{with} \quad n = 0, \pm 1, \pm 2, \pm 3, \ldots \quad . \tag{2.2 - 32}$$

The real subcritically damped or amplified oscillations $x(t)$ as general solutions of (2.2 - 28) can be written in various forms

α) as a linear combination of linearly independent functions with the constant real coefficients C_1 and C_2

$$x(t) = \{C_1 \cos\omega_0 t + C_2 \sin\omega_0 t\} \exp(-t/\tau) \quad . \tag{2.2 - 33}$$

β) with a real positive amplitude $A > 0$ and a real phase φ according to Figs. 2.2 - 6a&b

$$x(t) = A\,exp(-t / \tau)\,cos(\omega_0 t - \varphi) \quad . \tag{2.2 - 34}$$

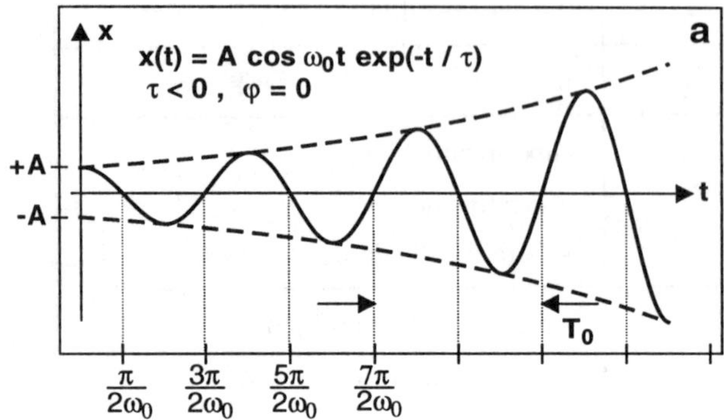

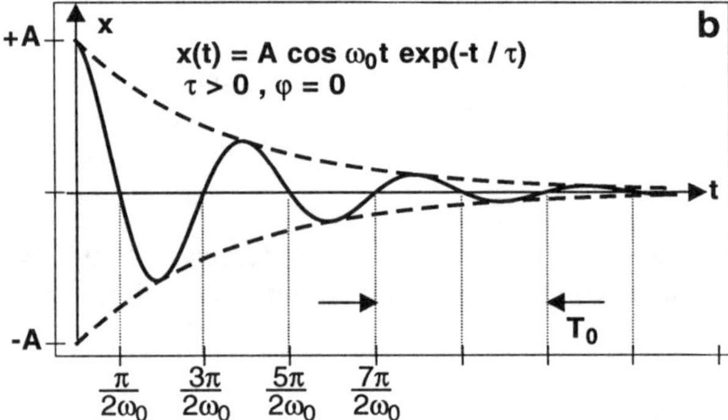

Fig. 2.2 - 6. Oscillations of harmonic oscillators a) with amplification ($\tau < 0$) and b) with damping ($\tau > 0$). The dashed lines indicate the exponential envelopes

γ) as a function of the state of the oscillator at time t_0 defined by $x(t_0)$ and $\dot{x}(t_0)$

$$x(t) = exp(-\Delta t / \tau)\left\{x(t_0)\cos\omega_0\Delta t + \left[\frac{x(t_0)}{\omega_0\tau} + \frac{\dot{x}(t_0)}{\omega_0}\right]\sin\omega_0\Delta t\right\} \tag{2.2 - 35}$$

with $\Delta t = t - t_0$.

This equation permits to determine the propagators $\boldsymbol{P}(t_0,t) = \boldsymbol{P}(\Delta t)$ of subcritically damped or amplified oscillators that relate the initial state at time t_0 with the state at time t according to (2.2 - 19a).

Fig. 2.2 - 7 shows examples of the phase diagrams of the damped (a), amplified (b) and undamped (c) oscillations described for $t > 0$ by the equations

$$x_a(t) = \{\cos t + 0.1 \cdot \sin t\}\,exp(-0.1t) \quad , \tag{2.2 - 36a}$$

$$x_b(t) = \{cos\,t - 0.1 \cdot sin\,t\} exp(+0.1t) \quad , \tag{2.2 - 36b}$$

$$x_c(t) = cos\,t \quad . \tag{2.2 - 36c}$$

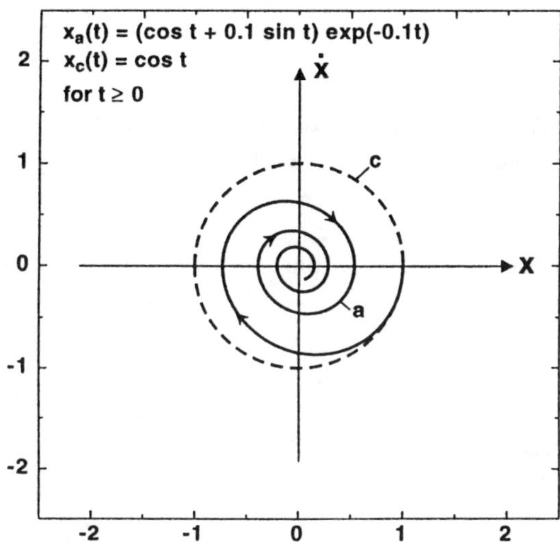

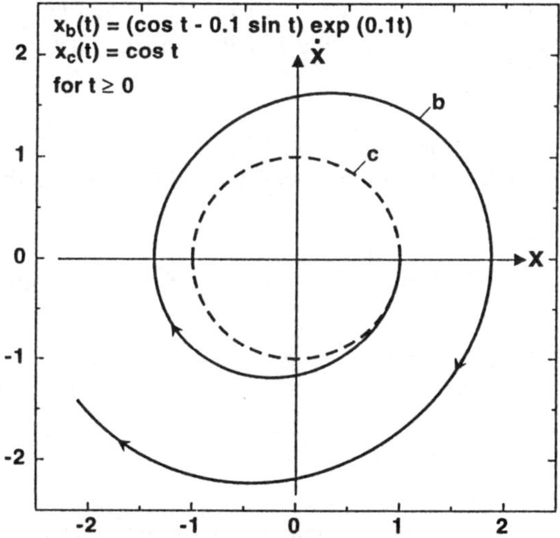

Fig. 2.2 - 7. Phase diagrams of the oscillations of harmonic oscillators a) with damping, b) with amplification and c) without damping or amplification

All oscillations (2.2 - 36a-c) start from the same *initial state* at time $t = 0$

$$x_a(0) = x_b(0) = x_c(0) = 1 \quad \text{and} \quad \dot{x}_a(0) = \dot{x}_b(0) = \dot{x}_c(0) = 0 \quad .$$

The phase diagrams of the damped and the amplified oscillations (2.2 - 36a&b) represent spirals, while that of the undamped oscillation (2.2 - 36c) is a circle.

b) Critical damping and amplification

Critical damping and amplification are defined by the equation

$$\Omega|\tau| = 1 \quad \text{or} \quad Q^2 = 1/4 \quad . \tag{2.2 - 37}$$

This condition determines the two limits where any apparent oscillation is suppressed. The motion $x(t)$ of a harmonic oscillator with critical damping or amplification can be described as

α) a linear combination of two specific basic functions with the constant real coefficients C_1 and C_2

$$x(t) = \{C_1 + C_2\, t\}\, exp(-t/\tau) \quad , \tag{2.2 - 38}$$

β) a function of a state of the oscillator at time t_0 that is defined by $x(t_0)$ and $\dot{x}(t_0)$

$$x(t) = \left\{ x(t_0)\left(1 + \frac{\Delta t}{\tau}\right) + \dot{x}(t_0)\,\Delta t \right\} exp(-\Delta t/\tau) \quad \text{with} \quad \Delta t = t - t_0 \quad . \tag{2.2 - 39}$$

This equation determines the propagators $\boldsymbol{P}(t_0,t) = \boldsymbol{P}(\Delta t)$ of critically damped or amplified oscillators which relate the initial state at time t_0 with the state at time t according to (2.2 - 19a).

γ) phase diagram in the phase plane $(x, \dot{x})$. For illustration Fig. 2.2 - 8a shows the phase diagram of the motion $x(t)$ of a critically damped oscillator described by the function

$$x_a(t) = (1 + t)\, exp(-t) \quad \text{for} \quad t > 0 \quad . \tag{2.2 - 40}$$

c) Supercritical damping and amplification

Supercritical damping and amplification are defined by the inequality

$$\Omega|\tau| < 1 \quad \text{or} \quad Q^2 < 1/4 \quad . \tag{2.2 - 41}$$

Under these circumstances the motions $x(t)$ of an harmonic oscillator are determined by two characteristic times τ_1 and τ_2. These can be calculated as follows

$$(1/\tau_{1,2}) = (1/\tau) \pm \left\{(1/\tau)^2 - \Omega^2\right\}^{1/2} \quad . \tag{2.2 - 42}$$

With these two parameters $\tau_{1,2}$ the motion $x(t)$ of supercritically damped and amplified oscillators can be represented as

α) a linear combination of two exponential functions with constant real coefficients C_1 and C_2

$$x(t) = C_1\, exp(-t / \tau_1) + C_2\, exp(-t / \tau_2) \quad . \tag{2.2 - 43}$$

β) a function of a state of the oscillator at time t_0 that is defined by $x(t_0)$ and $\dot{x}(t_0)$

$$x(t) = \frac{\{x(t_0) + \tau_2\, \dot{x}(t_0)\}}{\{1-(\tau_2 / \tau_1)\}} exp(-\Delta t / \tau_1) + \frac{\{x(t_0) + \tau_1\, \dot{x}(t_0)\}}{\{1-(\tau_1 / \tau_2)\}} exp(-\Delta t / \tau_2) \tag{2.2 - 44}$$

with $\Delta t = t - t_0$.

This equation determines the *propagators* $\boldsymbol{P}(t,t_0) = \boldsymbol{P}(\Delta t)$ of supercritically damped or amplified oscillators which relate the initial state at time t_0 with the state at time t according to (2.2 - 19a).

γ) phase diagram in the phase plane $(x, \dot{x})$. For illustration Fig. 2.2 - 8b shows the phase diagram of the motion $x(t)$ of a supercritically damped oscillator described by the function

$$x_b(t) = \{2exp(-t) - exp(-2t)\} \quad \text{for} \quad t \geq 0 \quad . \tag{2.2 - 45}$$

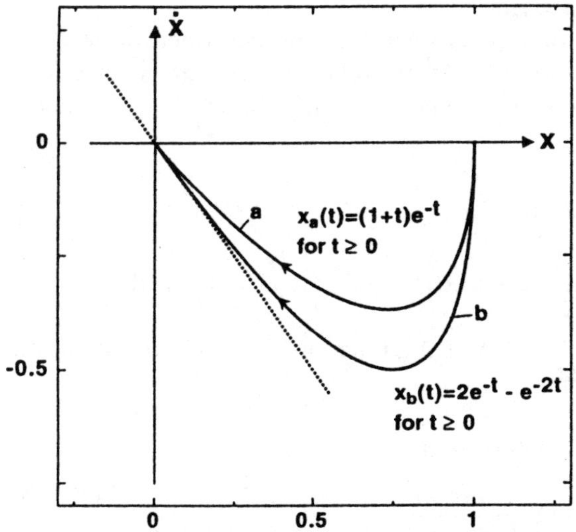

Fig. 2.2 - 8. Phase diagrams a) of a critically damped oscillation and b) of a supercritically damped oscillation

2.2.5 Frequency Spectra

The harmonic oscillation of an undamped harmonic oscillator is monochromatic and characterized by the circular frequency Ω. On the contrary an oscillation of a damped or amplified oscillator embraces a continuous frequency spectrum. Frequency spectra are evaluated with the aid of the Fourier transformation [Bracewell 1986 B, Campbell & Foster 1948 B, Carslaw 1930 B, Champeney 1973 B, Erdelyi et al. 1954 B] as demonstrated in the following.

a) Fourier transformation
The *Fourier transform* $\boldsymbol{F}\{x(t)\}$ of a function $x(t)$ is defined by the integral

$$\boldsymbol{F}\{x(t)\} = F(\omega) = \int_{-\infty}^{+\infty} x(t)\, exp(+i\omega t)\, dt \quad . \qquad (2.2 - 46a)$$

$F(\omega)$ is the Fourier or frequency spectrum of the function $x(t)$. The *inverse Fourier transform* of the frequency spectrum $F(\omega)$ corresponds to the following integral

$$\boldsymbol{F}^{-1}\{F(\omega)\} = x(t) = \frac{1}{2\pi} \int_{-\infty}^{+\infty} F(\omega)\, exp(-i\omega t)\, d\omega \quad . \qquad (2.2 - 46b)$$

This integral forms the basis of the *spectral analysis* of a motion $x(t)$. Fourier transforms of real functions are presented in Appendix A.2.

The subsequent consideration is restricted to linearly *damped harmonic oscillators*. These exhibit the following two types of frequency spectra.

b) Debye relaxation and attenuation
Harmonic oscillators with critical or supercritical damping described in Sections 2.2.4b&c exhibit the frequency spectra of the Debye relaxation and attenuation. This relaxation corresponds to an exponential decay of the variable $x(t)$ with the characteristic time $\tau > 0$

$$x(t) = A\, \tau^{-1} exp(-t/\tau) \quad \text{for} \quad t \geq 0 \quad . \qquad (2.2 - 47)$$

This function represents a special motion $x(t)$ of a critically damped harmonic oscillator (2.2 - 38) with $\Omega\tau = 2Q = 1$.

The Fourier transformation (2.2 - 46a) of the function (2.2 - 47) results in the following complex frequency spectrum

$$\begin{aligned} F(\omega) = \boldsymbol{F}\{x(t)\} &= A\, DR(\omega;\tau) + iA\, DA(\omega;\tau) \\ &= A \frac{1}{1+(\omega\tau)^2} + iA \frac{\omega\tau}{1+(\omega\tau)^2} \quad . \end{aligned} \qquad (2.2 - 48)$$

The Debye relaxation $DR(\omega;\ \tau)$ and the Debye attenuation $DA(\omega;\ \tau)$ are represented in Fig. 2.2 - 9. These functions find application in Section 7.5.2b.

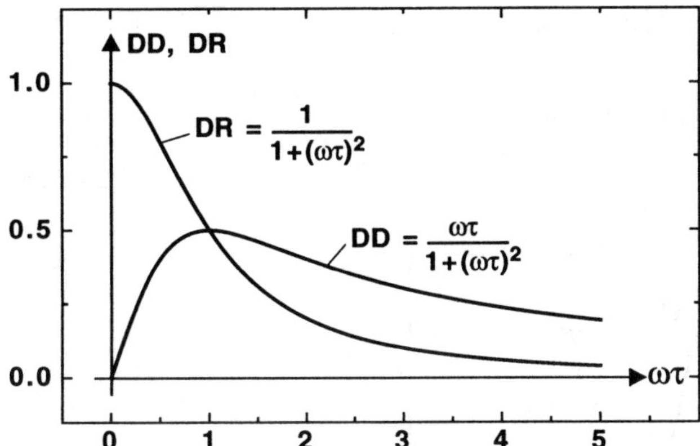

Fig. 2.2 - 9. Debye relaxation $DR\ (\omega;\ \tau)$ and Debye attenuation $DA\ (\omega;\ \tau)$ with the normalized frequency $\omega\tau$ and the characteristic time $\tau = 1$ of damping.

The Debye relaxation is a characteristic of *dielectric polar liquids* [Kneubühl 1989 J, 1994 B].

c) Lorentz line shape and dispersion

Harmonic oscillators with subcritical damping imply Lorentz line shapes and dispersion. These correspond to the subcritically damped harmonic oscillations of Section 2.2.4a with the quasi-circular frequency ω_0 and the characteristic time $\tau > 0$

$$x(t) = A\,\tau^{-1} \cos \omega_0 t\, exp(-t/\tau) \quad \text{for} \quad t \geq 0 \quad . \tag{2.2 - 49}$$

The Fourier transformation (2.2 - 46a) of the damped oscillation (2.2 - 49) results in

$$\begin{aligned} F(\omega) = \quad & A\,LL(\omega;\tau,\omega_0) - iA\,LD(\omega;\tau,\omega_0) \\ & + A\,LL(\omega;\tau,-\omega_0) - iA\,LD(\omega;\tau,-\omega_0) \quad \text{with} \end{aligned} \tag{2.2 - 50a}$$

$$LL(\omega;\tau,\omega_0) = \frac{1}{1+(\omega-\omega_0)^2\tau^2} \quad \text{and} \tag{2.2 - 50b}$$

$$LD(\omega;\tau,\omega_0) = -\frac{(\omega-\omega_0)\,\tau}{1+(\omega-\omega_0)^2\tau^2} \quad . \tag{2.2 - 50c}$$

LL indicates the *Lorentz line shape*, whilst *LD* represents the *Lorentz dispersion*. Both are illustrated in Fig. 2.2 - 10. These functions are used in Section 7.5.2c. In spectroscopy and quantum optics the Lorentz line shape is an indication for *homogeneously broadened spectral lines.* [Kneubühl & Sigrist 1989 B, Svelto &

Hanna 1989 B]. The spectral *half-width* $\Delta\omega$ of the Lorentz line shape that is defined by the equation

$$LL\left(\omega_0 \pm \frac{\Delta\omega}{2};\tau,\omega_0\right) = \frac{1}{2} LL(\omega_0;\tau,\omega_0) \qquad (2.2 - 51a)$$

is related to the characteristic time τ as follows

$$\Delta\omega\,\tau = 2 \quad . \qquad (2.2 - 51b)$$

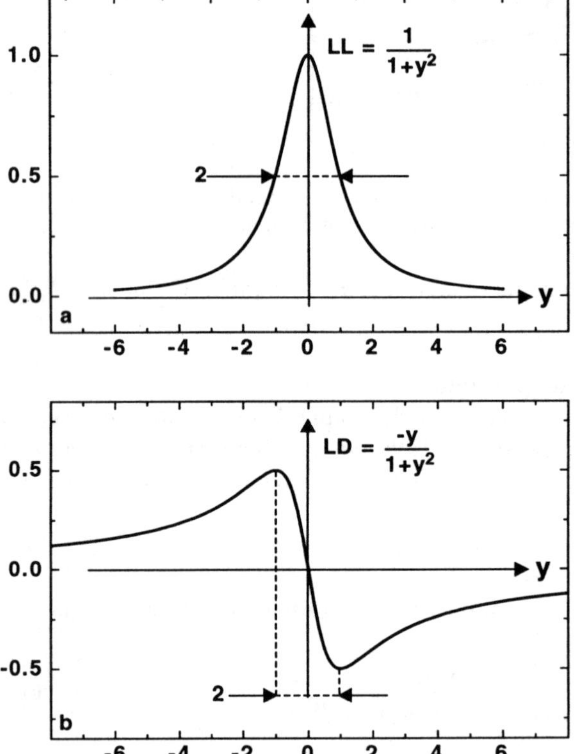

Fig. 2.2 - 10. Lorentz line shape LL (ω; τ, ω_0) and Lorentz dispersion LD (ω; τ, ω_0) as function of the normalized frequency $y = (\omega - \omega_0)\,\tau$ and $\tau = 1$

Of more interest is the *relative spectral half -width* given by

$$\Delta\omega / \omega_0 = 1 / Q_0 \quad , \qquad (2.2 - 52a)$$

where Q_0 indicates the quality factor of the oscillator

$$Q_0 = \frac{1}{2}\omega_0\,\tau \quad . \qquad (2.2 - 52b)$$

2.2.6 Linear Spring-Mass Systems

The undamped spring-mass system described in Section 2.2.1b and illustrated in Fig. 2.2 - 2 was previously characterized by the following equation of motion

$$\ddot{x} + \Omega^2 x = 0 \quad \text{with} \quad \Omega^2 = f / \mu \quad , \tag{2.2 - 4b}$$

where f [$Nm^{-1} = kg\ s^{-2}$] indicates the force constant and μ [kg] the mass of the system. Equation (2.2 - 4b) reveals that a spring-mass system represents a *harmonic oscillator*. An essential property of an undamped spring-mass system is the energy conservation discussed in the following.

a) Energy of undamped spring-mass systems

The *total energy* E [$J = kg\ m^2\ s^{-2}$] of a spring-mass system is the sum of the *kinetic energy* E_{kin} [$J = kg\ m^2\ s^{-2}$] and the *potential energy* V [$J = kg\ m^2\ s^{-2}$]. Kinetic and potential energy can be written as

$$E_{kin} = (\mu / 2)\dot{x}^2 \tag{2.2 - 53a}$$

$$V = (f / 2)x^2 \quad . \tag{2.2 - 53b}$$

The undamped spring-mass system described by (2.2 - 4b) obeys the law concerning the *conservation of the total energy*

$$E = E_{kin} + V = const \quad . \tag{2.2 - 53c}$$

This law can be derived in analogy to the nonlinear first-order differential equation (2.2 - 23) that determines the phase diagram of the undamped harmonic oscillator.

In all harmonic oscillators of physics the oscillation of a physical variable $x(t)$ with circular frequency Ω or ω_0 implies the *periodic exchange of two forms of energy* with the two-fold circular frequency 2Ω or $2\omega_0$. In the undamped spring-mass system this periodic exchange of energies occurs *between the kinetic and the potential energy.* It is described by the equations

$$x(t) = A\,cos(\Omega t - \varphi) \quad , \tag{2.2 - 54a}$$

$$V(t) = (f / 2)x^2(t) = \frac{E}{2}\{1 + cos\,2(\Omega t - \varphi)\} \quad , \quad \text{and} \tag{2.2 - 54b}$$

$$E_{kin}(t) = (\mu / 2)\dot{x}^2(t) = \frac{E}{2}\{1 - cos\,2(\Omega t - \varphi)\} \quad . \tag{2.2 - 54c}$$

b) Damped spring-mass systems

A spring-mass system can be damped by a *braking force* F_b proportional to the velocity υ

$$F_b = -b\upsilon = -b\dot{x} \quad \text{with} \quad b > 0 \quad . \qquad (2.2 - 55)$$

An example of such a force is the *drag* F_b [kg m/s^2] by a *laminar flow* of a fluid with the dynamic viscosity η [kg/ms] and the velocity υ [m/s] on a solid sphere of radius a [m]. According to *Stokes' law* this drag is [Kneubühl 1994 B, Prandtl & Tietjens 1975a&b B]

$$F_b = -6\pi\eta\, a\upsilon = -6\pi\eta\, a\dot{x} \quad . \qquad (2.2 - 56)$$

Taking into account both, the restoring force F_r and the braking force F_b, in Newton's second law of mechanics results in the following dynamic equation of a damped spring-mass system

$$\mu\ddot{x} = -fx - b\dot{x} \quad \text{or} \qquad (2.2 - 57a)$$

$$\ddot{x} + (2/\tau)\dot{x} + \Omega^2 = 0 \quad \text{with} \quad \Omega^2 = f/\mu \quad \text{and} \quad \tau = 2\mu/b \quad . \qquad (2.2 - 57b)$$

The second equation corresponds to the equation (2.2 - 28) characteristic of the damped oscillator.

The braking force F_b causes a continuous loss of the total energy E of the damped spring-mass system. The corresponding power loss P_b is according to the laws of mechanics determined by the relations

$$P_b = F_b\,\dot{x} = -b(\dot{x})^2 = -\frac{d}{dt}E \quad \text{with}$$
$$E = E_{kin} + V = (\mu/2)x^2 + (f/2)\dot{x}^2 \quad . \qquad (2.2 - 58)$$

Consequently, the total energy E of a damped spring-mass system is not conserved.

For a subcritically damped spring-mass system the decrease of the total energy E with time t can be evaluated by the combination of (2.2 - 49) and (2.2 - 58). For small coefficients b and correspondingly long characteristic times τ the decrease of the total energy E is approximately exponential

$$E(t) \approx E(0)\,exp(-2t/\tau) \quad . \qquad (2.2 - 59)$$

This equation permits the calculation of the relative energy decrease of a subcritically damped spring-mass system within one quasi-period $T_0 = 2\pi/\omega_0$ determined by (2.2 - 31)

$$\frac{\Delta E}{E} = \frac{E(T_0) - E(0)}{E(0)} \approx -2T_0 / \tau \approx -2\pi / Q_0 \quad . \qquad (2.2 - 60)$$

This decrease depends exclusively on the quality factor Q_0 defined by (2.2 - 52a&b).

2.2.7 Linear Electric Circuits

The linear passive electric *LC* circuit discussed in Section 2.2.1a and illustrated in Fig. 2.2 - 1 represents an example of an undamped harmonic oscillator. An undamped *LC* circuit can be damped by introduction of an ohmic resistance *R*.

a) *LCR* circuits

A common linear passive electric circuit including resistance consists of a *capacitor* with capacitance *C* [farad = F = As/V], a *coil* with self-inductance *L* [henry = H = Vs/A] and a *resistor* with resistance *R* [ohm = Ω = V/A] as illustrated in Fig. 2.2 - 11.

Because it is a closed circuit, the sum of the potential differences on the three elements vanishes according to *Kirchhoff's law.* This yields the relation

$$L\ddot{q} + R\dot{q} + C^{-1}q = 0 \qquad (2.2 - 61a)$$

that can be written in the form of (2.2 - 28)

$$\ddot{x} + (2/\tau)\dot{x} + \Omega^2 x = 0$$
$$\text{with} \quad x = q; \quad \Omega^2 = 1/LC; \quad \text{and} \quad \tau = 2L/R \quad . \qquad (2.2 - 61b)$$

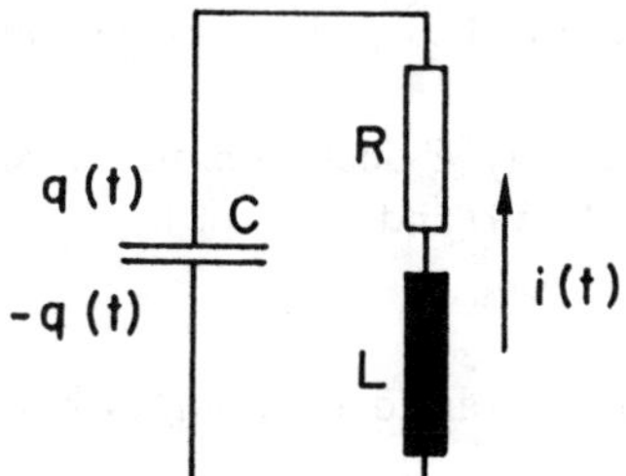

Fig. 2.2 - 11. Electric *LCR* circuit

Therefore, a passive electric circuit corresponds to a damped harmonic oscillator. For $R = 0$ that implies $\tau = \infty$ the electric circuit represents an undamped harmonic oscillator. The damping of the electric circuit is usually characterized by the *quality factor*

$$Q = \frac{1}{2}\Omega\tau = \left(L / CR^2\right)^{1/2} \quad . \qquad (2.2 - 62)$$

b) Energy in electric circuits

The *total energy* E of an electric circuit is the sum of the *electric energy*

$$E_{el} = (1/2C)q^2 \qquad (2.2 - 63a)$$

and the *magnetic energy*

$$E_{magn} = (L/2)\dot{q}^2 \quad . \qquad (2.2 - 63b)$$

For an undamped electric circuit the total energy obeys the law of *energy conservation*

$$E_{el} + E_{magn} = E = const \quad . \qquad (2.2 - 63c)$$

In an electric circuit oscillating at the frequency Ω or ω_0 the electric and the magnetic energy are exchanged periodically with the twofold circular frequency 2Ω or $2\omega_0$. In an undamped electric circuit this *energy exchange* is determined by the following equations

$$q(t) = q_0 \cos(\Omega t - \varphi) \quad , \qquad (2.2 - 64a)$$

$$E_{el}(t) = \frac{E}{2}\{1 + \cos 2(\Omega t - \varphi)\} \quad , \text{ and} \qquad (2.2 - 64b)$$

$$E_{magn}(t) = \frac{E}{2}\{1 - \cos 2(\Omega t - \varphi)\} \quad . \qquad (2.2 - 64c)$$

They are analogous to equations (2.2 - 54a-c) which describe the energy exchange in an undamped spring-mass system.

In a subcritically damped electric circuit the total energy E decreases in a first approximation exponentially with time t according to (2.2 - 59) and (2.2 - 60).

c) Feedback

Damping and amplification of an electric circuit can be varied by *feedback* as illustrated in Fig. 2.2 - 12.

For a simple feedback that is proportional to the current $i = \dot{q}$ the potential difference $V_R = Ri$ on the resistor with resistance R is amplified by a factor a and feeded back into the electric circuit. The corresponding relations are

$$L\ddot{q} + R\dot{q} + C^{-1}q = V \quad \text{and} \quad V = aV_R = aR\dot{q} \quad . \qquad (2.2 - 65)$$

These equations can be transformed into a modified equation of the harmonic oscillator with damping or amplification

$$\ddot{q}+(2/\tau_{\mathrm{f}})\dot{q}+\Omega^2 q=0 \tag{2.2 - 66a}$$

with the modified characteristic time

$$\tau_{\mathrm{f}} = \tau/(1-a) \tag{2.2 - 66b}$$

and the modified quality factor

$$Q_{\mathrm{f}} = \Omega\,\tau_{\mathrm{f}} = Q/(1-a) \quad . \tag{2.2 - 66c}$$

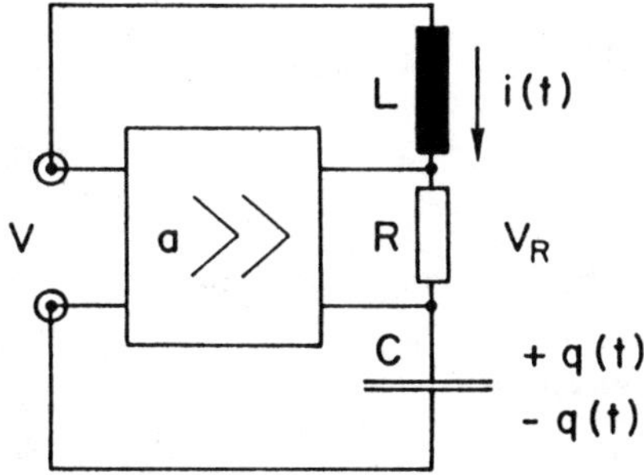

Fig. 2.2 - 12. Electric circuit with feedback

The feedback is called *positive* if $a > 0$ and *negative* if $a < 0$. Positive feedback decreases damping, while negative feedback enhances damping. The simplest *stabilization* of a system makes use of the negative feedback with $a < 0$ to stabilize by increasing damping. For a positive feedback with $a = 1$ the damping of an electric circuit is eliminated. As a consequence, the circuit oscillates harmonically with a constant amplitude.

2.2.8 Equivalent Nonlinear Oscillators

A harmonic oscillator described by the equation

$$\ddot{x}+(2/\tau)\dot{x}+\Omega^2 x=0 \tag{2.2 - 28}$$

fulfills the *linear superposition principle* of homogeneous linear differential equations which was mentioned in Section 2.2.2a. This requires that the superposition $x_{1+2}(t)$ of two solutions $x_1(t)$ and $x_2(t)$ of (2.2 - 28) equals their sum

$$x_{1+2}(t) = x_1(t) + x_2(t) \quad . \tag{2.2 - 8}$$

Equation (2.2 - 28) can be transformed into a nonlinear equivalent by a nonlinear transformation of the variable x into a new variable y. This new variable y represents a nonlinear oscillator that is equivalent to a harmonic oscillator [Ames 1972 B].

a) General nonlinear transformations

A common *nonlinear transformation* of the variable x into a new variable y is performed with a nonlinear function f and its inverse f^{-1}

$$x = f(y) \quad \text{and} \quad y = f^{-1}(x) \quad . \tag{2.2 - 67}$$

Thus, (2.2 - 28) that describes a harmonic oscillators is transformed into the equation

$$\ddot{y} + \{f''(y)/f'(y)\}(\dot{y})^2 + (2/\tau)\dot{y} + \Omega^2\{f(y)/f'(y)\} \quad . \tag{2.2 - 68}$$

The corresponding nonlinear *superposition principle* reads

$$y_{1+2}(t) = f^{-1}(f(y_1) + f(y_2)) \quad . \tag{2.2 - 69}$$

These general transformations can be illustrated by the following examples:

α) The *first example* to be considered is the transformation

$$x = x_0 \, exp(y/y_0) \quad \text{and} \quad y = y_0 \, \ell n(x/x_0) \quad . \tag{2.2 - 70}$$

The results is in a *Riccati differential equation* for $\dot{y}$ [Kamke 1956 B, Zwillinger 1989 B]:

$$\ddot{y} + \{1/y_0\}(\dot{y})^2 + (2/\tau)\dot{y} + \Omega^2 y_0 = 0 \tag{2.2 - 71}$$

and the nonlinear *superposition law*

$$exp(y_{1+2}/y_0) = exp(y_1/y_0) + exp(y_2/y_0) \quad . \tag{2.2 - 72}$$

β) The *second example* to be mentioned uses the transformation

$$x = x_0 y^n \quad \text{and} \quad y = (x/x_0)^{1/n} \quad \text{with} \quad n \neq 0,1 \quad . \tag{2.2 - 73}$$

This transformation yields

$$\ddot{y} + (n-1)\, y^{-1}(\dot{y})^2 + (2/\tau)\,\dot{y} + (1/n)y = 0 \tag{2.2 - 74}$$

and the nonlinear *superposition law*

$$y_{1+2}^n = y_1^n + y_2^n \quad . \tag{2.2 - 75}$$

b) Riccati transformation

A more fundamental aspect is revealed by the *Riccati transformation* [Zwillinger 1989 B]

$$y = -\dot{x} / x = -\frac{d}{dt} \ell n\, x \quad \text{and} \quad x = exp\left\{ -\int_0^t y\, dt \right\} \quad . \tag{2.2 - 76}$$

It transforms the homogeneous linear differential equation (2.2 - 28) of second order into a nonlinear differential equation of first order that is named *Riccati equation*

$$\dot{y} = y^2 - (2/\tau)\, y + \Omega^2 \quad . \tag{2.2 - 77}$$

The solution of this equation is

$$\int_{y_0}^{y} \left\{ y^2 - (2/\tau) y + \Omega^2 \right\}^{-1} dy = t \quad . \tag{2.2 - 78}$$

The *superposition law* corresponding to the Riccati transformation of the linear equation (2.2 - 28) is complicated

$$y_{1+2} = -\frac{d}{dt} \ell n \left[exp\left\{ -\int_0^t y_1 dt \right\} + exp\left\{ -\int_0^t y_2\, dt \right\} \right] \quad . \tag{2.2 - 79}$$

An *undamped harmonic oscillation* $x(t)$ and its Riccati transform $y(t)$ may serve as simple example

$$x(t) = x_0\, cos(\Omega t - \varphi) \quad \text{and} \quad y(t) = \Omega\, tan(\Omega t - \varphi) \quad . \tag{2.2 - 80}$$

These two variables fulfill the equations

$$\ddot{x} + \Omega^2 x = 0 \quad \text{and} \quad \dot{y} = y^2 + \Omega^2 \quad . \tag{2.2 - 81}$$

2.3 Modulated Linear Oscillators

This section deals with the free linear oscillators whose inherent properties vary with time. They show a considerable variety of important and complex phenomena. Of special interest are those oscillators whose characteristics vary periodically.

2.3.1 Representations

Free modulated linear oscillators can be represented by three different, yet equivalent differential equations:

a) General basic equation

Temporarily modulated linear oscillators are defined by the following basic homogeneous differential equation of second order

$$\ddot{x}+\{2/\tau(t)\}\dot{x}+\Omega^2(t)x=0 \tag{2.3 - 1}$$

with the time-dependent circular frequency $\Omega(t)$ and the time-dependent characteristic time $\tau(t)$, which implies amplification for $\tau(t) < 0$ and damping for $\tau(t) > 0$. The characteristics of harmonic oscillators with constant parameters $\Omega(t) = \Omega$ and $\tau(t) = \tau$ were already described in the previous Section 2.2.

Examples of temporarily modulated oscillators are the oscillators with a frequency chirp of Section 2.3.4, the oscillators with a step or pulse modulation of the frequency of Section 2.3.6 and the parametric oscillators of Section 2.3.7 such as the classical swing described in Section 2.3.7e and the passive LC circuit with a harmonically modulated capacitance $C(t)$ discussed in Section 2.3.7f.

b) Normal form

The basic equation (2.3 - 1) can be transformed into its normal form by the ansatz [Birkhoff & Rota, 1989 B]

$$x(t)=u(t)exp\left\{-\int\frac{dt}{\tau(t)}\right\} \quad . \tag{2.3 - 2}$$

The factor $u(t)$ obeys the basic differential equation in its *normal form*

$$\ddot{u}+\Omega_0^2(t)u=0 \tag{2.3 - 3}$$

with the term proportional to $\dot{u}$ missing. The modified circular frequency $\Omega_0(t)$ is determined by the equation

$$\Omega_0^2(t) =\Omega^2(t)-\{1/\tau(t)\}^2-\frac{d}{dt}\{1/\tau(t)\}=I(t) \quad . \tag{2.3 - 4}$$

In this equation $I(t)$ designates the *invariant* [Birkhoff & Rota 1989 B] of the original basic equation (2.3 - 1).

c) Riccati equation

The basic homogeneous differential equation (2.3 - 1) of second order and its normal form (2.3 -3) can be assigned an inhomogeneous quadratic differential equation of first order, called *Riccati differential equation* [Kamke 1956 B, Zwillinger 1989 B]. This is performed by the *Riccati* transformation [Birkhoff & Rota 1989 B] already introduced in Section 2.2.2b.

$$\upsilon = \dot{x} / x = \frac{d}{dt} \ell n\, x \quad \text{or} \tag{2.3 - 5a}$$

$$x = C\, exp \int \upsilon dt \quad . \tag{2.3 - 5b}$$

The result of the Riccati transformation of (2.3 - 1) is the Riccati equation

$$\dot{\upsilon} + \upsilon^2 + \{2 / \tau(t)\}\, \upsilon + \Omega^2(t) = 0 \quad . \tag{2.3 - 6}$$

The Riccati transformation (2.3 - 5a&b) reduces the problem of solving the basic oscillation equation (2.3 - 1) to the solution of the first-order nonlinear equation (2.3 - 6) and subsequent integration.

2.3.2 General Solutions

The solutions of (2.3 - 1) and (2.3 - 3) obey a series of general laws which are valid for all characteristic functions $\tau(t)$, $\Omega(t)$, or $\Omega_0^2(t) = I(t)$. Important are the following laws and relations:

a) Superposition principle
A linear combination or superposition of two solutions $x_1(t)$ and $x_2(t)$ of the homogeneous differential equations (2.3 - 1) and (2.3 - 3) is also a solution of this equation

$$x(t) = C_1\, x_1(t) + C_2\, x_2(t) \tag{2.3 - 7}$$

where C_1 and C_2 represent arbitrary constants.

b) Linear dependence and Wronski determinant
Two solutions $x_1(t)$ and $x_2(t)$ of the homogeneous differential equation (2.3 - 1) or (2.3 - 3) are linearly dependent if there exist two constants C_1 and C_2 such that

$$C_1\, x_1(t) + C_2\, x_2(t) = 0 \quad , \tag{2.3 - 8a}$$

$$C_1\, \dot{x}_1(t) + C_2\, \dot{x}_2(t) = 0 \quad . \tag{2.3 - 8b}$$

This linear system of equations for the determination of the constants C_1 and C_2 has nontrivial non-zero solutions if the *Wronski determinant* [Birkhoff & Rota 1989 B]

$$W(t) = \begin{vmatrix} x_1(t) & x_2(t) \\ \dot{x}_1(t) & \dot{x}_2(t) \end{vmatrix} = x_1(t)\, \dot{x}_2(t) - \dot{x}_1(t)\, x_2(t) \tag{2.3 - 9}$$

vanishes. This means two solutions $x_1(t)$ and $x_2(t)$ of (2.3 - 1) or (2.3 - 3) are *linearly dependent* (independent) if the Wronski determinant (2.3 - 9) vanishes (differs from zero).

c) Basis of solutions

Two solutions $x_1(t)$ and $x_2(t)$ of (2.3 - 1) or (2.3 - 3) form a basis of the general solution if they are linearly independent. This means that each solution $x(t)$ of (2.3 - 1) or (2.3 - 3) can be represented as a linear combination of the two linearly independent solutions $x_1(t)$ and $x_2(t)$ according to (2.3 - 7) with adequate constants C_1 and C_2.

d) Properties of the Wronski determinant

The Wronski determinant (2.3 - 9) of two arbitrary solutions $x_1(t)$ and $x_2(t)$ of (2.3 - 1) fulfills the relation

$$W(t) = W(0)\, exp\left(-2\int_0^t \frac{d\Theta}{\tau(\Theta)}\right) \quad . \qquad (2.3 - 10)$$

The proof of this equation is performed by differentiation of the Wronski determinant defined by (2.3 - 9)

$$\dot{W}(t) = x_1\ddot{x}_2 - \ddot{x}_1 x_2 = -\frac{2}{\tau}\left(x_1\dot{x}_2 - \dot{x}_1 x_2\right) = -\frac{2}{\tau}W(t) \quad . \qquad (2.3 - 11)$$

Equation (2.3 - 10) demonstrates that the Wronski determinant $W(t)$ of the solutions $x_1(t)$ and $x_2(t)$ of (2.3 -1) is *either always zero, positive or negative.*

In addition, (2.3 - 10) shows that the Wronski determinant of the solutions $x_1(t)$ and $x_2(t)$ of (2.3 - 3) is *always constant*

$$W(t) = W(0) = const \quad \text{for} \quad \tau(t) = \infty \quad . \qquad (2.3 - 12)$$

For *illustration* we consider the *harmonic oscillator* with $\Omega(t) = \Omega$ and $\tau(t) = \infty$. The Wronski determinant of the two linearly independent solutions and oscillations

$$x_1(t) = C_1 \cos\Omega t \quad \text{and} \quad x_2(t) = C_2 \sin\Omega t \qquad (2.3 - 13)$$

is the constant

$$W(t) = W(0) = \Omega C_1 C_2 \quad . \qquad (2.3 - 14)$$

By the combination of (2.3 - 9) and (2.3 - 10) it is possible to derive a linearly independent solution $x_2(t)$ from a given solution $x_1(t)$ of (2.3 - 1). For this purpose one evaluates the time derivative of the quotient $x_2(t)$ versus $x_1(t)$

$$\frac{d}{dt}\{x_2(t)/x_1(t)\}=x_1(t)^{-2}\,W(t) \tag{2.3 - 15}$$

This result permits to determine $x_2(t)$ when $x_1(t)$ is known

$$x_2(t)=x_1(t)\int^{t} d\vartheta\; x_2^{-2}(\vartheta)\,exp\left\{-2\int^{\vartheta}\frac{d\Theta}{\tau(\Theta)}\right\} \quad . \tag{2.3 - 16}$$

e) Initial value problems and propagator matrices

In many cases the main interest is in motions of an oscillator that occur for given initial conditions. These conditions may be the initial values of the variable $x(t)$ and its time derivative $\dot{x}(t)$ at time $t = 0$. The *existence* and *uniqueness* of the corresponding solutions of (2.3 - 1) are ruled by the *uniqueness theorem* [Birkhoff & Rota 1989 B]:

If the functions $\Omega(t)$ and $\{1 / \tau(t)\}$ of (2.3 - 1) are continuous, then there exists at most one solution $x(t)$ which fulfills the initial conditions $x(0) = x_0$ and $\dot{x}(0) = \dot{x}_0$.

The initial value problems of oscillations determined by (2.3 - 3) can be solved with *propagator matrices*. For this purpose one uses two linearly independent solutions $u_1(t)$ and $u_2(t)$ which fulfill the following initial conditions at time $t = 0$

$$u_1(0)=1\,;\quad \dot{u}_1(0)=0 \quad , \tag{2.3 - 17a}$$

$$u_2(0)=0\,;\quad \dot{u}_2(0)=1 \quad . \tag{2.3 - 17b}$$

If these solutions are known, every solution $u(t)$ of (2.3 - 3) with arbitrary initial values $u(0)$ and $\dot{u}(0)$ at time $t = 0$ can be represented with aid of a *propagator* [Zwillinger 1989 B] as follows

$$\begin{pmatrix} u(t) \\ \dot{u}(t) \end{pmatrix} = \boldsymbol{P}(0,t)\begin{pmatrix} u(0) \\ \dot{u}(0) \end{pmatrix} \quad \text{with } \boldsymbol{P}(0,t)=\begin{pmatrix} u_1(t) & u_2(t) \\ \dot{u}_1(t) & \dot{u}_2(t) \end{pmatrix} \quad . \tag{2.3 - 18a}$$

The determinant of the propagator (2.3 - 18a) is the Wronski determinant of $u_1(t)$ and $u_2(t)$. Since these two functions fulfill (2.3 - 3), this Wronski determinant is constant according to (2.3 - 12). Because of the initial conditions (2.3 - 17a&b) for $u_1(t)$ and $u_2(t)$ this Wronski determinant equals unity

$$det\,\boldsymbol{P}(0,t) = 1 \quad . \tag{2.3 - 18b}$$

Consequently, the propagator (2.3 - 18a) constitutes a *unimodular matrix*.

2.3.3 Zeros and Oscillatory Behavior

The *zeros* of the solutions of (2.3 - 1) and its normal form (2.3 - 3) give information on the *oscillatory behavior* of an oscillator. Oscillatory behavior is characterized by a finite or infinite number of sign changes or zeros of a solution $x(t)$ or $u(t)$. The appearance of zeros and their interrelations are governed by laws which have been originally derived by Ch. Sturm [Birkhoff & Rota 1989 B, Hairer et al. 1980 B].

a) Sturm's separation theorem

If $x_1(t)$ and $x_2(t)$ are linearly independent solutions of (2.3 - 1), then $x_2(t)$ *has a zero* at a time t_k^* between the times t_k and t_{k+1} of two successive zeros of $x_1(t)$ [Birkhoff & Rota 1989 B]. Consequently, the zeros of $x_1(t)$ and $x_2(t)$ are arranged according to the following scheme

$$\begin{aligned} &0=\ldots=x_1(t_{k-1})=x_1(t_k)=x_1(t_{k+1})=\ldots \quad , \\ &0=\ldots=x_2\left(t_{k-1}^*\right)=x_2(t_k^*)=x_2(t_{k+1}^*)=\ldots \quad , \\ &<<t_{k-1}<t_{k-1}^*<t_k<t_k^*<t_{k+1}<t_{k+1}^*<< \quad . \end{aligned} \tag{2.3 - 19}$$

As an *example* we consider the harmonic oscillator with $\Omega(t) = \Omega$, $\tau(t) = \infty$. Linearly independent solutions of the corresponding basic equation (2.3 - 1) and their zeros are

$$\begin{aligned} &x_1(t)=C_1 \cos\Omega t: \quad t_k = \ldots, \pi/2\Omega, 3\pi/2\Omega, \ldots \quad , \\ &x_2(t)=C_2 \sin\Omega t: \quad t_k^* = \ldots, 0, 2\pi/2\Omega, 4\pi/2\Omega, \ldots \quad . \end{aligned} \tag{2.3 - 20}$$

b) Sturm's comparison theorem

If $u(t)$ and $\upsilon(t)$ are nontrivial solutions of the two following oscillation equations in the normal form (2.3 - 3)

$$\ddot{u}+\Omega_1^2(t)u=0 \quad \text{and} \quad \ddot{\upsilon}+\Omega_2^2(t)\upsilon=0 \tag{2.3 - 21a}$$

with

$$\Omega_1^2(t)\geq\Omega_2^2(t) \quad , \tag{2.3 - 21b}$$

then there exists *at least one zero* of $u(t)$ between two arbitrary successive zeros of $\upsilon(t)$ except for $\Omega_1(t) \equiv \Omega_2(t)$ and $\upsilon(t) = C\,u(t)$ [Birkhoff & Rota 1989 B].

As example we consider two harmonic oscillators with

$$\Omega_1(t)=2\Omega>0, \tau(t)=\infty \quad \text{and} \quad \Omega_2(t)=\Omega>0, \tau_2(t)=\infty \quad . \tag{2.3. - 22a}$$

As nontrivial solutions of the corresponding oscillation equations (2.3 - 21) one can choose

$$u(t) = C_1 \, cos \, 2\Omega t \quad \text{and} \quad \upsilon(t) = C_2 \, sin \, \Omega t \quad . \tag{2.3 - 22b}$$

The zeros of these solutions are at

$$\begin{aligned} t_k &= \ldots, -\pi / 4\Omega, \pi / 4\Omega, 3\pi / 4\Omega, 5\pi / 4\Omega, \ldots \quad , \\ t_k^* &= \ldots, 0, 4\pi / 4\Omega, 8\pi / 4\Omega, \ldots \quad . \end{aligned} \tag{2.3 - 22c}$$

Thus, two zeros of $u(t)$ occur between two successive zeros of $\upsilon(t)$.

c) Absence of oscillation

A corollary of Sturm's comparison theorem permits to predict the absence of oscillation for the solutions of an oscillation equation in its normal form (2.3 - 3). If the invariant $I(t)$ of the normal form fulfills the condition

$$\Omega_0^2(t) = I(t) \leq 0 \tag{2.3 - 23}$$

then a nontrivial solution $u(t)$ of (2.3 - 3) has *at most one zero* [Birkhoff & Rota 1989 B]. Consequently, there occurs *no oscillation.*

An *example* is the oscillator that obeys the equation

$$\ddot{u} + \Omega_0^2(t) u = 0 \quad \text{with} \quad \Omega_0^2(t) = I(t) = -\alpha^2 < 0 \quad . \tag{2.3 - 24a}$$

The general solution of this equation

$$u(t) = u(0) \, cosh \, \alpha t + \left(\dot{u}(0) / \alpha \right) sinh \alpha t \tag{2.3 - 24b}$$

shows no oscillation. The only possible zero of $u(t)$ at a time t_1 is determined by the equation

$$tanh(\alpha t_1) = -\alpha \{ u(0) / \dot{u}(0) \} \quad . \tag{2.3 - 24c}$$

According to this equation there exists exactly one zero if the right-hand term is situated in the interval from –1 to +1. Otherwise, there exists no zero.

2.3.4 Chirp Oscillators

By definition chirp oscillators oscillate with a frequency that increases or decreases continuously with time t. In electronics and quantum electronics this phenomenon is called *up-chirp* if the frequency increases and *down-chirp* if the frequency decreases with time t. The following is devoted to the problem which of the time-dependent linear oscillators described by the normal form (2.3 - 3) exhibit a chirp.

a) Chirp and amplitude

An oscillation of a linear oscillator with chirp may be represented as

$$u(t) = exp - \{\vartheta(t) + i\varphi(t)\} \quad . \tag{2.3 - 25}$$

The combination of this ansatz with (2.3 - 3) yields the relations

$$\Omega_0^2 = \omega^2 - \dot{\vartheta}^2 + \ddot{\vartheta} \text{ with } \omega = \dot{\varphi} \quad , \quad \text{and} \tag{2.3 - 26a}$$

$$2\dot{\vartheta} = \dot{\omega} / \omega = \frac{d}{dt} \ell n \omega \quad . \tag{2.3 - 26b}$$

Equation (2.3 - 26b) demands a decreasing amplitude for an up-chirp with $\dot{\omega} > 0$ and an increasing amplitude for a down-chirp with $\dot{\omega} < 0$. This is demonstrated subsequently for a linear chirp.

b) Linear chirp

A linear chirp is defined by the ansatz

$$\varphi(t) = \varphi_0 + \omega_0 t \left[1 + (t / 2\Theta)\right] \quad , \tag{2.3 - 27a}$$

$$\omega(t) = \dot{\varphi}(t) = \omega_0 \left[1 + (t / \Theta)\right] \quad , \tag{2.3 - 27b}$$

$$\Theta > 0: \quad up-chirp; \qquad \Theta < 0: \quad down-chirp \quad . \tag{2.3 - 27c}$$

By introducing the ansatz (2.3 - 27b) in (2.3 - 26a&b), integration and separation of the real part (2.3 - 25) one finds in general

$$\Omega_0^2(t) = (\omega_0 / \Theta)^2 \left[t + \Theta\right]^2 - (3/4)\left[t + \Theta\right]^{-2} \quad . \tag{2.3 - 28}$$

One writes for the up-chirp with $\Theta > 0$, $t > 0$

$$u(t) = A\left[t + \Theta\right]^{-1/2} cos\left[\varphi_0 + \omega_0 t\left(1 + (t / 2\Theta)\right)\right] \quad , \tag{2.3 - 29a}$$

and for the down-chirp with $\Theta < 0$, $0 < t < |\Theta|$

$$u(t) = A\left[|\Theta| - t\right]^{-1/2} cos\left[\varphi_0 + \omega_0 t\left(1 - (t / 2|\Theta|)\right)\right] \quad . \tag{2.3 - 29b}$$

In the following other chirp oscillators are surveyed briefly with the aid of formulae describing $\omega(t)$, $\Omega_0^2(t)$ and $u(t)$. These can be derived by combining (2.3 - 3), (2.3 - 25) and (2.3 - 26a&b).

c) Euler down-chirp
Euler differential equation [Kamke 1956 B: equation 2.14]

$$\omega(t)=\dot{\varphi}(t)=\Phi / t \quad , \tag{2.3 - 30a}$$

$$\Omega_0^2(t)=\left[\Phi^2+(1/2)^2\right]t^{-2} \quad , \tag{2.3 - 30b}$$

$$u(t)=A\left(t/t_0\right)^{1/2} cos\left[\Phi \, \ell n(t/t_0)\right] \quad . \tag{2.3 - 30c}$$

General solution: A and t_0 arbitrary.

d) Down-chirp of the characteristic frequency
[Kamke 1956 B: equation 2.342]

$$\omega(t)=\dot{\varphi}(t)=\Theta / t^2 \quad , \tag{2.3 - 31a}$$

$$\Omega_0(t)=\Theta / t^2=\omega(t) \quad , \tag{2.3 - 31b}$$

$$u(t)=A\left(t/t_0\right) cos\left[\varphi_\infty-(\Theta / t)\right] \quad . \tag{2.3 - 31c}$$

General solution: A/t_0 and φ_∞ arbitrary.

Equation (2.3 - 31b) demonstrates that the effective circular frequency $\omega(t)$ of this oscillator equals the characteristic circular frequency $\Omega_0(t)$.

e) Exponential chirp

$$\omega(t)=\dot{\varphi}(t)=\omega_0 \, exp(t/\Theta) \quad , \tag{2.3 - 32a}$$

$$\Omega_0^2(t)=\omega_0^2 \, exp(2t/\Theta)-(1/2\Theta)^2 \quad , \tag{2.3 - 32b}$$

$$u(t)=A \, exp\left[-t/2\Theta\right] cos\left[\varphi_0+\omega\,\Theta\, exp(t/\Theta)\right] \quad . \tag{2.3 - 32c}$$

General solution: A and φ_0 arbitrary.

2.3.5 Aperiodically Modulated Linear Oscillators

For the chirp oscillators described in Section 2.3.4 the characteristic circular frequency $\Omega_0(t)$ of the oscillation equation in its normal form varies aperiodically. This is obvious for (2.3 - 28), (2.3 - 30b), (2.3 - 31b) and (2.3 - 32b). However, there exist many other linear oscillators whose characteristic circular frequencies $\Omega_0(t)$ exhibit an aperiodic modulation.

Oscillations of an aperiodically modulated oscillator described by (2.3 - 3) can be represented as linear combination of two linearly independent solutions $u_1(t)$ and $u_2(t)$ of this equation

$$u(t)=C_1\,u_1(t)+C_2\,u_2(t) \qquad (2.3 - 33)$$

in agreement with (2.3 - 7). C_1 and C_2 indicate arbitrary constants. Consequently, it suffices to use two linearly independent solutions $u_1(t)$ and $u_2(t)$ for the description of these oscillations. The following represents a *concise survey* on common aperiodically modulated linear oscillators. Listed are the characteristic circular frequencies $\Omega_0(t)$ and the solutions $u_k(t)$.

a) Airy oscillators
[Abramowitz & Stegun 1965 B: Chapter 10.4]

$$\Omega_0^2(t)=t \quad . \qquad (2.3 - 34a)$$

α) *Basic functions I:*

$$u_1(t)=Ai(-t) \quad , \qquad (2.3 - 34b)$$

$$u_2(t)=Bi(-t) \quad , \qquad (2.3 - 34c)$$

$Ai(t)$, $Bi(t)$ = Airy functions.

β) *Basic functions II:*

$$u_1(t)=Ac(t)=f(-t)=1-\frac{1}{3!}t^3+\frac{1\cdot 4}{6!}t^6-\frac{1\cdot 4\cdot 7}{9!}t^9+\;-\quad , \qquad (2.3 - 34d)$$

$$u_2(t)=As(t)=-g(-t)=t-\frac{2}{4!}t^4+\frac{2\cdot 5}{7!}t^7-\frac{2\cdot 5\cdot 8}{10!}t^{10}+\;-\quad , \qquad (2.3 - 34e)$$

$Ac(t)$, $As(t)$ = Airy cosine, Airy sine.

The functions $Ac(t)$ and $As(t)$ are illustrated in Fig. 2.3 - 1 for $t>0$. They are defined such that

$$Ac(0)=1,\ \frac{d}{dt}Ac(0)=0 \quad , \qquad (2.3 - 34f)$$

$$As(0)=0,\ \frac{d}{dt}As(0)=1 \quad . \qquad (2.3 - 34g)$$

Consequently they can be used for solving initial value problems with the propagator (2.3 - 18).

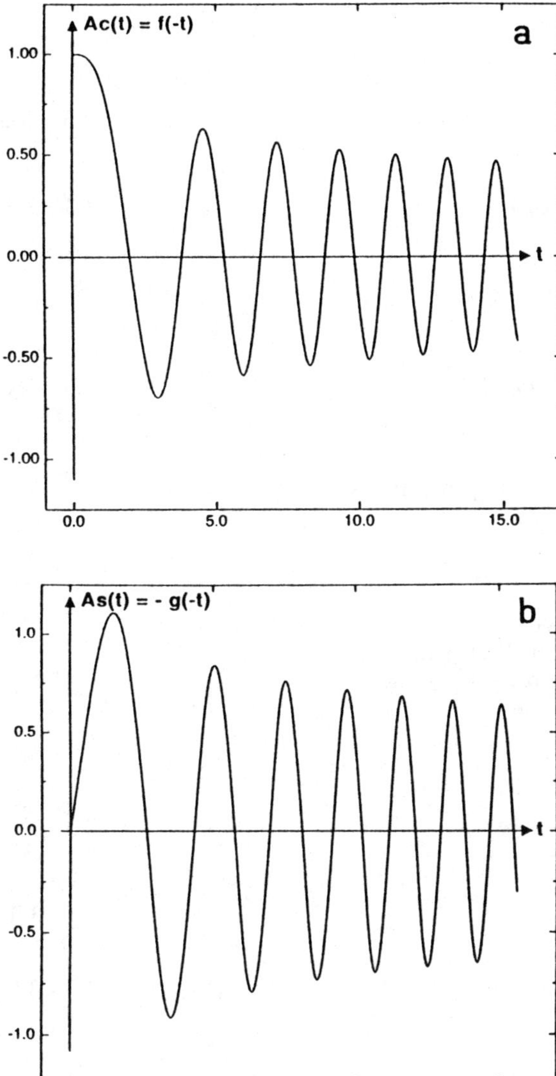

Fig. 2.3 - 1. Airy cosine (a) and Airy sine (b)

b) Weber oscillators
[Abramowitz & Stegun 1965 B, Pöschl 1956 B]

$$\Omega_0^2(t) = n + \frac{1}{2} - \frac{t^2}{4} \quad \text{with} \quad n = 0, 1, 2, 3, \ldots \quad , \tag{2.3 - 35a}$$

$$u_1(t) = H_n(t)\, exp\left[-t^2/2\right] \quad , \tag{2.3 - 35b}$$

$H_n(t)$ = Hermite polynomials .

c) Forsyth oscillators
[Kamke 1956 B: equation 2. 153]

$$\Omega_0^2(t)=\omega_\infty^2-n(n-1)t^{-2} \quad \text{with} \quad n = 0, 1, 2, 3, \ldots \quad , \tag{2.3 - 36a}$$

$$u(t)=A\,t^{n}\left[t^{-1}\frac{d}{dt}\right]^{n} \cos\left(\omega_\infty t-\varphi_0\right) \quad . \tag{2.3 - 36b}$$

General solution: A and φ_0 arbitrary.
Example: $n = 2$, $\varphi_0 = 0$,

$$u(t)=A\omega_\infty\left[t^{-1}\sin\omega_\infty t-\omega_\infty\cos\omega_\infty t\right] \quad . \tag{2.3 - 36c}$$

d) Bessel oscillators
[Abramowitz & Stegun 1965 B: equations 9.1.49 - 50 - 54]
[Kamke 1956 B: equation 2.162 (20)]

$n = 0, 1, 2, 3, \ldots$
$J_n(t)$ = Bessel functions of 1st order
$Y_n(t)$ = Bessel functions of 2nd order; Weber functions
$Z_n(t) = A\,J_n(t) + B\,Y_n(t)$ = general Bessel functions

α) *Oscillator type I*

$$\Omega_0^2(t)=\omega_\infty^2-\left(n^2-\frac{1}{4}\right)t^{-2} \quad , \tag{2.3 - 37a}$$

$$u_1(t)=t^{1/2}\,J_n\left(\omega_\infty t\right) \quad , \tag{2.3 - 37b}$$

$$u_2(t)=t^{1/2}\,Y_n\left(\omega_\infty t\right) \quad . \tag{2.3 - 37c}$$

β) *Oscillator type II*

$$\Omega_0^2(t)=(\eta/2)^2t^{-1}-\left(n^2-1\right)t^{-2} \quad , \tag{2.3 - 38a}$$

$$u_1(t)=t^{1/2}\,J_n\left(\eta t^{1/2}\right) \quad , \tag{2.3 - 38b}$$

$$u_2(t)=t^{1/2}\,Y_n\left(\eta t^{1/2}\right) \quad . \tag{2.3 - 38c}$$

γ) *Oscillator type III*

$$\Omega_0^2(t)=\left(\omega_0 e^{t}\right)^2-n^2 \quad , \tag{2.3 - 39a}$$

$$u_1(t)=J_n\left(\omega_0 e^t\right) \quad , \tag{2.3 - 39b}$$

$$u_2(t)=Y_n\left(\omega_0 e^t\right) \quad . \tag{2.3 - 39c}$$

δ) *Oscillator type IV*

$$\Omega_0^2(t)=t^{-4}\left[exp\left(2t^{-1}\right)-n^2\right] \quad , \tag{2.3 - 40a}$$

$$u_1(t)=t\,J_n\left(exp(1/t)\right) \quad , \tag{2.3 - 40b}$$

$$u_2(t)=t\,Y_n\left(\exp(1/t)\right) \quad . \tag{2.3 - 40c}$$

e) Laguerre oscillators
[Abramowitz & Stegun 1965 B: equations 22.6.17&18]

α) *Oscillator type I*

$$\begin{aligned}&\Omega_0^2(t)=-(1/4)+(2n+m+1)(2t)^{-1}+(1-m^2)(2t)^{-2}\\&\text{with}\quad n=1,2,3,\ldots\quad,\quad m=0,1,2,3,\ldots,n\quad ,\end{aligned} \tag{2.3 - 41a}$$

$$\begin{aligned}&u_1(t)=exp(-t/2)\,t^{(m+1)/2}L_n^m(t)\quad ,\\&L_n^m(t)=\text{generalized Laguerre polynomials}\quad .\end{aligned} \tag{2.3 - 41b}$$

β) *Oscillator type II*

$$\begin{aligned}&\Omega_0^2(t)=2(2n+m+1)-t^2+(1-m^2)(2t)^{-2}\\&\text{with}\quad n=1,2,3,\ldots\quad,\quad m=0,1,2,3,\ldots,n\quad ,\end{aligned} \tag{2.3 - 42a}$$

$$u_1(t)=exp(-t^2/2)\,t^{(2m+1)/2}\,L_n^m(t^2) \quad . \tag{2.3 - 42b}$$

f) Coulomb oscillators
[Abramowitz & Stegun 1965 B: Chapter 14]

$$\Omega_0^2(t)=1-2\eta t^{-1}-L(L+1)t^{-2} \quad \text{with} \quad L=0,\ 1,\ 2,\ 3,\ \ldots \quad , \tag{2.3 - 43a}$$

$$u_1(t)=F_L(\eta,t) \ = \text{regular Coulomb functions} \quad , \tag{2.3 - 43b}$$

$$u_2(t)=G_L(\eta,t) \ = \text{irregular (logarithmic) Coulomb functions} \quad . \tag{2.3 - 43c}$$

g) Whittaker oscillators
[Abramowitz & Stegun 1965 B: equations 13.1.31-33]

$$\Omega_0^2(t) = -\frac{1}{4} + \kappa t^{-1} + (\frac{1}{4} - \mu^2) t^{-2} \quad , \tag{2.3 - 44a}$$

$$u_1(t) = M_{\kappa,\mu}(t) \quad , \tag{2.3 - 44b}$$

$$u_2(t) = W_{\kappa,\mu}(t) \quad , \tag{2.3 - 44c}$$

$M_{x,\mu}(t)$, $W_{\kappa,\mu}(t)$ = Whittaker functions .

h) Modified Darboux oscillators
[Kamke 1956 B: e.g. equation 2.420]

$$\Omega_0^2(t) = \omega_\infty^2 - n(n-1)(\cosh t)^{-2} \quad \text{with} \quad n = 0, 1, 2, 3, \ldots \quad , \tag{2.3 - 45a}$$

$$u(t) = A\ (\cosh t)^{n} \left[\frac{1}{\cosh t} \frac{d}{dt} \right]^{n} \cos\left(\omega_\infty t - \varphi_0\right) \quad . \tag{2.3 - 45b}$$

General solution: A and φ_0 arbitrary.
Example: $n = 2$, $\varphi_0 = 0$,

$$u(t) = A\,\omega_\infty \left[\tanh t \ \sin \omega_\infty t - \omega_\infty \ \cos \omega_\infty t \right] \quad . \tag{2.3 - 45c}$$

2.3.6 Oscillators with Step or Pulse Modulation

The mathematical description of step and pulse modulation makes use of the Heaviside step function and the Dirac delta function which are introduced in the following. They correspond to infinitely rapid steps and infinitely short pulses.

a) Heaviside step function and Dirac delta function
The Heaviside step function $H(t)$ is defined as [Bracewell 1986 B]

$$H(t) = \begin{cases} 0 & \text{for } t < 0 \\ 1/2 & \text{for } t = 0 \\ 1 & \text{for } t > 0 \end{cases} \quad . \tag{2.3 - 46a}$$

This function represents a *unit step*. Furthermore, it is related to the *sign function*

$$H(t) = \frac{1}{2}(1 + sign\, t), \text{ where } sign\, t = \frac{d}{dt}|t| = \begin{cases} +1 & \text{for } t > 0 \\ 0 & \text{for } t = 0 \\ -1 & \text{for } t < 0 \end{cases} \quad . \qquad (2.3 - 46b)$$

The Heaviside step function $H(t)$ can also be represented with the aid of the *Dirac delta function* $\delta(t)$

$$H(t) = \int_{-\infty}^{t} \delta(t)\, dt \quad \text{and} \quad \delta(t) = \frac{d}{dt} H(t) = \frac{1}{2} \frac{d^2}{dt^2} |t| \quad . \qquad (2.3 - 46c)$$

The *derivatives* $\delta^{(n)}(t)$ of the Dirac delta function $\delta(t)$ can be written as (Kneubühl 1994 B)

$$\delta^{(n)}(t) = \frac{d^n}{dt^n} \delta(t) = n!(-t)^{-n} \delta(t) \quad , \quad \text{e.g. } \delta(t) = -t \frac{d}{dt} \delta(t) \quad . \qquad (2.3 - 46d)$$

The Dirac delta function $\delta(t)$ constitutes an *infinitely short unit impulse, pulse or shock*. It is characterized by the integral [Bracewell 1986 B, Schubert & Weber 1980 B]

$$\int_{t_1}^{t_2} \delta(t - t_0) f(t)\, dt = \begin{cases} f(t_0) & \text{if } \quad t_1 < t_0 < t_2 \\ 0 & \text{if } \quad t_0 < t_1 \quad \text{or} \quad t_0 > t_2 \end{cases} \quad . \qquad (2.3 - 47)$$

In addition, it can be represented as limits of continuous functions $\varphi(t, \varepsilon)$ which fulfill the condition

$$\int_{-\infty}^{+\infty} \varphi(t, \varepsilon)\, dt = 1 \quad \text{for} \quad 0 < \varepsilon < C \quad . \qquad (2.3 - 48a)$$

Examples are the limit of a *Lorentz function*

$$\delta(t) = \lim_{\varepsilon \to 0} \frac{\varepsilon}{\pi} \left[t^2 + \varepsilon^2 \right]^{-1} \quad \text{for} \quad 0 \leq \varepsilon \quad , \qquad (2.3 - 48b)$$

the limit of a *Gauss function*

$$\delta(t) = \lim_{\varepsilon \to 0} \frac{1}{\sqrt{2\pi}\, \varepsilon} exp\left[-t^2 / 2\varepsilon^2 \right] \quad \text{for} \quad 0 \leq \varepsilon \quad , \qquad (2.3 - 48c)$$

and the limit of a *pole*

$$\delta(t) = \lim_{\varepsilon \to 0} \begin{cases} 1/\varepsilon & \text{if } |t| < \varepsilon/2 \\ 0 & \text{if } |t| > \varepsilon/2 \end{cases} \quad \text{for } 0 \le \varepsilon \quad . \qquad (2.3 - 48d)$$

The Dirac delta function $\delta(t)$ is also significant with respect to the *Fourier transformation* (2.2 - 46a&b) because it is involved in the *Fourier transforms* of the following functions [Bracewell 1986 B, Pöschl 1956 B, Schubert & Weber 1980 B]

$$\boldsymbol{F}\{x(t) = 1\} = F(\omega) = 2\pi\,\delta(\omega) \quad , \qquad (2.3 - 49a)$$

$$\boldsymbol{F}\{x(t) = exp(-i\omega_0 t)\} = \mathrm{F}(\omega) = 2\pi\,\delta(\omega - \omega_0) \quad . \qquad (2.3 - 49b)$$

Similarly, the Dirac delta function $\delta(t)$ can also be represented by the *inverse Fourier transforms*

$$\boldsymbol{F}^{-1}\{F(\omega) = 1\} = x(t) = \delta(t) \quad , \qquad (2.3 - 49c)$$

$$\boldsymbol{F}^{-1}\{F(\omega) = exp(+i\omega\tau)\} = x(t) = \delta(t - \tau) \quad . \qquad (2.3 - 49d)$$

The *Laplace transforms* of $H(t)$, $\delta(t)$ and its derivatives are listed in Appendix A.3.2.

b) Step modulation

The following is dedicated to a harmonic oscillator whose characteristic circular frequency $\Omega_0(t)$ of (2.3 - 3) varies step by step as shown in Fig. 2.3 - 2. The steps k occur at successive times t_k such that $t_{k+1} > t_k$. In addition, the characteristic circular frequency $\Omega_0(t)$ does not vary between the steps and remains positive at all times t which implies

$$\begin{aligned} &\Omega_0(t_k \le t < t_{k+1}) = \Omega_k = const > 0 \quad , \\ &\text{with} \quad \Omega_k \ne \Omega_{k-1} \quad \text{and} \quad \Omega_k \ne \Omega_{k+1} \quad . \end{aligned} \qquad (2.3 - 50)$$

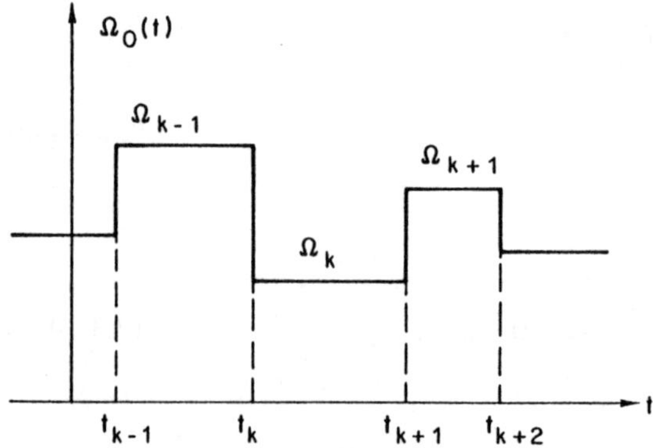

Fig. 2.3 - 2. Step modulation of the characteristic circular frequency $\Omega_0(t)$ [Kneubühl 1995 B]

With these assumptions $\Omega_0(t)$ can be represented as a sum over Heaviside step functions (2.3 - 46a-c)

$$\Omega_0(t) = \sum_k (\Omega_k - \Omega_{k-1}) H(t - t_k) \quad . \tag{2.3 - 51}$$

The oscillation $u(t)$ of the oscillator with a step modulation described by equation (2.3 - 51) and Fig. 2.3 - 2 can be determined with the aid of the propagators (2.3 - 18). If $u(t_0)$ and $\dot{u}(t_0)$ are the values of the variable $u(t)$ and its time derivative $\dot{u}(t)$ at time t_0, then $u(t)$ and $\dot{u}(t)$ are given by

$$\begin{pmatrix} u(t) \\ \dot{u}(t) \end{pmatrix} = \boldsymbol{P}(t_0, t) \begin{pmatrix} u(t_0) \\ \dot{u}(t_0) \end{pmatrix} \quad . \tag{2.3 - 52a}$$

For a time t in the interval between t_k and t_{k+1} the propagator of the oscillator with step modulation has the form

$$\boldsymbol{P}(t_0, t) = \boldsymbol{P}(t_k, t)\boldsymbol{P}(t_{k-1}, t_k)\ldots\boldsymbol{P}(t_1, t_2)\boldsymbol{P}(t_0, t_1) \quad . \tag{2.3 - 52b}$$

Each of these factors represents a matrix

$$\boldsymbol{P}(t_n, t) = \begin{pmatrix} \cos\Omega_n(t - t_n) & \dfrac{1}{\Omega_n}\sin\Omega_n(t - t_n) \\ -\Omega_n \sin\Omega_n(t - t_n) & \cos\Omega_n(t - t_n) \end{pmatrix} \text{where } t_n \le t < t_{n+1} \quad . \tag{2.3 - 52c}$$

These matrices (2.3 - 52c) obey the following laws:

α) They are *unimodular*. This means that their determinant is unity

$$\det \boldsymbol{P}(t_n, t) = |\boldsymbol{P}(t_n, t)| = 1 \quad . \tag{2.3 - 53a}$$

β) The traces of these matrices give information on the *phase shift* $\Omega_n(t - t_n)$ by the relation

$$tr\, \boldsymbol{P}(t_n, t) = 2 \cos \Omega_n (t - t_n) \quad . \tag{2.3 - 53b}$$

γ) The two *eigenvalues* $\gamma_{1,2}$ of these matrices are *conjugate complex*

$$\gamma_{1,2} = \Lambda(\boldsymbol{P}(t_n, t)) = \exp[\pm i\Omega_n (t - t_n)] \quad . \tag{2.3 - 53c}$$

The *propagator representation* (2.3 - 52a-c) and (2.3 - 53a-c) of the oscillations of harmonic oscillators with step modulation finds various *applications*, e.g.

α) the *digital computation* of the oscillations $u(t)$ of continuously modulated linear oscillators such as described in Sections 2.3 - 4&5. For this purpose the continuous time t is e.g. devided into equal intervals $t_{n+1} - t_n = \Delta t$. Each interval is assigned a constant circular frequency

$$\Omega_0 (t_n) = \Omega_n \quad \text{with} \quad t_n = t_0 + n\, \Delta t \quad . \tag{2.3 - 54}$$

The subsequent digital computation of the oscillation $u(t)$ is then performed with the propagators (2.3 - 52a-c).

β) the evaluation of oscillations of the *oscillators with periodic step modulations* which are discussed in Section 2.3.7d.

c) Pulse modulation

A harmonic oscillator with general pulse modulation can be described by a square of the characteristic circular frequency $\Omega_0(t)$ of the form

$$\Omega_0^2(t) = \Omega^2 + \sum_k p_k\, \delta(t - t_k) \tag{2.3 - 55}$$

where $\delta(t)$ indicates the Dirac delta function (2.3 - 46c).

The oscillations $u(t)$ of this oscillator can be determined with the aid of *propagators* by taking into account that

α) for times $t \neq t_k$ the propagator (2.2 - 20a-c) of the simple harmonic oscillator can be applied.

β) the intervals from $t_k - \delta t$ to $t_k + \delta t$ where $\delta t > 0$ are extremely short and therefore need special attention. Consequently, these intervals are assigned specific propagators $\boldsymbol{P}(t_k - \delta t, t_k + \delta t)$ which can be calculated as follows:

For simplicity one first assumes $t_n = 0$. Then one integrates the oscillator equation (2.3 - 3) with the characteristic circular frequency $\Omega_0(t)$ defined by (2.3 - 55) over the time t from $-\delta t$ to $+\delta t$

$$\int_{-\delta t}^{+\delta t} dt \left[\ddot{u} + u\, \Omega^2 + u \sum_k p_k\, \delta(t - t_k) \right] \approx \dot{u}(+\delta t) - \dot{u}(-\delta t) + 0\, u(0)\Omega^2 + u(0) p_n = 0 \quad .$$

For physical reasons one can assume that $u(t)$ is continuous. This implies

$$u(+\delta t) \approx u(-\delta t) \approx u(0) \quad \text{or} \quad \left[\dot{u}(+\delta t) - \dot{u}(-\delta t) + p_n u(-\delta t) \right] \approx 0 \quad .$$

From this equation one can deduce the propagators for the intervals $t_k - \delta t$ to $t_k + \delta t$ at the times t_k where the modulation pulses occur

$$P(t_k - \delta t,\, t_k + \delta t) \approx \begin{pmatrix} 1 & 0 \\ -p_k & 1 \end{pmatrix} \quad . \tag{2.3 - 56}$$

This is the *characteristic propagator of pulse modulation.*

2.3.7 Parametric Linear Oscillators

We define parametric linear oscillators [Nayfeh & Mook 1979 B] as periodically modulated linear oscillators described by a *Hill differential equation* of the form

$$\ddot{u}(t) + \Omega_0^2(t)u(t) = 0 \quad \text{with} \quad \Omega_0^2(t) = I(t) = I(t+T) = \Omega_0^2(t+T) \quad . \tag{2.3 - 57a}$$

This equation is characterized by a periodic invariant $I(t)$. The period T of modulation determines the *circular frequency* ω *of modulation*

$$\omega = 2\pi / T \tag{2.3 - 57b}$$

of the parametric oscillator. The following discussion is restricted to the well-known *real Hill differential equations* with real invariants $I(t)$. [Abramowitz & Stegun 1965 B, Jakubic & Starzinski 1975 B, Magnus & Winkler 1965 B, Nayfeh & Mook, 1979 B]. These equations are applied in solid state physics for the theory on energy bands of electrons in periodic potentials [Ashcroft & Mermin 1976 B, Blakemore 1974 B, Kachhava 1990 B, Kittel 1971 B, Kreher 1976 B, Kronig & Penney 1930 J, Ziman 1965 B], in electronics for the discussion of periodic filters [Brillouin 1946 B] and in microwave techniques for the characterization of periodic waveguides. In addition they can be applied to solve stability problems [Stoker 1957 B]. On the contrary, the complex Hill differential equations with complex invariants $I(t)$ are more difficult to understand and less known [Meiman 1977 J, Strutt 1949 J]. They find application e.g. in theories on distributed feedback lasers [Kneubühl 1993 B, Kneubühl & Sigrist 1995 B].

Parametric linear oscillators may exhibit resonant amplification and damping with respect to the modulation frequency ω. These are called *parametric resonances*. As an example the motion of a *swing* is discussed in Section 2.3.7d.

a) Floquet theory

Floquet's theory [Abramowitz & Stegun 1965 B, Magnus & Winkler 1965 B, Stoker 1957 B] provides general rules and a classification of the solutions $u(t)$ of the Hill differential equation (2.3 - 57a).

The characteristic propagator $\boldsymbol{P}(0, T)$ of a real Hill differential equation can be derived with the aid of two real linearly independent solutions $u_1(t)$ and $u_2(t)$ of the real Hill equation (2.3 - 57a), which fulfill the initial conditions

$$u_1(0) = 1, \quad \dot{u}_1(0) = 0 \quad , \tag{2.3 - 17a}$$

$$u_2(0) = 0,\ \dot{u}_2(0) = 1 \quad . \tag{2.3 - 17b}$$

In agreement with (2.3 - 18) the following propagator relates an arbitrary real solution *u*(*t*) of (2.3 - 57a) at time *t* with that one period *T* later [Stoker 1957 B, Zwillinger 1989 B]

$$\begin{pmatrix} u(t+T) \\ \dot{u}(t+T) \end{pmatrix} = \boldsymbol{P}(0,T) \begin{pmatrix} u(t) \\ \dot{u}(t) \end{pmatrix} \tag{2.3 - 58a}$$

with

$$\boldsymbol{P}(0,T) = \begin{pmatrix} u_1(T) & u_2(T) \\ \dot{u}_1(T) & \dot{u}_2(T) \end{pmatrix} \tag{2.3 - 58b}$$

and

$$det\, \boldsymbol{P}\,(0,\, T) = 1 \quad . \tag{2.3 - 58c}$$

A solution *u*(*t*) of equation (2.3 - 57a) is called a *Floquet solution*, *Bloch function* or normal [Stoker 1957 B] if it fulfills the relation

$$u(t + T) = \gamma\ u(t) \tag{2.3 - 59}$$

where γ indicates a constant factor.

Taking into account of (2.3 - 58a&b) we can satisfy (2.3 - 59) only if *u*(*t*) is a non-trivial solution of the following system of two linear equations

$$\begin{pmatrix} u_1(T) - \gamma & u_2(T) \\ \dot{u}_1(T) & \dot{u}_2(T) - \gamma \end{pmatrix} \begin{pmatrix} u(t) \\ \dot{u}(t) \end{pmatrix} = 0 \quad . \tag{2.3 - 60}$$

This solution requires that the determinant of the corresponding matrix vanishes. Because $\boldsymbol{P}(0, T)$ is unimodular according to equation (2.3 - 58c) this requirement yields the following *eigenvalue equation* for γ

$$\gamma^2 - \ tr\, \boldsymbol{P}(0,\, T)\, \gamma\ + 1 = 0 \quad . \tag{2.3 - 61}$$

Consequently, the sum and the product of the two eigenvalues of γ can be represented with the aid of the trace and determinant of the propagator matrix $\boldsymbol{P}(0, T)$ as follows

$$\gamma_1 + \gamma_2 = tr\, \boldsymbol{P}(0,T) = u_1(T) + \dot{u}_2(T) \quad , \tag{2.3 - 62a}$$

$$\gamma_1\, \gamma_2 = det\, \boldsymbol{P}(0,T) = 1 \quad . \tag{2.3 - 62b}$$

Equation (2.3 - 62b) permits the following ansatz for the two eigenvalues of γ

$$\gamma_{1,2}=exp\left[\pm\left(\frac{1}{\vartheta}+i\omega_S\right)T\right] \tag{2.3 - 63a}$$

or

$$\pm\left(\frac{1}{\vartheta}+i\omega_S\right)=\frac{1}{T}\,\ell n\;\gamma_{1,2} \quad . \tag{2.3 - 63b}$$

The meaning of ϑ and ω_S will be explained later with the aid of (2.3 - 65c). With the ansatz (2.3 - 63a) equation (2.3 - 62a) can be transformed into the relevant *Floquet equation*

$$cosh\left(\frac{T}{\vartheta}+i\omega_S T\right) = cos\left(\omega_S T-i\frac{T}{\vartheta}\right)=\frac{1}{2}\,tr\boldsymbol{P}(0,T)=\frac{1}{2}\left[u_1(T)+\dot{u}_2(T)\right] \ . \tag{2.3 - 64}$$

The Floquet solutions or Bloch functions defined by (2.3 - 59) exist only for the eigenvalues $\gamma_{1,2}$ of (2.3 - 60). These functions have a *characteristic form.* It can be derived with the aid of the two functions $u_{1,2}(t)$ and $w_{1,2}(t)$[Stoker 1957 B] related by

$$w_{1,2}(t)=u_{1,2}(t)\; exp\left[\mp\left(\frac{t}{\vartheta}+i\omega_S t\right)\right] \quad . \tag{2.3 - 65a}$$

The combination of (2.3 - 59), (2.3 - 63a) and (2.3 - 65a) leads to the conclusion that the functions $w_{1,2}(t)$ oscillate with the circular frequency $\omega = 2\pi/T$ of the modulation. Thus, $w_{1,2}(t)$ have the period T

$$w_{1,2}\,(t + T) = w_{1,2}(t) \quad \text{with} \quad T = 2\pi/\omega \quad . \tag{2.3 - 65b}$$

Therefore, a Floquet solution or Bloch function $u_{1,2}(t)$ can be represented as a product of a periodic function $w_{1,2}(t)$ and an exponential function as follows

$$u_{1,2}(t) = w_{1,2}(t)\,exp\left[\pm\left(\frac{t}{\vartheta}+i\omega_S t\right)\right] = w_{1,2}(t)\,exp\left[\pm\frac{t}{\vartheta}\right]exp\left[\pm i\omega_S t\right] \ . \tag{2.3 - 65c}$$

This equation also illustrates the meaning of ϑ and ω_S. The parameter ϑ is the *characteristic time of amplification or damping*, whilst ω_S represents a *beat frequency*, that originates in the modulation $w_{1,2}(t)$ of the oscillation.

b) Classification of Floquet solutions

The Floquet solutions or Bloch functions $u_{1,2}(t)$ of a real Hill differential equation (2.3 - 57a) can be classified into *five categories*. For this purpose we define the *Bragg circular frequency*

$$\omega_B = \omega\,/2 = \pi/T \tag{2.3 - 66}$$

and introduce an integer parameter $n = 0, \pm 1, \pm 2, \pm 3, \ldots$. Thus the *five categories* can be described as follows:

α) *Frequency gap* or *forbidden frequency range*
Pseudo-periodic solutions which grow or decay exponentially with the characteristic time ϑ

condition: $$cosh\left(\frac{T}{\vartheta}\right)=\frac{1}{2}tr\,\boldsymbol{P}(0,T)>+1 \quad . \qquad (2.3-67a)$$

parameters: $$\vartheta\neq\infty,\ \omega_S=k(even)\ \omega_B=n\omega \quad . \qquad (2.3-67b)$$

solutions: $$u_{1,2}(t)=w_{1,2}(t)\ exp\left[\pm\frac{t}{\vartheta}\right]\ exp(in\omega t) \quad . \qquad (2.3-67c)$$

β) *Band edge*
One periodic solution with the period T and the frequency $\omega = 2\pi / T$ [Nayfeh & Mook 1979 B]

condition: $$\frac{1}{2}tr\,\boldsymbol{P}(0,t)=+1 \quad . \qquad (2.3-68a)$$

parameters: $$\vartheta=\infty,\ \omega_S=k(even)\,\omega_B=n\omega \quad . \qquad (2.3-68b)$$

1st solution: $$u_1(t)=w_1(t)\,exp(in\omega t)=w_1(t)\ exp\left[ik(even)\,\omega_B t\right] \quad . \qquad (2.3-68c)$$

2nd solution: $$u_2(t)=\left[w_2(t)+(t/T)\,w_1(t)\right]exp(in\omega t) \quad . \qquad (2.3-68d)$$

γ) *Allowed frequency range*
Stationary stable oscillations

condition: $$-1< cos(\omega_S T)=\frac{1}{2}tr\,\boldsymbol{P}(0,T) <+1 \quad . \qquad (2.3-69a)$$

parameters: $$\vartheta=\infty,\ \omega_S\neq 0 \quad . \qquad (2.3-69b)$$

solutions: $$u_{1,2}(t)=w_{1,2}(t)\,exp\left[\pm i\omega_S t\right] \quad . \qquad (2.3-69c)$$

Under these circumstances the oscillations are stable. It should be noticed that the beat circular frequency ω_S is determined only to a multiple of the modulation circular frequency $\omega = 2\pi / T$

$$\omega_S = \omega_{SO} + n\omega \quad \text{with} \quad 0 \le \omega_{SO} < \omega \quad \text{and} \quad n = 0, \pm 1, \pm 2, \ldots \quad . \qquad (2.3-69d)$$

This equation justifies the introduction of *Brillouin zones* [Brillouin 1946 B, Kittel 1971 B] which can be numbered by the parameter n.

δ) *Band edge*
One periodic solution with the period $2T = 4\pi/\omega$ corresponding to the Bragg circular frequency $\omega_B = \omega/2$ [Nayfeh & Mook 1979 B] plus a non-periodic solution.

condition: $$\frac{1}{2} tr\, \boldsymbol{P}(0,t) = -1 \quad . \tag{2.3 - 70a}$$

parameters: $$\vartheta = \infty,\ \omega_S = k(odd)\,\omega_B = n\omega + \omega_B \quad . \tag{2.3 - 70b}$$

1st solution: $$\begin{aligned} u_1(t) &= w_1(t)\ exp\left[i(n\omega + \omega_B)t\right] \\ &= w_1(t)\ exp\left[ik(odd)\ \omega_B t\right] \quad . \end{aligned} \tag{2.3 - 70c}$$

2nd solution: $$u_2(t) = \left[w_2(t) + (t/T)\, w_1(t)\right] exp\left[i(n\omega + \omega_B)t\right] \quad . \tag{2.3 - 70d}$$

ε) *Frequency gap* or *forbidden frequency range*
Pseudo-periodic solutions which grow or decay exponentially with the characteristic time δ.

condition: $$-cosh\left(\frac{T}{\vartheta}\right) = \frac{1}{2} tr\boldsymbol{P}(0,t) < -1 \quad . \tag{2.3 - 71a}$$

parameters: $$\vartheta \neq \infty,\ \omega_S = k(odd)\,\omega_B = n\omega + \omega_B \quad . \tag{2.3 - 71b}$$

solutions: $$u_{1,2}(t) = w_{1,2}(t)\, exp\left(\pm \frac{t}{\vartheta}\right) exp\left[\ i(n\omega + \omega_B)t\right] \quad . \tag{2.3 - 71c}$$

In the categories α, β, δ and ε the parameter $k = 0, \pm 1, \pm 2, \pm 3, \ldots$ represents the *Bragg order*. In the categories β and γ where

$$\frac{1}{2} tr\ \boldsymbol{P}(0,T) = cos\ \omega_S T = (-1)^k \quad , \tag{2.3. - 72a}$$

there occur characteristic phenomena called *Bragg effects*.

In analogy to solid state physics, the categories α) and ε) are named *forbidden zones* or *forbidden frequency ranges*. In these ranges the oscillation of the oscillator is either amplified or damped by the modulation with the circular frequency ω depending on the phase lag between oscillation and modulation, respectively the initial conditions. According to (2.3 - 67b) and (2.3 - 71b) the circular frequencies ω_P of these oscillations are multiples of the Bragg circular frequency ω_B

$$k = 0, 1, 2, 3, \quad ; n = 0, \pm 1, \pm 2, \ldots \quad ; \qquad (2.3 - 72b)$$

$$\omega_P = (k + 2n)\, \omega_B = (k + 2n)\, \omega/2 \quad .$$

This phenomenon is named *parametric resonance* because it originates in the periodic modulation of a parameter that determines the frequency of the oscillator. An example is the *swing,* which is discussed in context with the square-wave modulation in Section 2.3.7d.

c) Periodic pulse modulation

The periodic pulse modulation of harmonic oscillators permits a good insight into the behavior of parametric linear oscillators, because it offers analytical solutions $u(t)$ of (2.3 - 3) with periodic invariants $I(t)$ and modified circular frequencies Ω_0 (2.3 - 57a). The periodic pulse modulation can be described with the aid of a sampling function [Bracewell 1986 B] as follows

$$I(t) = \Omega_0^2(t) = \Omega^2 + p \sum_{n=-\infty}^{+\infty} \delta(t - nT) \quad . \qquad (2.3 - 73)$$

It is illustrated in Fig. 2.3 - 3. The Dirac delta function $\delta(t)$ is defined by (2.3 - 46c).

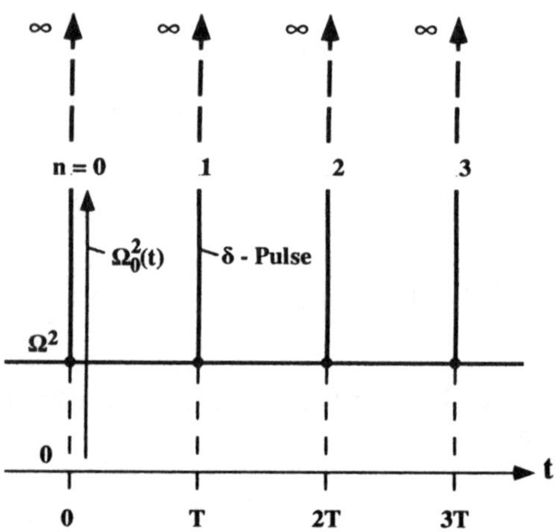

Fig. 2.3 - 3. Periodic pulse modulation

Because the periodically modulated invariant $I(t)$ of equation (2.3 - 73) exhibits Dirac delta functions $\delta(t)$ at the times $t = 0, \pm T, \pm 2T, \ldots$ one chooses the borders of the periodic intervals of duration T at times $t = -\delta t + nT$ with $n = 0, \pm 1, \pm 2, \ldots$ and $0 < \delta t << T$. Thus, one avoids Dirac delta functions at the borders. This assumption yields the following *Floquet equation* (2.3 - 64)

$$\cos\left(\omega_S T - i\frac{T}{\vartheta}\right) = \frac{1}{2} tr\, \boldsymbol{P}(-\delta t, T-\delta t) \quad . \qquad (2.3 - 74a)$$

The *propagator matrix* **P** $(-\delta t, T-\delta t)$ can be split into a product of simpler matrices

$$\boldsymbol{P}(-\delta t, T-\delta t) = \boldsymbol{P}(+\delta t, T-\delta t)\, \boldsymbol{P}(-\delta t, +\delta t) \quad . \qquad (2.3 - 74b)$$

The matrix on the left of the product describes the harmonic oscillation with the circular frequency Ω and therefore corresponds to the propagator matrix (2.2 - 20a)

$$\boldsymbol{P}(+\delta t, T-\delta t) \approx \boldsymbol{P}(0,T) = \begin{pmatrix} \cos \Omega T & \frac{1}{\Omega}\sin \Omega T \\ -\Omega \sin \Omega T & \cos \Omega T \end{pmatrix} \quad . \qquad (2.3 - 74c)$$

The matrix on the right includes the Dirac delta function $\delta(t)$. It has been evaluated in Section 2.3.6c on pulse modulation. According to (2.3 - 56) it can be written as

$$\boldsymbol{P}(-\delta t, +\delta(t)) = \begin{pmatrix} 1 & 0 \\ -p & 1 \end{pmatrix} \quad . \qquad (2.3 - 74d)$$

The combination of equations (2.3 - 74a-d) results in the *Floquet equation of the periodic pulse modulation*

$$\cos\left(\omega_S T - i\frac{T}{\vartheta}\right) = \cos \Omega T - \beta \sin \Omega T \qquad (2.3 - 74e)$$

with $\beta = p/2\,\Omega$; $\omega_S = \omega_{S0} + n\omega$; $0 < \omega_{S0} < \omega$; $n = 0, \pm 1, \pm 2, \ldots$.

For the *allowed frequency range* where $\vartheta = \infty$ the oscillations as Floquet solutions $u(t)$ exhibit neither damping nor amplification. They are characterized by the beat frequency ω_S. This beat frequency ω_S is not uniquely determined. A multiple of the modulation frequency $\omega = 2\pi/T$ can be added or substracted without influence on the solution $u(t)$.

The frequency gap or forbidden frequency range where $\vartheta \neq \infty$ exhibits parametric damping or amplification. It corresponds to the energy gap of solid state physics.

The extreme real values ± 1 of $\cos \omega_S T$ indicate the *borders* between allowed and forbidden frequency ranges. They fulfill the *Bragg conditions*

$$\cos \omega_S T = (-1)^k \quad \text{with} \quad k = 0,1,2,3,\ldots \quad \text{and} \quad \omega_S = k\omega_B = k\omega/2 \; , \qquad (2.3 - 75a)$$

where k indicates the *order* of the *Bragg effect.* By application of (2.3 - 75a) to (2.3 - 74e) and introduction of $\sin(\Omega T/2)$ and $\cos(\Omega T/2)$ one finds the *limit frequencies* $\omega_{1,2}$ corresponding to the Bragg conditions

$$\Omega / \omega_1 = k/2 \quad , \tag{2.3 - 75b}$$

$$\Omega / \omega_2 = k/2 - \frac{1}{\pi} arctan\, \beta \quad \text{with} \quad -\pi/2 \leq arctan\, \beta \leq \pi/2 \quad .$$

Fig. 2.4 - 4 illustrates the allowed and forbidden frequency ranges as well as their borders as functions of the modulation strength $\beta = p/2\ \Omega$. The forbidden frequency ranges or frequency gaps where $|cos\ (\omega_2 T - i\ T/\vartheta)| > 1$ have *widths* $\Delta(\Omega/\omega)$ which do not depend on the Bragg order k

$$\Delta(\Omega / \omega) = \frac{1}{\pi}\left| arctan\, \beta \right| \quad \text{with} \quad -\pi/2 \leq arctan\, \beta \leq \pi/2 \quad . \tag{2.3 - 75c}$$

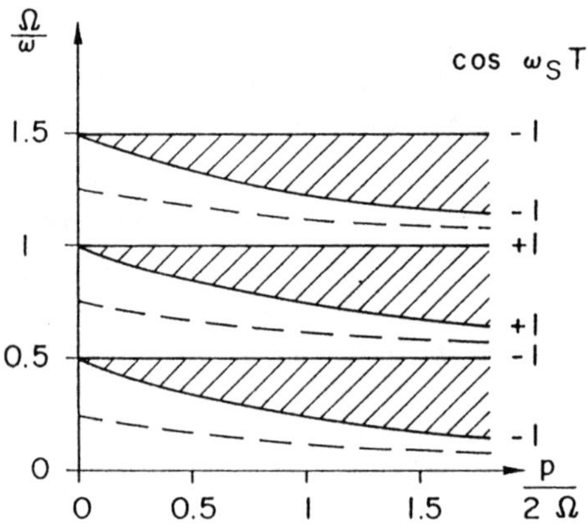

Fig. 2.3 - 4. Allowed (white) and forbidden (hatched) frequency ranges or zones of the periodic pulse modulation. The central positions of the allowed ranges are marked by dashed lines

In the *forbidden frequency ranges* or frequency gaps the parametric linear oscillators show *parametric resonances* caused by the modulation with the period $T = 2\pi/\omega$ of a harmonic oscillator with the circular frequency Ω. For small modulation strengths $|\beta| = |p/2\Omega| < 0.1$, equation (2.3 - 75b) demonstrates that the forbidden frequency ranges occur at $\Omega/\omega \approx k/2$, where k indicates the Bragg order. The circular frequencies ω_P of possible *parametric resonances* are determined by (2.3 - 72b). This yields

$$\omega_P = (k + 2n)\,\omega/2 \approx \left[1 + (2n/k)\right]\Omega \tag{2.3 - 76}$$

$$\text{with} \quad \omega \approx 2\Omega/k \quad \text{and} \quad k = 1,2,3,\ldots \quad ; \quad n = 0, \pm 1, \pm 2, \ldots \quad .$$

The *centers* of the *allowed frequency ranges* are defined by the relation

$$cos\ \omega_S T = 0 \tag{2.3 - 77a}$$

or $\omega_S = (k+\frac{1}{2})\,\omega_B = (k+\frac{1}{2})\,\omega/2 \quad ; \quad k = 0,1,2,3,\ldots$.

According to (2.3 - 74d) the corresponding circular frequencies ω are determined by the equation

$$(\Omega/\omega) = k/2 + \frac{1}{2\pi}\,arctan(1/\beta) \qquad \text{with} \quad -\pi/2 \le arctan(1/\beta) \le +\pi/2 \quad . \tag{2.3 - 77b}$$

They are illustrated in Fig. 2.3 - 4. Their approximations are

$$\text{for} \quad |\beta| \to 0: \ \Omega/\omega \approx \tfrac{1}{2}\left(k+\tfrac{1}{2}\right) \quad , \qquad \text{for} \quad |\beta| \to \infty: \ \Omega/\omega \approx \tfrac{1}{2}k \quad . \tag{2.3 - 77c}$$

The periodic pulse modulation is well suited to study the *influence of weak periodic perturbations on a linear harmonic oscillator*. This can be confirmed by various approximations of (2.3 - 74e).

For the *zero-order approximation* with $p = 0$ and $\vartheta = \infty$ one finds

$$\begin{aligned} &cos\,\Omega T = cos\,\omega_S T \quad , \\ &\Omega T = \pm\omega_S T - 2\pi n \quad , \\ &(\Omega/\omega) = \pm(\omega_S/\omega) - n \quad \text{with} \quad n = 0, \pm 1, \pm 2, \ldots \quad . \end{aligned} \tag{2.3 - 78}$$

In the $(\omega_S/\omega)-(\Omega/\omega)$ plane this approximation which is illustrated in Fig. 2.3 - 5 corresponds to two sets of parallel straight lines, whose intersections represent the Bragg conditions. For these conditions the effect of the periodic modulation reaches it maximum.

For the *first and higher-order approximations* of (2.3 - 74e) with $\vartheta = \infty$ one uses the ansatz

$$(\Omega/\omega) = \pm(\Omega/\omega_S) - n + \frac{1}{2\pi}\varepsilon(\omega_S/\omega); \quad n = 0, \pm 1, \pm 2, \pm 3, \ldots \quad . \tag{2.3 - 79a}$$

This ansatz transforms (2.3 - 74e) into

$$\begin{aligned} &cos(\pm\omega_S T)\left[1 - cos\,\varepsilon + \beta\,sin\,\varepsilon\right] \\ &\quad = -sin(\pm\omega_S T)\left[sin\,\varepsilon + \beta\,cos\,\varepsilon\right] \quad . \end{aligned} \tag{2.3 - 79b}$$

Exact solutions of this equation exist on the Bragg conditions

$$\varepsilon(\omega_S / \omega) = \varepsilon(k/2) = \begin{cases} 0 \\ -2 arctan\, \beta \end{cases} \qquad (2.3 - 79c)$$

with $-\pi/2 \leq arctan\, \beta \leq +\pi/2$ and $k = 0, 1, 2, 3, \ldots$

and for the centers of the allowed frequency ranges

$$\varepsilon(\omega_S / \omega) = \varepsilon\left((k + \frac{1}{2})/2 \right) = - arctan\, \beta \qquad (2.3 - 79d)$$

with $-\pi/2 \leq arctan\, \beta \leq +\pi/2$ and $k = 0, 1, 2, 3, \ldots$.

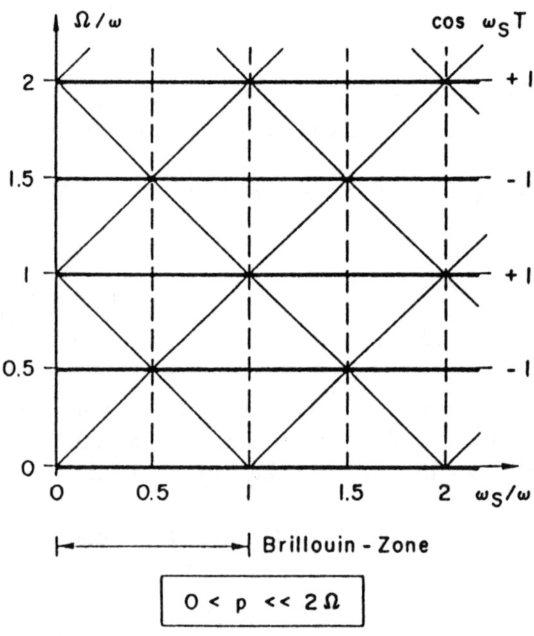

Fig. 2.3 - 5. $(\Omega/\omega) - (\omega_S/\omega)$ - relation for weak periodic pulse modulation with $|\beta| = |p/2\Omega| << 1$. The forbidden frequency zones are the horizontal lines with $\Omega/\omega = k/2$, $k = 0, 1, 2, \ldots$

For the other modulation frequencies ω the first approximation of (2.3 - 79b) yields

$$\varepsilon(\omega_S / \omega) \approx - \frac{\beta\, tan(\pm 2\pi\, \omega_S / \omega)}{\beta + tan(\pm 2\pi\, \omega_S / \omega)} \quad . \qquad (2.3 - 79e)$$

This demonstrates the effect of small modulation strengths $\beta = p/2\Omega$.

There exist explicit solutions of (2.3 - 74e) for the *strong periodic pulse modulations* with $\beta = p/2\Omega = \pm 1$. For $\beta = +1$ one finds

$$\cos\omega_S T = -\sqrt{2}\sin(\Omega T - \pi/4) \quad , \quad \text{or} \tag{2.3 - 80a}$$

$$(\Omega/\omega) = (1/8) - (1/2\pi)\arcsin\left[\left(1/\sqrt{2}\right)\cos\left(2\pi\omega_S/\omega\right)\right] \quad . \tag{2.3 - 80b}$$

As a consequence the allowed frequency ranges are situated between $\Omega/\omega = k/2$ and $(k + 1/2)/2$ and the forbidden ranges between $(k + 1/2)/2$ and $(k + 1)/2$, where $k = 0, 1, 2, \ldots$ indicates the Bragg order. These features are illustrated in Fig. 2.3 - 6.

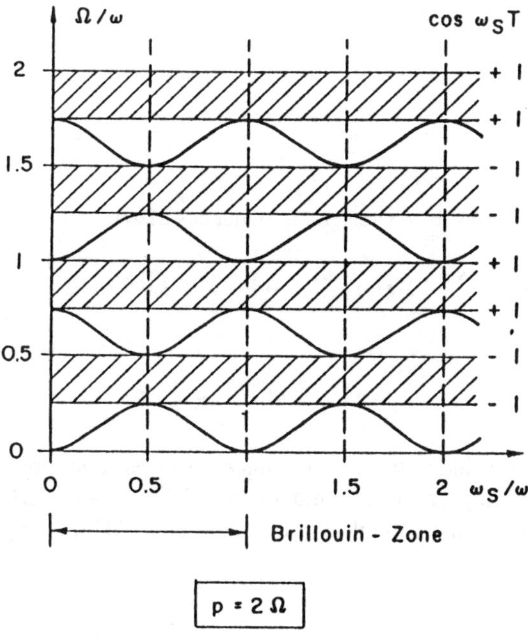

Fig. 2.3 - 6. $(\Omega/\omega) - (\omega_S/\omega)$ - relation for the strong periodic pulse modulation with $\beta = +1$. The forbidden frequency ranges correspond to the horizontal hatched bands

d) Square-wave modulation

The square-wave modulation of a harmonic oscillator is characterized by a characteristic circular frequency $\Omega_0(t)$ of the normal oscillation equation (2.3 - 3) that alternates at intervals $T/2 = \pi/\omega$ between the circular frequencies $\Omega - \Delta\Omega$ and $\Omega + \Delta\Omega$, with $0 < \Delta\Omega < \Omega$ as illustrated in Fig. 2.3 - 7. The corresponding modulation period is $T = 2\pi/\omega$. The circular frequency $\Omega_0(t)$ can be represented by

$$\Omega_0(t) = \Omega - \Delta\Omega\,\mathrm{sign}(\sin\omega t) \quad . \tag{2.3 - 81}$$

The propagator $\boldsymbol{P}(0, T)$ of the Floquet equation (2.3 - 64) of the square-wave modulation (2.3 - 81) can be derived from the propagator (2.3 - 52a-c) of the step modulation

$$\boldsymbol{P}(0,T)=\boldsymbol{P}(T/2,T)\boldsymbol{P}(0,T/2)=\prod_{m=1}^{2}\begin{pmatrix} \cos\varphi_m & (1/\Omega_m)\sin\varphi_m \\ -\Omega_m \sin\varphi_m & \cos\varphi_m \end{pmatrix} \qquad (2.3 - 82)$$

with $\Omega_m = \Omega \pm \Delta\Omega; \;\; \varphi_m = \Omega_m T/2$.

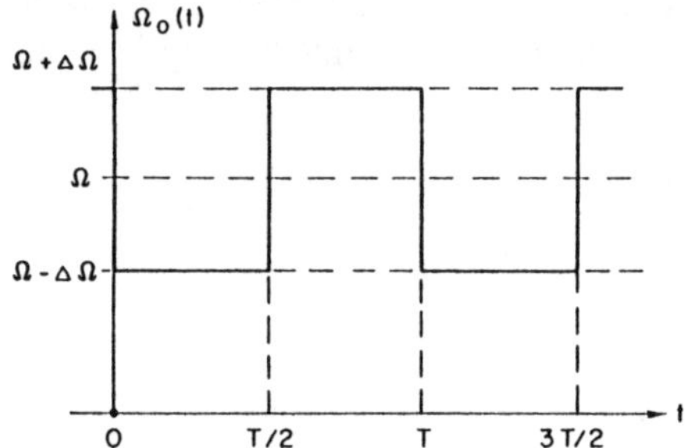

Fig. 2.3 - 7. Square-wave modulation of the characteristic frequency $\Omega_0(t)$

The application of the Floquet equation (2.3 - 64) to this propagator yields for $\vartheta = \infty$ the relation

$$\cos\omega_S T = \frac{\Omega^2 \cos\Omega T - \Delta\Omega^2 \cos\Delta\Omega T}{\Omega^2 - \Delta\Omega^2} \qquad . \qquad (2.3 - 83)$$

This relation determines allowed and forbidden frequency ranges (Ω/ω) as functions of the modulation strength $(\Delta\Omega/\Omega)$. They are illustrated in Fig. 2.3 - 8 and differ considerably from those of the periodic pulse modulation in Section 2.3.7d since they vary noticeably with the Bragg order k.

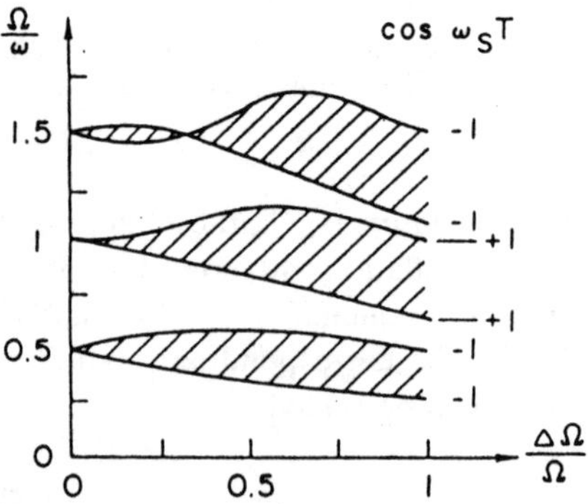

Fig. 2.3 - 8. Allowed (white) and forbidden (hatched) frequency ranges of the square-wave modulation and their variation with the modulation strength $(\Delta\Omega/\Omega)$

According to (2.3 - 83) a forbidden frequency range vanishes for modulation strengths ($\Delta\Omega/\Omega$) which obey the equations

$$\Delta\Omega/\Omega = 1 - (2n/k) \quad \text{with} \quad k = 1, 2, 3, \ldots \quad ; \quad n = 1, 2, 3, \ldots \qquad (2.3 - 84)$$

where k indicates the Bragg order. In Fig. 2.3 - 8 this occurs for $k = 3$; $\Omega/\omega = 1{,}5$; $n = 1$; $\Delta\Omega/\Omega = 1/3$.

The *limit frequencies* (Ω/ω) between allowed and forbidden zones for small modulation ($\Delta\Omega/\Omega$) can be estimated by a second-order approximation of (2.3 - 83). The result is

$$\text{for} \quad k \text{ odd:} \qquad (\Omega/\omega) = k/2 \pm \frac{1}{\sqrt{2}\pi}(\Delta\Omega/\Omega) \quad , \qquad (2.3 - 85a)$$

$$\text{for} \quad k \text{ even:} \qquad (\Omega/\omega) = k/2 \pm \frac{1}{2\sqrt{2}}(\Delta\Omega/\Omega)^2 \quad . \qquad (2.3 - 85b)$$

Consequently, the deviation from $k/2$ is linear in ($\Delta\Omega/\Omega$) for k odd and quadratic for k even.

Equations (23 - 85a&b) demonstrate that the *weak square-wave modulation* behaves similar to the *weak periodic pulse modulation* with respect to *parametric resonance.* The resonance circular frequencies ω_p of parametric amplification or damping fulfill the same condition (2.3 - 76).

Fig. 2.3 - 9 shows a *swing* as *classical example* of a parametric oscillator. Its parametric resonances can be evaluated with the aid of the square-wave modulation as first approximation. The swinging motion is parametrically amplified or damped by the periodic up and down of the person on the swing. Thus, the distance of the common center S of gravity of swing and person from the suspension A varies periodically. Under these circumstances the swing can be described by a mathematical pendulum with mass M and effective length $a(t)$ that is modulated with the period $T = 2\pi/\omega$.

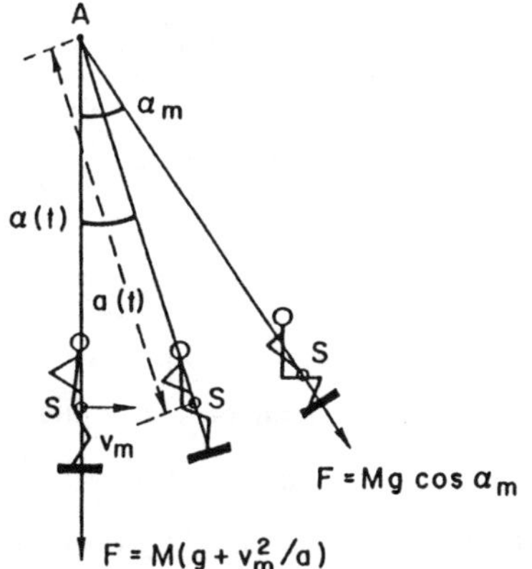

Fig. 2.3 - 9. Swing

For a constant effective length $a(t) = a$ and a small swing angle $\alpha(t)$ the motion of the swing and the corresponding mathematical pendulum is determined by the differential equation of the undamped harmonic oscillator

$$\ddot{\alpha} + \Omega^2 \alpha = 0 \quad \text{with} \quad \Omega = 2\pi / T_0 = (g / a)^{1/2} \tag{2.3 - 86a}$$

where g = 9.81... m/s^2 indicates the acceleration due to gravity. For the initial condition $\alpha(t = 0) = \alpha(0) = 0$ the above equation (2.3 - 86a) has the solution

$$\alpha(t) = \alpha_{\max} \, sin \; \Omega t \tag{2.3 - 86b}$$

that represents a stationary swinging motion.

For abrupt periodic up and downs of the person on the swing at identical time intervals $T/2=\pi/\omega$, the periodic variation of the effective length $a(t)$ of the pendulum corresponds to a *square-wave modulation* illustrated in Fig. 2.3 - 10

$$a(t) = a + \Delta a \; sign \; (sin \; (\omega t - \varphi)) \quad \text{with} \quad 0 < \Delta a \ll a \tag{2.3 - 87a}$$

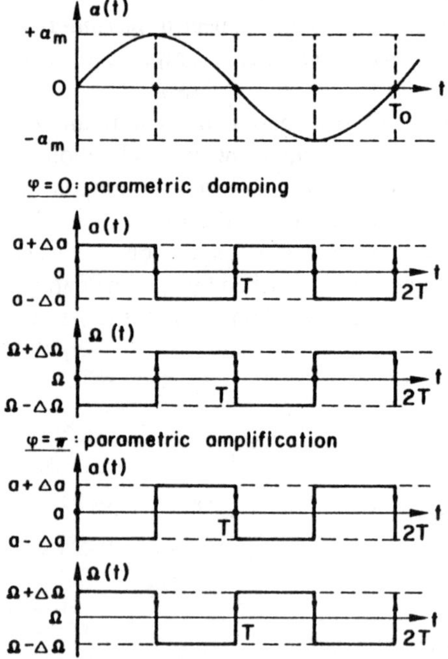

Fig. 2.3 - 10. Free swing motion and modulations which cause parametric damping and amplification of a swing

The person on the swing usually applies the *parametric resonance of the first Bragg order* to amplify or damp the swinging motion. Therefore, it modulates the effective length $a(t)$ of the pendulum with twice the circular frequency Ω (2.3 - 86a) of the free pendulum according to the relations (2.3 - 76)

$$k=1:\ \omega = 2\Omega;\quad T = T_0 / 2 \quad . \tag{2.3 - 87b}$$

The angle φ in equation (2.3 - 87a) measures the *phase shift* of the square-wave modulation of the length $a(t)$ of the pendulum versus the periodic motion of the swing as illustrated in Fig. 2.3 - 10.

The square-wave modulation of the length $a(t)$ of the pendulum according to (2.3 - 87a) causes a corresponding modulation of the characteristic circular frequency

$$\Omega_0(t) = \Omega - \Delta\Omega\, sign\big(sin(\omega t - \varphi)\big) \quad \text{with} \quad \Delta\Omega / \Omega = \Delta a / 2a \quad . \tag{2.3 - 87c}$$

This modulation is also included in Fig. 2.3 - 10.

The *phase shift* φ *decides whether the swing motion is amplified or damped in the parametric resonance* with $\omega = 2\Omega$. For the phase shifts $\varphi = 0$ and $\varphi = \pi$, which represent extreme damping and amplification, it is possible to evaluate the energy ΔE per period $T_0 = 2\pi/\Omega$ of the swinging motion that is added or subtracted to the energy E of the swing. If the phase shifts are $\varphi = 0$ or $\varphi = \pi$, the effective length $a(t)$ of the pendulum changes exactly at the swing angles $\alpha(t) = 0, \pm\, \alpha_{max}$. Consequently, it is possible to calculate the work to be performed to change the length $a(t)$ of the pendulum. In the vertical position of the swing when $\alpha(t) = 0$ the swing is subjected to the force of gravity as well as to the centrifugal force as illustrated in Fig. 2.3 - 9. Therefore, the *work performed by the person on the swing*, when it shortens the length $a(t)$ of the pendulum from $a + \Delta a$ to $a - \Delta a$ in the position $\alpha(t) = 0$, equals

$$\begin{aligned} W\big(\alpha = 0; a + \Delta a \rightarrow a - \Delta a\big) &= +2\Delta a\, M\big(g + \Omega^2\, \alpha_{max}^2\, a\big) \\ &= +2\Delta a\, Mg\big(1 + \alpha_{max}^2\big) \quad . \end{aligned} \tag{2.3 - 88a}$$

If the swing is in the positions $\alpha(t) = \pm\, \alpha_{max}$, its velocity is zero. Accordingly, the centrifugal force is absent. Therefore, the only force acting on the swing is gravity as illustrated in Fig. 2.3 - 9. Hence, during the increase of the effective length of the pendulum from $a - \Delta a$ to $a + \Delta a$ on this condition the *work performed by the swing* is

$$\begin{aligned} -W\big(\alpha = \pm\alpha_{max}\, ; a - \Delta a \rightarrow a + \Delta a\big) &= +2\Delta a\, Mg\, cos\, \alpha_{max} \\ &\approx +2\Delta a\, Mg\left(1 - \frac{1}{2}\alpha_{max}^2\right) \quad . \end{aligned} \tag{2.3 - 88b}$$

Thus, the energy ΔE gained by the swing per period $T_0 = 2\pi/\omega$ equals

$$\begin{aligned}\Delta E &\approx \pm 2W(\alpha = 0; a + \Delta a \rightarrow a - \Delta a)\\ &\pm 2W(\alpha = \pm\alpha_{\max}; a - \Delta a \rightarrow a + \Delta a)\\ &= \pm\, 6\Delta a\, Mg\, \alpha_{\mathrm{m}}^2 \quad .\end{aligned} \tag{2.3 - 88c}$$

According to Fig. 2.3 - 10 the energy gain ΔE is negative ($\Delta E < 0$) for $\varphi = 0$ and positive ($\Delta E > 0$) for $\varphi = \pi$. This means that the swing is parametrically damped for $\varphi = 0$ and amplified for $\varphi = \pi$.

e) Harmonic modulation

The standard periodic modulation of a harmonic oscillator is the harmonic modulation which is defined by the harmonic characteristic circular frequency

$$\Omega_0^2(t) = \Omega^2 + \Delta\Omega^2 cos\,\omega t \quad . \tag{2.3 - 89a}$$

The corresponding oscillation equation

$$\ddot{u} + \left[\Omega^2 + \Delta\Omega^2 cos\ \omega t\right] u = 0 \tag{2.3 - 89b}$$

represents a *Mathieu differential equation* [Abramowitz & Stegun 1965 B, Magnus & Winkler 1966 B, Nayfeh & Mook 1979 B, Stoker 1957 B]. Its solutions which are called *Mathieu functions* [Abramowitz & Stegun 1965 B, Gradshteyn & Ryzhik 1965 B] exhibit complicated features. The positive sign of $\Delta\Omega^2$ can be changed into a negative sign by a shift of time t by half the modulation period $T/2$, because

$$cos(\omega t - \pi) = -cos\,\omega t \quad .$$

An *example* of a parametric oscillator with harmonic modulation is the a *passive LC circuit with a harmonically modulated capacitance* $C(t)$. This device can be realized by making use of a parallel-plate capacitor with a constant plate area A and a harmonically modulated plate distance $d(t)$

$$C(t) = \varepsilon_0 A \,/\, d(t) \quad \text{with} \quad d(t) = d + \Delta d\ cos\,\omega t \quad . \tag{2.3 - 90a}$$

Hence, the oscillation equation of this LC circuit reads

$$L\ddot{q} + \frac{1}{C(t)} q = L\ddot{q} + \frac{d + \Delta d\, cos\,\omega t}{\varepsilon_0 A} q = 0 \quad , \tag{2.3 - 90b}$$

where q indicates the electric charge on the capacitor. This equation corresponds to (2.3 - 89b) and can therefore be written as

$$\ddot{q} + \left[\Omega^2 + \Delta\Omega^2\ cos\,\omega t\right] q = 0 \tag{2.3 - 90c}$$

with $\Omega^2 = d / \varepsilon_0 A L$ and $\Delta\Omega^2 = \Delta d / \varepsilon_0 A L$.

The harmonic modulation of oscillators yields the *same effects* as the periodic pulse modulation or the square-wave modulation, yet the calculation of the phenomena caused by harmonic modulation is more complicated and time consuming. Again the allowed and forbidden frequency ranges or zones are essential. The first exhibit stationary oscillations, whilst the second show parametric damping, amplification and resonances. The *borders between allowed and forbidden frequency ranges* are also determined by the relation

$$\frac{1}{2} tr\, \boldsymbol{P}(0,T) = cos\, \omega_S t = (-1)^k \quad , \tag{2.3 - 72a}$$

where k indicates the Bragg order. They are shown in Fig. 2.3 - 11a as functions of $(\Delta\Omega/\omega)^2$.

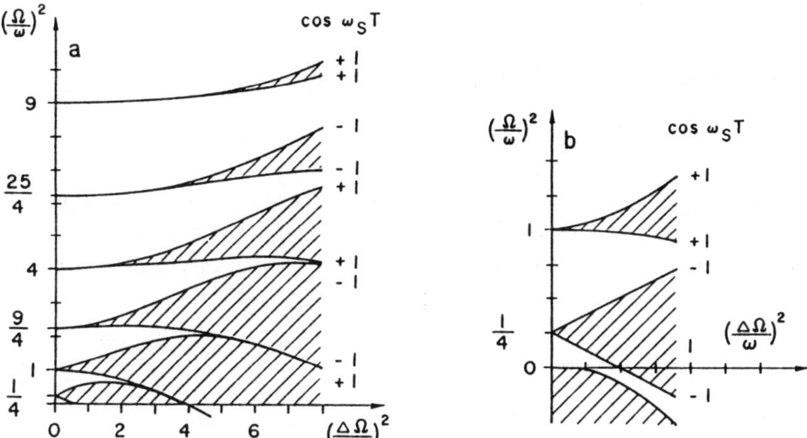

Fig. 2.3 - 11. $(\Omega/\omega) - (\Omega/\omega_S)$ relation of the harmonic modulation: a) for strong and b) for weak harmonic modulation

For *weak modulations* $(\Delta\Omega/\omega)^2$ the following approximations of the borders k can be applied for low Bragg orders [Stoker 1957 B]

$$k = 0: \quad (\Omega / \omega)^2 \approx -\frac{1}{2}(\Delta\Omega / \omega)^4 \quad , \tag{2.3 - 91a}$$

$$k = 1: \quad (\Omega / \omega)^2 \approx \frac{1}{4} \pm \frac{1}{2}(\Delta\Omega / \omega)^2 \quad , \tag{2.3 - 91b}$$

$$k=2:\quad (\Omega/\omega)^2 \approx 1-\frac{1}{12}(\Delta\Omega/\omega)^4 \quad ,$$
$$(\Omega/\omega)^2 \approx 1+\frac{5}{12}(\Delta\Omega/\omega)^4 \quad . \tag{2.3 - 91c}$$

These approximations are illustrated in Fig. 2.3 - 11b.
With respect to *parametric resonance* the harmonically modulated oscillator and the corresponding Mathieu differential equation behave as the weak periodic pulse modulation and the weak square-wave modulation. The parametric resonance of all these weak modulations occurs on the same frequency conditions (2.3 - 76).

The oscillator with harmonic modulation as well as the corresponding Mathieu differential equation are relevant in the theory of the *stabilization of inverted pendulums* by vibration of their support [Acheson 1993 J].

f) Darboux modulation
The Darboux modulation of an oscillator is defined by the following invariant in the normal form (2.3 - 3) of the basic oscillation equation

$$\begin{aligned} I(t) = \Omega_0^2(t) &= \Omega^2 - (\Delta\Omega)^2 \cos^{-2}(\omega t/2) \\ &= \Omega^2 - 2\left(\Delta\Omega^2\right)\left[1+\cos\omega t\right]^{-1} \end{aligned} \quad . \tag{2.3 - 92}$$

This unusual modulation is represented in Fig. 2.3 - 12. Little is known on this type of modulation. However, its corresponding oscillation equation (2.3 - 3) has simple analytical solutions $u(t)$ for the following discrete set of modulation strengths [Kamke 1965 B]:

$$(\Delta\Omega/\omega)^2 = n(n-1)/4, \ \text{with} \quad n = 1,2,3,\ldots \quad . \tag{2.3 - 93}$$

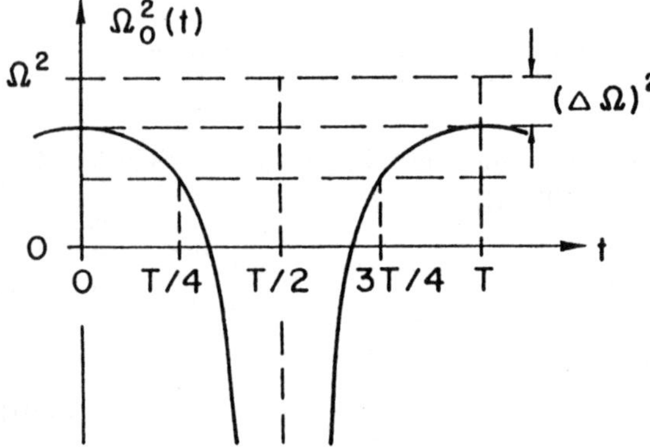

Fig. 2.3 - 12. Darboux modulation of the characteristic circular frequency

The simple solutions are

$$u(t,n) = A\cos^{n}(\omega t/2)\left(\frac{2/\omega}{\cos(\omega t/2)}D\right)^{n}\cos(\Omega t - \varphi) \tag{2.3 - 94a}$$

with $D = \frac{d}{dt}$.

In this equation A and φ represent arbitrary parameters.
A further implification occurs on the *Bragg conditions* where

$$(\Omega/\omega) = k/2: \quad k = 1, 2, 3, \ldots$$

$$u(t,n,k) = A\cos^{n}(\omega t/2)\left(\frac{2/\omega}{\cos(\omega t/2)}D\right)^{n}\cos((k\omega t/2) - \varphi) \quad . \tag{2.3 - 94b}$$

These solutions exhibit *poles* for times

$$t = (m + \frac{1}{2})T \quad \text{with} \quad m = 0, \pm 1, \pm 2, \ldots \quad . \tag{2.3 - 94c}$$

These poles can be avoided by special initial conditions or by a special choice of the phase φ. Thus one finds for $n = 2$, $\varphi = 0$ for k even and $\varphi = \pi/2$ for k odd

$$u(t,2,k) = -kA\left[k\cos((k\omega t/2) - \varphi) + \frac{\sin((k\omega t/2) - \varphi)}{\cos(\omega t/2)}\right] \quad . \tag{2.3 - 94d}$$

2.3.8 Singularities and Approximations

The solutions of the basic equation (2.3 - 1) of modulated linear oscillators are governed mainly by the presence and the characteristics of singularities. The behavior of a solution $x(t)$ can vary considerably with time t. Therefore, it is necessary to classify each point of time t_0 with respect to the basic differential equation (2.3 - 1) as explained in the following section [Birkhoff & Rota 1989 B, Kamke 1956 B, Zwillinger 1989 B]. Subsequently, the various approximations corresponding to the different classifications of the points of time are discussed. A further approximation, which deals with feedback and integral equations, is presented in Section 3.3.3.

a) Classification of the points of time

In order to simplify the following presentation the time scale has to be shifted from t to $(t - t_0)$ so that the point of time t_0 under consideration corresponds to the time $t = 0$. Secondly, the basic equation (2.3 - 1) of a modulated linear oscillator should be written as

$$\ddot{x} - S(t)\,\dot{x} + D(t)x = 0 \quad . \tag{2.3 - 95}$$

The point of time $t_0 = 0$ of the oscillation $x(t)$ determined by (2.3 - 95) is called

α) *ordinary or regular point of time,*
if the functions $S(t)$ and $D(t)$ are *analytic* at $t = 0$. Under these circumstances $S(t)$ and $D(t)$ can be represented by *Taylor series*

$$S(t) = \sum_{\mathrm{k}=0}^{\infty} S^{\mathrm{k}}(0)\frac{t^{\mathrm{k}}}{k!} = \sum_{\mathrm{k}=0}^{\infty} S_{\mathrm{k}} t^{\mathrm{k}} \quad , \tag{2.3 - 96a}$$

$$D(t) = \sum_{\mathrm{k}=0}^{\infty} D^{\mathrm{k}}(0)\frac{t^{\mathrm{k}}}{k!} = \sum_{\mathrm{k}=0}^{\infty} D_{\mathrm{k}} t^{\mathrm{k}} \quad . \tag{2.3 - 96b}$$

In this case the solution $x(t)$ of (2.3 - 95) is also analytic at $t = 0$. It can be written as a Taylor series. This is discussed in Section 2.3.8c.

As *example* one may consider the *linearly damped or amplified harmonic oscillator* of Section 2.2.3. This oscillator is characterized by the differential equation

$$\ddot{x} - S_0\,\dot{x} + D_0\,x = 0 \tag{2.3 - 97a}$$

with the parameters

$$S_0 = -1/\tau \quad , \tag{2.3 - 97b}$$

$$D_0 = (1/\tau)^2 + s\,\omega^2 \quad \text{with} \quad s = 0, \pm 1 \quad , \tag{2.3 - 97c}$$

and the solutions

$$\text{for} \quad s = +1: \quad x(t) = A\,exp(-t/\tau)sin(\omega t - \varphi) \quad , \tag{2.3 - 98a}$$

$$\text{for} \quad s = 0: \quad x(t) = [A + Bt]\,exp(-t/\tau) \quad , \tag{2.3 - 98b}$$

$$\text{for} \quad s = -1: \quad x(t) = [A\;exp(+\omega t) + B\;exp(-\omega t)]\;exp(-t/\tau) \quad . \tag{2.3 - 98c}$$

In these equations A, B and φ indicate free parameters. For *comparison* the function

$$x(t) = sin\;t \quad \text{with} \quad x(0) = 0, \quad \dot{x}(0) = 1 \tag{2.3 - 98d}$$

is plotted in Fig. 2.3 - 13 together with the function

$$x(t) = sin\;(t^2) \quad \text{with} \quad x(0) = 0, \quad \dot{x}(0) = 0 \quad .$$

Both functions exhibit an ordinary or regular point of time $t_0 = 0$.

β) *regular singular point of time,*
if the point is not regular and if the functions $S(t)$ and $D(t)$ can be represented by the *reduced Frobenius series*

$$S(t)=\sum_{k=-1}^{\infty} S_k t^k = t^{-1}\sum_{k=0}^{\infty} S_{k-1} t^k \qquad (2.3 - 99a)$$

$$D(t)=\sum_{k=2}^{\infty} D_k t^k = t^{-2}\sum_{k=0}^{\infty} D_{k-2} t^k \quad . \qquad (2.3 - 99b)$$

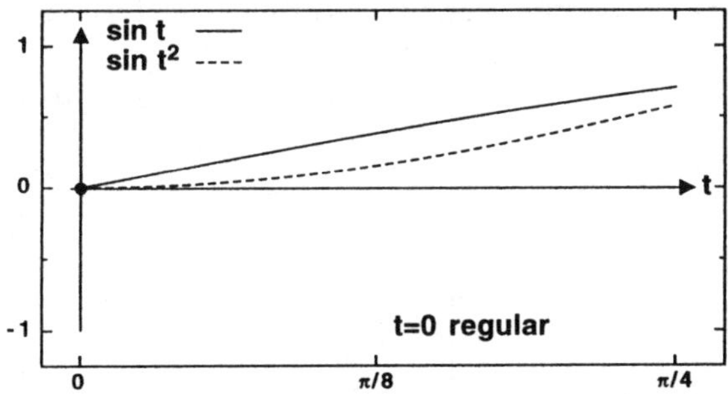

Fig. 2.3 - 13. Ordinary point of time $t_0 = 0$ of the functions *sin t* and *sin t²*

Because $t = 0$ is not an ordinary or regular point of time, at least one of the parameters S_{-1}, D_{-1} and D_{-2} differs from zero. On the other hand, one has to take into account that S_k $(k < -1) = 0$ and D_k $(k < -2) = 0$.

Here, it has to be noticed that at regular singular points the solutions $x(t)$ of (2.3 - 95) are not analytic. Consequently, poles occur. Characteristic of regular singular points of time are the *Frobenius series* described in Section 2.3.8d.

A classical *example* is *Euler's homogeneous differential equation* [Birkhoff & Rota 1989 B]

$$\ddot{x} - S_{-1} t^{-1} \dot{x} + D_{-2} t^{-2} x = 0 \quad \text{or} \quad t^2 \ddot{x} - S_{-1} t \dot{x} + D_{-2} x = 0 \qquad (2.3 - 100a)$$

with the parameters

$$S_{-1} = 2\alpha - 1 \qquad (2.3 - 100b)$$

$$D_{-2} = \alpha^2 + s\beta^2, \quad s = 0, \pm 1 \qquad (2.3 - 100c)$$

and the solutions

$$\text{for} \quad s=+1: \quad x(t) = t^{\alpha}\left[C t^{+i\beta} + D t^{-i\beta}\right]$$
$$= t^{\alpha}\left[C\,exp(i\beta\,\ell n\,t) + D\,exp(-i\beta\,\ell n\,t)\right] \tag{2.3 - 101a}$$
$$= A\,t^{\alpha}\,sin(\beta\,\ell n\,t - \varphi) \quad ,$$

$$\text{for} \quad s=0: \quad x(t) = t^{\alpha}\left[A + B\,\ell n\,t\right] \quad , \tag{2.3 - 101b}$$

$$\text{for} \quad s=1: \quad x(t) = t^{\alpha}\left[A t^{\beta} + B t^{-\beta}\right] \quad . \tag{2.3 - 101c}$$

In these equations A, B and φ indicate free parameters. For *illustration* the function

$$x(t) = sin(\ell n\,t) \tag{2.3 - 101d}$$

is shown in Fig. 2.3 - 14a.

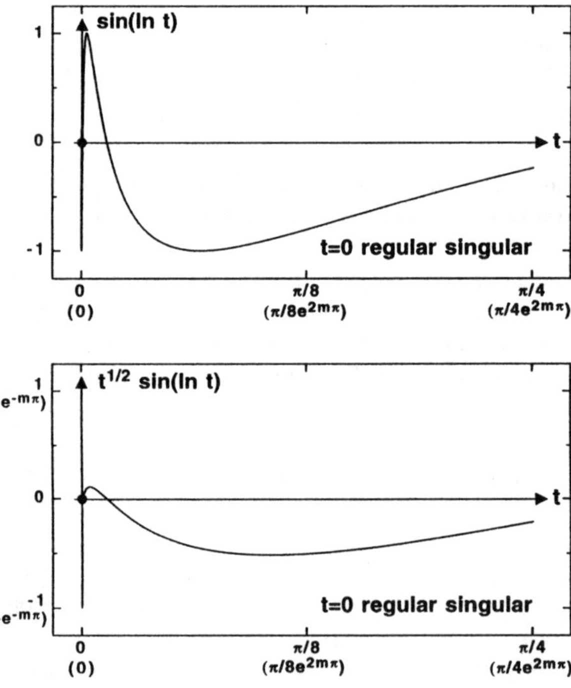

Fig. 2.3 - 14. Regular singular point $t_0 = 0$ for the functions a) $sin(\ell n t)$ and b) $t^{1/2}\ sin(\ell n t)$

This function exhibits a *restricted self similarity* or homogenuity [Magyari 1990 J] as manifested by the relation

$$sin[\ell n(t / exp\ m2\pi] = sin[\ell n\ t] \text{ with } m = 0, \pm 1, \pm 2, \ldots \quad . \tag{2.3 - 101e}$$

It demonstrates that time dilatations and contractions by the *similarity factors* *exp* ($m2\pi$) reproduce the function itself.

γ) *irregular singular point of time,*
if the point is neither ordinary nor regular singular. In this case the functions $S(t)$ and $D(t)$ can be described by *Laurent series*

$$S(t) = \sum_{k=-\infty}^{+\infty} S_k\, t^k \quad , \tag{2.3 - 102a}$$

$$D(t) = \sum_{k=-\infty}^{+\infty} D_k\, t^k \tag{2.3 - 102b}$$

where at least one of the parameters $S_k(k < -1)$ and $D_k(k < -2)$ differs from zero.

At an irregular singular point of time $t_0 = 0$ the solution $x(t)$ of (2.3 - 95) can also be represented by a *Laurent series*

$$x(t) = \sum_{k=-\infty}^{+\infty} x_k\, t^k \quad . \tag{2.3 - 103}$$

As *example* one may consider the differential equation

$$\ddot{x} + \sigma t^{-1}\, \dot{x} + \beta^2 t^{-2\sigma} x = 0 \quad \text{with} \quad \sigma > 1 \quad , \tag{2.3 - 104a}$$

the parameters

$$S_{-1} = -\sigma \,,\; S_k(k \neq -1) = 0 \quad , \tag{2.3 - 104b}$$

$$D_{-2\sigma} = \beta^2 \,,\quad D_k(k \neq -2\sigma) = 0 \quad , \tag{2.3 - 104c}$$

and the solution

$$x(t) = A\ sin\left[\frac{\beta}{\sigma - 1} t^{1-\sigma} - \varphi\right] \quad . \tag{2.3 - 105a}$$

For *illustration* the solution $x(t)$ for the parameters $\sigma = 2$, $\beta = 1$, $A = 1$, $\varphi = 0$, which has the form

$$x(t) = sin(1/t) = t^{-1} - \frac{t^{-3}}{3!} + \frac{t^{-5}}{5!} - + - \tag{2.3 - 105b}$$

is shown in Fig. 2.3 - 15a.

For $\sigma = 0$ the differential equation (2.3 - 104a) corresponds to that of the *undamped harmonic oscillator* with an ordinary point of time $t_0 = 0$ and with the typical solution (2.3 - 98d) for $\beta = 1$, $A = 1$, $\varphi = \pi$.

On the contrary (2.3 - 104a) exhibits for $\sigma = 1$ a regular singular point of time $t_0 = 0$ and the typical solution (2.3 - 101d) for $\beta = 1$, $A = 1$, $\varphi = 0$.

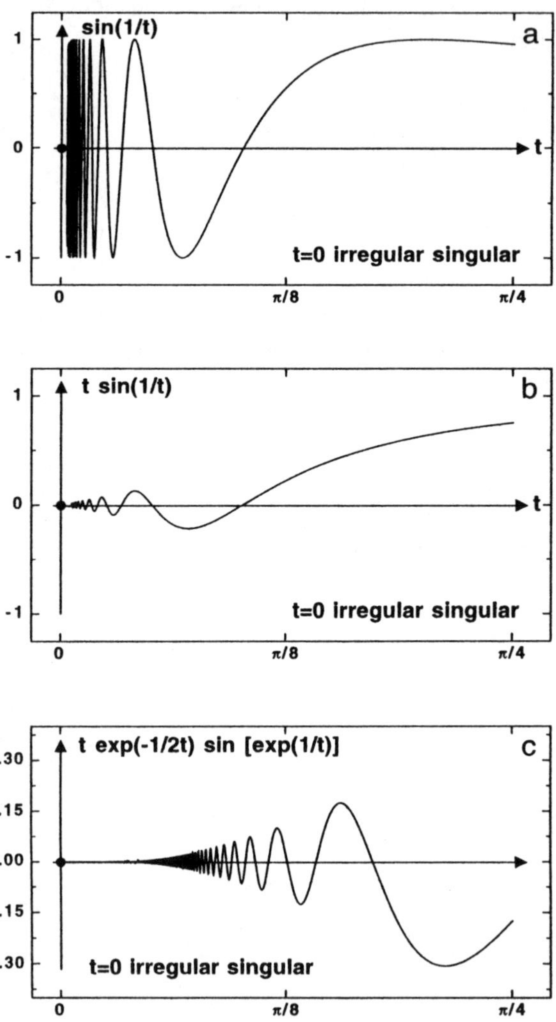

Fig. 2.3 - 15. Irregular singular point of time $t_0 = 0$ of the functions a) $sin(1/t)$, b) $t\ sin(1/t)$ and c) $t\ exp(-1/2\ t)\ sin\ [exp(1/t)]$

b) Normalized basic oscillation equation

The solutions $u(t)$ of the oscillator equation

$$\ddot{u} + \Omega_0^2(t)u = 0 \qquad (2.3 - 3)$$

with the specific characteristic circular frequency

$$\Omega_0(t) = \Omega\, t^{-(1+\gamma)} \tag{2.3 - 106}$$

exhibit all types of the point of time $t_0 = 0$ for different parameters γ. Equation (2.3 - 3) with the circular frequency $\Omega_0(t)$ of (2.3 - 106) can be solved analytically for all γ. The points of time $t_0 = 0$ of the solutions $u(t)$ differ essentially for $\gamma = 0$ and $\gamma \neq 0$:

α) In the case where $\gamma = 0$ the point of time $t_0 = 0$ is regular singular. In this case the normalized basic equation (2.3 - 3) equals

$$\ddot{u} + (\Omega / t)^2 u = 0 \tag{2.3 - 107a}$$

with the parameter

$$\Omega^2 = 1/4 + s\beta^2 \quad \text{where} \quad s = 0, \pm 1 \quad . \tag{2.3 - 107b}$$

The solutions $u(t)$ of this equation form *three classes* [Kamke 1956 B] characterized by the parameter s

$$\text{for} \quad s = +1: \; u(t) = A\, t^{1/2} \sin(\beta\, \ell n\, t - \varphi) \quad , \tag{2.3 - 108a}$$

$$\text{for} \quad s = 0: \quad u(t) = t^{1/2}(A + B\, \ell n\, t) \quad , \tag{2.3 - 108b}$$

$$\text{for} \quad s = -1: \; u(t) = A\, t^{(1/2)+\beta} + B\, t^{(1/2)-\beta} \quad . \tag{2.3 - 108c}$$

In these equations A, B and φ represent free parameters. For *illustration* of (2.3 - 108a) the function

$$u(t) = t^{1/2} \sin(\ell n\, t) \tag{2.3 - 108d}$$

is plotted in Fig. 2.3 - 14b. It represents the solution $u(t)$ defined by the parameters $s = +1$, $\beta = 1$, $A = 1$, $\varphi = 0$. Also this function exhibits a *restricted self similarity* or homogenuity [Magyari 1990 J] according to the relation

$$\begin{gathered} exp(m\pi)(t / exp\, 2m\pi)^{1/2} \sin\{\ell n(t / exp\, 2m\pi)\} = \sin(\ell n\, t) \\ \text{with} \quad m = 0, \pm 1, \pm 2, \ldots \quad . \end{gathered} \tag{2.3 - 108e}$$

According to this relation the function (2.3 - 108d) reproduces itself if t is multiplied by the similarity factor $exp\,(2m\pi)$, while the variable x is simultaneously multiplied by the similarity factor $exp\,(m\pi)$ with $m = \pm 1, \pm 2, \pm 3, \ldots$.

β) Solutions $u(t)$ defined by specific discrete non-zero values of $\gamma \neq 1$ can be represented with the formulas of A.R. Forsyth and W. Jacobsthal [Kamke 1965 B: equation 2.14].
With the assumptions

$$D = d/dt \quad \text{and} \quad m = -\frac{1+\gamma}{2\gamma} = 0, \pm 1, \pm 2, \pm 3, \dots \tag{2.3 - 109a}$$

the solutions $u(t)$ for $\gamma < 0$ can be written as

$$u(t) = A\,t\left(t^{1+2\gamma} D\right)^{m+1} sin\left(\frac{\Omega}{\gamma} t^{-\gamma} - \varphi\right)$$
$$\text{with} \quad \gamma = -1, -1/3, -1/5, -1/7, \dots \;\; ; \quad m = 0, 1, 2, 3, \dots \; , \tag{2.3 - 109b}$$

whilst for $\gamma > 0$ the solutions $u(t)$ are of the form

$$u(t) = A\left(t^{1+2\gamma} D\right)^{-m} sin\left(\frac{\Omega}{\gamma} t^{-\gamma} - \varphi\right)$$
$$\text{with} \quad \gamma = 1, 1/3, 1/5, 1/7, \dots \;\; ; \quad m = -1, -2, -3, -4, \dots \; . \tag{2.3 - 109c}$$

In these equations A and φ indicate free parameters.

The solution $u(t)$ with the parameters $m = -1$, $\gamma = 1$, $\Omega = 1$, $A = 1$, $\varphi = 3\pi/2$, has the form

$$u(t) = t\,sin(1/t) = 1 - \frac{1}{3!t^2} + \frac{1}{5!t^4} - + - \; . \tag{2.3 - 109d}$$

It is plotted in Fig. 2.3 - 15b for illustration.

On the contrary to the function defined by (2.3 - 108d) this function exhibits no self similarity.

Equations (2.3 - 109b&c) demonstrate that the point of time $t_0 = 0$ is ordinary or regular for $\gamma < 0$ and irregular singular for $\gamma > 0$. As mentioned before this point of time is regular singular for $\gamma = 0$.

A further general analytical representation of the solutions $u(t)$ for $\gamma \neq 0$ is based on *Bessel functions* $Z_\mu(z)$ [Abramowitz & Stegun 1965 B, Gradshteyn & Ryzhik 1965 B, Kamke 1956 B]. They are defined by *Bessel 's differential equation*

$$z^2 Z_\mu'' + z Z_\mu' + \left(z^2 - \mu^2\right) Z_\mu = 0 \; . \tag{2.3 - 110a}$$

The Z_μ can be represented as a linear combination of Bessel functions of the first kind $J_\mu(z)$ and of the second kind $Y_\mu(z)$

$$Z_\mu(z) = A\,J_\mu(z) + B\,Y_\mu(z) \qquad (2.3 - 110b)$$

with the coefficients A and B as free parameters. With these functions the solutions $u(t)$ for $\gamma \neq 0$ can be written as

$$u(t) = t^{1/2}\, Z_{-\frac{1}{2\gamma}}\left(-\frac{\Omega}{\gamma}t^{-\gamma}\right) \quad . \qquad (2.3 -111)$$

If the Bessel functions $J_\mu(z)$ and $Y_\mu(z)$ are represented as series with respect to $z = 0$ [Abramowitz & Stegun 1965 B], then the point of time $t_0 = 0$ is regular for $\mu < 0$.

c) Approximation by Taylor series

The solutions of the basic equations of modulated linear oscillators in general (2.3 - 1) or normalized (2.3 - 3) form can be approximated by Taylor series at ordinary points of time t_0. This section deals with the solution $u(t)$ of the normalized basic equation

$$\ddot{u} + \Omega_0^2(t)\,u = 0 \qquad (2.3 - 3)$$

that has an ordinary point of time at $t_0 = 0$. According to the definition of Section 2.3.8a the invariant $I(t)$ can then be represented as a Taylor series with respect to $t = 0$

$$\Omega_0^2(t) = I(t) = \sum_{k=0}^{\infty} I^{(k)}(0)\frac{t^k}{k!} = \sum_{k=0}^{\infty} I_k\, t^k \quad . \qquad (2.3 - 112)$$

Consequently, the solution $u(t)$ can also be approximated by a Taylor series at $t = 0$

$$u(t) = \sum_{k=0}^{\infty} u^{(k)}(0)\frac{t^k}{k!} = \sum_{k=0}^{\infty} u_k\, t^k \quad . \qquad (2.3 - 113)$$

The combination of (2.3 -3), (2.3 - 112) and (2.3 - 113) yields the relations

$$(m+1)(m+2)\,u_{m+2} + \sum_{k=0}^{m} I_{m-k}u_k = 0 \quad \text{with} \quad m = 0, 1, 2, 3, \ldots \quad . \qquad (2.3 - 114)$$

These equations permit to evaluate the constant coefficients u_k of the Taylor series (2.3 - 113).

The initial conditions

$$u(0) = u_0 \;,\; \dot{u}(0) = u_1 \qquad (2.3 - 115a)$$

yield the equations

$$1\cdot 2\cdot u_2 = -I_0\, u_0$$
$$2\cdot 3\cdot u_3 = -I_1\, u_0 - I_0\, u_1$$
$$3\cdot 4\cdot u_4 = \left(-I_2 + \frac{1}{2} I_0^2\right) u_0 - I_1 u_1 \qquad (2.3 - 115b)$$
$$4\cdot 5\cdot u_5 = \left(-I_3 + \frac{2}{3} I_0 I_1\right) u_0 + \left(-I_2 + \frac{1}{6} I_0^2\right) u_1$$
$$5\cdot 6\cdot u_6 = \ \ldots \ .$$

For illustration one may assume the parameters $I_{k\neq 1} = 0$, $I_1 = 1$, $u_0 = 1$, $u_1 = 0$. With these parameters equations (2.3 - 115b) yield the Airy Cosine of Section 2.3.5a

$$u(t) = Ac(t) = f(-t) = 1 - \frac{1}{3!} t^3 + \frac{1\cdot 4}{6!} t^6 - + \quad . \qquad (2.3 - 34d)$$

d) Approximation by Frobenius series

At regular singular points $t_0 = 0$ the functions $S(t)$ and $D(t)$ of (2.3 - 95) can be represented by special Frobenius series according to (2.3 - 99a&b). Therefore, the oscillation equation (2.3 - 95) can be written as

$$\ddot{x} - \dot{x}\left[\sum_{k=-1}^{\infty} S_k t^k\right] + x\left[\sum_{k=-2}^{\infty} D_k t^k\right] = 0 \quad . \qquad (2.3 - 116)$$

The solutions $x(t)$ of these equations also contain *Frobenius series* [Kamke 1956 B, Zwillinger 1989 B]. In order to calculate these series it is necessary to determine the *indicial equation*. For these purpose one starts with the ansatz

$$x = t^{\alpha} \qquad (2.3 - 117a)$$

to solve (2.3 - 116). This results in the indicial equation

$$\alpha^2 - \alpha(1 + S_{-1}) + D_{-2} = 0 \qquad (2.3 - 117b)$$

for the exponent α in $t^{\alpha-2}$. The roots α_1 and α_2 of (2.3 - 117b), which are called *exponents of the singularity* at $t = 0$, determine the type of series which approximate the solutions $x(t)$. The following *survey* is restricted to positive times $t > 0$.

α) If $\alpha_1 = \alpha_2 = \alpha$, then there exist two linearly independent solutions [Zwillinger 1989 B] of the form

$$x_1(t) = t^{\alpha}\left[1 + \sum_{k=1}^{\infty} a_k\, t^k\right] \quad , \qquad (2.3 - 118a)$$

$$x_2(t) = x_1(t)\ell n\ t + t^{\alpha}\left[\sum_{k=0}^{\infty} b_k t^k\right] \quad . \tag{2.3 - 118b}$$

β) If $\alpha_{1,2} = \alpha \pm i\beta$ with α and β real, then there exist two linearly independent solutions of the form

$$x_1(t) = t^{\alpha} sin(\beta\ \ell n\ t)\left[1 + \sum_{k=1}^{\infty} c_k t^k\right] \quad , \tag{2.3 - 119a}$$

$$x_2(t) = t^{\alpha} cos(\beta\ \ell n\ t)\left[1 + \sum_{k=1}^{\infty} d_k t^k\right] \quad . \tag{2.3 - 119b}$$

γ) If $\alpha_1 \neq \alpha_2$ with α_1 as well as α_2 real, and when the difference $\alpha_1 - \alpha_2 = \Delta\alpha$ represents *no integer*, then there exist two linearly independent solutions of the form

$$x_1(t) = t^{\alpha_1}\left[1 + \sum_{k=1}^{\infty} e_k t^k\right] \quad , \tag{2.3 - 120a}$$

$$x_2(t) = t^{\alpha_2}\left[1 + \sum_{k=1}^{\infty} f_k t^k\right] \quad . \tag{2.3 - 120b}$$

δ) If $\alpha_1 \neq \alpha_2$ with α_1 as well as α_2 real, and when the difference $\alpha_1 - \alpha_2 = \Delta\alpha = M > 0$ represents a positive integer M, then there exist two linearly independent solutions of the form

$$x_1(t) = t^{\alpha_1}\left[1 + \sum_{k=1}^{\infty} g_k t^k\right] \quad , \tag{2.3 - 121a}$$

$$x_2(t) = t^{\alpha_2}\left[\sum_{k=0}^{\infty} h_k t^k\right] + C x_1(t)\ell n\ t \tag{2.3 - 121b}$$

where the constant C in (2.3 - 121b) may be zero.

e) Wentzel-Kramers-Brillouin approximation

The Wentzel-Kramers-Brillouin or WKB approximation [Blochinzew 1966 B, Flügge 1990 B, Kamke 1956 B, Kramers 1926 J, Landau & Lifschitz 1965 B, Pauli 1950 B, Schubert & Weber 1980 B, Wentzel 1926 J] works in the neighborhood of an *irregular singular point* as well at other singularities of solutions of ordinary linear differential equations. It finds application in optics as well as in wave mechanics.

In order to understand the WKB approximation one should start with *Planck's law*

$$E = \hbar\Omega = h\Omega / 2\pi \qquad (2.3 - 122)$$

where E indicates the energy of a particle, Ω the corresponding circular frequency while

$$h \approx 6{,}626 \cdot 10^{-34}\ \mathrm{Ws}^2 \approx 2\pi \cdot 10^{-34}\mathrm{Ws}^2 \approx (2/3) \cdot 10^{-33}\ \mathrm{Ws}^2$$

represents *Planck's constant.* Since $\hbar$ is extremely small, the circular frequency Ω is usually high. As example one should mention an optical phonon with $\Omega \approx 3 \cdot 10^{14}$ s^{-1} or a free non-relativistic particle of mass $m = 1$ g and velocity $\upsilon = 2$ cm/s with $\Omega = 3 \cdot 10^{26}\ \mathrm{s}^{-1}$. Consequently, one can write the normalized basic equation (2.3 - 3) as

$$\ddot{u}(t) + \frac{1}{\hbar^2} E^2(t)u(t) = 0 \qquad (2.3 - 123a)$$

$$\text{with} \quad \Omega_0(t) = E(t)/\hbar \quad , \qquad (2.3 - 123b)$$

where $\hbar$ represents an extremely small quantity. In order to solve this equation one makes the ansatz

$$u(t) = exp\left(\frac{i}{\hbar}\right)\sum_{k=0}^{\infty} S_k(t)\left(\frac{\hbar}{i}\right)^k \quad . \qquad (2.3 - 124a)$$

Because it is required that this ansatz converges in the time interval $t_1 < t < t_2$, the functions $S_k(t)$ have to fulfill the following inequalities [Zwillinger 1989B]

$$\left|\hbar^k S_{k+1}(t)\right| << 1 \quad \text{for} \quad \hbar \to 0 \quad , \qquad (2.3 - 124b)$$

$$\left|S_{k+1}(t) / S_k(t)\right| \leq C \quad . \qquad (2.3 - 124c)$$

On these conditions the application of the ansatz (2.3 - 124a) to (2.3 -123a) results in

$$\begin{aligned}
E^2 &= \dot{S}_0^{\,2} \\
0 &= 2\,\dot{S}_0\,\dot{S}_1 + \ddot{S}_0 \\
0 &= 2\,\dot{S}_0\,\dot{S}_2 + \dot{S}_1^{\,2} + \ddot{S}_1 \\
0 &= 2\,\dot{S}_0\,\dot{S}_3 + 2\,\dot{S}_1\,\dot{S}_2 + \ddot{S}_2 \\
0 &= 2\,\dot{S}_0\,\dot{S}_4 + 2\,\dot{S}_1\,\dot{S}_3 + \dot{S}_2^{\,2} + \ddot{S}_3 \\
&\text{etc.}
\end{aligned} \qquad (2.3 - 125a)$$

These equations permit to evaluate successively the functions $S_0(t)$, $S_1(t)$, $S_2(t)$, etc.
This procedure yields e.g.

$$S_0(t) = \pm\hbar \int_{\tau}^{t} \Omega_0(t')\,dt' \quad , \tag{2.3 - 125b}$$

$$S_1(t) = -\frac{1}{2}\ell n[\Omega_0(t)/\Omega_0(\tau)] \quad . \tag{2.3 - 125c}$$

These two functions determine the first approximation also called *geometrical-optics approximation* [Zwillinger 1989 B]

$$u(t) = A\,sin\left[\int_{\tau}^{t} \Omega_0(t')\,dt' - \varphi\right] \quad , \tag{2.3 - 126a}$$

as well as the second approximation also named *physical-optics approximation* [Zwillinger 1989 B]

$$u(t) = A[\Omega_0(\tau)/\Omega_0(t)]^{1/2} sin\left[\int_{\tau}^{t} \Omega_0(t')\,dt' - \varphi\right] \quad . \tag{2.3 - 126b}$$

In (2.3 - 126a&b) A and φ represent free parameters. For most problems the second approximation (2.3 - 126b) suffices. *Higher approximations* can be calculated with the aid of the functions $S_2(t)$, $S_3(t)$, etc.

An *example* is the oscillation $u(t)$ of an oscillator with the modulated circular frequency

$$\Omega_0(t) = t^{-2} exp(1/t) \quad . \tag{2.3 - 127a}$$

This oscillation $u(t)$ exhibits an *irregular singular* point of time $t_0 = 0$. The second or physical-optics approximation (2.3 - 126b) yields at $t = 0$

$$u(t) = A\,t\,exp(-1/2t)\,sin[exp(1/t) - \varphi] \quad . \tag{2.3 - 127b}$$

As illustration this function is plotted in Fig. 2.3 - 15c for the parameters $A = 1$ and $\varphi = 0$.

2.4 Oscillators with Relays

Linear oscillators become nonlinear when *relays* and *switches* are introduced. Relays and switches are relevant in electrical engineering and computer science. In order to determine the motion of an oscillator with relays or switches one usually takes advantage of phase diagrams as demonstrated in the following.

2.4.1 Switching Functions

Relays and switches are nonlinear elements characterized by *discontinuous switching functions* $a = \sigma(e)$ which relate the input signal e with the output signal a. Examples are illustrated in Fig. 2.4 - 1. The switching functions can be either single-valued as that for switching on (Fig 2.4 - 1a), or multi-valued as that of the hysteresis (Fig. 2.4 - 1e). The following basic switching functions are of interest

α) *Switching on* (Fig. 2.4 - 1a)

$$a = \sigma(e) = H(e) \quad , \qquad (2.4 - 1a)$$

where $H(e)$ indicates the Heaviside step function (2.3 - 46a-c).

β) *Switching off* (Fig. 2.4 - 1b)

$$a = \sigma(e) = 1 - H(e) \quad . \qquad (2.4 - 1b)$$

γ) *Change of polarity* (Fig. 2.4 - 1c)

$$a = \sigma(e) = sign\ (e) = 2H(e) - 1 \quad . \qquad (2.4 - 1c)$$

δ) *Change of polarity with dead zone* (Fig. 2.4 - 1d)

$$a = \sigma(e) = \frac{1}{2}\left[sign(e-1) + sign(e+1)\right] \quad . \qquad (2.4 - 1d)$$

ε) *Hysteresis* (Fig. 2.4 - 1e)

$$ea = e\sigma_{\mathrm{H}}(e) \geq -1 \quad \text{with} \quad a = \sigma_{\mathrm{H}}(e) = \pm 1 \quad . \qquad (2.4 - 1e)$$

In accordance with this inequality one can write

$$a = \sigma_{\mathrm{H}}(e) = \begin{array}{lll} +1 & \text{for} & +1 < e \\ \pm 1 & \text{for} & -1 \leq e \leq +1 \\ -1 & \text{for} & e < -1 \end{array} \quad .$$

Switching occurs when

$$ea = e\sigma_{\mathrm{H}}(e) = -1 \quad .$$

As a consequence, there exist two transitions

$$a = \sigma_H(e) = -1 \to +1 \quad \text{for} \quad e = +1 \quad ,$$
$$a = \sigma_H(e) = +1 \to -1 \quad \text{for} \quad e = -1 \quad .$$

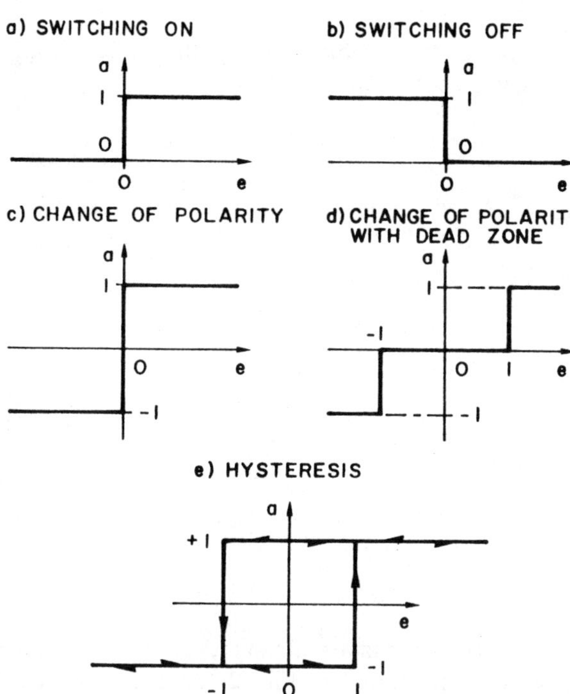

Fig. 2.4 - 1. Switching functions $a = \sigma(e)$

2.4.2 Oscillators with Dry Friction

Dry friction or *Coulomb friction* involves a *braking force* characterized by the equation

$$\vec{F}_C = -R\frac{\vec{v}}{v} \quad \text{with} \quad R > 0 \tag{2.4 - 2a}$$

where $\vec{v}$ indicates the velocity. If the motion occurs along the x axis, this vector relation is reduced to the scalar equation

$$F_C = -R\,sign(v) = -R\,sign(\dot{x}) \quad . \tag{2.4 - 2b}$$

Consequently, a *harmonic oscillator* with the circular frequency $\Omega_0 = 1$ under the influence of the dry friction (2.4 - 2b) with $R = 1$ is described by the equation

$$\ddot{x} + x + sign(\dot{x}) = 0 \tag{2.4 - 3a}$$

with the switching function

$$\sigma(e) = \sigma(\upsilon) = sign(\upsilon) = sign(\dot{x}) \quad . \tag{2.4 - 3b}$$

The equation of the *trajectories in the phase plane*, which describe the motion of this oscillator, can be determined by taking into account the relations

$$x = \frac{1}{2}\frac{d}{dx}\left(x^2\right) \quad \text{and} \tag{2.4 - 4a}$$

$$\ddot{x} = \frac{d\dot{x}}{dx}\frac{dx}{dt} = \dot{x}\frac{d\dot{x}}{dx} = \frac{1}{2}\frac{d}{dx}\left(\dot{x}^2\right) \quad . \tag{2.4 - 4b}$$

Thus, one finds

$$\frac{d}{dx}\left[\dot{x}^2 + x^2 + 2\,sign(\dot{x})\right] = 0 \tag{2.4 - 5a}$$

and after integration over x

$$\dot{x}^2 + \left[x + sign(\dot{x})\right]^2 = x_0^2 \quad . \tag{2.4 - 5b}$$

According to this equation the trajectories in the phase plane $(x,\ \dot{x})$ are *semicircles* with the centers at $\left(x_M = \pm 1, \dot{x}_M = 0\right)$ as illustrated in Fig. 2.4 - 2. A trajectory passes from a semicircle into the following at the straight switching line

$$S\colon\ \dot{x} = 0 \tag{2.4 - 5c}$$

except in the section

$$-1 \le x \le +1,\ \ \dot{x} = 0 \tag{2.4 - 5d}$$

where the oscillator stops its motion. This section covers the positions of rest of the oscillator for sufficient static friction. Attention should be paid to the fact that the position of rest of the oscillator with dry friction depends on its initial conditions.

The equation (2.4 - 3a) that describes the motion of a normalized harmonic oscillator with dry friction can be solved completely with the aid of the integral

$$t - t_0 = \int_{x_0}^{x} \frac{dx}{\dot{x}} = f(x) - f\left(x_0\right) \quad \text{where} \quad \dot{x} = \dot{x}(x) \quad . \tag{2.4 - 6}$$

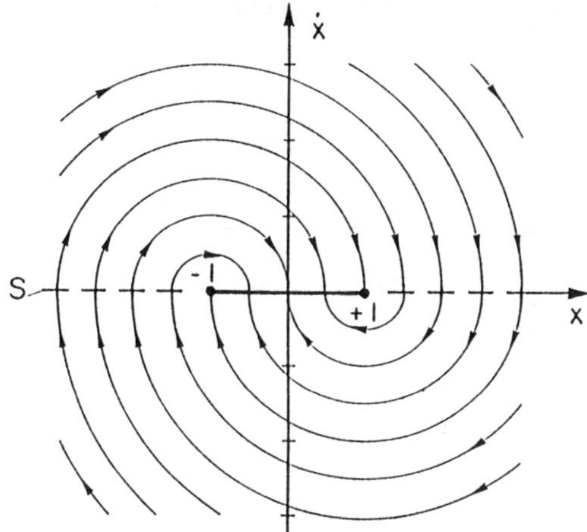

Fig. 2.4 - 2. Phase diagram of an oscillator with dry friction

The equation (2.4 - 3a) that describes the motion of a normalized harmonic oscillator with dry friction can be solved completely with the aid of the integral

$$t - t_0 = \int_{x_0}^{x} \frac{dx}{\dot{x}} = f(x) - f(x_0) \quad \text{where} \quad \dot{x} = \dot{x}(x) \quad . \tag{2.4 - 6}$$

For trajectories in the phase plane $(x, \dot{x})$ that form segments of circles with centers on the x axis ($\dot{x} = 0$) according to Fig. 2.4 -2, the integral (2.4 - 6) implies that the time $\Delta t = t - t_0$ needed to travel along this segment corresponds to its segment angle φ. Consequently, the time required for a semicircle is $\Delta t = \pi$.

The *solution of* (2.4 - 3a) has the form

$$x(t) = -sign[\dot{x}(t)] + A(t)cos(t - \varphi) \tag{2.4 - 7a}$$

$$\text{with} \quad 0 \le \varphi = arctg[\dot{x}(0) / x(0)] \le \pi \quad , \tag{2.4 - 7b}$$

$$A(0) = \left\{ [x(0) + sign[\dot{x}(0)]]^2 + [\dot{x}(0)]^2 \right\}^{1/2} > 0 \quad \text{and} \tag{2.4 - 7c}$$

$$A(t) = A(0) - 2\sum_{n=0}^{\infty} H(t - \varphi - n\pi) > 0 \quad . \tag{2.4 - 7d}$$

For long times t these equations result in

$$\lim_{t \to \infty} x(t) = x_\infty = const \quad \text{with} \quad |x_\infty| \le 1 \quad . \tag{2.4 - 7e}$$

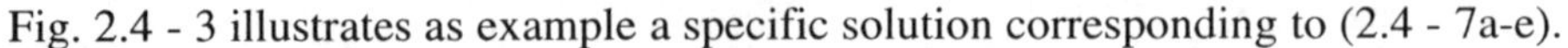
Fig. 2.4 - 3 illustrates as example a specific solution corresponding to (2.4 - 7a-e).

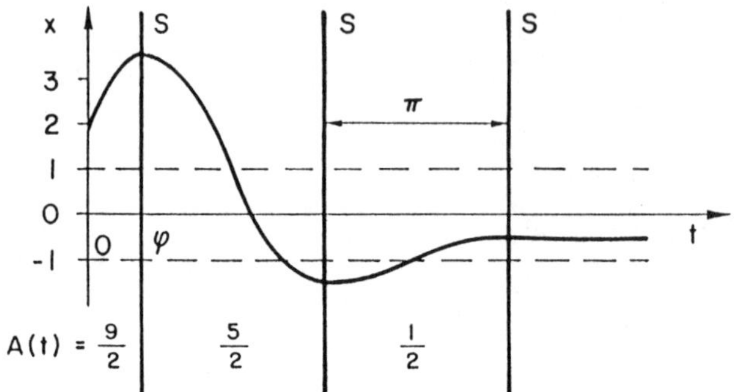

Fig. 2.4 - 3. Example of the motion $x(t)$ of an oscillator with dry friction

2.4.3 Oscillators with Drag

Drag is a frequent phenomenon in fluid dynamics [Prandtl & Tietjens 1957 B, Bohl 1980 B]. It designates the force $\vec{F}_{\mathrm{D}}$ on a solid body moving in a fluid, which is proportional to the square of the relative velocity $\vec{\upsilon}$. As a consequence the drag is dominant for high velocities and negligible for low velocities. In addition, it is proportional to the density ρ of the fluid and to the maximum cross section A of the solid body

$$\vec{F}_{\mathrm{D}} = -c_{\mathrm{D}}\, A\left(\frac{1}{2}\rho\, \upsilon\right)\vec{\upsilon} \quad . \tag{2.4 - 8a}$$

The *drag coefficient* c_{D} depends on the shape of the body in the fluid.

For *one-dimensional motions* along the x axis the drag $\vec{F}_{\mathrm{D}}$ can be described by the equation

$$F_{\mathrm{D}} = -c_{\mathrm{D}} A\left(\frac{1}{2}\rho\, \upsilon^2\right) sign(\upsilon) \tag{2.4 - 8b}$$

where $\upsilon = \dot{x}$ indicates the velocity along the x axis.

The motion of a *harmonic oscillator with drag,* when normalized by the assumptions $\Omega_0 = 1$ and $c_{\mathrm{D}}\rho\, A = 1$, obeys the basic oscillation equation

$$\ddot{x} + x + \frac{1}{2}\,\dot{x}^2 sign(\dot{x}) \tag{2.4 - 9a}$$

with the switching function

$$\sigma(e) = \sigma(\upsilon) = sign(\upsilon) = sign(\dot{x}) \quad . \tag{2.4 - 9b}$$

The trajectories of motions of this oscillator with drag in the phase plane $(x,\ \dot{x})$ are determined by

$$\frac{d}{dx}\left[\dot{x}^2\right]+\left[\dot{x}^2\right]sign(\dot{x})=-2x \tag{2.4 - 10a}$$

which can be derived with the aid of (2.4 - 4a&b). This relation represents a differential equation of first order for $\dot{x}^2$ as a function of x. Its solution is

$$2x\,sign(\dot{x})=\left[(\dot{x}(0))^2-2\right]exp\left[-x\,sign(\dot{x})\right]-\left[(\dot{x}(x))^2-2\right] \tag{2.4 - 10b}$$

with the straight switching line

$$S:\ \dot{x}=0 \quad . \tag{2.4 - 10c}$$

The trajectories $\dot{x}(x)$ described by (2.4 - 10b) are illustrated in Fig. 2.4 - 4 for various initial values $\dot{x}(0)$. These trajectories represent semi-parabolas

$$x=\pm\left[1-\frac{1}{2}\dot{x}^2\right] \quad \text{if} \quad \dot{x}(0)=\pm\sqrt{2} \quad . \tag{2.4 - 10d}$$

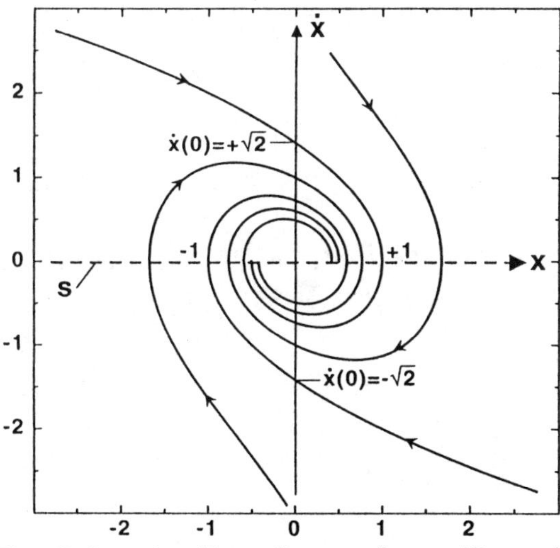

Fig. 2.4 - 4. Phase diagram of an oscillator with drag

In these two cases, the time t as running parameter of these semi-parabolas can be determined with (2.4 - 6). The result is

$$t=\pm\,\dot{x} \quad . \tag{2.4 - 11a}$$

Under these circumstances the motion $x(t)$ of the harmonic oscillator with drag can be evaluated by combination of (2.4 - 10d) and (2.4 - 11a). The result is

$$x(t) = \pm\left[1 - \frac{1}{2}t^2\right] \quad \text{for} \quad -\infty < t \leq 0 \quad . \tag{2.4 - 11b}$$

The limit values on the two semi-parabolas are $x(0) = \pm 1$ and $x(-\infty) = \mp\infty$.

2.4.4 Oscillators with Constant Restoring Force

A constant central restoring force in the two- or three-dimensional space is defined by the formula

$$\vec{F}_Z = -F_Z \frac{\vec{r}}{r} \quad \text{with} \quad F_Z = const > 0 \quad . \tag{2.4 - 12a}$$

If the motion is restricted to the one-dimensional x axis this force is described by

$$F_R = -F_Z \; sign\,(x) \quad \text{with} \quad F_Z = const > 0 \quad . \tag{2.4 - 12b}$$

An *oscillator* that consists of a point with mass $m = 1$, which is subjected to a *constant restoring force* F_R with $F_Z = 1$, obeys the oscillation equation

$$\ddot{x} + sign(x) = 0 \tag{2.4 - 13a}$$

including the switching function

$$\sigma(e) = \sigma(x) = sign\;(x) \quad . \tag{2.4 - 13b}$$

The *trajectories* which represent the motion of this oscillator in the phase space $(x, \dot{x})$ can be evaluated with (2.4 - 4a&b) and integration. This procedure yields

$$x = \frac{1}{2}\, sign(x)\left[(\dot{x}(0))^2 - \dot{x}^2\right] \quad . \tag{2.4 - 14a}$$

The corresponding straight *switching line* is defined by

$$S: \quad x = 0 \quad . \tag{2.4 - 14b}$$

Trajectories and switching line are illustrated in Fig. 2.4 - 5.

The *period* T of the oscillations of the oscillator described by (2.4 - 13a) depends on its amplitude x_{max}

$$T = 4\,\dot{x}(0) = \left[32\, x_{max}\right]^{1/2} \quad . \tag{2.4 - 15}$$

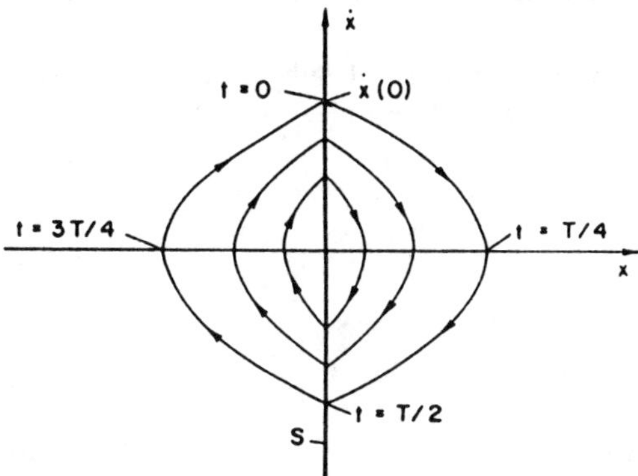

Fig. 2.4 - 5. Phase diagram of an oscillator with constant restoring force

This can be demonstrated by consideration of the following *explicit solution* of (2.4 - 13a)

$$\begin{aligned} &\text{for} \quad 0 \le t \le T/2: \quad x(t) = -\frac{1}{2}t^2 + \frac{1}{4}Tt \quad , \\ &\text{for} \quad T/2 \le t \le T: \quad x(t) = +\frac{1}{2}\left(t - \frac{1}{2}T\right)^2 - \frac{1}{4}T\left(t - \frac{1}{2}T\right) \quad . \end{aligned} \qquad (2.4 - 16)$$

This solution $x(t)$ is plotted in Fig. 2.4 - 6. In a first approximation $x(t)$ corresponds to a harmonic oscillation.

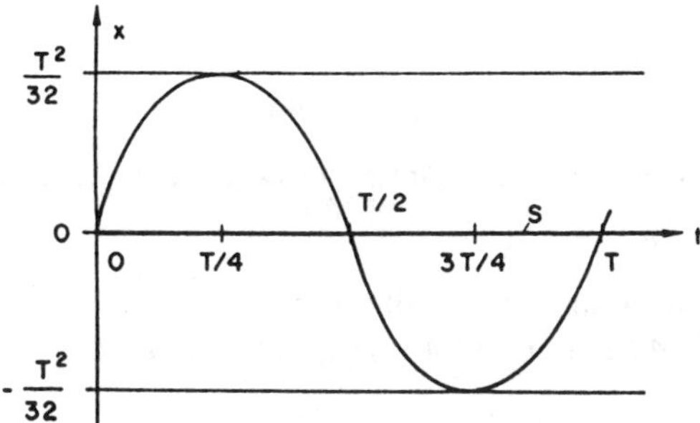

Fig. 2.4 - 6. Example of a motion $x(t)$ of an oscillator with constant restoring force

2.4.5 Oscillators with Constant Restoring Force and Dead Zone

According to (2.4 - 1d) and Fig. 2.4 - 1d a *constant restoring force* F_R *with a dead zone* is described by the equation

$$F_R = -\frac{1}{2} F_Z \left[sign(x - x_0) + sign(x + x_0) \right] \quad \text{with} \quad F_Z \; const > 0 \quad . \qquad (2.4 - 17)$$

The *dead zone* is in the range $|x| < x_0$, whilst the *active zone* extends over the range $|x| > x_0$.

An *oscillator* which consists of a point of mass $m = 1$ subjected to a constant restoring force with a dead zone characterized by $F_Z = 1$ and $x_0 = 1$ oscillates according to the relation

$$\ddot{x} + \frac{1}{2}\left[sign(x-1) + sign(x+1)\right] = 0 \qquad (2.4 - 18a)$$

with the switching function

$$\sigma(e) = \sigma(x) = \frac{1}{2}\left[sign(x-1) + sign(x+1)\right] \quad . \qquad (2.4 - 18b)$$

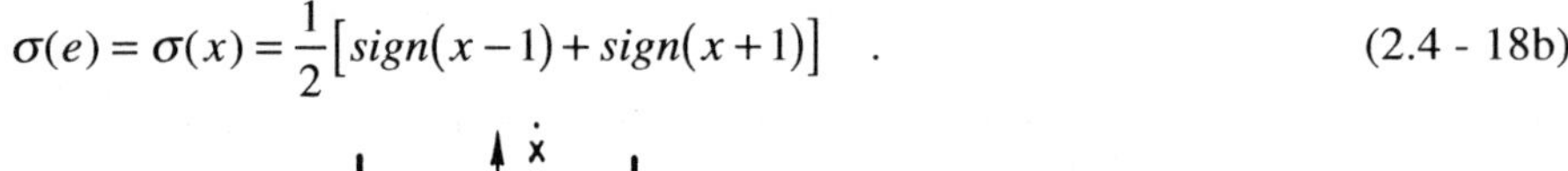

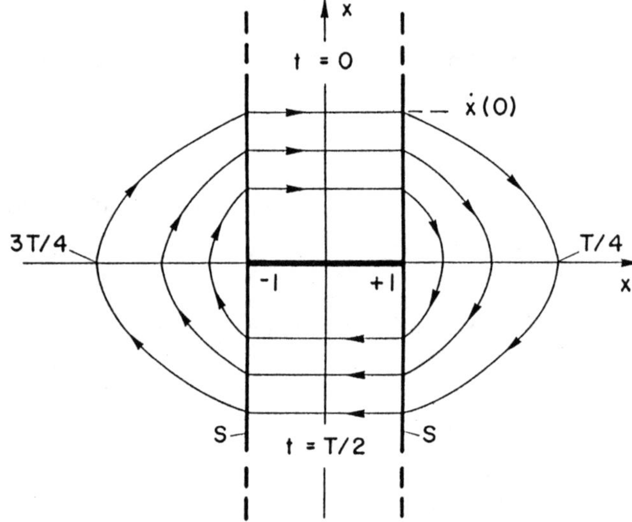

Fig. 2.4 - 7. Phase diagram of an oscillator with constant restoring force and a dead zone

The *trajectories* of the motion of this oscillator in the phase plane $(x, \dot{x})$ can be determined by the combination of (2.4 - 4a&b) and (2.4 - 18a). They obey the two following equations:

α) in the dead zone: $|x| \le 1$,

$$\dot{x}(x) = \dot{x}(-1) = \dot{x}(0) = \dot{x}(+1) = const \quad , \qquad (2.4 - 19a)$$

β) in the active zone: $|x| \ge 1$,

$$x = sign(x) \cdot \left\{1 + \frac{1}{2}\left[(\dot{x}(0))^2 - \dot{x}^2\right]\right\} \quad . \qquad (2.4 - 19b)$$

The corresponding two straight *switching lines* are defined by

$$S: \quad x = \pm 1 \quad . \tag{2.4 - 19c}$$

These two straight switching lines and trajectories (2.4 - 19a&b) are shown in Fig. 2.4 - 7.

The *period* T of the oscillations of the oscillator defined by (2.4 - 18a) depends on the initial condition $\dot{x}(x=0)$ according to

$$T = 4\left[\dot{x}(0) + \frac{1}{\dot{x}(0)}\right] \quad . \tag{2.4 - 20a}$$

Consequently, this period T as a function of $\dot{x}(0)$ exhibits a minimum

$$T_{\min} = 2 \quad \text{for} \quad \dot{x}(0) = 1 \quad . \tag{2.4 - 20b}$$

The relation (2.4 - 20a) between $\dot{x}(0)$ and T can be derived by the explicit solution of (2.4 - 18a). This yields

$$\begin{aligned}
&\text{for} \quad -T_1 \le t \le T_1: \\
&\qquad x = t / T_1 \quad , \\
&\text{for} \quad T_1 \le t \le T_1 + 2T_2: \\
&\qquad x = -\tfrac{1}{2}(t - T_1)^2 + T_2(t - T_1) + 1 \quad , \\
&\text{for} \quad T_1 + 2T_2 \le t \le 3T_1 + 2T_2: \\
&\qquad x = (2T_1 + 2T_2 - t) / T_1 \quad , \\
&\text{for} \quad 3T_1 + 2T_2 \le t \le T - T_1: \\
&\qquad x = \tfrac{1}{2}(t - 3T_1 - 2T_2)^2 - T_2(t - 3T_1 - 2T_2) - 1 \quad , \\
&\text{with} \quad T_2 = \dot{x}(0) = 1 / T_1 \quad \text{and} \quad T = 4(T_1 + T_2) \quad .
\end{aligned} \tag{2.4 - 21}$$

This solution is illustrated in Fig. 2.4 - 8.

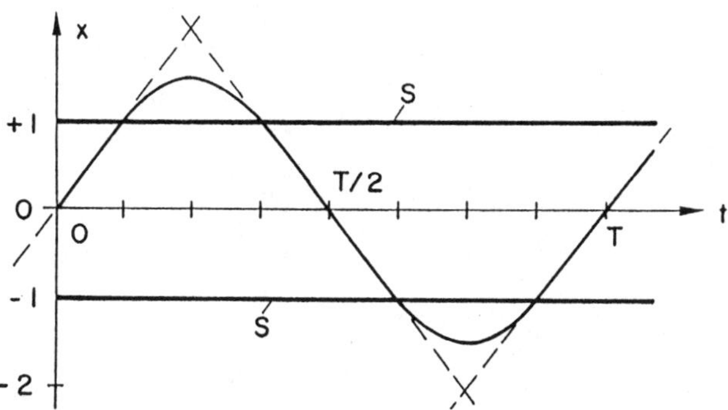

Fig. 2.4 - 8. Example of a motion $x(t)$ of an oscillator with a constant restoring force and a dead zone

2.4.6 Oscillators with Hysteresis

Among the most interesting nonlinear oscillators are those that contain a switch with hysteresis. An *example* is the oscillator defined by the equation

$$\ddot{x} + x - \sigma_H(x) = 0 \quad , \tag{2.4 - 22}$$

where $\sigma_H(x)$ represents the *switch with hysteresis* characterized by (2.4 - 1e) and Fig. 2.4 - 1e.

The *trajectories* of the motion of this oscillator in the phase plane $(x, \dot{x})$ can be determined by the combination of (2.4 - 4a&b) and (2.4 - 22). They fulfill the relation

$$[\dot{x}(x)]^2 + [x - \sigma_H(x)]^2 = [\dot{x}(\sigma_H(x))]^2 \tag{2.4 - 23a}$$

and exhibit the straight *switching lines* described by

$$S: \ \dot{x} > 0, x = 1 \quad \text{and} \quad \dot{x} < 0, x = -1 \quad . \tag{2.4 - 23b}$$

Trajectories and switching lines are shown in the phase diagram of Fig. 2.4 - 9. This diagram elucidates the effect of the hysteresis e.g. in point A, where two circular trajectories with the different centers (± 1, 0) meet and continue as a new trajectory with the center (+ 1, 0). In addition, it demonstrates that for long times t the solutions $x(t)$ of (2.4 - 22) emerge as harmonic oscillations in the form

$$\lim_{t\to\infty} (t) = \pm 1 + A \sin(t - \varphi) \quad \text{with} \quad 0 \le A < 2 \quad . \tag{2.4 - 24a}$$

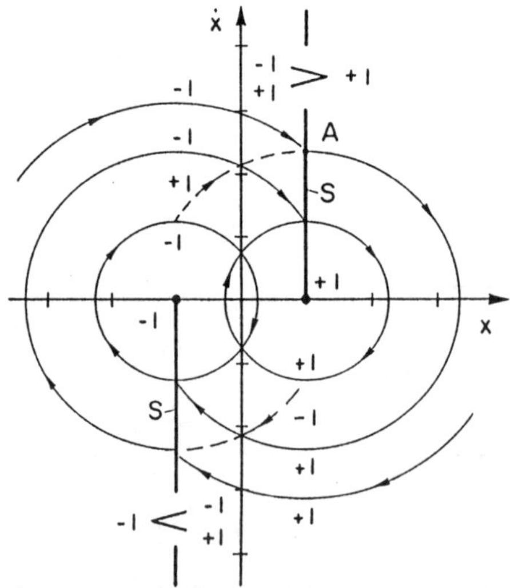

Fig. 2.4 - 9. Phase diagram of an oscillator with hysteresis

This phenomenon is recognized in the following particular solution of (2.4 - 22)

$$
\begin{aligned}
&\text{for} \quad 0 \le t \le t_1 \quad : x_0(t) = -1 + 4 \sin t \\
&\text{with} \quad t_1 = \arcsin \frac{1}{2} = \pi / 6 \quad , \\
&\text{for} \quad t_1 \le t \le t_2 \quad : x_1(t) = +1 + 2\sqrt{3} \sin(t - t_1) \\
&\text{with} \quad t_2 = t_1 + \pi + \arcsin\left(1/\sqrt{3}\right) \quad , \\
&\text{for} \quad t_2 \le t \le t_3 \quad : x_2(t) = -1 - 2\sqrt{2} \sin(t - t_2) \\
&\text{with} \quad t_3 = t_2 + \pi + \arcsin\left(1/\sqrt{2}\right) = t_2 + 5\pi/4 \quad , \\
&\text{for} \quad t_3 \le t \le t_4 \quad : x_3(t) = 1 + 2 \sin(t - t_3) \\
&\text{with} \quad t_4 = t_3 + 3\pi/2 \quad , \\
&\text{for} \quad t_4 < t \qquad : x_4(t) = -1 \quad .
\end{aligned}
\qquad (2.4 - 24b)
$$

This solution is illustrated in Fig. 2.4 - 10.

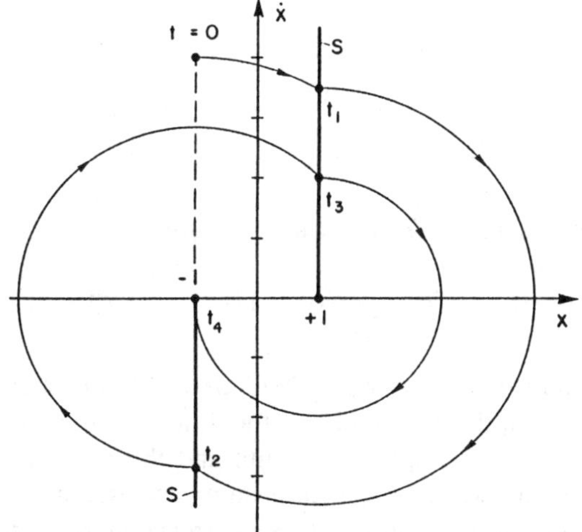

Fig. 2.4 - 10. Phase diagram of an oscillator with hysteresis according to the solution $x(t)$ described by (2.4 - 24b)

2.5 Nonlinear Liénard Oscillators

Liénard oscillators are defined by the *oscillator equation*

$$\ddot{x} - S(x)\dot{x} + D(x)x = 0 \quad . \qquad (2.5 - 1a)$$

This differential equation is linear, if both, $S(x)$ and $D(x)$, are constant. Otherwise, this equation and the corresponding oscillator are *nonlinear.*

The best known Liénard oscillators are discussed in the following chapters. They include the *Duffing* oscillator, the *simple pendulum,* the *Smith* oscillator and the *van der Pol* oscillator.

2.5.1 General Properties

The oscillation equation (2.5 - 1a) can be assigned to an *autonomous system* of two first order differential equations of the type described in Section 4.2. This system has the form

$$\begin{aligned} \dot{x} &= u = -U_x + H_y = -U_x(x) + y \\ \dot{y} &= \upsilon = -U_y - H_x = -H_x(x) \end{aligned} \qquad (2.5 - 1b)$$

with the potential U and the Hamiltonian H. The indices x and y indicate partial differentiations. U and H are the functions

$$U = U(x) \quad \text{and} \quad H = H(x,y) = \frac{1}{2} F(x^2) + \frac{1}{2} y^2 \quad . \qquad (2.5 - 1c)$$

They are related to the coefficients of (2.5 - 1a) by

$$S(x) = -U_{xx}(x) \quad \text{and} \quad D(x) = dF(x^2)/dx^2 \quad . \qquad (2.5 - 1d)$$

The system (2.5 - 1b) of equations exhibits a *singular point* at its *origin*

$$\dot{x}(x=0, y=0) = \dot{y}(x=0, y=0) = 0 \quad \text{if} \quad U_x(0) = H_x(0) = 0 \quad . \qquad (2.5 - 1e)$$

A large variety of Liénard oscillators can perform *stationary oscillations.* Some show stationary oscillations with arbitrary amplitudes, such as the linear undamped harmonic oscillator, the nonlinear undamped simple pendulum, and the nonlinear undamped Duffing oscillator. Other Liénard oscillators perform stationary oscillations only for well defined amplitudes and forms of oscillations. These correspond to stable *limit cycles* in the phase plane $(x, \dot{x})$ or in the xy plane.

The existence of a *single stable limit cycle* in the phase plane $(x, \dot{x})$ of a Liénard oscillator is determined by the *theorem of Levinson and Smith* [Birkhoff & Rota 1989 B]. This theorem involves the coefficients of (2.5 - 1a). It requires

$$\alpha) \quad D(-x) = D(+x) \quad , \quad S(-x) = S(x) \quad , \qquad (2.5 - 2a)$$

$$\beta) \quad D(x) > 0 \quad , \qquad (2.5 - 2b)$$

γ) $S(x) > 0$ for $x^2 < a^2 < \infty$, $S(x) < 0$ for $x^2 > a^2$, (2.5 - 2c)

δ) $\int_0^\infty S(x)\,dx = -\infty$. (2.5 - 2d)

On these conditions the following two statements are valid:

α) There exists a *single stable limit cycle* in the phase plane $(x, \dot{x})$ of the solutions $x(t)$ of the Liénard equation (2.5 - 1a).

β) Each solution $x(t)$ of (2.5 - 1a) either *converges in the phase plane* $(x, \dot{x})$ *towards this limit cycle* for long times $t \to \infty$, or it corresponds to this limit cycle itself.

2.5.2 Duffing Oscillators

Undamped Duffing oscillators [Bogoljubow & Mitropolski 1965 B, McLachlan 1950 B, Nayfeh & Mook 1979 B, Reitman 1996 B, Stoker 1957 B, Zwillinger 1989 B] are characterized by the basic nonlinear differential equation

$$\Omega^{-2}\,\ddot{x} + x + \varepsilon\,x^3 = 0 \quad . \qquad (2.5 - 3a)$$

Consequently, this oscillator equation represents a special nonlinear Liénard oscillator whose basic equation (2.5 - 1a) has the coefficients

$$S(x) = 0 \quad \text{and} \quad D(x) = \Omega^2\left(1 + \varepsilon\,x^2\right) \quad . \qquad (2.5 - 3b)$$

For $\varepsilon = 0$ the Duffing oscillators correspond to the *linear harmonic oscillators* defined by (2.2 - 1).

a) Associated Hamiltonian systems

The associated autonomous two-dimensional systems of first-order differential equations

$$\begin{aligned} \dot{x} &= u = H_y(x, y) = +y \\ \dot{y} &= v = -H_x(x, y) = -\Omega^2 x\left[1 + \varepsilon\,x^2\right] \end{aligned} \qquad (2.5 - 4a)$$

are *Hamiltonian systems* with a zero potential U and a specific Hamiltonian H as described in Sections 4.2.3d and 4.2.5

$$\begin{aligned} U &= U(x, y) = 0 \quad , \\ H &= H(x, y) = \frac{1}{2}y^2 + \frac{1}{2}\Omega^2\left[x^2 + \frac{1}{2}\varepsilon\,x^4\right] \quad . \end{aligned} \qquad (2.5 - 4b)$$

In *Hamilton mechanics* the Hamiltonian $H(x, y)$ of a conservative system corresponds to the total energy E represented by position and momentum coordinates x and $\dot{x}$ in the phase space. For the system (2.5 - 4ab) this yields

$$H(x,\dot{x}) = E(x,\dot{x}) = E_{\text{kin}}(\dot{x}) + V(x)$$

$$\text{with} \quad E_{\text{kin}}(\dot{x}) = \frac{1}{2}\dot{x}^2 \quad \text{and} \quad V(x) = \frac{1}{2}\Omega^2 x^2\left[1 + \frac{1}{2}\varepsilon x^2\right] \quad . \tag{2.5 - 5a}$$

$E_{\text{kin}}(\dot{x})$ signifies the *kinetic energy* of a particle with mass $m = 1$, whilst the existence of the *potential energy* $V(x)$ reveals that the undamped Duffing oscillators represent *conservative systems.* The potential energies $V(x)$ for $\varepsilon = 0, \pm 1$ are illustrated in Fig. 2.5 - 1.

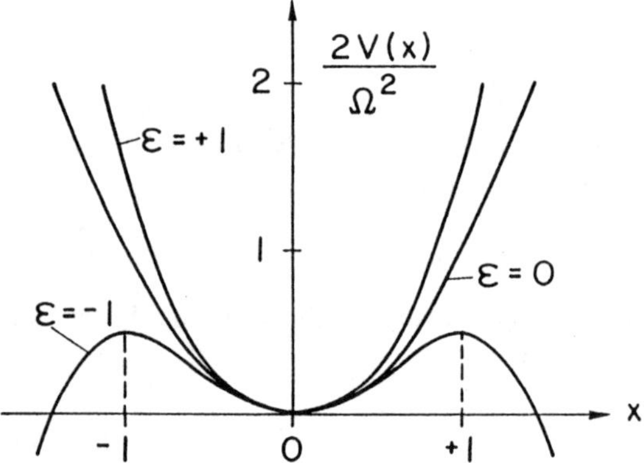

Fig. 2.5 - 1. Potential energies $V(x)$ of Duffing oscillators

According to the relations (4.2 - 11g-i) the Hamiltonian $H(x,\dot{x})$ determines the trajectories in the phase space $(x,\dot{x})$. As a consequence the *trajectories* of (2.5 - 3a) and (2.5 - 4a) in the phase space are defined by

$$H(x,\dot{x}) = \frac{1}{2}\dot{x}^2 + \frac{1}{2}\Omega^2 x^2\left(1 + \frac{1}{2}\varepsilon x^2\right) = H = E = const \quad , \tag{2.5 - 5b}$$

where E represents the total energy of the Duffing oscillator.

b) Solutions with elliptic functions

The solution of the oscillator equation (2.5 - 3a) of the undamped Duffing oscillator is relatively simple on the initial conditions

$$x(0) = x_0 \quad \text{and} \quad \dot{x}(0) = 0 \quad . \tag{2.5 - 6a}$$

The potential energy $V(x)$ described by (2.5 - 5a) and Fig. 2.5 - 1 exhibits a minimum for all ε and two maxima for $\varepsilon < 0$. Therefore, an undamped Duffing oscillator oscillates

$$\begin{aligned} &\text{for} \quad x_0 \; \textit{arbitrary} \quad &&\text{if} \quad \varepsilon \geq 0 \;, \\ &\text{for} \quad |x_0| < +\sqrt{-1/\varepsilon} \quad &&\text{if} \quad \varepsilon < 0 \;. \end{aligned} \tag{2.5 - 6b}$$

In most circumstances one is interested mainly in these oscillations.

In order to solve the oscillation equation (2.5 - 3a) one starts with (2.5 - 5b) and changes the time scale by introducing

$$\tau = \Omega t \quad . \tag{2.5 - 7a}$$

The combination of this relation with (2.5 - 5b) and (2.5 - 6a) results in

$$2\Omega^{-2} E = (dx/d\tau)^2 + x^2\left(1 + \frac{1}{2}\varepsilon x^2\right) = x_0^2\left(1 + \frac{1}{2}\varepsilon x_0^2\right) \quad \text{or} \tag{2.5 - 7b}$$

$$\begin{aligned} (dx/d\tau)^2 &= \left[x_0^2 - x^2\right]\left[1 + \frac{1}{2}\varepsilon\left(x_0^2 + x^2\right)\right] \\ &= \frac{1}{2}\varepsilon\left[x_0^2 - x^2\right]\left[x_0^2 + (2/\varepsilon) + x^2\right] \\ &= -\frac{1}{2}\varepsilon\left[x_0^2 - x^2\right]\left[(-2/\varepsilon) - x_0^2 - x^2\right] \quad . \end{aligned} \tag{2.5 - 7c}$$

The solutions of this equation involve *Legendre's elliptic integrals* and *Jacobi's elliptic functions* [Abramowitz & Stegun 1965 B, Erdelyi et al. 1952-1954 B, Jahnke & Emde 1952 B, Milne-Thomson 1931 B]. These solutions depend on the sign of the parameter ε:

α) *ε positive*

$$x(t) = x_0 \, cn\left(\left[1 + \varepsilon x_0^2\right]^{1/2} \Omega t \,|\, m\right) \quad \text{with} \quad m = \frac{1}{2}\left[1 + \left(1/\varepsilon x_0^2\right)\right]^{-1} \quad . \tag{2.5 - 8}$$

The *Jacobian elliptic cosine* $cn(u \mid m)$ illustrated in Fig. 2.5 - 3 is periodic in u. Its period equals

$$T_{\text{cn}} = 4K(m) = 4\int_0^{\pi/2} \left[1 - m\sin^2\Theta\right]^{-1/2} d\Theta \tag{2.5 - 9a}$$

where $K(m)$ represents the *complete elliptic integral of the first kind* shown in Fig. 2.5 - 2. It can be approximated by the following *series* [Abramowitz & Stegun 1965 B]

$$\frac{2}{\pi}K(m) = 1 + \left(\frac{1}{2}\right)^2 m + \left(\frac{1\cdot 3}{2\cdot 4}\right)^2 m^2 + \left(\frac{1\cdot 3\cdot 5}{2\cdot 4\cdot 6}\right)^2 m^3 + \dots \quad . \tag{2.5 - 9b}$$

Accordingly, the period $T(\varepsilon, x_0)$ of an undamped Duffing oscillator with positive ε is

$$\begin{aligned} &T(\varepsilon, x_0) = T_0\left(1+\varepsilon x_0{}^2\right)^{-1/2} \frac{2}{\pi}K(m) \\ &\text{with} \quad m = \frac{1}{2}\left[1+\left(1/\varepsilon x_0^2\right)\right]^{-1} \\ &\text{and} \quad T(0,x_0) = T_0 = 2\pi/\Omega\,, \quad T(\infty, x_0) = 0 \quad . \end{aligned} \tag{2.5 - 9c}$$

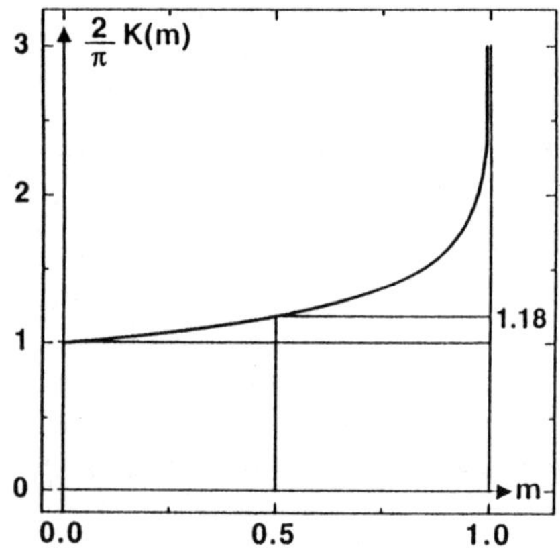

Fig. 2.5 - 2. The complete elliptical integral $K(m)$ reduced by the factor $\pi/2$

The *Jacobian elliptic cosine* $cn(u \mid m)$ that defines the oscillation of a Duffing oscillator with positive ε is characterized by the following *properties* [Abramowitz & Stegun 1965 B]

$$\begin{aligned} &cn(u+4K(m) \mid m) = +cn(u \mid m) \quad , \\ &cn(u+2K(m) \mid m) = -cn(u \mid m) \quad , \\ &cn(-u \mid m) \qquad\quad = +cn(u \mid m) \quad , \end{aligned} \tag{2.5 - 10a}$$

$$\begin{aligned} &cn(0 \mid m) = 1 \quad , \quad cn(2K(m) \mid m) = -1 \quad , \\ &cn(K(m) \mid m) = cn(3K(m) \mid m) = 0 \quad . \end{aligned} \tag{2.5 - 10b}$$

$$cn(u \mid 0) = \cos u \text{ and } cn(u \mid 1) = \operatorname{sech} u \quad . \tag{2.5 - 10c}$$

Fig. 2.5 - 3 illustrates the functions $cn(u \mid m)$ for $m = 0$; 0.5 ; 0.95 and 1. For $m < 1/2$ the Jacobian cosines look similar to the trigonometric function $\cos u$

$$cn(u, m) \approx \cos u \quad \text{for} \quad m < 1/2 \quad . \tag{2.5 10d}$$

β) *ε negative and* $x_0 \le \sqrt{-1/\varepsilon}$

$$x(t) = x_0\, cd\left(\left[1 + \frac{1}{2}\varepsilon x_0^2\right]^{1/2} \Omega t \mid m\right) \quad \text{with} \quad m = -\left[1 + \left(2/\varepsilon x_0^2\right)\right]^{-1} \quad . \tag{2.5 - 11}$$

The *Jacobian elliptic function* $cd(u \mid m)$ has the same period as the Jacobian cosine $cn(u \mid m)$

$$T_{\text{cd}} = 4\, K(m) = T_{\text{cn}} \quad . \tag{2.5 - 12a}$$

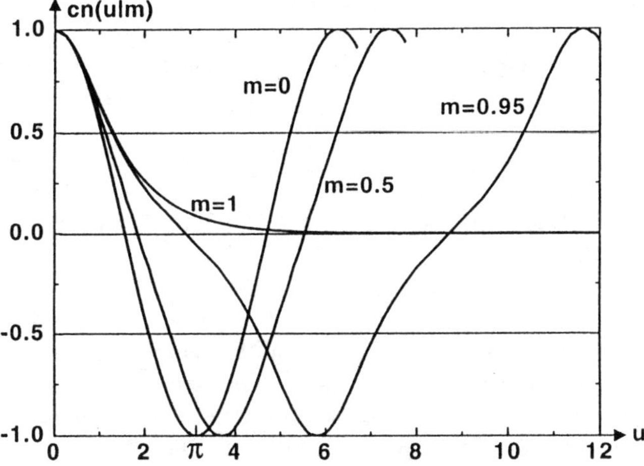

Fig. 2.5 - 3. Jacobi's elliptical cosine $cn(u \mid m)$ for $m = 0$; 0.5; 0.95; 1

The corresponding period $T(\varepsilon, x_0)$ of an undamped Duffing oscillator with negative ε equals

$$T(\varepsilon, x_0) = T_0\left(1 + \frac{1}{2}\varepsilon x_0^2\right)^{-1/2} \frac{2}{\pi} K(m)$$

$$\text{with} \quad m = -\left[1 + \left(2/\varepsilon x_0^2\right)\right]^{-1} \tag{2.5 - 12b}$$

$$\text{and} \quad T(0, x_0) = T_0 = 2\pi/\Omega \quad , \quad T(\varepsilon, \sqrt{-1/\varepsilon}) = \infty \quad .$$

The *Jacobian function* $cd(n \mid m)$ shows properties similar to those of the Jacobian cosine $cn(u \mid m)$ [Abramowitz & Stegun 1965 B]

$$\begin{aligned} cd\big(u+4K(m) \mid m\big) &= +cd\big(u \mid m\big) \ , \\ cd\big(u+2K(m) \mid m\big) &= -cd\big(u \mid m\big) \ , \\ cd\big(-u \mid m\big) &= +cd\big(u \mid m\big) \ , \end{aligned} \qquad (2.5 - 13a)$$

$$\begin{aligned} &cd\big(0 \mid m\big) = 1 , \quad cd\big(2K(m) \mid m\big) = -1 \ , \\ &cd\big(K(m) \mid m\big) = cd\big(3K(m) \mid m\big) = 0 \ . \end{aligned} \qquad (2.5 - 13b)$$

The *limit functions* of $cd(u \mid m)$ are

$$cd\big(u \mid 0\big) = cos\, u \ \text{ and } \ cd\big(u \mid 1\big) = 1 \ . \qquad (2.5 - 13c)$$

The second limit function $cd(u \mid 1) = 1$ corresponds to the constant solutions $x(t) = \pm\, x_0 = \pm \sqrt{-1/\varepsilon}$ with the infinite period $T(\varepsilon, \sqrt{-1/\varepsilon}\,) = \infty$. The Jacobian functions $cd(u \mid m)$ with $m < 1/2$ behave approximately as the trigonometric function $cos\, u$

$$cd(u \mid m) \approx cos\, u \quad \text{for} \quad m < 1/2 \ . \qquad (2.5 - 13d)$$

The Jacobian function $cd(u \mid 1/2)$ is illustrated in Fig. 16.3 of the reference [Abramowitz & Stegun 1965 B].

γ) $|\varepsilon|$ *small*

There exist *approximative solutions* of (2.5 - 3a) for small $|\varepsilon|$. They are valid for both signs of ε. In most circumstances, the main interest is in *periodic solutions* that can be represented as *Fourier series.* Because the period T of such a solution depends on ε and on the initial amplitude x_0, the time scale needs to be modified in the course of improving the approximation. This is often accomplished by the *approach of Poincaré-Lindstedt-Lighthill* [Andersen & Geer 1982 J, Hairer et al. 1987 B, Lindstedt 1883 J, McLachlan 1950 B, Verhulst 1990 B, Zwillinger 1989 B]. On the initial conditions (2.5 - 6a) this approximation starts with the ansatz

$$\begin{aligned} &\Omega t = \tau = s\big[1 + \varepsilon c_1 + \varepsilon^2 c_2 + \ldots\big] \ , \\ &x(s) = x_0(s) + \varepsilon x_1(s) + \varepsilon^2 x_2(s) + \ldots \\ &\text{with} \quad x(0) = x_0(0) = x_0 \ ; \ x_1(0) = x_2(0) = \ldots = 0 \\ &\text{and} \quad x'(0) = x'_0(0) = x'_1(0) = \ldots = 0 \ . \end{aligned} \qquad (2.5 - 14a)$$

Application of this ansatz in (2.5 - 3a) and subsequent comparison of the terms with the same powers of ε, result in a sequence of differential equations. These are solved successively on the condition that the *solution is periodic*. For the approximate solution of (2.5 - 3a) the first terms are

$$\Omega\, t = \tau = s\left[1 - \frac{3}{8}\varepsilon\, x_0^2 + \ldots\right] \quad ,$$
$$x(s) = x_0\left[1 - \frac{\varepsilon}{32} x_0^2\right] \cos s + \frac{\varepsilon}{32} x_0^3 \cos 3s + \ldots \quad . \tag{2.5 - 14b}$$

This solution has the period

$$T(\varepsilon, x_0) = T_0\left[1 - \frac{3}{8}\varepsilon\, x_0^2 + \ldots\right] \quad \text{with} \quad T(0, x_0) = T_0 = 2\pi / \Omega \quad . \tag{2.5 - 14c}$$

c) Cubic Duffing oscillators

A cubic undamped Duffing oscillator obeys the oscillator equation

$$\Omega^{-2}\ddot{x} + x^3 = 0 \quad . \tag{2.5 - 15a}$$

The corresponding Liénard differential equation (2.5 - 1a) has the coefficients

$$S(x) = 0 \text{ and } D(x) = \Omega^2\, x^2 \quad . \tag{2.5 - 15b}$$

In this situation the potential energy $V(x)$ of a particle with the mass $m = 1$ is of the form

$$V(x) = \frac{1}{4}\Omega^2\, x^4 \text{ with } V_{xx}(0) = 0 \quad . \tag{2.5 - 15c}$$

Starting with the initial conditions

$$x(0) = x_0 \;\; ; \;\; \dot{x}(0) = 0 \tag{2.5 - 6a}$$

one finds the periodic solution

$$x(t) = x_0\, cn\left(x_0 \Omega t \,|\, 1/2\right) \tag{2.5 - 16a}$$

with the *Jacobian cosine* $cn(u\,|1/2)$ [Abramowitz & Stegun 1965 B] described by (2.5 - 10a-c) and Fig. 2.5 - 2. This solution has the period

$$T(\Omega, x_0) = T_0 \frac{2}{\pi} K(1/2) \frac{1}{x_0}$$
$$\text{with} \quad T_0 = 2\pi / \Omega \quad \text{and} \quad T(\Omega, 0) = \infty \quad . \tag{2.5 - 16b}$$

This period increases with decreasing $|x_0|$ towards infinity because $V_{xx}(0) = 0$ according to (2.5 - 15c).

d) Damped Duffing oscillators

Damped Duffing oscillators can be described by the oscillator equation

$$\Omega^{-2}\ddot{x} + (\Omega Q)^{-1}\dot{x} + x + \varepsilon x^3 = 0 \qquad (2.5 - 17a)$$

where the quality factor Q is defined as usual

$$\Omega\tau = 2Q \quad . \qquad (2.2 - 29)$$

The oscillator equation (2.5 - 17a) represents also a Liénard differential equation (2.5 - 1a). Its coefficients are

$$S(x) = -(\Omega / Q) \quad \text{and} \quad D(x) = \Omega^2\left(1 + \varepsilon x^2\right) \quad . \qquad (2.5 - 17b)$$

For the initial conditions

$$x(0) = x_0 \quad \text{and} \quad \dot{x}(0) = 0 \qquad (2.5 - 6a)$$

and *weak damping* defined by $Q >> 0$ there exists the following approximate solution [McLachlan 1950 B] of (2.5 - 17a)

$$x(t) = x_0\, exp(-t/\tau) cos\left\{\Omega t\left[1 + \frac{3}{4}\varepsilon x_0^2\, exp(-2t/\tau)\right]^{1/2}\right\} \quad . \qquad (2.5 - 18)$$

2.5.3 Simple Pendulum

The undamped *simple or mathematical pendulum* is introduced in Section 2.2.1c and illustrated in Fig. 2.2 - 3. It is characterized by its mass μ and its length a. Its motion is usually described with the angle α between the pendulum and the vertical as variable. In Section 2.2.1c the equation of motion (2.2 - 6b) is derived on the basis of the relation between the time derivate of the angular momentum L and the torque T exerted on the pendulum. In the following the equation of motion is derived with the law of energy conservation.

Without friction the simple pendulum represents a conservative mechanical system with the total energy

$$\begin{aligned} E_{tot} &= H(\alpha, \dot{\alpha}) = E_{kin}(\dot{\alpha}) + V(\alpha) = \frac{1}{2}\mu a^2\,\dot{\alpha}^2 + \mu g a(1 - cos\,\alpha) \\ &= H(\alpha, L) = \frac{1}{2}\left(\mu a^2\right)^{-1} L^2 + \mu g a(1 - cos\,\alpha) \end{aligned} \qquad (2.5 - 19a)$$

where L indicates the angular momentum, $E_{kin}(\dot{\alpha})$ the kinetic energy, $V(\alpha)$ the potential energy, $H(\alpha, L)$ the Hamiltonian and $g \approx 9.81$ m s^{-2} the acceleration of free fall due to gravity. The stable equilibrium with $\alpha = 0$ is characterized by the minimum potential energy $V(0) = 0$. The unstable equilibrium with $\alpha = \pi$ corresponds to the maximum potential energy with $V(\pi) = 2$ mga. The total energy E_{tot} does not vary with time. Therefore, it is possible to derive the oscillator differential equation by differentiation of (2.5 - 19a) with respect to time t. The resulting *oscillator equation* is in agreement with (2.2 - 6b)

$$\Omega^{-2}\ddot{\alpha} + sin\ \alpha = 0 \text{ with } \Omega^2 = g / a \quad . \tag{2.5 - 19b}$$

Thus, the simple pendulum represents a Liénard oscillator according to (2.5 - 1a) with the coefficients

$$S(x) = 0 \quad \text{and} \quad D(x) = \Omega^2 \frac{sin\, x}{x} \quad . \tag{2.5 - 19c}$$

In order to deduce the angular motion α of the simple pendulum it is of advantage to devide (2.5 - 19a) by 2 mgL which yields

$$\Omega^{-2}(\dot{\alpha}/2)^2 + sin^2(\alpha/2) = const \quad . \tag{2.5 - 19d}$$

a) Oscillations with large amplitudes

Of most interest are the two *angular oscillations* $\alpha(t)$ initiated by the following two sets of conditions:

α) The first initial conditions

$$\alpha(0) = \alpha_0 \quad \text{and} \quad \dot{\alpha}(0) = 0 \tag{2.5 - 20a}$$

imply

$$\alpha(t) = 2\, arc\, sin\left[sin(\alpha_0/2)\, cd\left(\Omega t \mid m = sin^2(\alpha_0/2)\right)\right] \tag{2.5 - 20b}$$

with the *Jacobi's elliptic function* $cd(u \mid m)$ [Abramowitz & Stegun 1965 B] characterized by (2.5 - 13a-c). The solution (2.5 - 20b) of (2.5 - 19d) has the *period*

$$T(\alpha_0) = T_0 \frac{2}{\pi} K\left(m = sin^2\alpha_0\right) \quad \text{with} \quad T(0) = T_0 = 2\pi/\Omega \quad . \tag{2.5 - 20c}$$

β) The second initial conditions

$$\alpha(0) = 0 \quad \text{and} \quad \dot{\alpha}(0) = \dot{\alpha}_0 \tag{2.5 - 21a}$$

yield

$$\alpha(t) = 2\,arc\,sin\left[\left(\dot{\alpha}_0 / 2\Omega\right) sn\left(\Omega t \mid m = \left(\dot{\alpha}_0 / 2\Omega\right)^2\right)\right] \qquad (2.5 - 21b)$$

with the period

$$T(\dot{\alpha}_0) = T_0 \frac{2}{\pi} K\left(m = \left(\dot{\alpha}_0 / 2\Omega\right)^2\right) \quad \text{with} \quad T(0) = T_0 = 2\pi / \Omega \quad . \qquad (2.5 - 21c)$$

Consequently, the initial conditions of the motion described by (2.5 - 21b) are limited to $\dot{\alpha}_0^2 \leq 4\Omega^2$.

The *Jacobian elliptic sine* $sn(u| m)$ [Abramowitz & Stegun 1965 B] of (2.5 - 21b) has the following properties:

$$\begin{aligned} sn\left(u + 4K(m) \mid m\right) &= +sn\left(u \mid m\right) \quad , \\ sn\left(u + 2K(m) \mid m\right) &= -sn\left(u \mid m\right) \quad , \\ sn\left(-u \mid m\right) &= -sn\left(u \mid m\right) \quad , \end{aligned} \qquad (2.5 - 22a)$$

$$\begin{aligned} &sn\left(0 \mid m\right) = sn\left(2K(m) \mid m\right) = 0 \quad , \\ &sn\left(K(m) \mid m\right) = -sn\left(3K(m) \mid m\right) = 1 \quad . \end{aligned} \qquad (2.5 - 22b)$$

The limit functions of $sn(u \mid m)$ are

$$sn\left(u \mid 0\right) = sin\,u \quad \text{and} \quad sn\left(u \mid 1\right) = tanh\,u \quad . \qquad (2.5 - 22c)$$

For $m < 1/2$ the Jacobian elliptic sine $sn(u \mid m)$ is similar to the trigonometric function $sin\,u$

$$sn\left(u \mid m\right) \approx sin\,u \quad \text{for} \quad m < 1/2 \quad . \qquad (2.5 - 22d)$$

The Jacobian elliptic sine $sn(u \mid 1/2)$ with $m = 1/2$ is illustrated in Fig. 16.1 of the reference [Abramowitz & Stegun 1965 B].

Furthermore the Jacobian elliptic sine $sn(u \mid m)$ is related to the other Jacobian elliptic function $cd(u \mid m)$ by a *phase shift* as follows

$$sn\left(u \pm K(m) \mid m\right) = \pm cd\left(u \mid m\right) \quad . \qquad (2.5 - 22e)$$

b) Weak oscillations

For motions of the simple pendulum with *small angles* $\alpha(t)$ it can be approximated by a *Duffing oscillator* defined by

$$\Omega^{-2}\ddot{\alpha}+\alpha+\varepsilon\alpha^{3}\approx 0 \quad \text{with} \quad \varepsilon=-1/3!=-1/6$$
$$\text{because} \quad \sin\alpha\approx\alpha-\frac{1}{3}\alpha^{3} \quad . \tag{2.5 - 23a}$$

On these circumstances the *method of Poincaré, Lindstedt and Lighthill* (2.5 - 14a&b) yields an *approximate oscillation* determined by the relations

$$\Omega\, t=\tau=s\left[1+\frac{1}{16}\alpha_0^2+\ldots\right] \quad ,$$
$$\alpha(s)=\alpha_0\left[1+\frac{1}{192}\alpha_0^2\right]\cos s-\frac{1}{192}\alpha_0^3\cos 3s+\ldots \quad . \tag{2.5 - 23b}$$

The *period* of this oscillation corresponds to those of (2.5 - 9c) and (2.5 - 20c)

$$\alpha(s)=\alpha_0\left[1+\frac{1}{192}\alpha_0^2\right]\cos s-\frac{1}{192}\alpha_0^3\cos 3s+\ldots \quad ,$$
$$T(\alpha_0)=T_0\frac{2}{\pi}K\left(m=\sin^2(\alpha_0/2)\right)$$
$$\approx T_0\left[1+\frac{1}{4}\sin^2(\alpha_0/2)+\ldots\right]\approx T_0\left[1+\frac{1}{16}\alpha_0^2+\ldots\right] \tag{2.5 - 23c}$$
$$\text{with} \quad T(0)=T_0=2\pi/\Omega \quad .$$

2.5.4 Smith Oscillators

Smith oscillators are defined by the basic differential equation [Bellman 1966 B, Smith 1961 J]

$$\ddot{x}+\left[(n+2)b\,x^{n}-2a\right]\dot{x}+\left[\omega^2+\left(bx^{n}-n\right)^2\right]x=0$$
$$\text{with} \quad a>0 \quad , \quad b>0 \quad , \quad n=2,4,6,\ldots \quad . \tag{2.5 - 24a}$$

Consequently, it represents a Liénard differential equation (2.5 - 1a) with the coefficients

$$S(x)=2a-(n+2)b\,x^{n} \quad \text{and} \quad D(x)=\omega^2+\left(b\,x^{n}-n\right)^2 \quad . \tag{2.5 - 24b}$$

These coefficients fulfill the conditions (2.5 - 2a-d) required by the *theorem by Smith and Levinson*. As a consequence the Smith oscillator has *exactly one limit cycle* in the phase plane ($x, \dot{x}$). Limit cycles will be discussed in detail in Section 4.5.

With respect to mathematics, the Smith oscillators have the advantage, that their basic equation (2.5 - 24a) possesses *simple analytical solutions* $x(t)$ determined by

$$x^{n}(t)=\frac{\cos^{n}(\omega t-\varphi)\exp nat}{\left[\frac{\cos\varphi}{x(0)}\right]^{n}+nb\int_{0}^{t}\exp(na\,\theta)\cos^{n}(\omega\theta-\varphi)\,d\theta} \quad . \tag{2.5 - 25}$$

For $n = 2$ and $\varphi = 0$ the solution $x(t)$ fulfills the relation

$$\begin{aligned}\left(\frac{\cos\omega t}{x(t)}\right)^{2}&=\frac{b}{2a}\left[1+\frac{a^{2}}{a^{2}+\omega^{2}}\cos 2\omega t+\frac{a\omega}{a^{2}+\omega^{2}}\sin 2\omega t\right]\\&+\exp(-2at)\left[\left(\frac{1}{x(0)}\right)^{2}-\frac{b}{2a}\left(1+\frac{a^{2}}{a^{2}+\omega^{2}}\right)\right] \quad .\end{aligned} \tag{2.5 - 26a}$$

For long times $t \rightarrow +\infty$ this solution becomes a *stationary oscillation* corresponding to a limit cycle in the phase plane $(x, \dot{x})$. The *period* of this oscillation is

$$T_{\mathrm{Sm}} = 2\pi / \omega \quad . \tag{2.5 - 26b}$$

Fig. 2.5 - 4 shows the limit cycle of the Smith oscillator for $n = 2$, $\varphi = 0$, $2a = b = \omega = 1$. The motion $x(t)$ of this oscillator is

$$x(t)=\pm\sqrt{2}\cos t\left[2+\cos 2t+\sin 2t+\left(\frac{2}{x(0)^{2}}-3\right)\exp(-t)\right]^{-1/2} \quad . \tag{2.5 - 26c}$$

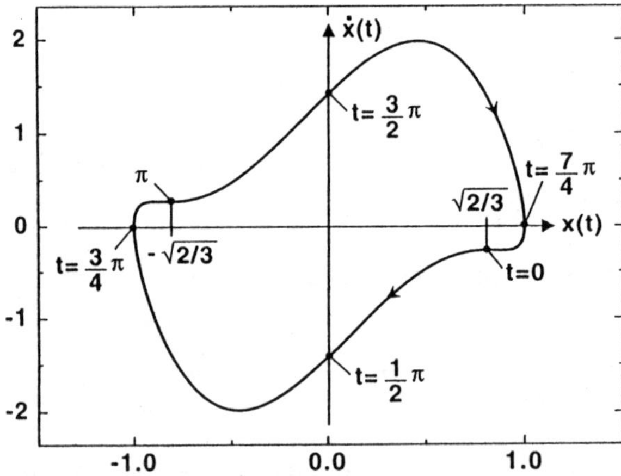

Fig. 2.5 - 4. Limit cycle of the Smith oscillator with $n = 2$, $\varphi = 0$, $2a = b = \omega = 1$ represented in the phase space $(x, \dot{x})$

For the initial condition $x(0) = \pm\sqrt{2/3}$ or for long times $t \to \infty$ this motion represents a stationary oscillation with the period $T_{\mathrm{Sm}} = 2\pi$. This oscillation corresponds to the limit cycle.

2.5.5 Van der Pol Oscillators

A *van der Pol oscillator* [van der Pol 1926 J, 1934 J] consists of a *LCR* circuit with a resistor whose resistance R depends on the current I as follows

$$R(I) = R_0\left[\frac{1}{3}(I/I_0)^2 - 1\right] \tag{2.5 - 27a}$$

$$\text{with} \quad R(I) > 0 \text{ for } I^2 > 3I_0^2 \quad \text{and} \quad R(I) < 0 \text{ for } I^2 < 3I_0^2 \quad .$$

Therefore, the oscillation of this *LCR* circuit will be amplified for small currents I with $I^2 < 3\,I_0^2$ and damped for large currents I with $I^2 > 3\,I_0^2$. Its basic oscillation equation is according to (2.2 - 61a)

$$L(dI/dt) + I\,R(I) + C^{-1}q = 0 \quad \text{with} \quad I = dq/dt \quad , \tag{2.5 - 27b}$$

where q indicates the electric charge on the capacitor with capacitance C. L is the inductance. By the introduction of the new parameters and variables

$$\Omega = +(L\,C)^{-1/2}\,, \quad \varepsilon = +R_0(C/L)^{1/2}\,, \quad \tau = \Omega t \quad , \quad y = (\Omega/I_0)\,q \tag{2.5 - 27c}$$

equation (2.5 - 27b) can be reduced to the *Rayleigh equation* [Nayfeh & Mook 1979 B]

$$y'' + \varepsilon\left[\frac{1}{3}(y')^2 - 1\right]y' + y = 0 \quad \text{with} \quad y' = dy/d\tau \quad . \tag{2.5 - 28a}$$

By the differentiation $d/d\tau$ and the definition x=y' one finds the *van der Pol equation* [Bellman 1966 B, Guckenheimer and Homes 1983 B, Hairer et al 1987 B, Liénard 1928 J, McLachlan 1950 B, van der Pol 1926 J, 1934 J, Verhulst 1990 B, Zwillinger 1989 B]

$$x'' + \varepsilon\left(x^2 - 1\right)x' + x = 0 \quad \text{with} \quad x' = dx/d\tau \quad . \tag{2.5 - 28b}$$

This equation represents a Liénard equation (2.5 - 1a) with the coefficients

$$S(x) = \varepsilon\left(1 - x^2\right) \quad \text{and} \quad D(x) = 1 \quad . \tag{2.5 - 28c}$$

These coefficients fulfill the conditions (2.5 - 2a-d) required by the *theorem of Levinson and Smith* for all $\varepsilon > 0$. Consequently, the van der Pol equation (2.5 - 28b) possesses *one stable limit cycle* in the phase space $(x, \dot{x})$ for $\varepsilon > 0$. Therefore the van der Pol oscillator with $\varepsilon > 0$ exhibits *one stable stationary oscillation.*

a) Van der Pol system of equations

The van der Pol equation (2.5 - 28b) and the Rayleigh equation (2.5 - 28a) correspond to an *autonomous two-dimensional system* of first-order differential equations characteristic of Liénard oscillators

$$\begin{aligned} \dot{x} &= u = -U_x(x) + y \\ \dot{y} &= \upsilon = -H_x(x) \end{aligned} \qquad (2.5 - 1b)$$

with the following potential $U(x)$ and Hamiltonian $H(x, y)$

$$U = U(x) = \frac{1}{12}\varepsilon\left(x^4 - 6x^2\right) \quad \text{with}$$

$$U_x(x) = \frac{1}{3}\varepsilon\left(x^3 - 3x\right) \quad \text{and} \quad U_{xx}(x) = \varepsilon\left(x^2 - 1\right) \quad , \qquad (2.5 - 28d)$$

$$H = H(x, y) = +\frac{1}{2}\left(x^2 + y^2\right) \quad .$$

These functions define the system

$$\begin{aligned} x' &= u = -\frac{1}{3}\varepsilon\left(x^3 - 3x\right) + y \\ y' &= \upsilon = -x \quad . \end{aligned} \qquad (2.5 - 28e)$$

In this system the variable x fulfills the van der Pol equation (2.5 - 28b) whilst the variable y fulfills the Rayleigh equation (2.5 - 28a).

b) Approximations of solutions

The solution of the van der Pol equation (2.5 - 28b) is difficult. Therefore, this apparently simple equation often serves for a test of new approximation techniques created to solve nonlinear ordinary differential equations.

α) For $\varepsilon = 0$ the van der Pol equation (2.5 - 28b) corresponds to the linear second-order differential equation (2.2 - 3) of the *undamped harmonic oscillator.* Its general solution

$$x = A\cos(\tau - \varphi) = A\cos(\Omega t - \varphi) \qquad (2.5 - 29)$$

contains as *arbitrary parameters* the phase φ and the amplitude A. They are determined by the *initial conditions*. This is characteristic of *linear* differential equations. As soon as ε differs slightly from zero, however, the van der Pol equation (2.5 - 28b) becomes nonlinear. Then the amplitude A is determined by the equation and, as a consequence, cannot be chosen arbitrarily. This is typical for *nonlinear* differential equations as demonstrated in the following.

β) For small $|\varepsilon| << 1$ the *method by averaging* [Verhulst 1990 B, Zwillinger 1989 B] yields the following approximation of the solution $x(t)$ of (2.5 - 28b)

$$0<|\varepsilon|<<1:$$

$$x(t)=A(t)\cos(\Omega t-\varphi) \quad \text{with} \tag{2.5 - 30}$$

$$A(t)\cong 2\left[1+\left[4A(0)^{-2}-1\right]exp(-\varepsilon\Omega t)\right]^{-1/2} \text{ and } \quad A(\infty)=2 \quad .$$

Consequently, there exists a *stable limit cycle* and a *stable stationary oscillation* with the amplitude $A \approx 2$ for small positive $\varepsilon << 1$.

γ) For $0 < \varepsilon < 1/4$ the *approximation technique of Poincaré, Lindstedt and Lighthill* [Andersen & Geer 1982 J, Hairer et al. 1987 B, Lindstedt 1883 J, McLachlan 1950 B, Verhulst 1990 B, Zwillinger 1989 B] yields on the initial condition

$$\dot{x}(0)=0 \tag{2.5 - 31a}$$

an approximate *periodic solution* that corresponds to the *limit cycle*

$$0<\varepsilon<1/4:$$

$$x(s)=2\cos s+\varepsilon\left[\frac{3}{4}\sin s-\frac{1}{4}\sin 3s\right]$$

$$+\varepsilon^2\left[-\frac{1}{8}\cos s+\frac{3}{16}\cos 3s-\frac{5}{96}\cos 5s\right]+\ldots \tag{2.5 - 31b}$$

$$\text{with} \quad s=\Omega t\left[1-\frac{1}{16}\varepsilon^2+\frac{17}{3072}\varepsilon^4+\ldots\right] \quad .$$

The *period* $T(\varepsilon)$ of this solution is determined by taking $s = 2\pi$. The result is

$$T(\varepsilon)=T_0\left[1+\frac{1}{16}\varepsilon^2-\frac{5}{3072}\varepsilon^4+\ldots\right] \quad \text{with} \quad T(0)=T_0=2\pi/\Omega \quad . \tag{2.5 - 31c}$$

The behavior of a van der Pol oscillator with $\varepsilon = 0.1$ that fulfills the condition $0 < \varepsilon < 1/4$ is illustrated in Fig. 2.5 - 5. It shows the phase diagram $(x,\dot{x})$ with the limit cycle C as well as an oscillation $x(t)$.

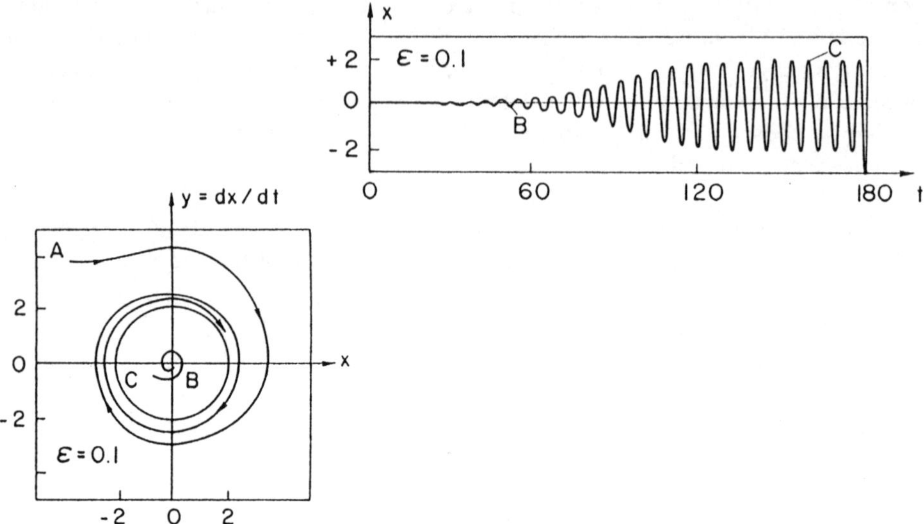

Fig. 2.5 - 5. Oscillation $x(t)$, phase diagram $(x, \dot{x} = y)$ and limit cycle C of the van der Pol oscillator with $\varepsilon = 0.1$

δ) A better technique is the *approximation of J. Shohat* [Bellman 1966 B, Shohat 1944 J]. It works for almost all positive ε, i.e. for $0 < \varepsilon < \infty$. For the initial condition

$$\dot{x}(0) = 0 \tag{2.5 - 31a}$$

one applies a modification of Shohat's method by introducing a parameter r defined by

$$r = \frac{\varepsilon}{1+\varepsilon} < 1 \quad \text{and} \quad \varepsilon = \frac{r}{1-r} < \infty \tag{2.5 - 31b}$$

and by assuming the following relations

$$\begin{aligned} &s = \Omega t (1-r) g(r) \quad , \\ &g(r) = 1 + r g_1 + r^2 g_2 + \dots \quad , \\ &x(s) = x_0(s) + r\, x_1(s) + r^2\, x_2(s) + \dots \quad , \\ &\dot{x}(s) = \dot{x}_0(s) = \dot{x}_1(s) = \dots = 0 \quad . \end{aligned} \tag{2.5 - 31c}$$

The *relevant property* of the parameter r is the fact that it varies from zero to unity when ε varies from zero to infinity. This property guarantees the good convergence of the series introduced for the Sohat approximation.

With the initial condition (2.5 - 31a) the described Shohat technique yields the following *approximative periodic solution* corresponding to the *limit cycle*

$0 < \varepsilon < \infty$:

$$x(s) = 2\cos s + \left(r + r^2\right)\left[\frac{3}{4}\sin s - \frac{1}{4}\sin 3s\right]$$

$$+ r^2\left[-\frac{1}{8}\cos s + \frac{3}{16}\cos 3s - \frac{5}{96}\cos 5s\right] + \dots \quad , \qquad (2.5 - 32a)$$

with $s = \Omega t\left[1 - \frac{1}{16}r^2 - \frac{1}{8}r^3 + \dots\right]$

and $g(r) = 1 + r + \frac{15}{16}r^2 + \frac{13}{16}r^3 + \dots$.

The *period* $T(r)$ of this solution is obtained for $s = 2\pi$

$$T(r) = T_0\left[1 + \frac{1}{16}r^2 + \frac{1}{8}r^3 + \dots\right] \quad \text{with} \quad T(0) = T_0 = 2\pi / \Omega \quad . \qquad (2.5 - 32b)$$

For large $\varepsilon \to \infty$ and $r \to 1$ the Shohat method yields

$$T(r \to 1) \approx \frac{T_0}{(1-r)\,g(1)} \approx \frac{T_0\varepsilon}{3{,}75} \qquad (2.5 - 32c)$$

instead of the more precise value [Verhulst 1990 B]

$$T(\varepsilon \to \infty) \approx \frac{(3 - 2\,\ell n 2)}{2\pi} T_0\varepsilon \approx \frac{T_0\varepsilon}{3{,}89} \quad . \qquad (2.5 - 32d)$$

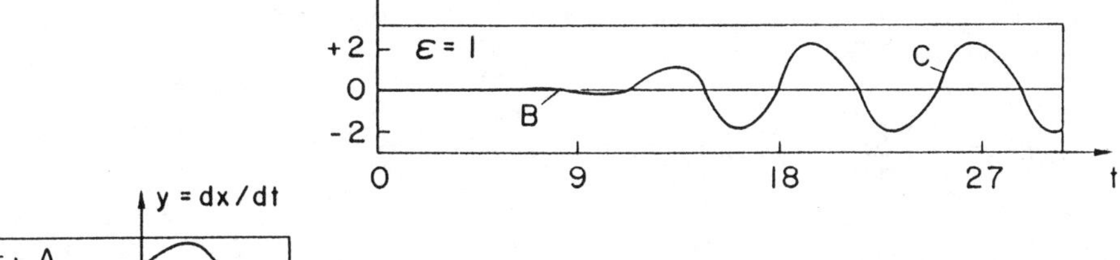

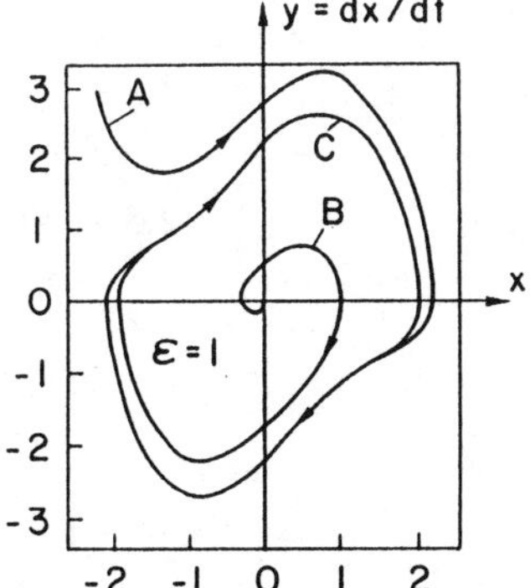

Fig. 2.5 - 6. Oscillation $x(t)$, phase diagram ($x, \dot{x} = y$) and limit cycle C of the van der Pol oscillator with $\varepsilon = 1$

Fig. 2.5 - 6 shows the phase diagram $(x, \dot{x})$ with the limit cycle C and a motion $x(t)$ of a van der Pol oscillator with $\varepsilon = 1$ and $r = 1/2$. This oscillator is well suited for the approximation of Shohat.

c) Relaxation oscillations

For large $\varepsilon > 1$ the van der Pol oscillator performs *relaxation oscillations* [Verhulst 1990 B]. Relaxation oscillations are characterized by intermittent long and short intervals within a period. During the long intervals the oscillator stays practically at rest and within the short intervals it moves rapidly. In electronics an extreme relaxation oscillation is performed by a *square-wave generator.*

α) In order to determine the behavior of a van der Pol oscillator with large $\varepsilon > 10$ during a *long interval of effective rest* the time scale must be shortened to detect any motion. This is achieved by introducing the new time variable

$$\theta = \tau / \varepsilon = \Omega t / \varepsilon \quad . \tag{2.5 - 33a}$$

This *time contraction* transforms the van der Pol equation (2.5 - 28b) into

$$\varepsilon^{-2}\left(d^2x / d\theta^2\right) + (dx / d\theta)\left[x^2 - 1\right] + x = 0 \quad . \tag{2.5 - 33b}$$

Since ε^2 is extremely large, it is permitted to omit the first term in this equation with the result

$$dx / d\theta = x\left[1 - x^2\right]^{-1} \quad \text{or} \quad d\theta = \left[x^{-1} - x\right]dx \quad . \tag{2.5 - 33c}$$

The integration of the second equation with the initial condition $x(t_0) = x_0$ yields [Hairer et al. 1987 B]

$$\ell n\left(x / x_0\right) - \frac{1}{2}\left[x^2 - x_0^2\right] = (\Omega / \varepsilon)(t - t_0) \quad . \tag{2.5 - 33d}$$

On the assumptions

$$x_0 = x(t_0) = \frac{3}{2} \quad \text{and} \quad x = \frac{3}{2}(1 + \Delta) \quad \text{with} \quad 0 < \Delta << 1 \tag{2.5 - 34a}$$

one finds the approximate solution

$$\begin{aligned} &t(x) \approx t_0 - \Delta(5\varepsilon / 4\Omega) \quad , \quad \text{and} \\ &\frac{dx}{dt}(t_0) = \frac{dx}{dt}\left(x_0 = \frac{3}{2}\right) = -4\Omega / 5\varepsilon \quad . \end{aligned} \tag{2.5 - 34b}$$

Equations (2.5 - 33d) and (2.5 - 34b) show that $x(t)$ changes little with time.

β) In order to understand the behavior of the van der Pol oscillator with large $\varepsilon > 10$ during a *short interval of rapid changes* the time scale must be extended to see the details of motion. This is accomplished with the aid of the new time variable

$$\vartheta = \varepsilon \Omega t \quad . \tag{2.5 - 35a}$$

This *time dilatation* transforms the van der Pol equation (2.5 - 28b) into

$$\left(d^2 x / d\vartheta^2\right) + \left(dx / d\vartheta\right) \cdot \left[x^2 - 1\right] + \varepsilon^{-2} x = 0 \quad . \tag{2.5 - 35b}$$

Since ε^2 is extremely large the last term of this equation can be omitted. This results in

$$\frac{d}{dx}\left(dx / d\vartheta\right) = 1 - x^2 \quad \text{and} \quad dx / d\vartheta = x - \frac{1}{3} x^3 + C \quad . \tag{2.5 - 35c}$$

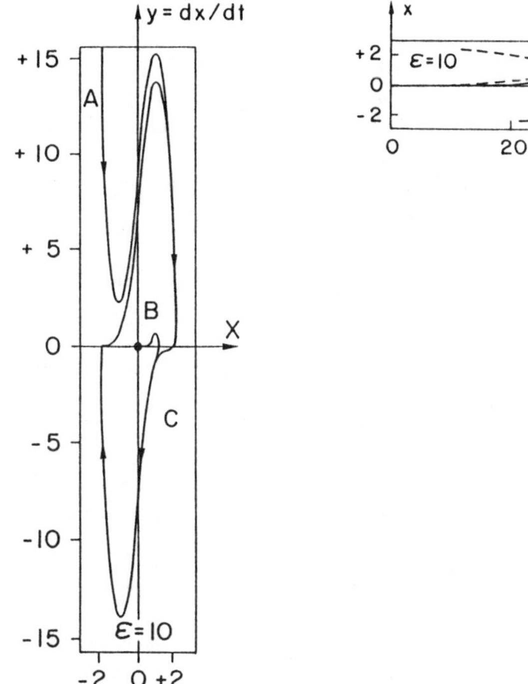

Fig. 2.5 - 7. Relaxation oscillation $x(t)$ and its phase diagram $(x, \dot{x} = y)$ with the limit cycle C of the van der Pol oscillator characterized by ε = 10. The plot of $x(t)$ also shows the time-contraction approximation (2.5 - 33d)

The second equation can be transformed into

$$\dot{x} - \varepsilon\Omega\left(x + \frac{1}{3}x^3\right) = \dot{x}(x_0) - \varepsilon\Omega\left(x_0 + \frac{1}{3}x_0^3\right) \quad . \tag{2.5 - 36a}$$

For small $|x_0|$ and $|x|$ and on the initial condition $\dot{x}(x_0) \neq 0$ the differential equation (2.5 - 36a) has the approximate solution

$$x(t) = x_0 + \frac{1}{\varepsilon\Omega}\dot{x}(x_0)\left[exp(\varepsilon\Omega t) - 1\right]$$
$$\text{for } |x| << 1 \text{ and } \dot{x}(x_0) \neq 0 \quad . \tag{2.5 - 36b}$$

Thus $x(t)$ varies rapidly with time t.

As illustration Fig. 2.5 - 7 shows the phase diagram $(x, \dot{x})$ with the limit cycle C and the corresponding relaxation oscillation of the van der Pol oscillator with $\varepsilon = 10$.

3. Forced Oscillations

The free oscillators described in the previous chapter represent isolated systems shielded from external forces. On the contrary this chapter deals with the effects of these forces or excitations on the oscillators. In this context transient as well as continuous periodic forces are taken into account. A *characteristic* of forced oscillations is *resonance*, a phenomenon well known in physics and engineering. In addition forced *nonlinear* oscillations may also show bistability, frequency mixing, frequency multiplication and division as well as deterministic chaos.

3.1 External Forces

The free oscillators of Chapter 2 oscillate undisturbed by external forces. Frequently, their oscillations are determined by equations of the form

$$\Phi(x, \dot{x}, \ddot{x}, t) = 0 \quad . \tag{3.1 - 1}$$

Oscillators can, however, be subjected to *external excitations or forces* $F(t)$ varying with time t. Thus they can be forced to oscillate. These *forced oscillations* are frequently solutions of inhomogeneous differential equations such as

$$\Phi(x, \dot{x}, \ddot{x}, t) = F(t) \quad . \tag{3.1 - 2}$$

There exist various *types of external excitations* of oscillators. They are defined by characteristic external forces:

α) *Harmonic excitation* is either described by the *real* periodic force

$$F(t) = F \cos \omega t \quad \text{with} \quad \omega = 2\pi / T \quad \text{and a real} \quad F > 0 \tag{3.1 - 3a}$$

or by the *complex* periodic force

$$F(t) = F \exp(-i\omega t) \quad \text{with} \quad \omega = 2\pi / T \quad \text{and } F \text{ complex} \quad . \tag{3.1 - 3b}$$

In these two equations ω and T indicate circular frequency and period of the excitation.

β) *Impulse or shock excitation* [Bracewell 1986 B] is represented by an infinitely short and infinitely strong force of the form

$$F(t) = K\,\delta(t) \tag{3.1 - 4}$$

where $\delta(t)$ signifies the Dirac delta function (2.3 - 46c). If $K = 1$ this excitation is called *unit impulse*.

γ) A *step excitation* is accomplished by switching on an external force at time $t = 0$. This force can be described as follows

$$F(t<0) = 0 \quad \text{and} \quad F(t>0) = f(t) \quad \text{or} \quad F(t) = H(t)f(t) \tag{3.1 - 5a}$$

where $H(t)$ represents the Heaviside step function (2.3 - 46a-c) and $f(t)$ an arbitrary real function.

A special type of this excitation is the *step excitation* defined by a constant $f(t)$

$$F(t<0) = 0 \text{ and } F(t>0) = f(t) = f \text{ or } F(t) = f\,H(t) \quad . \tag{3.1 - 5b}$$

δ) For both, impulse excitation (3.1 - 4) and step excitation (3.1 - 5a), the oscillation of an oscillator on the initial conditions

$$x(t<0) = 0 \quad \text{and} \quad \dot{x}(t<0) = 0 \tag{3.1 - 6}$$

is determined uniquely by the inherent properties of the oscillator. Therefore, an oscillator under these conditions is called a *causal system*. These conditions imply that the oscillator is at rest for times $t < 0$.

3.2 Excitation of Harmonic Oscillators

The forced oscillations $x(t)$ of harmonic oscillators obey the following linear *inhomogeneous* differential equation

$$\ddot{x} + \frac{2}{\tau}\dot{x} + \Omega^2 x = F(t) \quad . \tag{3.2 - 1a}$$

The undamped harmonic oscillator is characterized by $\tau = \infty$ and the damped by $0 < \tau < \infty$. With respect to forced oscillations the harmonic oscillator with amplification characterized by $\tau < 0$ is of little interest. $F(t)$ represents the external force of excitation. When it is missing the harmonic oscillator is described by the linear *homogeneous* differential equation of the free harmonic oscillators

$$\ddot{x}+\frac{2}{\tau}\dot{x}+\Omega^2 x=0 \quad . \tag{3.2 - 1b}$$

As discussed in general in the following sections the forced oscillation $x(t)$ of a harmonic oscillator is the sum of the *stationary oscillation* $x_p(t)$ and the *transient* $x_h(t)$

$$x(t)=x_p(t)+x_h(t) \quad . \tag{3.2 - 2}$$

The stationary oscillation $x_p(t)$ is a particular solution of the inhomogeneous equation (3.2 - 1a), whereas the transient $x_h(t)$ corresponds to the general solution of the homogeneous equation (3.2 - 1b).

3.2.1 Transients

In order to describe a transient $x_h(t)$ as a solution of the homogeneous equation (3.2 - 1b) one defines ω_0 by the relation

$$\Omega^2 = \tau^{-2} + s\,\omega_0^2 \quad \text{with} \quad s=0,\pm 1 \quad . \tag{3.2 - 3a}$$

Thus, one finds in agreement with (2.2 - 25), (2.2 - 30) and (2.2 - 35) the general solution

for *subcritical damping* with $s=+1, \Omega\tau>1$

$$x_h(t)=\left[a\,cos\;\omega_0 t+b\,sin\;\omega_0 t\right]exp(-t/\tau) \quad , \tag{3.2 - 3b}$$

for *critical damping* with $s=0, \Omega\tau=1$

$$x_h(t)=\left[a+bt\right]exp(-t/\tau) \quad , \tag{3.2 - 3c}$$

and for *supercritical damping* with $s=-1, \Omega\tau<1$

$$x_h(t)=\left[a\,cosh\;\omega_0 t+b\,sinh\;\omega_0 t\right]exp(-t/\tau) \quad . \tag{3.2 - 3d}$$

These equations demonstrate that for $0<\tau<\infty$ the transient *decays* with time.

The parameters a and b in (3.2 - 3b-d) are determined by the *initial conditions* $x(0)$ and $\dot{x}(0)$ of the forced oscillation $x(t)$.

3.2.2 Real Harmonic Excitation

The real harmonic excitation of a harmonic oscillator is represented by the inhomogeneous differential equation

$$\ddot{x}+\frac{2}{\tau}\dot{x}+\Omega^2 x = F\cos\omega t \quad \text{with real positive} \quad F>0 \quad . \tag{3.2 - 4}$$

This equation originates in (3.1 - 3a) and (3.2 - 1a). The *stationary forced oscillation* of the harmonic oscillator with harmonic excitation is the particular solution $x_p(t)$ of this equation. This solution depends on the parameters F, Ω, τ and ω as follows

$$x_p(t) = G(\omega;\Omega;\tau)\,F\cos[\omega t - \varphi(\omega;\Omega,\tau)] \tag{3.2 - 5a}$$

$$\text{with} \quad G(\omega;\Omega,\tau) = +\left[\left(\Omega^2-\omega^2\right)^2 + (2\omega/\tau)^2\right]^{-1/2} \geq 0 \tag{3.2 - 5b}$$

$$\text{and} \quad \varphi(\omega;\Omega,\tau) = \arctan\frac{2\omega/\tau}{\Omega^2-\omega^2} \quad . \tag{3.2 - 5c}$$

Equation (3.2 - 5a) demonstrates that the stationary forced oscillation exhibits the *circular frequency* ω of the exciting force $F(t)$ instead of the circular frequency ω_0 of the free harmonic oscillator determined by (3.2 - 3a).

The gain function $G(\omega;\,\Omega,\,\tau)$ and the phase lag $\varphi(\omega;\,\Omega,\,\tau)$, are illustrated in Figs. 3.2 - 1&2.

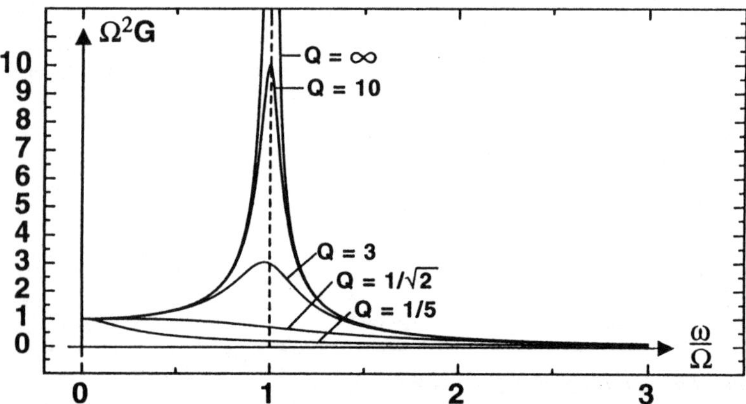

Fig. 3.2 - 1. Gain function $G(\omega;\Omega,\tau)$ in normalized representation with $\Omega^2 G$ as a function of ω/Ω for the quality factors $Q = 1/5, 1/\sqrt{2}, 3, 10, \infty$

The *gain function* $G(\omega;\,\Omega,\,\tau)$ shows a maximum at

$$\Omega_{res} = \Omega\left[1-\frac{1}{2}Q^{-2}\right] \tag{3.2 - 6a}$$

$$\text{with} \quad Q = \Omega\,\tau/2 \tag{3.2 - 6b}$$

where Q represents the quality factor (2.2 - 21). According to these equations, the stationary forced oscillation exhibits a *resonance* for $Q > 1/\sqrt{2}$ with a maximum amplitude at the *resonance circular frequency* ω_{res}. The quality factor Q is a measure for the *resonance width* $\Delta\omega$ shown in Fig. 3.2 - 3. This width $\Delta\omega$ is measured where the gain function $G(\omega;\, \Omega,\, \tau)$ takes the value of $1/\sqrt{2}$ of the maximum at the resonance circular frequency ω_{res}. This yields for $Q >> 1$

$$G\left(\Omega \pm \frac{\Delta\omega}{2};\Omega,\tau\right) = \frac{1}{\sqrt{2}}\; G(\Omega;\,\Omega,\tau) \quad , \quad \text{and} \tag{3.2 - 6c}$$

$$\Delta\omega / \Omega = 1/Q \quad \text{for} \quad Q >> 1 \quad . \tag{3.2 - 6d}$$

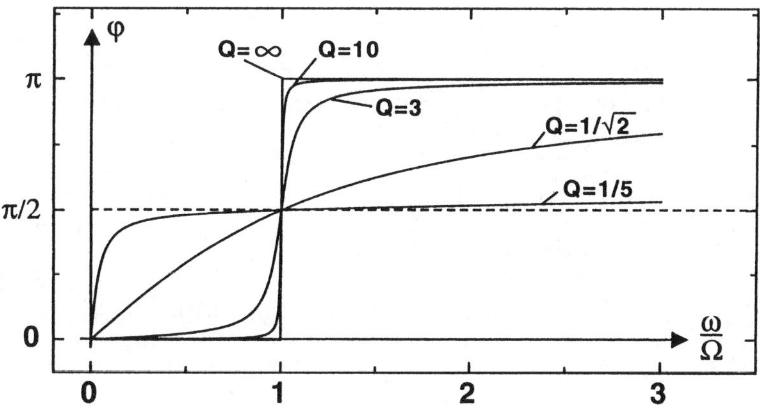

Fig. 3.2 - 2. Phase lag $\varphi(\omega;\Omega,\tau)$ in normalized representation with φ as a function of ω/Ω for the quality factors $Q = 1/5, 1\sqrt{2}, 3, 10, \infty$

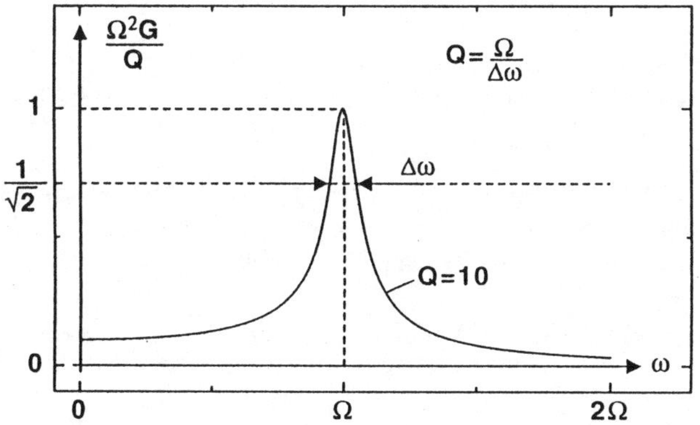

Fig. 3.2 - 3. Determination of a high quality factor Q from the resonance data. The resonance is represented by the normalized gain function $Q^{-1}\Omega^2 G(\omega;\Omega,\tau)$ as a function of ω for the quality factor $Q = \Omega/\Delta\omega = 10$

The *phase lag* $\varphi(\omega;\, \Omega,\, \tau)$ illustrated in Fig. 3.2 - 2 varies from $\varphi = 0$ at $\omega = 0$ to $\varphi = \tau$ at $\omega = \infty$. Close to the resonance it is $\varphi = \tau/2$. In general the phase lag obeys the relations

$$\begin{aligned} &\varphi(\omega = 0;\Omega,\tau) = 0 \quad , \\ &\varphi(\omega = \Omega;\Omega,\tau) = \pi/2 \quad , \\ &\varphi(\omega = \infty;\Omega,\tau) = \infty \quad . \end{aligned} \tag{3.2 - 7}$$

The harmonic excitation of *undamped harmonic oscillators* characterized by $\tau = \infty$ shows particularities. First, the solution $x_h(t)$ of the homogeneous equation (3.2 - 1b) is an undamped oscillation if $\tau = \infty$. Thus, it does not represent a transient. Secondly, in resonance the gain function $G(\omega;\, \Omega,\, \tau)$ grows to infinity

$$G(\omega = \Omega;\Omega,\tau = \infty) = \infty \quad . \tag{3.2 - 8a}$$

In this case (3.2 - 5a-c) do not represent a particular solution $x_p(t > 0)$ of the basic equation (3.2 - 4). A particular solution is then

$$\begin{aligned} \tau = \infty,\ \omega = \Omega: \quad x_p(t) &= (F/2\Omega)\, t\, cos(\Omega t - \pi/2) \\ &= (F/2\Omega)\, t\, sin\, \Omega t \quad . \end{aligned} \tag{3.2 - 8b}$$

In agreement with (3.2 - 8a) this solution grows to infinity with a phase shift $\varphi = \pi/2$.

3.2.3 Complex Harmonic Excitation

According to (3.1 - 1a) and (3.2 - 3b) the complex harmonic excitation of harmonic oscillators can be described by the basic equation

$$\ddot{x} + \frac{2}{\tau}\dot{x} + \Omega^2 x = F\, exp(-i\omega t) \text{ with } F \text{ real or complex} \quad . \tag{3.2 - 9}$$

Also in case of this excitation the transient $x_h(t)$ corresponds to the solutions (3.2 - 3b-d) of the homogeneous equation (3.2 - 1b) with $0 < \tau < \infty$. On this condition, however, the constant parameters a and b in these solutions can be complex.

The particular solution $x_p(t)$ of (3.2 - 9) that corresponds to the stationary oscillation is

$$x_p(t) = Z(\omega;\Omega,\tau)\ F\, exp(-i\omega t) \tag{3.2 - 10a}$$

$$\text{with} \quad Z(\omega;\Omega,\tau) = \left[\Omega^2 - \omega^2 - i(2\omega/\tau)\right]^{-1} \quad . \tag{3.2 - 10b}$$

The complex *transfer function* $Z(\omega; \Omega, \tau)$ can be split in two ways

$$Z(\omega;\Omega,\tau) = X(\omega;\Omega,\tau) + iY(\omega;\Omega,\tau) \quad \text{or} \tag{3.2 - 10c}$$

$$Z(\omega;\Omega,\tau) = G(\omega;\Omega,\tau)\,exp\left[+i\varphi(\omega;\Omega,\tau)\right] \quad . \tag{3.2 - 10d}$$

In these equations $X(\omega;\Omega,\tau)$ and $Y(\omega;\Omega,\tau)$ indicate real and imaginary part of the transfer function, whereas $G(\omega;\Omega,\tau)$ and $\varphi(\omega;\Omega,\tau)$ represent the gain function and the phase lag of (3.2 - 5b&c).

In electrical engineering the transfer function $Z(\omega;\Omega,\tau)$ is plotted in the complex $Z = X + iY$ plane as a function of the excitation circular frequency ω for constant parameters Ω and τ [Birkhoff & Rota 1989 B]. This plot is called *Nyquist diagram.* Two examples are shown in Fig. 3.2 - 4.

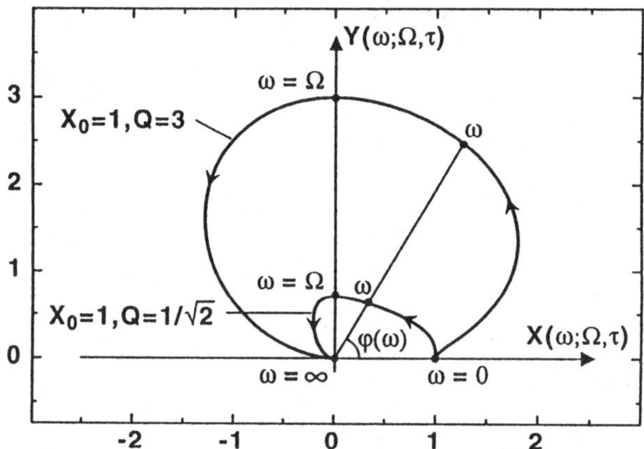

Fig. 3.2 - 4. Nyquist diagrams of two harmonically excited harmonic oscillators with the quality factors $Q = 1/\sqrt{2}$ and $Q = 3$

3.2.4 Exceptional Sub- and Ultraharmonics

Sub- and ultraharmonics are characteristic *nonlinear phenomena.* In general, linear oscillators show no sub- and ultraharmonics. However, there exist exotic exceptions.

After the decay of the transient $x_h(t)$ (3.2 - 3b-d) of a damped harmonic oscillator with $0 < \tau < \infty$ the stationary forced oscillation $x_p(t)$ has the same circular frequency ω as the exciting harmonic force $F(t)$ of (3.1 - 3a). On the contrary, *nonlinear oscillators* may behave quite differently. Their oscillations may exhibit circular frequencies or periods different from those of the exciting force $F(t)$. An example represent the subharmonics whose periods $T_r = rT$ are integer multiples of the period $T = 2\pi/\omega$ of the harmonic exciting force $F(t)$.

As an exotic exception [Stoker 1957 B] subharmonics and ultraharmonics can also be excited in a *linear harmonic oscillator* if it is undamped ($\tau = \infty$). Prerequisites are an adequate ratio of the characteristic circular frequency Ω of the oscillator versus the circular frequency ω of excitation as well as proper initial conditions. The resulting oscillations give hints concerning the phenomena that occur in nonlinear oscillators [Stoker 1957 B].

The harmonically excited undamped harmonic oscillator is described by (3.2 - 4) when $\tau = \infty$. The forced oscillations of this oscillator are according to (3.2 - 3b) and (3.2 - 5a-c)

$$x(t) = F\left[\Omega^2 - \omega^2\right]^{-1} + a\cos\Omega t + b\sin\Omega t \quad \text{for} \quad \omega \neq \Omega \quad . \tag{3.2 - 11a}$$

By a special choice of the parameters of this equation one can select among forced oscillations corresponding to: α) the pure harmonic with the circular frequency ω of the harmonic exciting force $F(t)$; β) a combination of the harmonic and a subharmonic; γ) a combination of the harmonic with an ultraharmonic; and δ) a combination of a subharmonic with an ultraharmonic. For comparison of these four types of oscillations one chooses the common initial conditions

$$x(0) = 1, \quad \dot{x}(0) = 0 \quad . \tag{3.2 - 11b}$$

Thus, one finds for the

α) pure harmonic

$$\begin{aligned} &F = \Omega^2 - \omega^2;\ a = b = 0 \quad ; \\ &x(t) = \cos \omega t \\ &\text{with the period } T_{max} = T = 2\pi/\omega \quad . \end{aligned} \tag{3.2 - 12a}$$

β) combination of *harmonic and subharmonic*

$$\begin{aligned} &F = (1-a)\left(\Omega^2 - \omega^2\right); \quad b = 0; \quad n = 0,1,2,\ldots \quad ; \\ &x(t) = (1-a)\cos\omega t + a\cos\left(\frac{\omega}{2+n}t\right) \\ &\text{with the period } T_{max} = (2+n)T \quad . \end{aligned} \tag{3.2 - 12b}$$

γ) combination of *harmonic* and *ultraharmonic*

$$\begin{aligned} &F = (1-a)\left(\Omega^2 - \omega^2\right); \quad b = 0; \quad m = 0,1,2,\ldots \quad ; \\ &x(t) = (1-a)\cos\omega t + a\cos(2+m)\omega t \\ &\text{with the period } T_{max} = T \quad . \end{aligned} \tag{3.2 - 12c}$$

δ) combination of *subharmonic* and *ultraharmonic*

$$F = (1-a)\left(\Omega^2 - \omega^2\right); \quad b = 0; \quad m,n = 0,1,2,\ldots \; ;$$
$$x(t) = (1-a)\cos\,\omega t + a\,\cos\left(\frac{2+m}{2+n}\right)\omega t \tag{3.2 - 12d}$$

with the period $T_{max} = (2+n)\,T$ and the *irreducible fraction* $(2+m)/(2+n)$, e.g. 2/3, 3/2, 3/4, 4/3, etc.

The four types (3.2 - 12a-d) of oscillations are illustrated in Fig. 3.2 - 5.

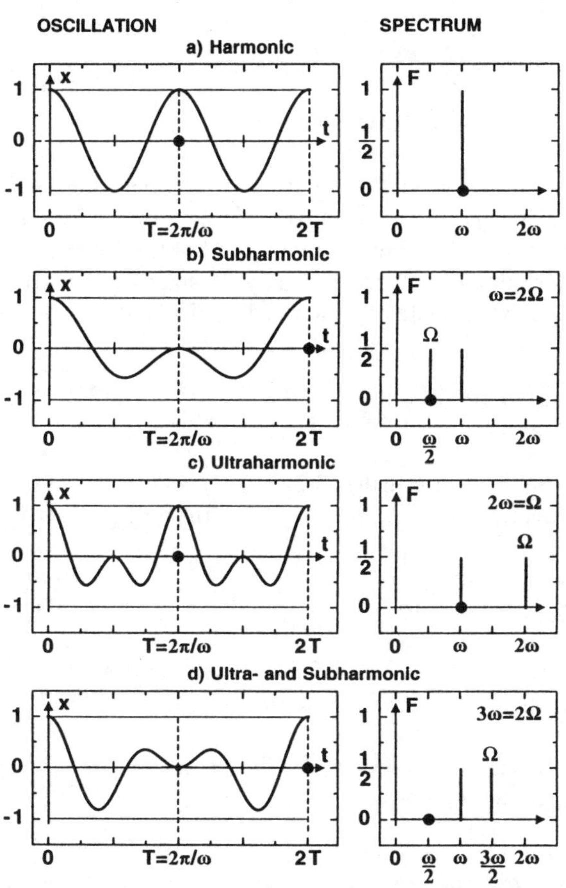

Fig. 3.2 - 5. Excitation of sub- and ultraharmonics of the undamped harmonic oscillator for $\omega = 1$ and $a = 1/2$

a) $x(t) = \cos\, t$

b) $x(t) = \dfrac{1}{2}\cos\, t + \dfrac{1}{2}\cos\,(t/2)$ for $n = 0$

c) $x(t) = \frac{1}{2} \cos t + \frac{1}{2} \cos 2t$ for $m = 0$

d) $x(t) = \frac{1}{2} \cos t + \frac{1}{2} \cos (3t/2)$ for $m = 1, n = 0$

3.2.5 General Periodic Excitation

A periodic excitation $F(t)$ of an oscillator can usually be represented by a *Fourier series*. Consequently, the corresponding basic equation of the forced oscillation is the following

$$\ddot{x} + \frac{2}{\tau}\dot{x} + \Omega^2 x = F(t) = F(t+T) = \sum_{m=-\infty}^{+\infty} F_m \, exp(-im\,\omega\, t) \tag{3.2 - 13}$$

with $\omega = 2\pi / T$ and $F_{-m} = F_m^*$ for $F(t)$ real .

The terms of the sum in (3.2 - 13) correspond to individual complex harmonic excitations according to (3.2 - 9). Because of the linearity of (3.2 - 9) and (3.2 - 13) the particular solution $x_p(t)$ of (3.2 - 13) can be represented as a linear superposition of those of the individual complex excitations characterized by F_m and (3.2 - 9). Thus, the periodically excited oscillation of a harmonic oscillator is described by

$$x(t) = \sum_{m=-\infty}^{+\infty} Z(m\omega;\Omega,\tau)\; F_m \; exp(-im\omega\, t) + x_h(t) \tag{3.2 - 14a}$$

where $x_h(t)$ designates the transient according to (3.2 - 3a-d) and $Z(m\omega;\,\Omega,\,\tau)$ indicates the complex transfer function of (3.2 - 10a-d). This function fulfills the relation

$$Z(-m\,\omega;\Omega,\tau) = Z^*(+m\,\omega;\Omega,\tau) \quad . \tag{3.2 - 14b}$$

3.2.6 Broad-Band Excitation

A non-periodic excitation $F(t)$ with the *frequency spectrum* $F(\omega)$ acting on a harmonic oscillator is described by the basic equation

$$\ddot{x} + \frac{2}{\tau}\dot{x} + \Omega^2 x = F(t) = \frac{1}{2\pi}\int_{-\infty}^{+\infty} F(\omega)\, exp(-i\omega\, t)\, d\omega \tag{3.2 - 15a}$$

$$\text{with} \quad F(\omega) = \int_{-\infty}^{+\infty} F(t)\, exp(+i\omega\, t)\, dt \tag{3.2 - 15b}$$

$$\text{and} \quad F(-\omega) = F^*(\omega) \quad \text{for} \quad F(t) \text{ real} \tag{3.2 - 15c}$$

according to the rules of the *Fourier transformation* (2.2 - 46a&b). Because of the linearity of the basic equation (3.2 - 15a) its particular solution $x_p(t)$ can be constructed by superposition of the effects of the partial excitations

$$dF(\omega t)=\frac{1}{2\pi}F(\omega)\,exp(-i\omega t)d\omega \quad . \qquad (3.2 - 15d)$$

The result is the following forced oscillation

$$x(t)=\frac{1}{2\pi}\int_{-\infty}^{+\infty} Z(\omega;\Omega,\tau)\;F(\omega)\,exp(-i\omega t)\,d\omega\;+x_h(t) \quad . \qquad (3.2 - 16a)$$

Similar to (3.2 - 14a) this oscillation is composed of the transient $x_h(t)$ described by (3.2 - 3a-d) and the stationary oscillation determined by complex transfer function $Z(\omega;\,\Omega,\,\tau)$ of (3.2 - 10a-d) with

$$Z(-\omega;\Omega,\tau)=Z^{*}(\omega;\Omega,\tau) \quad . \qquad (3.2 - 16b)$$

3.2.7 Impulse or Shock Excitation

The oscillation of a harmonic oscillator initiated by an impulse or shock excitation (3.1 - 4) at time $t = 0$ is determined by the basic equation

$$\ddot{x}+\frac{2}{\tau}\dot{x}+\Omega^2 x=F(t)=K\,\delta(t) \qquad (3.2 - 17)$$

where $\delta(t)$ indicates the Dirac delta function defined by (2.3 - 47) and (2.3 - 48a-d). When $K = 1$ this excitation is named *unit impulse* or *unit shock.*

a) Boundary conditions at the impulse

The essential task when solving (3.2 - 17) is to calculate the state of the oscillator at the time $t = +0$ just after the impulse starting from the given state at time $t = -0$ just before the impulse.

For physical reasons $x(t)$ has to be continuous also at $t = 0$. This implies

$$x(+0)=x(0)=x(-0) \quad . \qquad (3.2 - 18a)$$

This statement, however, is not valid for $\dot{x}(t)$. The integration of (3.2 - 17) over the time t from $t = -\delta t$ to $t = +\delta t$ with $\delta t > 0$ gives

$$0=\int_{-\delta t}^{+\delta t}\left[\ddot{x}+\frac{2}{\tau}\dot{x}+\Omega^2 x-K\,\delta(t)\right]dt$$

$$=\left[\dot{x}(+\delta t)-\dot{x}(-\delta t)\right]+\frac{2}{\tau}\left[x(+\delta t)-x(-\delta t)\right]+\Omega^2\,2x(0)\delta t-K \quad .$$

For vanishing $\delta t \to 0$ this relation can be written as

$$\dot{x}(+0) = \dot{x}(-0) + K \quad . \tag{3.2 - 18b}$$

For times $t > 0$ the basic equation (3.2 - 17) represents the free harmonic oscillator since it is equivalent to

$$\ddot{x} + \frac{2}{\tau}\dot{x} + \Omega^2 x = 0 \quad \text{for} \quad t > 0 \tag{3.2 - 1b}$$

that is subjected to the specific boundary conditions

$$x(+0) = x(-0) \quad \text{and} \quad \dot{x}(+0) = \dot{x}(-0) + K \quad \text{for} \quad t \approx 0 \quad . \tag{3.2 - 18 a\&b}$$

b) Response of causal systems

A characteristic of the impulse or shock excitation is the behavior or the harmonic oscillator as *causal system* defined by (3.1 - 6). The combination of this equation with (3.2 - 18a&b) yields

$$x(-0) = x(t < 0) = 0 \quad \text{and} \quad \dot{x}(-0) = \dot{x}(t < 0) = 0 \quad , \tag{3.2 - 19a}$$

$$x(+0) = 0 \quad \text{and} \quad \dot{x}(+0) = K \quad . \tag{3.2 - 19b}$$

The solution of the homogeneous differential equation (3.2 - 1b) of the free harmonic oscillator with the boundary conditions (3.2 - 19a&b) represents the *impulse response* or *shock response $x(t)$ of the causal system.* If

$$\Omega^2 = \tau^{-2} + s\,\omega_0{}^2, \quad \text{with} \quad s = 0, \pm 1 \tag{3.2 - 20a}$$

this response is

for *subcritical* damping with $s = +1; \quad \Omega\tau > 1$

$$x(t) = K\,x_{\text{shock}}(t) = K\,\omega_0^{-1}\,\sin\,\omega_0 t\;\exp(-t/\tau) \quad , \tag{3.2 - 20b}$$

for *critical* damping with $s = 0; \quad \Omega\tau = 1$

$$x(t) = K\,x_{\text{shock}}(t) = K\,t\;\exp(-t/\tau) \quad , \tag{3.2 - 20c}$$

for *supercritical* damping with $s = 0; \quad \Omega\tau < 1$

$$x(t) = K\,x_{\text{shock}}(t) = K\,\omega_0{}^{-1}\,\sinh\,\omega_0 t\;\exp(-t/\tau) \quad . \tag{3.2 - 20d}$$

In these equations $x_{\text{shock}}(t)$ indicates the response of the causal system to the unit impulse or unit shock $\delta(t)$ when $K = 1$.

c) Response of non-causal systems
The impulse or shock excitation of a harmonic oscillator as *non-causal system* with

$$x(-0) \neq 0 \text{ and/ or } \dot{x}(+0) \neq 0 \qquad (3.2 - 21a)$$

can be calculated in two ways. Either one solves (3.2 - 17) on the conditions (3.2 - 18a&b), or one considers

$$x(t) = K\, x_{\text{shock}}(t) + x_{\text{h}}(t) \quad \text{for} \quad t > 0 \qquad (3.2 - 21b)$$

$$\text{with} \quad x_{\text{h}}(+0) = x(-0)^{\cdot} \text{ and } \quad \dot{x}_{\text{h}}(+0) = \dot{x}(-0) \quad . \qquad (3.2 - 21c)$$

The function $x_{\text{h}}(t)$ indicates the solution of the homogeneous equation (3.2 - 1b). It does not depend on the impulse or shock excitation.

3.2.8 Transients after Switching on

The transients after *switching on an external force* are solutions of the basic equation

$$\ddot{x} + \frac{2}{\tau}\dot{x} + \Omega^2 x = F(t) = H(t)\, f(t) \quad . \qquad (3.2 - 22)$$

This represents the combination of (3.1 - 5a) and (3.2 - 1a). $H(t)$ is the Heaviside step function (2.3 - 46a-c) and $f(t)$ an arbitrary continuous function.

a) The Laplace transformation
The transients $x(t)$ for times $t \geq 0$ as solutions of the basic equation (3.2 - 22) can be usually determined with the aid of the *Laplace transformation* [Abramowitz & Stegun 1965 B, Bracewell 1986 B, Doetsch 1970 B, Erdelyi et al. 1954 B, Pöschl 1956 B, Poularikas 1995 B, Zayed 1996 B]. The Laplace transform $F(p)$ of a function $F(t)$ is defined by the *Laplace integral*

$$L\{F(t)\} = F(p) = \int_0^{\infty} F(t)\, exp(-pt)\, dt = \int_{-\infty}^{+\infty} H(t)\, f(t)\, exp(-pt) dt \quad . \qquad (3.2 - 23a)$$

The inverse Laplace transformation makes use of the *Bromwich-Wagner integral*

$$L^{-1}\{F(p)\} = F(t) = H(t)\, f(t) = \frac{1}{2\pi i} \int_{q-i\infty}^{q+i\infty} F(p)\, exp(pt)\, dp \qquad (3.2 - 23b)$$

with a real $q > 0$.

There exists numerous *tables* on the Laplace transformation and its inverse [Abramowitz & Stegun 1965 B, Bracewell 1986 B, Bronstein et al. 1993 B, Doetsch 1971/73 B, Erdelyi et al. 1954 B]. *Rules* of the Laplace transformation are listed in Appendix A.3.1.

The last equation of Appendix A.3.1 involves the Laplace transform of the *convolution* [Bracewell 1989 B] defined by

$$f_1(t) * f_2(t) = \int_{-\infty}^{+\infty} f_1(s) f_2(t-s)\,ds \quad \text{or} \tag{3.2 - 24a}$$

$$F_1(t) * F_2(t) = \int_0^t F_1(s) F_2(t-s)\,ds \quad \text{if} \quad F_{1,2}(t) = H(t) f_{1,2}(t) \quad . \tag{3.2 - 24b}$$

The convolution is *commutative*

$$f_1(t) * f_2(t) = f_2(t) * f_1(t) = \int_{-\infty}^{+\infty} f_2(s) f_1(t-s)\,ds \quad \text{or} \tag{3.2 - 24c}$$

$$F_1(t) * F_2(t) = F_2(t) * F_1(t) = \int_0^t F_2(s) F_1(t-s)\,ds \quad \text{if} \quad F_{1,2}(t) = H(t) f_{1,2}(t) \quad . \tag{3.2 - 24d}$$

Convolutions are listed in Appendix A.4.

The calculation of transients of the harmonic oscillator after impulse excitation or switching on requires the Laplace transforms of the Dirac delta function $\delta(t)$ and the Heaviside step function $H(t)$. These functions are described in Section 2.3.6a, whilst their Laplace transforms are listed in Appendix A.3.2. Laplace transforms of simple functions are summarized in Appendix A.3.3.

b) Calculation of transients

The application of the Laplace transformation (3.2 - 23a) to equation (3.2 - 22) in order to calculate the transients of a harmonic oscillator after switching on an external force yields

$$x(p)\left[p^2 + \frac{2}{\tau}p + \Omega^2\right] - p\,x(0) - \left[\frac{2}{\tau}x(0) + \dot{x}(0)\right] = F(p) \quad \text{with} \quad x(p) = \mathbf{L}\{x(t)\} \quad \text{and} \quad F(p) = \mathbf{L}\{F(t)\} \quad . \tag{3.2 - 25a}$$

In this equation it is of importance that the reciprocal factor of $x(p)$ corresponds to the Laplace transform $x_{\text{shock}}(p)$ of the response $x_{\text{shock}}(t)$ of the harmonic oscillator to the unit impulse $\delta(t)$. This response $x_{\text{shock}}(t)$ is described by (3.2 - 20a-d) when $K = 1$. Its Laplace transform is

$$x_{\text{shock}}(p) = L\{x_{\text{shock}}(t)\} = \left[p^2 + \frac{2}{\tau}p + \Omega^2\right]^{-1} \quad . \qquad (3.2 - 26)$$

Consequently, (3.2 - 25a) can also be written as

$$x(p) = x_{\text{shock}}(p)\left[F(p) + x(0)p + \left\{\frac{2}{\tau}x(0) + \dot{x}(0)\right\}\right] \quad . \qquad (3.2 - 25b)$$

This equation permits one to calculate the Laplace transform $x(p)$ of the transient response $x(t)$ of the harmonic oscillator for any kind of force $F(t)$ switched on at time $t = 0$. The real-time transient response $x(t)$ is subsequently evaluated with the aid of the inverse Laplace transform (3.2 - 23b) and Appendix A.3.1

$$x(t) = F(t) * x_{\text{shock}}(t) + x(0)\frac{d}{dt}x_{\text{shock}}(t) + \left[\frac{2}{\tau}x(0) + \dot{x}(0)\right]x_{\text{shock}}(t) \quad . \qquad (3.2 - 25c)$$

c) Causal systems

If the harmonic oscillator represents a *causal system* described by

$$x(t<0) = 0 \quad \text{and} \quad \dot{x}(t<0) = 0 \quad , \qquad (3.1 - 6)$$

then (3.2 - 25b&c) are reduced to the equations

$$x(p) = F(p)x_{\text{shock}}(p) \quad , \qquad (3.2 - 27a)$$

$$x(t) = F(t) * x_{\text{shock}}(t) = \int_0^t F(s)\, x_{\text{shock}}(t-s)ds \quad . \qquad (3.2 - 27b)$$

These two relations demonstrate the significance of the response $x_{\text{shock}}(t)$ to the unit impulse $\delta(t)$.

An *example* is the transient $x(t)$ of a harmonic oscillator as *causal system* (3.1 - 6) initiated by a *Heaviside* or *unit step excitation* according to (3.1 - 5b)

$$F(t) = H(t) \quad \text{with} \quad F(p) = p^{-1} \quad . \qquad (3.2 - 28a)$$

This system replies as follows

$$x(t) = \int_0^t x_{\text{shock}}(s)ds \quad \text{with} \quad x(p) = x_{\text{shock}}(p)p^{-1} \quad . \qquad (3.2 - 28b)$$

Therefore, this response $x(t)$ represents the integral of the response $x_{shock}(t)$ to a unit impulse $\delta(t)$ over time t.

3.3 Excitation of Modulated Linear Oscillators

The excitation of modulated linear oscillators by external forces can be studied on the basis of the *inhomogeneous* linear differential equation

$$\ddot{x} + \frac{2}{\tau(t)}\dot{x} + \Omega^2(t)\,x = F(t) \qquad (3.3 - 1)$$

that represents a special form of the general equation (3.1 - 2).

The forced oscillations of modulated linear oscillators correspond to the general solution of the inhomogeneous equation (3.3 - 1)

$$x(t) = x_{\mathrm{p}}(t) + x_{\mathrm{h}}(t) \qquad (3.3 - 2)$$

where $x_{\mathrm{p}}(t)$ represents a particular solution of the inhomogeneous differential equation (3.3 - 1), whilst $x_{\mathrm{h}}(t)$ signifies the general solution of the corresponding homogeneous differential equation of the free oscillator

$$\ddot{x} + \frac{2}{\tau(t)}\dot{x} + \Omega^2(t)\,x = 0 \quad . \qquad (2.3 - 1)$$

According to (2.3 - 7) the general solution of this equation can be written as a linear combination of two linearly independent solutions $x_1(t)$ and $x_2(t)$

$$x_{\mathrm{h}}(t) = a\,x_1(t) + b\,x_2(t) \qquad (3.3 - 3)$$

where a and b are arbitrary constants. If a particular solution $x_{\mathrm{p}}(t)$ of (3.3 - 1) is known, these constants can be determined from the initial conditions

$$x(t_0) = x_0 \quad \text{and} \quad \dot{x}(t_0) = \dot{x}_0 \quad . \qquad (3.3 - 4)$$

The initial conditions (3.3 - 4) and the constants a and b of (3.3 - 3) are related by

$$a = W\big(x(t_0) - x_{\mathrm{p}}(t_0), x_2(t_0)\big) \,/\, W\big(x_1(t_0), x_2(t_0)\big) \quad \text{and} \qquad (3.3 - 5a)$$

$$b = W\big(x_1(t_0), x(t_0) - x_{\mathrm{p}}(t_0)\big) \,/\, W\big(x_1(t_0), x_2(t_0)\big) \quad . \qquad (3.3 - 5b)$$

The function $x_p(t)$ represents a particular solution of (3.3 - 1) that does not depend on the initial conditions (3.3 - 4). $W(x_i(t), x_k(t))$ indicates the *Wronski determinant*

$$W(x_i(t), x_k(t)) = x_i(t)\dot{x}_k(t) - \dot{x}_i(t)x_k(t) \quad . \qquad (3.3 - 6)$$

If the functions $x_1(t)$ and $x_2(t)$ in the linear combination (3.3 - 3) are linearly independent, then the Wronski determinant differs from zero

$$W(x_1(t), x_2(t)) \neq 0 \qquad (3.3 - 7)$$

as discussed in Section 2.3.2b. The Wronski determinant (3.3 - 6) is the common denominator of equations (3.3 - 5a&b). If the free oscillation $x_h(t)$ of (3.3 - 3), which is essentially determined by the initial conditions, is damped, then it constitutes a *transient*. In this case the particular solution $x_p(t)$ of (3.3 - 1) describes the *stationary or long-time motion* of the oscillator.

The above considerations demonstrate that the description of the forced oscillations of a linear oscillator requires knowledge of a particular solution $x_p(t)$ of (3.3 - 1). This is elucidated in the following section.

3.3.1 Stationary Motion and Green Functions

As mentioned above, the stationary motion of a linear oscillator with a damped free oscillation or transient corresponds to a particular solution $x_p(t)$ of (3.3 - 1). By taking into account relations (3.3 - 5a&b), which permit to derive the constants a and b of $x_h(t)$ in (3.3 - 3) from the initial conditions (3.3 - 4), one comes to the conclusion that it suffices to find a particular solution $x_p(t)$ of (3.3 - 1) with the initial conditions

$$x_p(0) = 0 \quad \text{and} \quad \dot{x}_p(0) = 0 \quad . \qquad (3.3 - 8)$$

This solution $x_p(t)$ can be evaluated with the aid of the *integral transformation* [Birkhoff & Rota 1989 B]

$$x_p(t) = \int_0^t F(s)G(t,s)ds \quad . \qquad (3.3 - 9)$$

In this equation $F(t)$ indicates the external force, and the *kernel* $G(t,s)$ represents a *Green function*. This function fulfills the following conditions

$$0 \leq t \leq s\text{:} \quad G(t,s) = 0 \quad , \qquad (3.3 - 10a)$$

$$0 \leq s = t\text{:} \; G(t,s) = 0; \frac{\partial}{\partial t}G(t,s) = 0 \quad , \qquad (3.3 - 10b)$$

$$0 \le s < t: \quad \frac{\partial^2}{\partial t^2} G(t,s) + \frac{2}{\tau(t)} \frac{\partial}{\partial t} G(t,s) + \Omega^2(t)\, G(t,s) = 0 \quad . \tag{3.3 - 10c}$$

For a *non-modulated harmonic oscillator* with $\tau(t) = \tau$ and $\Omega(t) = \Omega$, which is described in Section 3.2, the Green function $G(t,s)$ corresponds to the transient $x_{\text{shock}}(t)$ after a unit impulse $\delta(t)$

$$G(t,s) = x_{\text{shock}}(t-s) \tag{3.3 - 11}$$

in accordance with (3.2 - 27b) and (3.3 - 9).

In general the Green function $G(t,s)$ can be calculated if two linearly independent solutions $x_1(t)$ and $x_2(t)$ of the homogeneous basic equation (2.3 - 1) or (2.3 - 3) of the free oscillator are known. Then, $G(t,s)$ is determined by

$$G(t,s) = \frac{x_1(s)x_2(t) - x_1(t)x_2(s)}{x_1(s)\dot{x}_2(s) - \dot{x}_1(s)x_2(s)} = \frac{x_2(t)x_1(s) - x_1(t)x_2(s)}{W[x_1(s),\, x_2(s)]} \quad . \tag{3.3 - 12a}$$

This equation can be derived e.g. by the "variation of parameters" [Bellman 1966 B, Zwillinger 1989 B].

According to (2.3 - 10) the Wronski determinant $W[x_1(t), x_2(t)]$ of the solutions $x_1(t)$ and $x_2(t)$ of the homogeneous differential equation (2.3 - 1) can be replaced by an exponential function. This yields the equation [Bellman 1966 B]

$$G(t,s) = \frac{x_2(t)x_1(s) - x_1(t)x_2(s)}{W[x_1(0), x_2(0)]}\, exp\left(2\int_0^t \frac{d\theta}{\tau(\theta)}\right) \quad . \tag{3.3 - 12b}$$

If the homogeneous part of the inhomogeneous basic equation (3.3 - 1) is in its normalized form

$$\ddot{x} + \Omega_0^2(t)\, x = 0 \quad , \tag{2.3 - 3}$$

then the Wronski determinant $W[x_1(t), x_2(t)]$ is constant according to (2.3 - 12). Under these circumstances, (3.3 - 12b) is reduced to

$$G(t,s) = \frac{x_2(t)\, x_1(s) - x_1(t)\, x_2(s)}{W[x_1(0),\, x_2(0)]} \quad . \tag{3.3 - 12c}$$

3.3.2 Excitation of Specific Modulated Oscillators

The application of Green functions to the calculation of particular solutions $x_p(t)$ of inhomogeneous linear differential equations (3.3 - 1) with the initial conditions (3.3 - 8) is common knowledge. In spite of this fact only few particular solutions

$x_p(t)$ derived with the aid of Green functions have been published. Particular solutions $x_p(t)$ derived by other means are discussed in the following.

a) Chirp oscillator

Chirp oscillators and their free oscillations were discussed in Section 2.3.4. For specific excitations the forced oscillations of these chirp oscillators can be described in a simple manner. A corresponding inhomogeneous oscillation equation in its normal form is

$$\ddot{x} + \Omega_0^2(t)x = F\dot{\Omega}_0(t)\cos\int_{t_0}^{t}\Omega_0(s)\,ds \tag{3.3 - 13a}$$

with the particular solution

$$x_p(t) = F\sin\int_{t_0}^{t}\Omega_0(s)\,ds = F\cos\left[\int_{t_0}^{t}\Omega_0(s)\,ds - \frac{\pi}{2}\right] \quad . \tag{3.3 - 13b}$$

An *example* is the oscillator with a down-chirp of the characteristic frequency $\Omega_0(t)$ described by (2.3 - 31a-c). Its inhomogeneous oscillation equation with the corresponding specific excitation has the form

$$\ddot{x} + \left(\theta^2 / t^4\right)x = F\left(\theta / t^3\right)\cos\theta\left(\frac{1}{t_0} - \frac{1}{t}\right) \quad . \tag{3.3 - 14a}$$

The general solution of this equation is

$$x(t) = -F\sin\theta\left(\frac{1}{t_0} - \frac{1}{t}\right) + A\ (t / t_0)\cos\left(\varphi_\infty - \frac{\theta}{t}\right) \tag{3.3 - 14b}$$

where the free parameters A and φ_∞ are determined by the initial conditions $x(t_0) = x_0$ and $\dot{x}(t_0) = \dot{x}_0$.

b) Aperiodically modulated oscillators

The excitations of aperiodically modulated oscillators whose forced oscillations have simple analytical representations can be derived from the following general inhomogeneous oscillation equation

$$\ddot{x} + \Omega_0^2(t)x = \ddot{x} + \left[\omega^2 - \frac{\dot{F}(t)}{\int_0^t F(s)\,ds}\right]x = F(t)\cos\omega t \quad . \tag{3.3. - 15a}$$

For *causal systems* defined by the conditions

$$x(t < 0) = 0 \quad \text{and} \quad \dot{x}\ (t < 0) \tag{3.1 - 6}$$

this equation has the solution

$$x(t) = \frac{\sin \omega t}{2\omega} \int_0^t F(s) ds \quad . \tag{3.3 - 15b}$$

An *example* constitutes the inhomogeneous oscillation equation

$$\ddot{x} + \left[\omega^2 - t^{-2}\right] x = F\, t \cos \omega t \tag{3.3 - 16a}$$

The corresponding causal system oscillates according to

$$x(t) = \frac{1}{2} F\, t^2 \sin \omega t \quad . \tag{3.3 - 16b}$$

c) Periodically modulated oscillators

A harmonic excitation of a periodically modulated linear oscillator can generate *subharmonics*. *Examples* to be mentioned are the harmonically excited and modulated oscillators that exhibit

α) *period doubling:* The oscillation equation

$$\ddot{x} + \left[\Omega^2 + 2\Delta\Omega \cos \Omega t\right] x = F \sin \omega t \tag{3.3 - 17a}$$

defines an oscillator with the modulation circular frequency $\Omega = 2\pi/T_0$ and a harmonic excitation with the circular frequency $\omega = 2\pi/T$. This oscillator oscillates with *half the excitation circular frequency* $\omega/2$ if $\omega = 2\Omega$. This condition yields the following particular solution $x_p(t)$ of (3.3 - 17a)

$$x_p(t) = \frac{F}{\Delta\Omega} \sin\left(\frac{\omega}{2} t\right) \quad \text{with} \quad \Omega = \omega/2 \quad \text{and} \quad T_p = 2T = T_0 \; . \tag{3.3 - 17b}$$

where T_p indicates the apparent oscillation period.

β) *period tripling:* The oscillation equation

$$\ddot{x} + \left[(\Omega/2)^2 + \Delta\Omega(1 + 2 \cos \Omega t)\right] x = F \sin \omega t \tag{3.3 - 18a}$$

represents the harmonic excitation with the circular frequency $\omega = 2\pi/T$ of an oscillator modulated with the circular frequency $\Omega = 2\pi/T_0$. The oscillators defined by (3.3 - 17a) and (3.3 - 18a) are essentially the same. The oscillator (3.3 - 18a) oscillates with one *third* $\omega/3$ *of the excitation circular frequency* $\omega = 3\Omega/2$. The corresponding particular solution $x_p(t)$ is

$$x_p(t) = \frac{F}{\Delta\Omega} \sin\left(\frac{\omega}{3}t\right) \quad \text{with} \quad \Omega/2 = \omega/3 \quad \text{and} \quad T_p = 3T = 2T_0 \ , \qquad (3.3 - 18b)$$

where T_p designates the apparent oscillation period.

Finally, it should be noticed that (3.3 - 17a&b) and (3.3 - 18a&b) are based on the decomposition of $\sin 2\alpha$ and $\sin 3\alpha$ [Gradshteyn & Ryzhik 1965 B].

3.3.3 Feedback

The *feedback in harmonic oscillators* was already discussed with regard to linear electric circuits in Section 2.2.7c and illustrated in Fig. 2.2 - 12. Equation (2.2 - 65), which represents this feedback, can be generalized to include the feedback in modulated linear oscillators. The result is

$$\ddot{x} + \frac{2}{\tau(t)}\dot{x} + \Omega^2(t)x = \Psi(x, \dot{x}, \ddot{x}, t) \quad . \qquad (3.3 - 19a)$$

The feedback function Ψ should depend at least on one of the variables x, $\dot{x}$ or $\ddot{x}$. However, it is not necessary that Ψ depends explicitly on time t. A feedback independent of time t is characterized by

$$\frac{\partial\Psi}{\partial t} = 0 \quad \text{or} \quad \Psi = \Psi(x, \dot{x}, \ddot{x}) \quad . \qquad (3.3 - 19b)$$

By comparison of (3.3 - 1) and (3.3 - 19a) one finds that in (3.3 - 19a) the external excitation $F(t)$ is replaced by the feedback function $\Psi(x, \dot{x}, \ddot{x}, t)$ that is determined by the momentary state of the oscillator, i.e. by the variables x, $\dot{x}$ and $\ddot{x}$.

An *approximation scheme* exists for the calculation of free oscillations of modulated linear oscillations [Bellman 1966 B] that makes use of a *virtual feedback*.

An *example* is the free oscillator described by the oscillation equation

$$\ddot{u} + \Omega_0^2(t)u = 0 \qquad (2.3 - 3)$$

with the characteristic circular frequency

$$\Omega_0(t) = \Omega + \varepsilon\,\Delta\Omega(t) \quad \text{where} \quad |\Delta\Omega(t)| \le \Omega \quad \text{and} \quad 0 < \varepsilon << 1 \quad . \qquad (3.3 - 20a)$$

In a first approximation this oscillation equation can be written as

$$\ddot{u}+\left[\Omega^2+2\varepsilon\Delta\Omega(t)\right]u=0 \quad . \qquad (3.3 - 20b)$$

Since ε is small one may expect that the free oscillation $u(t)$ can be approximated by various procedures. Thus, one writes (3.3 - 20b) in the form of (3.3 - 19a) that is characteristic for feedback in an oscillator

$$\ddot{u}+\Omega^2 u=\Psi(u,t)=-2\varepsilon\Delta\Omega(t)u \quad . \qquad (3.3 - 20c)$$

The solution of this equation is performed with the *Green function* introduced in Section 3.3.1. This equation is homogeneous for ε = 0. Then the normalized solutions of this equation are

$$u_c(t)=\cos\Omega t \quad \text{and} \quad u_s(t)=\frac{1}{\Omega}\sin\Omega t \qquad (3.3 - 21a)$$

with the corresponding Wronski determinant

$$W\left[u_c(t),u_s(t)\right]=1 \quad . \qquad (3.3 - 21b)$$

Application of (3.3 - 12c) to the functions of (3.3 - 21a) and to the Wronski determinant (3.3 - 21b) yields the Green function

$$G(t,s)=\frac{1}{\Omega}\sin\Omega(t-s) \quad . \qquad (3.3 - 21c)$$

The oscillation $u(t)$ following the initial conditions $u(0) = u_0$ and $\dot{u}(0) = \dot{u}_0$ can be evaluated with the *Volterra integral equation* [Bellman 1966 B]

$$u(t)=w(t)+\varepsilon\int_0^t u(s)G(t,s)ds \quad \text{with} \qquad (3.3 - 22a)$$

$$w(t)=u_0\cos\Omega t+\left(\dot{u}_0/\Omega\right)\sin\Omega t \quad . \qquad (3.3 - 22b)$$

This equation can be solved by *successive substitution*. The result is the *Liouville-Neumann series* [Bellman 1966 B]

$$\begin{aligned} u(t)&=w(t)+\varepsilon\int_0^t G(t,\alpha)w(\alpha)\,d\alpha \\ &+\varepsilon^2\int_0^t G(t,\alpha)\left[\int_0^\alpha G(\alpha,\beta)w(\beta)d\beta\right]d\alpha+\dots \quad . \end{aligned} \qquad (3.3 - 23)$$

3.4. Excitation of Nonlinear Liénard Oscillators

The *free* oscillations of nonlinear Liénard oscillators have been described in Section 2.5. *Forced* oscillations $x(t)$ of nonlinear Liénard oscillators as a result of an external time-dependent excitation $F(t)$ obey the basic equation

$$\ddot{x} - S(x)\dot{x} + D(x)x = F(t) \tag{3.4 - 1}$$

in accordance with (2.5 - 1a) and (3.1 - 2). Equation (3.4 - 1) is linear if both, $S(x)$ and $D(x)$, are constant. Otherwise (3.4 - 1) represents an *inhomogeneous nonlinear Liénard equation.*

3.4.1 Lagrange Stability

A relevant aspect of forced oscillations $x(t)$ of Liénard oscillators as real solutions of (3.4 - 1) is the *Lagrange stability* [LaSalle & Lefschetz 1961 B, 1967 B]. According to definition, the Lagrange stability of a real solution $x(t)$ of (3.4 - 1) means that for times t after a certain point of time t_0, e.g. $t_0 = 0$, this solution is bounded

$$|x(t)| \leq C < \infty \quad \text{for} \quad t \geq t_0 = 0 \quad . \tag{3.4 - 2}$$

The following conditions guarantee that the real solutions $x(t)$ of (3.4 - 1) exhibit Langrange stability and thus fulfill inequality (3.4 - 2) for times $t \geq t_0 = 0$.

$$\left| \int_0^t F(\Theta)\, d\Theta \right| \leq C_F \quad \text{for all } t \geq 0 \quad , \tag{3.4 - 3a}$$

$$-sign(x) \int_0^x S(y)dy \geq C_S > C_F \quad \text{for } |x| \geq a > 0 \quad , \tag{3.4 - 3b}$$

$$|x| \cdot D(x) \geq C_D \quad \text{for } |x| \geq a > 0 \quad . \tag{3.4 - 3c}$$

These conditions can be derived [LaSalle & Lefschetz 1961 B, 1967 B] with the aid of the Lyapunov functions described in Section 4.6.

For a harmonic excitation with a periodic external force

$$F(t) = F \cos \omega t \quad \text{with} \quad \omega = 2\pi / T \quad \text{and} \quad F > 0 \tag{3.1 - 3a}$$

the relation (3.4 - 3a) yields

$$C_F = F / \omega \quad . \tag{3.4 - 4}$$

If, in addition, the forced Liénard oscillator (3.4 - 1) includes a *linear damping* with

$$S(x) = -2/\tau \quad \text{and} \quad \tau > 0, \tag{3.4 - 5a}$$

then (3.4 - 3b) is reduced to the form

$$a > (\tau / 2\omega) F \quad . \tag{3.4 - 5b}$$

This inequality can be applied to (3.4 - 3c) and thus yields the condition

$$|x| D(x) \geq C_D \quad \text{for} \quad |x| \geq a > (\tau / 2\omega) F > 0 \quad . \tag{3.4 - 5c}$$

3.4.2 Periodic Excitation of Duffing Oscillators

The *free* oscillations of Duffing oscillators were already discussed in Section 2.5.2. The *forced* oscillations of Duffing oscillators as a result of the periodic real harmonic excitations (3.1 - 3a) correspond to the real solutions $x(t)$ of the inhomogeneous Duffing equation [Duffing 1918 B, Hale 1969 B, Korsch & Jodl 1994 B, Minorsky 1974 B, McLachlan 1950 B, Nayfeh & Mook 1979 B, Plaschko & Brod 1995 B, Stoker 1957 B]

$$\ddot{x} + (2/\tau)\dot{x} + \Omega^2 x + \varepsilon \Omega^2 x^3 = F \cos \omega t \ \text{ with real } \ F > 0 \quad . \tag{3.4 - 6a}$$

For $\varepsilon > 0$, $\tau > 0$ the real solutions $x(t)$ of this equation show Lagrange *stability*. Thus, they are bonded for times $t > t_0$. This can be proved with the inequalities (3.4 - 5a-c).

The *stationary periodic solutions* $x(t)$ of the inhomogeneous Duffing equation (3.4 - 6a) are of special interest. These solutions include periodic motions with the excitation circular frequency ω as well as those with the frequencies 3ω and $\omega/3$. Thus, there exist the *harmonic*, an *ultraharmonic* and a *subharmonic* with *frequency and period tripling*. This is the effect of the cubic term in the Duffing equation (3.4 - 6a). The following is dedicated to the oscillations with the three frequencies mentioned. In order to simplify the presentation one introduces the parameters

$$\omega^* = \omega / \Omega \quad \text{and} \quad Q = \Omega \tau / 2 \tag{3.4 - 6b}$$

where Q indicates the *quality factor.*

a) Harmonic oscillations and bistability

The *stationary harmonic oscillations* with the excitation circular frequency ω can be determined with the ansatz

$$x(t) = A_1 \cos(\omega t - \varphi_1) \quad \text{with} \quad A_1 \geq 0 \quad \text{and} \quad 0 \leq \varphi_1 \leq \pi \quad . \tag{3.4 - 7a}$$

The relation

$$cos^3\alpha = \frac{3}{4} cos\ \alpha + \frac{1}{4} cos\ 3\alpha$$

in combination with the ansatz (3.4 - 7a) in the inhomogeneous Duffing equation (3.4 - 6a) yields the *amplitude-frequency relation*

$$\left[1+\frac{3}{4}\varepsilon A_1^2 - \omega^{*2}\right]^2 + [\omega^*/Q]^2 = \left(F/A_1\Omega^2\right)^2 = \left[\Omega^2 G(\omega;\Omega,\tau,A_1)\right]^{-2} \quad (3.4 - 7b)$$

as well as the phase-frequency relation

$$tan\ \varphi_1 = (\omega^*/Q)\left[1+\frac{3}{4}\varepsilon A_1^2 - \omega^{*2}\right]^{-1}$$
$$\text{with}\quad \varphi_1 = \pi/2 \quad \text{for} \quad \omega^{*2} = 1+\frac{3}{4}\varepsilon A_1^2 \quad . \qquad (3.4 - 7c)$$

In (3.4 - 7b) the *gain function*

$$G(\omega;\Omega,\tau,A_1) = A_1/F \quad , \qquad (3.4 - 7d)$$

which was initially defined by (3.2 - 5a&b) for the forced linear harmonic oscillator, depends generally on the amplitude A_1 of the oscillation. Similarly, (3.4 - 7c) shows that the phase φ_1 also depends on this amplitude. This is characteristic for forced oscillations of nonlinear oscillators.

The relations (3.4 - 7b&c) are illustrated in Figs. 3.4 - 1a-c for $\varepsilon > 0$, $\varepsilon = 0$ and $\varepsilon < 0$.

For $\varepsilon = 0$ the two relations (3.4 - 7b&c) describe the forced oscillations and resonances of the *damped linear harmonic oscillator* discussed in Section 3.2.2. In this case the gain function $G(\omega;\Omega,\tau)$ as well as the phase φ_1 are independent of the amplitude A_1.

In the case of the *undamped nonlinear Duffing oscillator* with $\varepsilon \neq 0$ and $\tau = Q = \infty$ the relations (3.4 - 7b&c) can be simply written as

$$\left|1+\frac{3}{4}\varepsilon A_1^2 - \omega^{*2}\right| = \left(F/A_1\Omega^2\right) = [\Omega^2 G(\omega;\Omega,\tau,A_1)]^{-1} \quad , \qquad (3.4 - 8a)$$

$$\varphi_1 \begin{cases} = 0 \\ = \pi/2 \\ = \pi \end{cases} \text{for} \quad 1+\frac{3}{4}\varepsilon A_1^2 - \omega^{*2} \begin{cases} > 0 \\ = 0 \\ < 0 \end{cases} \quad . \qquad (3.4 - 8b)$$

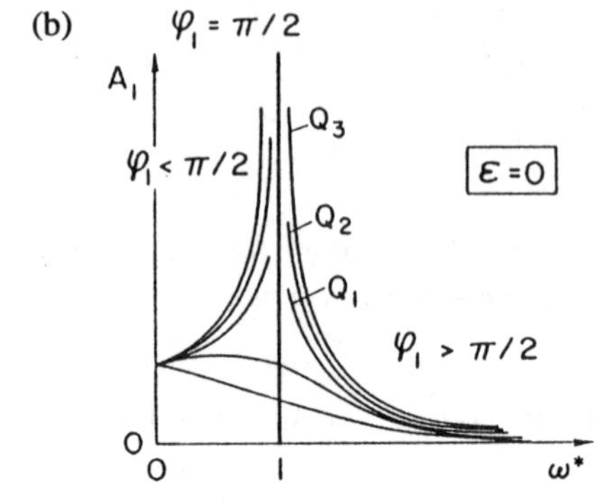

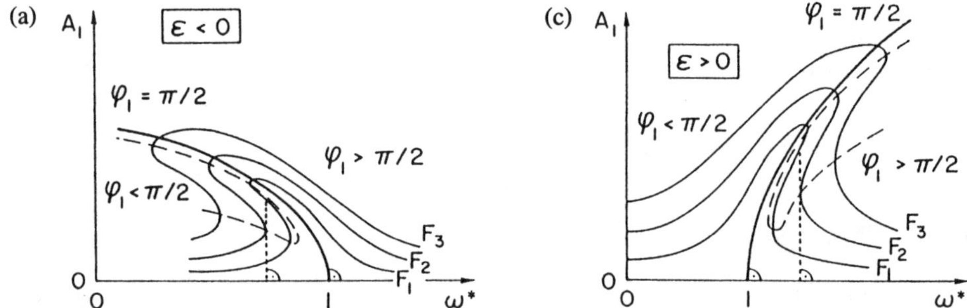

Fig. 3.4 - 1. Amplitude-frequency relations of forced oscillations of the Duffing oscillator for a) $\varepsilon < 0$, b) $\varepsilon = 0$, c) $\varepsilon > 0$

Figs. 3.4 - 1 a-c demonstrate a *further essential difference* between linear ($\varepsilon = 0$) and nonlinear ($\varepsilon \neq 0$) Duffing oscillators. Fig. 3.4 - 1b shows that in case of the damped linear Duffing oscillator with $\varepsilon = 0$ each frequency ω or ω^* is assigned a single amplitude A_1 of the stationary forced oscillation $x(t)$. On the contrary, a nonlinear Duffing oscillator with $\varepsilon \neq 0$ exhibits ranges of excitation frequencies ω or ω^*, where three different amplitudes A_1 are assigned to each frequency ω or ω^*. These frequency ranges have borders defined by the condition $d\omega/dA_1 = d\omega^*/dA_1 = 0$. These borders are indicated in Figs. 3.4 - 1 a&c by dashed curves. Also indicated in these figures are the curves corresponding to $\varphi_1 = \pi/2$. These full curves form the back bones of all the resonance curves. For $\varepsilon > 0$ these resonance curves bend toward high frequencies ω or ω^*, whilst for $\varepsilon < 0$ they bend towards low frequencies. The nonlinearity of Duffing oscillators is the origin of the phenomenon called *bistability* or *jump phenomenon.* Bistability occurs during variations of the excitation circular frequency ω or ω^* and affects the amplitude A_1 of the forced oscillation. It is located in the frequency range where each ω or ω^* is assigned three different amplitudes A_1 by the amplitude-frequency relation.

Fig. 3.4 - 2 demonstrates the bistability of a forced oscillation of a Duffing oscillator with $\varepsilon > 0$. For this demonstration it is assumed that the excitation frequency ω^* is varied slowly, while the amplitude $F > 0$ of excitation is kept constant.

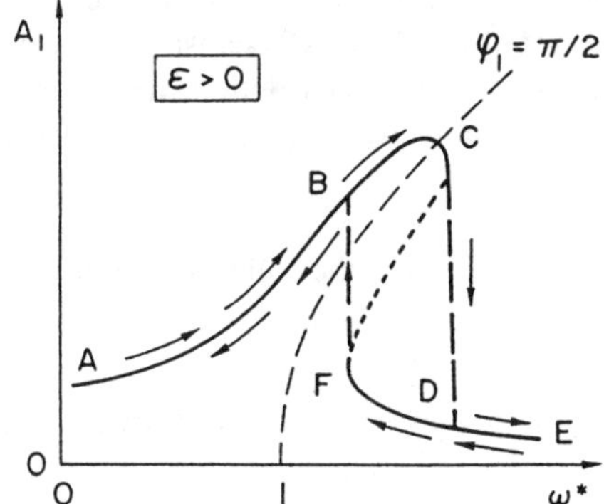

Fig. 3.4 - 2. Bistability of the forced harmonic oscillation of a Duffing oscillator with $\varepsilon > 0$. This bistability is elucidated by the amplitude-frequency relation illustrated in this figure

In order to explain bistability one starts in Fig. 3.4 - 2 with $\omega^* < 1$ at A and then increases ω^*. Thus, one stays on the upper curve of the amplitude-frequency relation and moves via B to C, where $d\omega^*/dA_1 = 0$. If one further increases ω^*, the positive *amplitude* A_1 *jumps* from the large value at C to the small value at D and then decreases in the direction of E. In order to return from E to A one now decreases the frequency ω^* and thus moves on the lower curve of the amplitude-frequency relation via D to F, where also $d\omega^*/dA = 0$. A further decrease of the excitation frequency ω^* causes the positive *amplitude* A_1 *to jump* from the low value at F to the high value at B and, subsequently, to decrease versus A.

The bistable behavior of the amplitude A_1 during the variation of the excitation circular frequency ω^* shows that the central section CF of the amplitude-frequency relation corresponds to *instable oscillations.* On the contrary, the sections AC and EF represent *stable oscillations.*

The bistability that manifests in the nonlinear Duffing oscillators [Duffing 1918 B] is a frequent phenomenon in *science and technology.*

b) Ultraharmonic oscillations

The excitation of the third harmonic [Mc Lachlan 1950 B, Stoker 1957 B], i.e. an oscillation with the threefold excitation circular frequency 3ω or $3\omega^*$, can be described most simply in case of the undamped nonlinear Duffing oscillator with $\varepsilon \neq 0$ and $\tau = Q = \infty$. Under these circumstances the solution of (3.4 - 6a) is attempted with the aid of the ansatz

$$x(t) = A_1 \cos \omega t + A_3 \cos 3\omega t \quad . \tag{3.4 - 9}$$

This ansatz is missing a phase φ between the harmonic and the third harmonic because one expects this phase difference to be either 0 or π. The latter can be taken into account by a negative A_3. Common approximation schemes on the basis of the ansatz (3.4 - 9) were developed by Duffing and by Poincaré, Lindstedt and Lighthill

[Stoker 1957 B]. The result of both of these types of approximation is the excitation circular frequency ω or ω^* as a function of an amplitude A_1 or A because the inverse relation is multivalued in wide frequency ranges. This was already demonstrated in the previous Section 3.4.2a that describes forced harmonic oscillations of the Duffing oscillator.

α) The *approximation by Duffing* [Stoker 1957 B] results in the following stationary forced oscillation

$$x(t) \approx A_1 \cos \omega t + \frac{\varepsilon}{32\,\omega^{*2}} A_1^3 \cos 3\omega t \qquad (3.4 - 10a)$$

with the amplitude-frequency relation

$$\omega^{*2} = (\omega / \Omega)^2 = 1 + \frac{3}{4} \varepsilon A_1^2 - \left(F / A_1 \Omega^2\right) \qquad (3.4 - 10b)$$

that is in agreement with the amplitude-frequency relation (3.4 - 8a).

β) The *approximation by Poincaré-Lindstedt-Lighthill* [Lindstedt 1883 J, Verhulst 1990 B, Zwillinger 1989 B] yields [Stoker 1957 B]:

$$x(t) \approx A \cos \omega t + \frac{\varepsilon}{32} A^3 \left(-\cos \omega t + \cos 3\omega t\right) \qquad (3.4 - 11a)$$

with the amplitude-frequency relation

$$\omega^* = \omega / \Omega \approx 1 + \frac{3}{8} \varepsilon A^2 - \frac{1}{2}\left(F / A \Omega^2\right) \quad . \qquad (3.4 - 11b)$$

c) Subharmonic oscillations

The occurrence of the subharmonic [McLachlan 1950 B, Stoker 1957 B] with a circular frequency $\omega/3$ or $\omega^*/3$ of one third of the excitation frequency ω or ω^* in an *undamped Duffing oscillator* with $\varepsilon \neq 0$ and $\tau = Q = \infty$ can be demonstrated with the aid of the ansatz [Stoker 1957 B]:

$$x(t) = A_{1/3} \cos \frac{1}{3} \omega t + A_1 \cos \omega t + A_{5/3} \cos \frac{5}{3} \omega t + + \quad . \qquad (3.4 - 12a)$$

This ansatz in combination with the inhomogeneous Duffing equation (3.4 - 1a) for $\varepsilon \neq 0, \tau = Q = \infty$ and with the mathematical relations

$$\cos^3(\alpha / 3) = \frac{3}{4} \cos(\alpha / 3) + \frac{1}{4} \cos \alpha \quad ,$$

$$cos^2(\alpha/3)\cos\alpha = \frac{1}{4}\cos(\alpha/3) + \frac{1}{2}\cos\alpha + + \quad ,$$

$$\cos(\alpha/3)\cos^2\alpha = \frac{1}{2}\cos(\alpha/3) + + \quad , \quad \text{etc.}$$

yields the following amplitude-frequency relations

$$\left(1-\omega^{*2}\right)A_1 + \frac{1}{4}\varepsilon\left(A_{1/3}^3 + 6A_{1/3}^2 A_1 + 3A_1^3\right) = F/\Omega^2 \quad , \qquad (3.4 - 12b)$$

$$\left(9-\omega^{*2}\right)A_{1/3} + \frac{27}{4}\varepsilon\left(A_{1/3}^3 + A_{1/3}^2 A_1 + 2A_{1/3}A_1^2\right) = 0 \quad . \qquad (3.4 - 12c)$$

The evaluation of these two coupled equations [Stoker 1957 B] results in the following remarkable rules on subharmonics that imply to *period tripling:*

α) A subharmonic oscillation with a circular frequency $\omega/3$ and an apparent period $T_p = 3T$ exists only if

$$+3\left[1+\frac{21}{1024}\varepsilon\Omega^{-4}F^2\right]^{1/2} \begin{cases} < \omega^* & \text{for} \quad \varepsilon > 0 \\ > \omega^* & \text{for} \quad \varepsilon < 0 \end{cases} \quad . \qquad (3.4 - 13)$$

Consequently, there exists no subharmonic with the circular frequency $\omega/3 = \Omega$ for $\varepsilon \neq 0$.

β) There exist ranges of A_1 and $\omega^* = \omega/\Omega$ where no subharmonics with the circular frequency $\omega/3$ occur. These are characterized by $A_1 \neq 0$, $A_3 = 0$.

In these ranges the amplitude A_1 of the harmonic is determined by the amplitude-frequency relation (3.4 - 10b). The formation of the subharmonic with the circular frequency $\omega/3$ occurs by *bifurcation* of the harmonic with the frequency ω or ω^*, a phenomenon that is described in Section 6.1. This bifurcation occurs at the amplitude [McLachlan 1950 B, Stoker 1957 B]

$$A_1 = A_B \approx -\frac{1}{8}\Omega^{-2}F\left[1-0{,}02\cdot\varepsilon\Omega^{-4}F^2\right] \qquad (3.4 - 14a)$$

and at the normalized frequency

$$\omega^* = \omega^*_B \approx 3\left[1+\frac{21}{1024}\varepsilon\Omega^{-4}F^2\right] \quad . \qquad (3.4 - 14b)$$

According to the inequalities (3.4 - 13) the *subharmonics with period tripling exist* in the frequency ranges

$$\begin{aligned} &\omega^* \geq \omega_B{}^* > 3 \quad \text{for} \quad \varepsilon > 0 \quad , \\ &\omega^* \leq \omega_B \quad < 3 \quad \text{for} \quad \varepsilon < 0 \quad . \end{aligned} \tag{3.4 - 14c}$$

γ) The subharmonic with period tripling exists *alone* [McLachlan 1950 B], when the amplitudes A_1 and A_3 fulfill the conditions

$$A_1 = 0 \quad \text{and} \quad A_{1/3} = 2\sqrt{F / \varepsilon \Omega^2} = 2\sqrt{\frac{1}{27\varepsilon}\left(\omega^{*2} - 9\right)} \quad . \tag{3.4 - 15a}$$

The corresponding excitation frequency $\omega_{1/3}$ is determined.by the equation

$$\left[\omega_{1/3}^* / 3\right]^2 = 1 + 3\left[\frac{\varepsilon F^2}{4\Omega^4}\right]^{1/3} \quad . \tag{3.4 - 15b}$$

The *influence of damping* on the formation of the subharmonic with period tripling by harmonic excitation of the nonlinear Duffing oscillator has also been investigated and published [McLachlan 1950 B, Stoker 1957 B].

d) Chaotic response

Computer simulated solutions of the harmonically excited Duffing oscillator [3.4 - 6a] have uncovered an unforeseen complexity of response including parameter ranges that display chaos [Korsch & Jodl 1994 B, Pain 1993 B, Parlitz & Lauterborn 1985 J, Ueda 1979 J, 1980 J]. Even with identical parameters Ω, τ and ε the long-term behavior of the oscillator is found to depend critically on the initial or starting values $x(0)$ and $\dot{x}(0)$.

e) Elliptic excitation and oscillation

An *analytic solution of the inhomogeneous Duffing equation* exists [Zwillinger 1989 B] if the undamped Duffing oscillator is excited periodically in the form of *Jacobi's elliptic cosine* $cn(u \mid m)$ instead of a harmonic cosine $\cos u$. This type of excitatiom is described by the equation

$$\ddot{x} + \Omega^2 x + \varepsilon \Omega^2 x^3 = F\, cn(\frac{2}{\pi} K(m)\, \omega t \,|\, m) \quad . \tag{3.4 - 16a}$$

The complete elliptic integral $K(m)$ and Jacobi's elliptic cosine $cn(u \mid m)$ [Abramowitz & Stegun 1965 B, Erdelyi et al. 1952-1954 B, Jahnke & Emde 1952 B, Milne-Thompson 1931 B] are described by (2.5 - 9a&b), (2.5 - 10a-c) and Figs. 2.5 - 3&4. For $m < 1/2$ the function $cn(u \mid m)$ is similar to $\cos u$.

For this excitation of the undamped Duffing oscillator, the excitation and the corresponding forced oscillation have the same form. The particular solution of (3.4 - 16a) is

$$x(t) = A\, cn(\frac{2}{\pi} K(m)\, \omega t | m) \tag{3.4 - 16b}$$

with the amplitude-frequency relation

$$\omega^{*2} = \frac{\varepsilon}{2}\left[\frac{2}{\pi} K(m)\right]^{-2} (A/m)^2 \quad . \tag{3.4 - 16c}$$

In these equations A and m are related by the equation

$$m^2 = \frac{\varepsilon}{2}\left[\varepsilon + A^{-2} - F\Omega^{-2} A^{-3}\right]^{-1} \quad . \tag{3.4 - 16d}$$

The three variables ω^*, A and m are interdependent. One of these can be chosen arbitrarily. The form of the excitation as well as of the oscillation is determined by the parameter m since it defines the shape of $cn(u \mid m)$.

3.4.3 Harmonic Excitation of the van der Pol Oscillator

The periodic excitation of the van der Pol oscillator involves complicated phenomena because this oscillator can oscillate permanently without external excitation as discussed in Section 2.5.5. This free permanent or stationary oscillation is characterized by a stable limit cycle. The following discussion reveals a range of excitation frequencies ω or ω^*, where the forced oscillation suppresses this free permanent oscillation. This range is called the *silent zone.*

a) Basic oscillation equation

The periodic harmonic excitation of the van der Pol oscillator [McLachlan 1950 B, Nayfeh & Mook 1979 B, Stoker 1957 B, van der Pol 1927 J,] is described by the following *inhomogeneous van der Pol equation*

$$\ddot{x} + \varepsilon\Omega\left(x^2 - 1\right)\dot{x} + \Omega^2 x = F\cos(\omega t + \varphi) \quad \text{with real} \quad F > 0 \quad , \tag{3.4 - 17a}$$

where φ represents the phase shift of the resulting forced oscillation versus the harmonic excitation. This phase shift is determined in the course of the solution of this equation. Equation (3.4 - 17a) can be reduced to the normal or *standard form*

$$\begin{aligned} &x'' + \varepsilon\left(x^2 - 1\right)x' + x = f\cos(\omega^* \tau + \varphi); f > 0 \\ &\text{with} \quad f > 0 \quad \text{and} \quad x' = dx/d\tau \quad . \end{aligned} \tag{3.4 - 17b}$$

This reduction can be performed by the following transformations

$$\tau = \Omega t, \;\; \omega^* = \omega / \Omega, \;\; f = F / \Omega^2 \quad . \tag{3.4 - 17c}$$

For $f = 0$ the standard form (3.4 - 17b) corresponds to the normalized homogeneous van der Pol equation (2.5 - 28b) describing the free oscillations of the van der Pol oscillator.

The solutions $x(t)$ and $x(\tau)$ of the inhomogeneous van der Pol equations (3.4 - 17a&b) show *Lagrange stability* for $\varepsilon > 0$, i.e. they are bonded for times $t > t_0$. This can be proven with the aid of the relations (3.4 - 3b&c) and (3.4 - 4).

b) Harmonic and free oscillations

In a first approximation the inhomogeneous van der Pol equation in the standard form (3.4 - 17b) can be solved with the ansatz [McLachlan 1950 B]

$$x(\tau) = a\cos\tau + A\cos\omega^*\tau \quad . \tag{3.4 - 18a}$$

This ansatz includes the circular frequency Ω of the free oscillation as well as the circular frequency $\omega = \Omega\omega^*$ of the excitation. Ultra- and subharmonics are neglected.

The combination of the ansatz (3.4 - 18a) with (3.4 - 17b) yields the following equations for the coefficients of the terms with $\sin\tau$, $\cos\tau$, $\sin\omega^*\tau$ and $\cos\omega^*\tau$

$$\begin{aligned}
&\sin\tau - \text{Term:} && 0 = a\left(1 - \frac{1}{4}a^2 - \frac{1}{2}A^2\right) \quad , \\
&\cos\tau - \text{Term:} && 0 = -a + a \quad , \\
&\sin\omega^*\tau - \text{Term:} && f\sin\varphi = -\varepsilon\,\omega^* A\left(1 - \frac{1}{2}a^2 - \frac{1}{4}A^2\right) \quad , \\
&\cos\omega^*\tau - \text{Term:} && f\cos\varphi = A\left(1 - \omega^{*2}\right) \quad .
\end{aligned} \tag{3.4 - 18b}$$

Higher terms are neglected. The following discussion makes use of the excitation E, the energies S_e and S_f of the forced and the free oscillation, the total oscillator energy S and the detuning ϑ which are determined by the equations

$$\begin{aligned}
&\vartheta = \left(\omega^{*2} - 1\right)/\varepsilon\,\omega^* \quad , \qquad E = (f/2\varepsilon\omega^*)^2 \quad , \\
&S_e = \frac{1}{4}A^2 \quad , \qquad S_f = \frac{1}{4}a^2 \quad , \qquad S = S_e + S_f \quad .
\end{aligned} \tag{3.4 - 18c}$$

The combination of these parameters and of the $\sin\omega^*\tau$ and $\cos\omega^*\tau$ terms (3.4 - 18b) yields the following equations

$$\tan\varphi = (1/\vartheta)(1 - S_e - 2S_f) \quad , \quad \text{and} \tag{3.4 - 18d}$$

$$E = S_e\left[(1 - S_e - 2S_f)^2 + \vartheta^2\right] \quad . \tag{3.4 - 18e}$$

The $sin\tau$ term (3.4 - 18b) gives rise to the following relations among the various parameters:

$$\begin{aligned}
&\alpha)\quad E=0, A=0, a=2;\quad S_e=0, S=S_f=1\quad .\\
&\beta)\quad E>0, A>0, a=0;\quad S=S_e, S_f=0\quad .\\
&\gamma)\quad E>0, A>0, a>0;\quad S=1-S_e=(1+S_f)/2;\quad 2S_e+S_f=1\quad .
\end{aligned}\qquad (3.4 - 18f)$$

The relations α, β and γ define *three different categories* of solutions of the normalized inhomogeneous van der Pol equation (3.4 - 17b):

α) *Pure free oscillation without excitation* characterized by
$E = 0$, $S_e = 0$, $S = S_f = 1$.

In a first approximation this solution corresponds to the oscillation and to the limit cycle of a free van der Pol oscillator with a small parameter ε according to Section 2.5.5

$$x = 2\cos\tau = 2\cos\Omega t\quad . \qquad (3.4 - 19)$$

β) *Pure forced oscillation in the silent zone* characterized by
$E > 0$, $A \geq \sqrt{2}$, $S = S_e \geq 1/2$, $a = 0$, $S_f = 0$.

In this case the forced oscillation suppresses the free oscillation. Because $S_f = 0$ equations (3.4 - 18d-e) are reduced to the relations

$$tan\,\varphi = (1/\vartheta)(1-S)\quad ,\quad \text{and} \qquad (3.4 - 20a)$$

$$E = S\left[(1-S)^2+\vartheta^2\right]\quad . \qquad (3.4 - 20b)$$

Equation (3.4 - 20b) is illustrated in Fig. 3.4 - 3. It shows the total oscillator energy $S = S_e$ as a function of the detuning ϑ, while E characterizes the different curves.

In order to avoid a free oscillation with $a \neq 0$, $S_f > 0$, one requires

$$S_f = \frac{1}{4}a^2 = 1-2S_e \leq 0\quad \text{or}\quad S_e = S \geq 1/2\quad \text{and}\quad A \geq \sqrt{2}\quad . \qquad (3.4 - 20c)$$

As a consequence the curves for $S < 1/2$ in Fig. 3.4 - 3 represent *instable* oscillations.

The pure forced oscillations described by (3.4 - 20a&b) become also instable if $dS/d\vartheta = \infty$ or $d\vartheta/dS = 0$. The differentiation of (3.4 - 20b) shows that this condition is fulfilled on an *ellipse* in the ϑ–S plane defined by the equation

$$3\vartheta^2 + 9\left(S-\frac{2}{3}\right)^2 = 1\quad . \qquad (3.4 - 20d)$$

The principal axes of this ellipse are $1/\sqrt{3}$ and 1/3, whilst its center is situated at $\vartheta = 0$, $S = 2/3$. The ellipse is also plotted in Fig. 3.4 - 3. In this figure the dashed curves correspond to *instable* oscillations mentioned above.

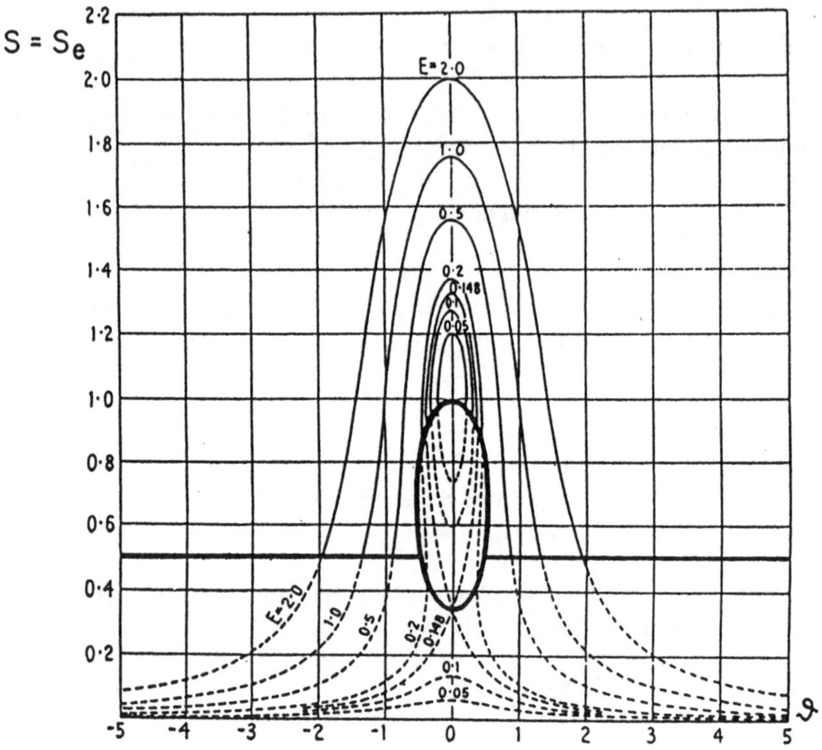

Fig. 3.4 - 3. Amplitude-frequency relations of the pure forced oscillations of a harmonically excited van der Pol oscillator in the quiet zone. The figure shows the energy $S_e = S$ of the pure forced oscillation as a function of the detuning ϑ. The dashed curves represent instable oscillations

γ) *Simultaneous free and forced oscillations* characterized by
$E > 0, A > 0, S_e > 0, a > 0, S_f > 0.$

The relevant relations between the oscillation energies S, S_e and S_f are part of the relations γ of (3.4 - 18f)

$$S = 1 - S_e = \frac{1}{2}(1 + S_f) \quad \text{and} \quad 2S_e + S_f = 1 \quad . \tag{3.4 - 21a}$$

The introduction of these relations in (3.4 - 18d&e) results in

$$tan\,\varphi = (1/\vartheta)(2 - 3S) \quad , \quad \text{and} \tag{3.4 - 21b}$$

$$E = (1-S)\left[(2-3S)^2 + \vartheta^2\right] \quad . \tag{3.4 - 21c}$$

Fig. 3.4 - 4 shows S as a function of ϑ where E labels the different curves. This figure contains the pure forced oscillation described by (3.4 - 20b) as well as the simultaneous forced and free oscillations according to (3.4 - 21c). For an identical excitation E the corresponding curves intersect outside the ellipse (3.4 - 20d) at

$$\begin{aligned} &S = 1/2, \ \vartheta_0^2 = 2E - 1/4 > 1/4 \\ &\text{with} \quad E \geq 1/4 \quad \text{and} \quad dS/d\vartheta = \infty \quad . \end{aligned} \tag{3.4 - 21d}$$

This condition determines the transition from the pure forced oscillation β to the simultaneous free and forced oscillations γ). It defines the *border of the silent zone.* In Fig. 3.4 - 4 this zone is marked for E = 0.5.

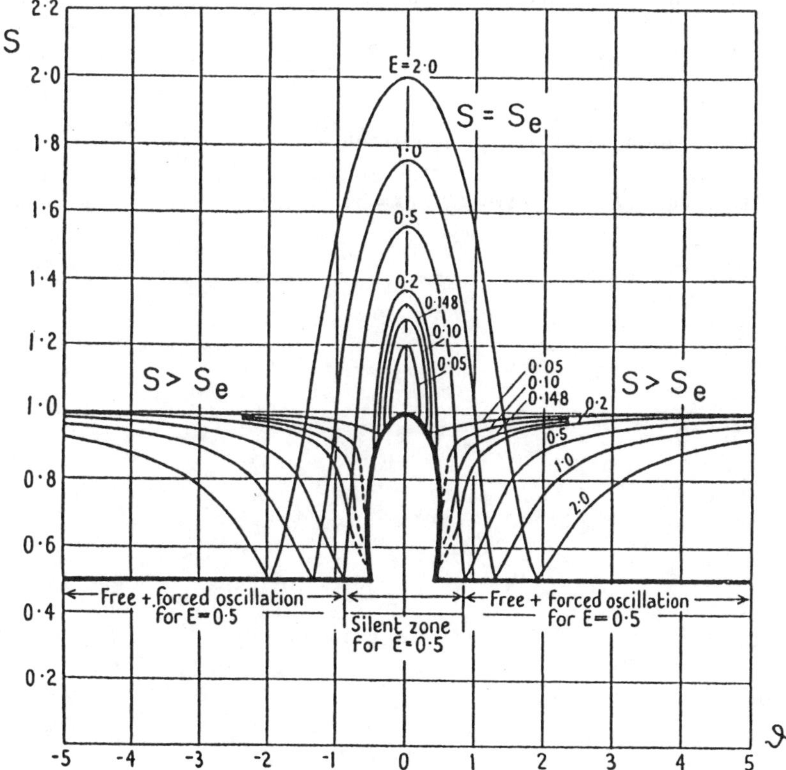

Fig. 3.4 - 4. Amplitude-frequency relations of the simultaneous forced and free oscillations of the harmonically excited van der Pol oscillator. This figure shows the total oscillator energy $S = S_e + S_f$ as a function of the detuning ϑ as well as the silent zone for $E = 0.5$

For large detuning ϑ (3.4 - 21a-d) yield the data

$$S(\vartheta=\pm\infty)=S_f(\vartheta=\pm\infty)=1\,,\; S_e(\vartheta=\pm\infty)=0 \quad ,$$
$$a(\vartheta=\pm\infty)=2,\; A(\vartheta=\pm\infty)=0,\; \varphi(\vartheta=\pm\infty)=\pi \quad . \qquad (3.4 - 21e)$$

These equations demonstrate that for large detuning ϑ the harmonic excitation looses its effect on a van der Pol oscillator.

c) Subharmonic oscillation

The generation of ultraharmonics with the triple excitation frequency 3ω and subharmonics with the triple excitation period $T_p = 3T = 6\pi/\omega$ by harmonic excitation of the nonlinear Duffing oscillator is discussed in Section 3.4.2. The same effect is caused by harmonic excitation of a van der Pol oscillator.

An *example* is the excitation of a *pure subharmonic* with the frequency $\Omega = \omega/3$ according to the inhomogeneous van der Pol equation

$$\ddot{x}+\varepsilon\Omega\left(x^2-1\right)\dot{x}+\Omega^2 x = F\,cos\,3\Omega t \quad \text{with} \quad F=2\,\varepsilon\,\Omega^2 \quad . \qquad (3.4 - 22a)$$

The solution of this equation shows *pure period tripling*

$$x(t)=-2\,sin\,\Omega t \quad \text{and} \quad T_p=2\pi/\Omega=6\pi/\omega=3T \quad . \qquad (3.4 - 22b)$$

4. Kinematics of Systems

This chapter concerns oscillations and other motions of *systems of ordinary differential equations* also called *coupled differential equations.* These systems are analogous to the velocity fields of fluid dynamics whose concepts are of great help in the following considerations.

4.1 Survey

The complexity of the systems of ordinary differential equations demands a systematic survey before discussion and interpretation. This survey requires first the standardization of these systems.

4.1.1 Standard Systems

The *standard form* of the systems of ordinary differential equations to be investigated is represented by a system of n explicit ordinary differential equations for n variables x_j, j = 1, 2, ... n as functions of time t [e.g. Kamke 1956 B, Hubbard & West 1995 B]

$$dx_j / dt = \dot{x}_j = v_j(t, x_1, \ldots x_k, \ldots x_n) \quad \text{with} \quad j = 1, 2, \ldots, n; k = 1, 2, \ldots, n. \quad (4.1 - 1a)$$

This system of differential equations can also be written as a *vector equation* in the n-dimensional space

$$\dot{\vec{r}} = \vec{v}(t, \vec{r}) \quad \text{with} \quad \vec{r} = [x_1, \ldots. x_k, \ldots.. x_n], \; \vec{v} = [v_1, \ldots. v_j, \ldots v_n] \quad . \qquad (4.1 - 1b)$$

In *fluid dynamics* this vector equation describes an *instationary flow*. This aspect will be discussed in more detail in Section 4.10.

The standard system (4.1a) can also be represented by a *differential dt* of the time t [Madelung 1956 B]

$$dt = \frac{dx_1}{\upsilon_1(t, x_1, .. x_k, .., x_n)} = \ldots = \frac{dx_j}{\upsilon_j(t, x_1, .. x_k, .., x_n)}$$
$$= \ldots = \frac{dx_n}{\upsilon_n(t, x_1, .. x_k, .., x_n)} \tag{4.1 - 1c}$$
with $j = 1, 2, \ldots, n; k = 1, 2, \ldots, n$.

A system of n ordinary differential equations of *higher order* can usually be transformed into a standard system (4.1 - 1a) in spite of the fact that the latter includes only differential equations of first order [Madelung 1943 B]. If, for example, the original system includes the second time derivative $\ddot{x}_k$ of the variable x_k, one sets $\dot{x}_{n+k} = \ddot{x}_k$. As a consequence the standard system deduced by this method has a higher dimension $n^* > n$ than that of the original system.

An *example* is the two-dimensional *acceleration field* defined by the system

$$\ddot{x}_1 = a_1(t, x_1, x_2, \dot{x}_1, \dot{x}_2)$$
$$\ddot{x}_2 = a_2(t, x_1, x_2, \dot{x}_1, \dot{x}_2) \tag{4.1 - 2a}$$

that corresponds to the vector equation

$$\ddot{\vec{r}} = \vec{a}(t, \vec{r}) \quad . \tag{4.1 - 2b}$$

The transformation of (4.1 - 2a) into a standard system (4.1 - 1a) by the method described increases the dimension from $n = 2$ to $n^* = 4$ as follows

$$\dot{x}_1 = x_3$$
$$\dot{x}_2 = x_4$$
$$\dot{x}_3 = a_1(t, x_1, x_2, x_3, x_4)$$
$$\dot{x}_4 = a_2(t, x_1, x_2, x_3, x_4) \quad . \tag{4.1 - 2c}$$

4.1.2 Autonomous Systems

A standard system defined by (4.1 - 1a-c) is called *time-invariant* or *autonomous* [Birkhoff & Rota 1989 B, Verhulst 1989 B, Zwillinger 1990 B] when the velocity $\vec{\upsilon}$ is explicitly independent of time t. On this condition the standard system has the form

$$dx_j / dt = \dot{x}_j = \upsilon_j(x_1, \ldots, x_k, \ldots, x_n) \quad \text{with} \quad j = 1, 2, \ldots, n; k = 1, 2, \ldots, n. \tag{4.1 - 3a}$$

It corresponds to the vector equation

$$\dot{\vec{r}} = \vec{\upsilon}(\vec{r}) \quad \text{with} \quad \frac{\partial}{\partial t} \vec{\upsilon} = \vec{0} \quad , \tag{4.1 - 3b}$$

where $\partial / \partial t$ indicates the partial differentiation with respect to time t. This equation describes a *stationary flow* by a *velocity field independent of time t*.

The solution of autonomous standard systems of differential equations is relatively simple because the time t can be separated from the other variables x_k, $k = 1, 2, ..., n$. This is revealed by the equations

$$\begin{aligned} dt &= \frac{dx_1}{\upsilon_1(x_1,..x_k,..x_n)} = \dots = \frac{dx_j}{\upsilon_j(x_1,..x_k,..x_n)} \\ &= \dots = \frac{dx_n}{\upsilon_n(x_1,..x_k,..x_n)} \qquad (4.1\text{ - }3c) \\ \text{with} \quad & j = 1,2,\dots,n \quad ; \quad k = 1,2,\dots,n \quad . \end{aligned}$$

That can be derived from (4.1 - 3a). The separation of dt in these equations demonstrates that *autonomous systems are invariant with respect to a shift of the time scale* from t to $t - t_0$, where t_0 represents an arbitrary reference time.

Characteristic of autonomous systems are the *fixed singular or critical points.* They correspond to vectors $\vec{r}_S = [x_{S1},\dots x_{Sk},\dots x_{Sn}]$, which fulfill the condition

$$\upsilon_j(x_{S1},..x_{Sk},..x_{Sn}) = 0 \quad \text{with} \quad j = 1,2,\dots,n \quad ; \quad k = 1,2,\dots,n \qquad (4.1\text{ - }4a)$$

$$\text{or} \quad \vec{\upsilon}(\vec{r}_S) = \vec{0} \quad . \qquad (4.1\text{ - }4b)$$

4.1.3 Linear Systems

General standard linear systems have the form

$$dx_j / dt = \dot{x}_j = \upsilon_j = \sum_{k=1}^{n} a_{jk}(t)\, x_j + b_j(t) \quad . \qquad (4.1\text{ - }5a)$$

They are equivalent to the vector equation

$$\begin{aligned} &\dot{\vec{r}} = \vec{\upsilon} = \boldsymbol{A}(t)\,\vec{r} + \vec{b}(t) \\ &\text{with} \quad \boldsymbol{A}(t) = \{a_{jk}(t)\} \quad \text{and} \quad \vec{b}(t) = [b_1(t),..b_j(t),..b_n(t)] \quad . \end{aligned} \qquad (4.1\text{ - }5b)$$

In this equation $\boldsymbol{A}(t)$ designates the *characteristic matrix* and $\vec{b}(t)$ the *perturbation vector*. A linear system is called *homogeneous* if

$$b_j(t) = 0 \quad \text{or} \quad \vec{b}(t) = \vec{0} \quad , \qquad (4.1\text{ - }6a)$$

and inhomogeneous, if

$$b_j(t) \neq 0 \quad \text{or} \quad \vec{b}(t) \neq \vec{0} \quad . \tag{4.1 - 6b}$$

Autonomous linear systems [Kamke 1956 B], which are also called *d'Alembert systems* [Kamke 1956 B], are the most simple standard systems (4.1 - 1a) of differential equations of first order. *Homogeneous* d'Alembert systems are written as

$$dx_j / dt = \dot{x}_j = \upsilon_j = \sum_{k=1}^{n} a_{jk}\, x_k \quad \text{with} \quad j = 1,2,\ldots,n \quad . \tag{4.1 - 7a}$$

They correspond to the vector equation

$$\dot{\vec{r}} = \vec{\upsilon} = \boldsymbol{A}\,\vec{r} \quad \text{with} \quad \boldsymbol{A} = \left\{a_{ij}\right\}, \tag{4.1 - 7b}$$

where the characteristic matrix does not vary with time t.

A homogeneous d'Alembert system possesses a *single critical point in the origin:*

$$\vec{r}_S = \vec{0} \quad \text{with} \quad \vec{\upsilon}(\vec{r}_S) = \vec{\upsilon}\left(\vec{0}\right) = \vec{0} \quad . \tag{4.1 - 8}$$

Inhomogeneous d'Alembert systems can be transformed into homogeneous d'Alembert systems by a simple shift of the origin according to the following scheme:

Starting with the vector equation of an inhomogeneous d'Alembert system

$$\dot{\vec{r}} = \boldsymbol{A}\,\vec{r} + \vec{b} \tag{4.1 - 9a}$$

one introduces a vector $\vec{r}_0$ defined by

$$\boldsymbol{A}\,\vec{r} + \vec{b} = \boldsymbol{A}\left(\vec{r} - \vec{r}_0\right) \quad .$$

This results in

$$\vec{r}_0 = -\boldsymbol{A}^{-1}\,\vec{b} \tag{4.1 - 9b}$$

By a shift of the origin with the vector $\vec{r}_0$ the inhomogeneous d'Alembert system described by (4.1 - 9a) is transformed into a vector equation of a homogeneous d'Alembert system

$$\frac{d}{dt}\left(\vec{r} - \vec{r}_0\right) = \boldsymbol{A}\left(\vec{r} - \vec{r}_0\right) \tag{4.1 - 9c}$$

with the characteristic constant matrix $\boldsymbol{A}$. Consequently, it suffices to investigate the properties and solutions of the homogeneous d'Alembert systems.

4.1.4 Linearization of Autonomous Systems

Autonomous systems (4.1 - 3a&b) can be characterized mainly by their behavior at their fixed critical points $\vec{r}_S$ defined by (4.1 - 4ab). In most cases the behavior can be deduced from the linear approximation of these systems in the immediate neighborhood of their critical points $\vec{r}_S$. The linear approximation can be derived from (4.1 - 3a&b) and (4.1 - 4a&b). It yields the following linear autonomous systems

$$\frac{d}{dt}\left(x_j - x_{Sj}\right) = \sum_k J_{jk}\left(x_k - x_{Sk}\right) \tag{4.1 - 10a}$$

or

$$\frac{d}{dt}\left(\vec{r} - \vec{r}_S\right) = \boldsymbol{J}\left(\vec{r} - \vec{r}_S\right) \tag{4.1 - 10b}$$

with the *Jacobi matrix*

$$\boldsymbol{J} = \left\{J_{jk}\right\} = \left\{\frac{\partial}{\partial x_k} v_j\left(x_k = x_{Sk}\right)\right\} \tag{4.1 - 10c}$$

and $\quad j = 1, 2, \ldots, n \quad ; \quad k = 1, 2, \ldots, n \quad .$

This linear approximation represents a *homogeneous d'Alembert system* (4.1 - 7a&b). It suffices for the study of the behavior of the autonomous system at the singular point $\vec{r}_S$ if

$$det\,\boldsymbol{J} \neq 0 \quad . \tag{4.1 - 11a}$$

Two-dimensional homogeneous d'Alembert systems are discussed extensively in Section 4.3, while higher-dimensional homogeneous d'Alembert systems are the topic of Section 4.10.5.

The linear approximation (4.1 - 10a-c) of an autonomous system at the singular point $\vec{r}_S$ is insufficient for the description of its behavior at this point if

$$det\,\boldsymbol{J} = 0 \quad . \tag{4.1 - 11b}$$

In this case quadratic or higher-order approximations have to be taken into account. The corresponding *two-dimensional quadratic autonomous systems* are presented in Section 4.4.

4.2 Two-dimensional Autonomous Systems in General

The two-dimensional autonomous systems [Hubbard & West 1995 B] are relatively simple yet of fundamental interest. For example they describe two-dimensional flows of incompressible fluids.

4.2.1 Description in Cartesian Coordinates

In order to understand and to classify the two-dimensional autonomous systems it is of advantage to use the following *standard representation*

$$\begin{aligned}\dot{x} &= u(x,y) = -U_x(x,y) + H_y(x,y)\\ \dot{y} &= \upsilon(x,y) = -U_y(x,y) - H_x(x,y)\end{aligned} \qquad (4.2 - 1)$$

where $U(x, y)$ represents the *potential* and $H(x, y)$ the *Hamiltonian*. The indices x and y indicate the partial differentiations $\partial / \partial x$ and $\partial / \partial y$.

In *fluid kinematics* the two-dimensional autonomous systems (4.2 - 1) correspond to *stationary planar flows of incompressible fluids* [Prandtl & Tietjens 1957ab B]. They can be described by the vector equation

$$\dot{\vec{r}} = \vec{\upsilon}(\vec{r}) \quad \text{with} \quad \vec{r} = [x,y,z] \quad \text{and} \quad \vec{\upsilon}(\vec{r}) = [u(x,y), \upsilon(x,y), 0] \quad . \qquad (4.2 - 2)$$

In such a stationary velocity field the velocity $\vec{\upsilon}\left[\mathrm{m\,s^{-1}}\right]$ is a function of the position vector $\vec{r}\,[\mathrm{m}]$, yet independent of time t [s].

The flows of an incompressible fluid are usually classified on the basis of the following characteristics

$$div\,\vec{\upsilon} = q = u_x + \upsilon_y = -\Delta U = -\left(U_{xx} + U_{yy}\right) \quad , \qquad (4.2 - 3a)$$

$$curl\,\vec{\upsilon} = 2\vec{\omega} = [0,0,\upsilon_x - u_y] = [0,0,-\Delta H] = \left[0,0,-\left(H_{xx} + H_{yy}\right)\right] \quad , \qquad (4.2 - 3b)$$

$$\omega = \frac{1}{2}\left[curl\,\vec{\upsilon}\right]_z = -\frac{1}{2}\Delta H = -\frac{1}{2}\left(H_{xx} + H_{yy}\right) \quad . \qquad (4.2 - 3c)$$

The divergence $div\,\vec{\upsilon}$ of the velocity $\vec{\upsilon}$ at the position $\vec{r}$ corresponds to the local *source or sink strength* $q(\vec{r})\left[\mathrm{s^{-1} = m^2 s^{-1} / m^2}\right]$. This strength q is positive for a source and negative for a sink. The rotation *curl* $\vec{\upsilon}$ or *rot* $\vec{\upsilon}$ of the velocity $\vec{\upsilon}$ at the position $\vec{r}$ determines the angular velocity $\omega(\vec{r})$ and the axis of the *local rotation of a fluid particle*. In a planar flow described by (4.2 - 2) the axis of rotation is the vertical z-axis. In (4.2 - 3a) Δ indicates the two-dimensional Laplace operator. A more detailed description of these characteristics is offered in Section 4.10 devoted to three-dimensional flows.

4.2.2 Description in Polar Coordinates

In the study of two-dimensional flows the Cartesian coordinates x,y are often replaced by the polar coordinates r,φ. These are defined by the equations

$$r = +\left(x^2 + y^2\right)^{1/2} \quad , \quad \varphi = arctan\,(y/x) \quad , \tag{4.2 - 4a}$$

$$x = r\,cos\,\varphi \quad , \quad y = r\,sin\,\varphi \quad . \tag{4.2 - 4b}$$

The corresponding radial velocity $\dot{r}$ and angular velocity $\dot{\varphi}$ are related to the velocity components u and υ as follows

$$\begin{aligned} \dot{r} &= \upsilon_r = \quad u\,cos\,\varphi + \upsilon\,sin\,\varphi \\ r\dot{\varphi} &= \upsilon_\varphi = -u\,sin\,\varphi + \upsilon\,cos\,\varphi \quad , \end{aligned} \tag{4.2 - 5a}$$

$$\begin{aligned} u &= \dot{r}\,cos\,\varphi - (r\dot{\varphi})sin\,\varphi \\ \upsilon &= \dot{r}\,sin\,\varphi + (r\dot{\varphi})cos\,\varphi \quad . \end{aligned} \tag{4.2 - 5b}$$

In (4.2 - 5a) υ_r indicates the radial velocity and υ_φ the transverse velocity. As a consequence, the two-dimensional autonomous system (4.2 - 1) corresponds in polar coordinates to the system

$$\begin{aligned} \dot{r} &= \upsilon_r = -U_r(r,\varphi) + \frac{1}{r}H_\varphi(r,\varphi) \\ r\dot{\varphi} &= \upsilon_\varphi = -\frac{1}{r}U_\varphi(r,\varphi) - H_r(r,\varphi) \end{aligned} \tag{4.2 - 6}$$

where $U(r,\varphi)$ and $H(r,\varphi)$ represent potential and Hamiltonian in polar coordinates.

The characteristics of a planar flow of an incompressible fluid in polar coordinates are

$$div\,\vec{\upsilon} = q = -\Delta U = -U_{rr} - r^{-1}U_r - r^{-2}U_{\varphi\varphi} \quad , \tag{4.2 - 7a}$$

$$\left[curl\,\vec{\upsilon}\right]_z = 2\omega = -\Delta H = -H_{rr} - r^{-1}H_r - r^{-2}H_{\varphi\varphi} \quad . \tag{4.2 - 7b}$$

4.2.3 Classification of Systems and Flows

In this section the two-dimensional systems and flows are classified with respect to rotation and divergence as well as to the corresponding potential and Hamiltonian. Most remarkable are the ideal systems of flows since they can be described in a complex representation. They will be discussed in Section 4.2.4.

a) Irrotational systems or flows
These two-dimensional systems are defined by the condition

$$\left[curl\ \vec{\upsilon}\right]_z = 2\omega = 0 \quad . \qquad (4.2 - 8a)$$

According to (4.2 - 3a-c) this implies that the Hamiltonian $H = H(x,y) = H(r,\varphi)$ fulfills the Laplace equation:

$$\Delta H = H_{xx} + H_{yy} = H_{rr} + r^{-1}H_r + r^{-2}H_{\varphi\varphi} = 0 \quad . \qquad (4.2 - 8b)$$

The *source and sink strength* q of irrotational systems or flows is determined by (4.2 - 7a).

b) Gradient systems or flows

Gradient systems are characterized by a zero Hamiltonian:

$$H = H(x,y) = H(r,\varphi) = 0 \quad . \qquad (4.2 - 9a)$$

According to (4.2 - 8a) the gradient systems and flows defined by this equation are *irrotational* because they fulfill the relations (4.2 - 8a&b).

These systems are called *gradient fields* since the velocity $\vec{\upsilon}$ equals the negative gradient of the potential U

$$\vec{\upsilon}(\vec{r}) = -grad\ U(\vec{r}) \quad . \qquad (4.2 - 9b)$$

For two-dimensional gradient systems and flows this equation can be represented in Cartesian coordinates by

$$\begin{aligned} \dot{x} &= u = -U_x(x,y) \\ \dot{y} &= \upsilon = -U_y(x,y) \end{aligned} \qquad (4.2 - 9c)$$

and in polar coordinates by

$$\begin{aligned} \dot{r} &= \upsilon_r = -U_r(r,\varphi) \\ r\dot{\varphi} &= \upsilon_\varphi = -r^{-1}U_\varphi(r,\varphi) \quad . \end{aligned} \qquad (4.2 - 9d)$$

Equations (4.2 - 9c) can also be written as differential dt of the time t

$$dt = -U_x^{-1}dx = -U_y^{-1}dy \quad . \qquad (4.2 - 9e)$$

The second equation implies that the vector product of the gradient of U and the infinitesinal displacement vector $d\vec{r}$ vanishes

$$grad\, U \times d\vec{r} = \vec{0} \text{ with } grad\, U = \left[U_x, U_y, 0\right] \text{ and } d\vec{r} = \left[dx, dy, dz\right] \quad . \qquad (4.2 - 9f)$$

This yields the geometrical relations

$$d\vec{r} \| grad\, U \quad \text{and} \quad d\vec{r} \perp \; (U = const) \quad . \qquad (4.2 - 9g)$$

They imply that the *streamlines* $\vec{r}(t) = [x(t), y(t), z]$ of *a steady gradient flow* as solutions of (4.2 - 9c) or (4.2 - 9d) are *perpendicular to the equipotential lines*

$$U(x,y) = U(r,\varphi) = U = const \quad . \qquad (4.2 - 9h)$$

As in Section 4.2.3a the source and sink strength q of the gradient systems is given by

$$q = -\Delta U = -\left(U_{xx} + U_{yy}\right) = -U_{rr} - r^{-1}U_r - r^{-2}U_{\varphi\varphi} \quad . \qquad (4.2 - 3a)$$

c) Systems and flows without sources and sinks
A system or flow has *neither sources nor sinks* if it fulfills the condition

$$div\, \vec{\upsilon} = q = u_x + \upsilon_y = \upsilon_{r,r} + r^{-1}\upsilon_r + r^{-1}\upsilon_{\varphi,\varphi} = 0 \quad . \qquad (4.2 - 10a)$$

This condition and (4.2 - 3a) require that the potential $U=U(x,y)=U(r,\varphi)$ is determined by the *Laplace equation*

$$\Delta U = U_{xx} + U_{yy} = U_{rr} + r^{-1}U_r + r^{-2}U_{\varphi\varphi} = 0 \quad . \qquad (4.2 - 10b)$$

The local rotation ω of these systems and flows is determined by

$$\omega = -\frac{1}{2}\Delta H = -\frac{1}{2}\left(H_{xx} + H_{yy}\right) = -\frac{1}{2}\left(H_{rr} + r^{-1}H_r + r^{-2}H_{\varphi\varphi}\right) \quad . \qquad (4.2 - 3c)$$

d) Hamiltonian systems and flows
The *potential U* of Hamiltonian systems and flows is zero by definition

$$U = U(x,y) = U(r,\varphi) = 0 \quad . \qquad (4.2 - 11a)$$

According to (4.2 - 3a) the Hamiltonian systems fulfill (4.2 - 10a) and, consequently, possess *neither sources nor sinks.*

In addition, Hamiltonian systems and flows represent pure *vortex velocity fields* which can be represented as

$$\vec{\upsilon}(\vec{r}) = curl\, \vec{A}(\vec{r}) \quad . \qquad (4.2 - 11b)$$

Two-dimensional Hamiltonian systems and flows are determined by a *vector potential* $\vec{A}$ in the *z*-direction vertical to the *xy* plane

$$\vec{A} = [0, 0, H] \quad \text{with} \quad H = H(x, y) = H(r, \varphi) \quad . \tag{4.2 - 11c}$$

Therefore, they can be represented in Cartesian coordinates as

$$\begin{aligned} \dot{x} &= u = +H_y(x, y) \\ \dot{y} &= \upsilon = -H_x(x, y) \end{aligned} \tag{4.2 - 11d}$$

and in polar coordinates as

$$\begin{aligned} \dot{r} &= \upsilon_r = r^{-1} H_\varphi(r, \varphi) \\ r\dot{\varphi} &= \upsilon_\varphi = -H_r(r, \varphi) \quad . \end{aligned} \tag{4.2 - 11e}$$

Equations (4.2 - 11d) can also be written as a differential *dt* of the time *t*

$$dt = +H_y^{-1} dx = -H_x^{-1} dy \quad . \tag{4.2 - 11f}$$

The second equation implies that the scalar product of the gradient of *H* and the infinitesinal displacement vector $d\vec{r}$ vanishes

$$\begin{aligned} &grad\ H \cdot d\vec{r} = 0 \\ &\text{with} \quad grad\ H = [H_x, H_y, 0] \quad \text{and} \quad d\vec{r} = [dx, dy, 0] \quad . \end{aligned} \tag{4.2 - 11g}$$

This requirement yields the geometrical relations

$$d\vec{r} \perp grad\ H \quad \text{and} \quad d\vec{r} \parallel H = const \quad . \tag{4.2 - 11h}$$

From these statements we can conclude that the *trajectories and streamlines* $\vec{r} = [x(t), y(t), z]$ *of a stationary Hamiltonian flow* as solutions of (4.2 - 11d) or (4.2 - 11e) *correspond to the lines of a constant Hamiltonian*

$$H(x, y) = H(r, \varphi) = H = const \quad . \tag{4.2 - 11i}$$

Therefore, the Hamiltonian *H* represents a *stream function.*

4.2.4 Ideal Flows

Systems and flows can be represented with the aid of *complex analytical functions* $w = f(z)$ if they are *irrotational* according to Section 4.2.3a and possess *neither sources nor sinks* according to Section 4.2.3c. They are called *ideal flows* [Chung 1988 B] and form the basis of fluid kinematics [Anderson 1988 B, Chung 1989 B, Lighthill 1989 B].

a) Basic concepts

An irrotational two-dimensional flow of an incompressible fluid is characterized by zero rotation

$$[curl\ \vec{\upsilon}]_z = 2\omega = 0 \quad , \tag{4.2 - 8a}$$

and a Hamiltonian H that fulfills the Laplace equation

$$\Delta H = H_{xx} + H_{yy} = H_{rr} + r^{-1}H_r + r^{-2}H_{\varphi\varphi} = 0 \quad . \tag{4.2 - 8b}$$

A two-dimensional flow of an incompressible fluid without sources and sinks has a zero divergence

$$div\ \vec{\upsilon} = q = 0 \quad , \tag{4.2 - 10a}$$

and a potential U that also fulfills the Laplace equation

$$\Delta U = U_{xx} + U_{yy} = U_{rr} + r^{-1}U_r + r^{-2}U_{\varphi\varphi} = 0 \quad . \tag{4.2 - 10b}$$

Thus, both, the potential U and the Hamiltonian H, obey the Laplace equation in the case of an irrotiational flow of an incompressible fluid without sources and sinks. A relation between the potential U and the Hamiltonian H can be derived by taking into account that in a potential velocity field the gradient of U is parallel to the streamlines according to (4.2 - 9g) whilst in a Hamiltonian system the gradient of H is perpendicular to the streamlines according to (4.2 - 11h). As a consequence, the gradients of U and H in our two-dimensional flow are perpendicular. This demands

$$grad\ U \cdot grad\ H = U_x H_x + U_y H_y = 0 \quad . \tag{4.2 - 12a}$$

This relation is valid if U and H obey the *Cauchy-Riemann conditions*

$$H_x(x,y) = +U_y(x,y) \quad \text{and} \quad H_y(x,y) = -U_x(x,y) \quad . \tag{4.2 - 12b}$$

These conditions yield the following standard representation (4.2 - 1) of the irrotational flows without sources or sinks

$$\begin{aligned} \dot{x} &= u(x,y) = -2U_x(x,y) = +2H_y(x,y) \\ \dot{y} &= \upsilon(x,y) = -2U_y(x,y) = -2H_x(x,y) \quad . \end{aligned} \tag{4.2 - 12c}$$

In polar coordinates this system of equations has the form

$$\begin{aligned} \dot{r} &= v_{\mathrm{r}}(r,\varphi) = -2U_{\mathrm{r}}(r,\varphi) = +\frac{2}{r}H_{\varphi}(r,\varphi) \\ r\dot{\varphi} &= v_{\varphi}(r,\varphi) = -\frac{2}{r}U_{\varphi}(r,\varphi) = -2H_{\mathrm{r}}(r,\varphi) \quad . \end{aligned} \tag{4.2 - 12d}$$

This system can either be considered as a pure *potential flow* with the potential $2U$ or as a proper *Hamiltonian system* with the Hamiltonian $2H$.

b) Complex representations

The Cauchy-Riemann conditions (4.2 - 12b) are essential in the theory of analytic complex functions of a complex variable [Arbenz & Wohlhauser 1986 B, Churchill 1948 B, Convay 1973 B, Hurwitz & Courant 1929 B, Levinson & Redheffer 1970 B, Titchmarsh 1932 B]. In this theory the xy plane is replaced by the *Argand diagram* $z = x + iy$ [Margenau & Murphy 1956 B] that consists of a real axis along x and an imaginary axis along y. A complex function $w = f(z)$ of the complex variable z is called *analytic* in the point $z_0 = x_0 + iy_0$ when it can be represented by the *Taylor series*

$$w = f(z) = \sum_{\mathrm{n}=0}^{\infty} \frac{1}{n!} f^{(\mathrm{n})}(z_0)(z - z_0)^{\mathrm{n}} \quad . \tag{4.2 - 13a}$$

This requires, that *all derivatives* $f^{(\mathrm{n})}(z)$ *exist* at $z = z_0$.

A complex function $f(z)$ of the complex variable z can be split into a real and an imaginary part

$$w = f(z) = f(x + iy) = -U(x,y) + i\,H(x,y) \quad . \tag{4.2 - 13b}$$

This function is *analytic* in a point or a range of the Argand diagram if *U and H fulfill the Cauchy-Riemann conditions* (4.2 - 12b) and, consequently, the *Laplace equation* (4.2 - 10b), respectively (4.2 - 8c). Therefore, it is possible to *describe the irrotational flow of an incompressible fluid without sources and sinks by the analytic complex function* (4.2 - 13a&b) by taking $U(x,y)$ as *potential* and $H(x,y)$ as *Hamiltonian.*

The potential U and the Hamiltonian H can be *exchanged* by a simple multiplication of the analytic function $f(z)$ with the imaginary unit i

$$f_{\mathrm{e}}(z) = if(z) = -H(x,y) - i\,U(x,y) = -U_{\mathrm{e}}(x,y) + i\,H_{\mathrm{e}}(x,y) \quad . \tag{4.2 - 13c}$$

An important aspect of analytic complex functions is *conformal mapping* [Betz 1964 B, Bieberbach 1986 B, Convay 1973 B, Hurwitz & Courant 1929 B, Ivanov & Trubetskov 1995 B, Nehari 1952 B, Titchmarsh 1932 B]. An analytic complex function $w = f(z)$ maps the $z = x + iy$ plane onto the $w = -U + i\,H$ plane without changing angles between crossing lines, except for the sign. Thus, in the $z = x + iy$

plane the equipotential lines $U = const$ cross the streamlines $H = const$ at right angles as described before. Since the streamlines are determined by $H = const$ the imaginary part H of the complex function (4.2 - 13b) acts as *stream function* as well as Hamiltonian. Conformal mapping is a common tool in fluid kinematics [Kober 1975 B, Prandtl & Tietjens 1957a&b B].

In our presentation the complex velocity $\upsilon_c = u + i\upsilon$ in the Argand diagram $z = x + iy$ is related to the corresponding analytic function $f(z)$ by

$$\upsilon_c = u + i\upsilon = 2df / dz^* \qquad (4.2 - 13c)$$

where the asterisk* indicates the complex conjugate. This is the result of the following calculation that makes use of (4.2 - 12bc) and (4.2 - 13b)

$$df = f_x dx + f_y dy = (-U_x + i H_x)dx + (-U_y + i H_y)dy$$

$$= (-U_x - U_y)dx + (-U_y + i U_x)dy = \frac{1}{2}(u + i\upsilon)(x - i\,dy) \quad .$$

c) Rectilinear flow

A simple application of conformal mapping is the complex representation of the rectilinear flow shown in Fig. 4.2 - 1. It is determined by the analytic complex function

$$w = f(z) = \frac{\upsilon_0}{2} z = -U + i H = \frac{\upsilon_0}{2} x + \frac{i\,\upsilon_0}{2} y \quad . \qquad (4.2 - 14a)$$

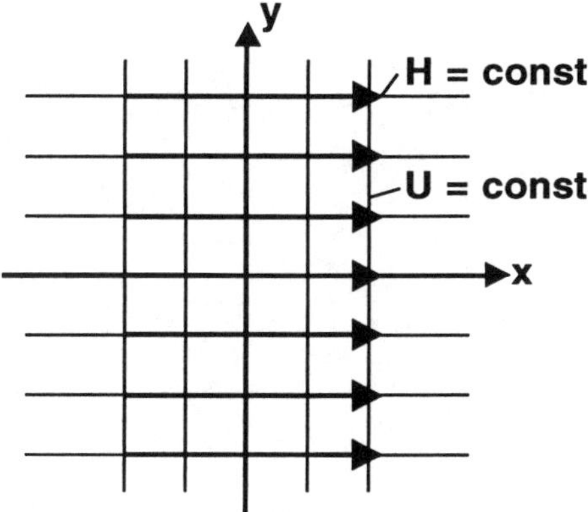

Fig. 4.2 - 1. Ideal rectilinear flow (4.2 - 14a-d)

According to (4.2 - 12c) this function is related to the *system*

$$\dot{x} = u = -2U_x(x,y) = +2H_y(x,y) = \upsilon_0$$
$$\dot{y} = \upsilon = -2U_y(x,y) = -2H_x(x,y) = 0 \qquad (4.2 - 14b)$$

with the *solutions*

$$x(t) = \upsilon_0 t + x_0 \quad \text{and} \quad y(t) = y_0 = const \quad . \tag{4.2 - 14c}$$

The *streamlines* are determined by the Hamiltonian H acting as stream function

$$y = 2H / c = y_0 = const \quad \text{because } H = const \tag{4.2 - 14d}$$

in accordance with the second equation of (4.2 - 14c).

d) Two-dimensional flow against a wall

As a second simple application of conformal mapping concerns the stationary two-dimensional flow against a plane wall. This flow, which is shown in Fig. 4.2 - 2, can be described by the analytic complex function

$$w = f(z) = \frac{z^2}{2\tau} = -U + i\,H = -\frac{1}{2\tau}(y^2 - x^2) + \frac{i}{\tau}xy \quad . \tag{4.2 - 15a}$$

According to (4.2 - 12c) this function corresponds to the system

$$\begin{aligned} \dot{x} &= x / \tau \\ \dot{y} &= -y / \tau \end{aligned} \tag{4.2 - 15b}$$

with the solution

$$x(t) = x_0\, exp\,(t / \tau) \quad \text{and} \quad y(t) = y_0\, exp\,(-t / \tau) \quad . \tag{4.2 - 15c}$$

The streamlines of this system are the hyperbolas

$$xy = H\tau = const \quad \text{because} \quad H = const \quad . \tag{4.2 - 15d}$$

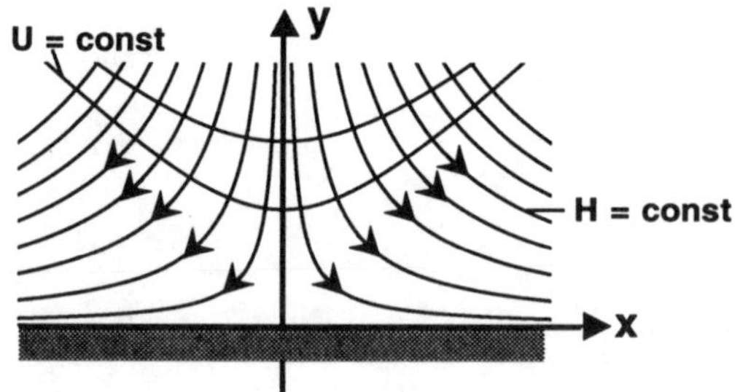

Fig. 4.2 - 2. Two-dimensional flow against a wall (4.2 - 15a-d)

e) Rectilinear source or sink of zero diameter

This section deals with a source or sink in the form of a straight line where the incompressible fluid exits or enters radially as shown in Fig. 4.2 - 3. If the straight line coincides with the real z axis the flow from this rectilinear source or sink can be represented with the aid of the analytic complex function

$$w = f(z) = \frac{Q}{4\pi} \ell n\, z = -U + i\, H = \frac{Q}{4\pi} \ell n\, r + \frac{i\, Q}{4\pi} \varphi \qquad (4.2 - 16a)$$

where r and φ indicate the polar coordinates defined by (4.1 - 4a&b) and by the *Euler relation*

$$z = x + i\, y = r\, exp(i\, \varphi) = r[cos\, \varphi + i\, sin\, \varphi] \quad . \qquad (4.2 - 16b)$$

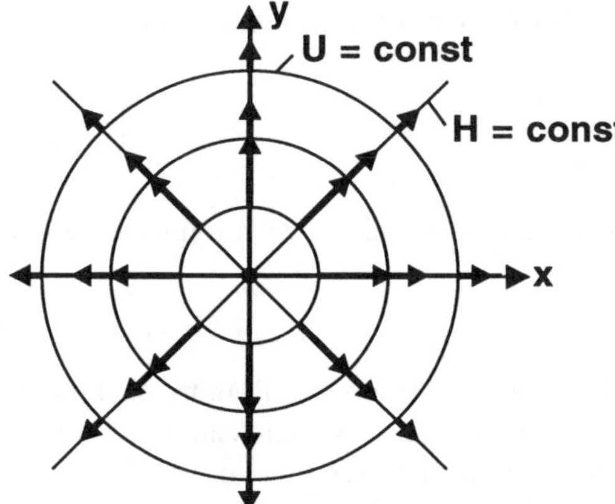

Fig. 4.2 - 3. Rectilinear source (4.2 - 16a-h)

According to (4.2 - 12d) the function (4.2 - 16a) corresponds to the following system in polar coordinates

$$\dot{r} = \upsilon_r = \frac{Q}{2\pi\, r} \quad \text{and} \quad r\dot{\varphi} = \upsilon_\varphi = 0 \quad . \qquad (4.2 - 16c)$$

The solution of this system, which determines the motion of an incompressible fluid at a rectilinear source or sink, is

$$r(t) = \left[r_0^2 + \frac{Qt}{\pi} \right]^{1/2} \quad \text{and} \quad \varphi = \varphi_0 = const \quad . \qquad (4.2 - 16d)$$

The *streamlines* of this flow are determined by

$$\varphi = 4\pi\, H \,/\, Q = \varphi_0 = const \quad \text{because} \quad H = const \qquad (4.2 - 16e)$$

and in accordance with the second equation of (4.2 - 16d).

The *strength* Q of the source or sink is given by the integral

$$\int_0^{2\pi} \upsilon_r \, r\, d\varphi = \int_0^{2\pi} \dot{r}\, r\, d\varphi = \int_0^{2\pi} \frac{Q}{2\pi r} r\, d\varphi = Q = const \quad . \tag{4.2 - 16f}$$

The *strength* Q is positive for a *source* and negative for a *sink*.

The application of the *two-dimensional Gauss law* to $\upsilon_r(r)$ yields

$$Q = \int_0^{2\pi} \upsilon_r(r)\, r d\varphi = 2\pi r\, \upsilon_r(r) = \int_0^r [div\, \vec{\upsilon}] 2\pi r dr = 2\pi \int_0^r q(r) r dr \quad . \tag{4.2 - 16g}$$

Since Q is a non-zero constant and (4.2 - 16g) is valid for all r the divergence and strength $q(r)$ takes the values

$$q(r>0) = 0 \quad \text{and} \quad q(r=0) = \pm\infty \quad . \tag{4.2 - 16h}$$

Consequently, the *divergence* of a rectilinear source or sink shows a *singularity* on its axis.

f) Rectilinear vortex

A rectilinear vortex has a straight axis as vortex filament [Prandtl & Tietjens 1957a&b B]. If this straight axis coincides with the real z axis one can represent this flow that is illustrated in Fig. 4.2 - 4 by the analytic complex function

$$w = f(z) = \frac{i\Gamma}{4\pi} \ell n\, z = -U + iH = -\frac{\Gamma\varphi}{4\pi} + i\frac{\Gamma}{4\pi} \ell n\, r \quad , \tag{4.2 - 17a}$$

where r and φ indicate the polar coordinates and Γ designates the *circulation* or *vortex strength*. The comparison shows that this analytic function equals the analytic function (4.2 - 16a) of the rectilinear source or sink multiplied by the imaginary unit i. According to (4.2 - 13c) the corresponding potentials U and Hamiltonians H are exchanged as well as the equipotential lines and the streamlines.

From (4.2 - 12d) it can be concluded that the analytic function (4.2 - 17a) is equivalent to the following system in polar coordinates

$$\dot{r} = \upsilon_r = 0 \quad \text{and} \quad r\dot{\varphi} = \upsilon_\varphi = \frac{\Gamma}{2\pi r} \quad . \tag{4.2 - 17b}$$

The solution of this system that describes the flow of an incompressible fluid around a rectilinear vortex is

$$r = r_0 = const \quad \text{and} \quad \varphi = \varphi_0 + \frac{\Gamma}{2\pi r_0^2} t \quad . \tag{4.2 - 17c}$$

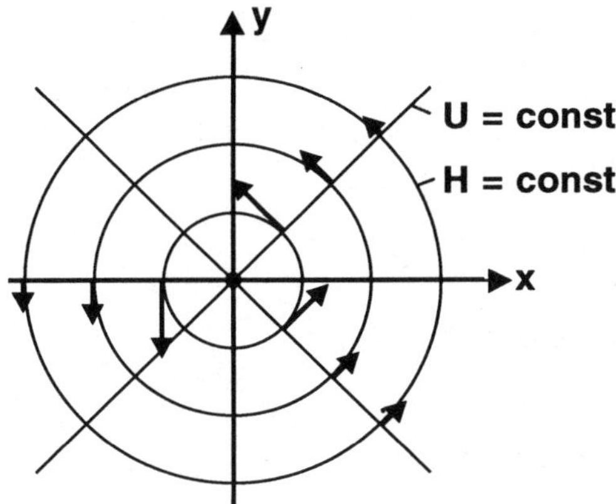

Fig. 4.2 - 4. Rectilinear vortex (4.2 - 17a-g)

The *streamlines* of this flow fulfill the equation

$$r = exp(4\pi H / \Gamma) = r_0 = const \quad \text{because} \quad H = const \tag{4.2 - 17d}$$

in accordance with the first equation of (4.2 - 17c).

The *vortex strength* or *circulation* Γ is given by the integral

$$\int_0^{2\pi} \upsilon_\varphi r \, d\varphi = \int_0^{2\pi} r\dot{\varphi} \, r d\varphi = \int_0^{2\pi} \frac{\Gamma}{2\pi r} r d\varphi = \Gamma = const \quad . \tag{4.2 - 17e}$$

The sign of Γ determines the direction of circulation.

The application of the *two-dimensional Stokeslaw* to $\upsilon_\varphi(r)$ yields

$$\Gamma = \int_0^{2\pi} \upsilon_\varphi(r) r d\varphi = 2\pi r \upsilon_\varphi(r) = \int_0^{r} [curl\, \vec{\upsilon}]_z 2\pi r dr = 4\pi \int_0^{r} \omega(r) r dr \quad . \tag{4.2 - 17f}$$

Since Γ is a non-zero constant and this equation is valid for all r the rotation $\omega(r)$ takes the values

$$\omega(r > 0) = 0 \quad \text{and} \quad \omega(r = 0) = \pm\infty \quad . \tag{4.2 - 17g}$$

In conclusion, the *rotation* of a rectilinear vortex exhibits a *singularity on the axis.*

g) Flow around a straight circular cylinder

The flow around a straight circular cylinder of radius R shown in Fig. 4.2 - 5 can be represented by the analytic complex function [Prandtl & Tietjens 1957a&b B]

$$w = f(z) = \frac{u_0}{2}\left[z + \frac{R^2}{z}\right]$$
$$= -U + iH = \frac{u_0}{2} r\,cos\varphi\left[1 + \frac{R^2}{r^2}\right] + i\frac{u_0}{2} r\,sin\varphi\left[1 - \frac{R^2}{r^2}\right] \quad . \qquad (4.2 - 18a)$$

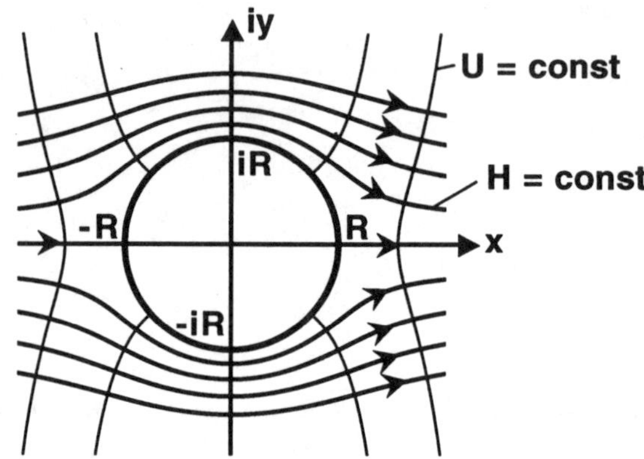

Fig. 4.2 - 5. Flow around a straight circular cylinder (4.2 - 18a-g)

This function is related to the *Joukowski transformation* [Zwillinger 1989 B]

$$s = p + iq = \frac{1}{2}\left(z + \frac{R^2}{z}\right) = \frac{R}{2}\left(\frac{z}{R} + \frac{R}{z}\right) \quad . \qquad (4.2 - 18b)$$

It maps the circle of radius R with the center in the origin of the $z = x + iy$ plane onto the straight line of length $2R$ on the real p axis of the $s = p + iq$ plane from $s = -R$ to $s = +R$ as follows

$$\begin{aligned} &\text{circle in } z \text{ plane:} \quad z = R\,exp(i\varphi) \quad , \\ &\text{line in } s \text{ plane:} \quad s = p = R\,cos\varphi \quad . \end{aligned} \qquad (4.2 - 18c)$$

The *flow around the circular cylinder* of radius R is well manifested by the *streamlines* defined by the zero Hamiltonian $H = 0$ of (4.2 - 18a)

$$\begin{aligned} &H = 0 \quad , \\ &y = 0, z = x \quad \text{for} \quad x^2 \geq R^2 \quad \text{and} \quad r = R, z = Re^{i\varphi} \quad \text{for} \quad x^2 \leq R^2 \quad . \end{aligned} \qquad (4.2 - 18d)$$

The components of the velocity $\vec{v}$ on the x axis outside the cylinder where $x^2 > R^2$ are

$$u(z=x) = -2U_x(z=x) = u_0\left[1 - R^2x^2(x^2+y^2)^{-2}\right]$$
$$\upsilon(z=x) = 0 \quad . \tag{4.2 - 18e}$$

Thus, the velocity of the flow far from the cylinder is $u(z=\pm\infty)=u_0$. The flow comes to a *stop* at the intersections of the x axis with the cylinder because $u(z=\pm R)=0$. The transverse velocity υ_φ on the cylinder is determined by

$$\upsilon_\varphi(z=Re^{i\varphi}) = -\frac{2}{R}U_\varphi(R,\varphi) = -2u_0 sin\varphi \quad . \tag{4.2 - 18f}$$

The *maximal velocity* $2u_0$ is reached at the intersections $z=\pm iR$ of the y axis with the cylinder according to

$$u(\pm iR) = \mp\upsilon_\varphi(\pm iR) = 2u_0 \quad . \tag{4.2 - 18g}$$

This velocity is parallel to the x-axis.

h) Flow around a plate at right angles

The flow around a plane plate of width $2R$ at right angles to a stream of velocity u_0 shown in Fig. 4.2 - 6 is represented by the analytic complex function $w=f(z)$ defined by the relation

$$w^2 = \frac{1}{4}u_0^2(R^2+z^2)$$
$$\text{with} \quad z = x+iy \quad \text{and} \quad w = f(z) = -U+iH \quad . \tag{4.2 - 19a}$$

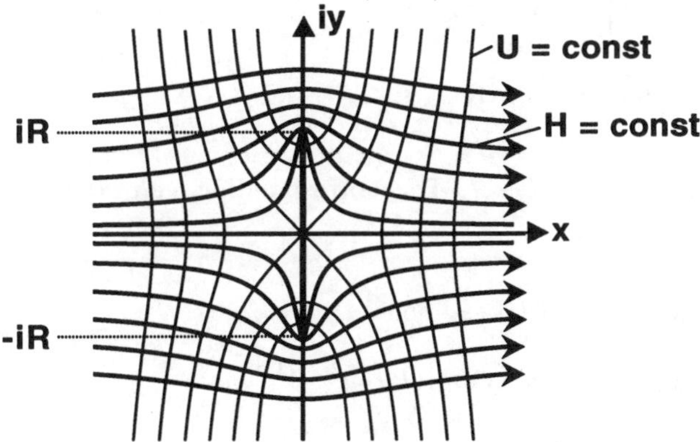

Fig. 4.2 - 6. Flow around a plate at right angles (4.2 - 19a-h)

The *streamlines* with $H = const$ and the *equipotential lines* with $U = const$ of this flow can be evaluated by the separation of real and imaginary part of (4.2 - 19a). These streamlines and trajectories are determined by

$$y = y(x,H) = \pm(2H/u_0)\left[\frac{R^2 + x^2 + (2H/u_0)^2}{x^2 + (2H/u_0)^2}\right]^{1/2} \quad , \tag{4.2 - 19b}$$

whilst the equipotential lines are given by

$$y = y(x,U) = \pm(2U/u_0)\left[\frac{R^2 + x^2 - (2U/u_0)^2}{(2U/u_0)^2 - x^2}\right]^{1/2} \quad . \tag{4.2 - 19c}$$

In order to discuss the flow (4.2 - 19a) in more detail one considers the complex function $w = f(z)$ on the real x axis and on the imaginary y axis

$$\begin{aligned} w(z = x) &= -U(x,0) + iH(x,0) = \pm\frac{1}{2}u_0\left[R^2 + x^2\right]^{1/2} \quad \text{and} \\ w(z = iy) &= -U(0,y) + iH(0,y) = \pm\frac{1}{2}u_0\left[R^2 - y^2\right]^{1/2} \quad . \end{aligned} \tag{4.2 - 19d}$$

The *streamline* defined by the zero Hamiltonian $H = 0$ includes the x axis as well as the straight line of length $2R$ on the y axis

$$y = 0 \quad \text{and} \quad x = 0 \quad , \quad y^2 \le R^2 \quad \text{for} \quad H = 0 \quad . \tag{4.2 - 19e}$$

According to equation (4.2 - 19d) the *potential* U on this streamline fulfills the equations

$$\begin{aligned} U(x,0) &= -sign\, x\frac{1}{2}u_0\left[R^2 + x^2\right]^{1/2} \quad \text{and} \\ U(\varepsilon \to 0, y) &= -sign\, \varepsilon\frac{1}{2}u_0\left[R^2 - y^2\right]^{1/2} \quad \text{for} \quad y^2 \le R^2 \quad . \end{aligned} \tag{4.2 - 19f}$$

This potential is illustrated in Fig. 4.2 - 7. The corresponding *velocities* are determined by

$$\begin{aligned} u(x,0) &= -2U_x(x,0) = +u_0\left[(R/x)^2 + 1\right]^{-1/2} \quad \text{and} \\ \upsilon(\varepsilon \to 0, y) &= -2U_y(\varepsilon \to 0, y) = -sign(\varepsilon y)\, u_0\left[(R/y)^2 - 1\right]^{-1/2} \quad . \end{aligned} \tag{4.2 - 19g}$$

Consequently, the velocity far from the plate is $u(\pm\infty, 0) = u_0$. It is zero at both sides of the intersection of the x-axis with the plate since $u(0,0) = \upsilon(\varepsilon \to 0,0) = 0$. At the sharp edges $z = \pm iR$ of the plate the velocity grows to infinity according to

$$\upsilon(\varepsilon \to 0, y = \pm R) = -sign(\varepsilon y) \cdot \infty \quad . \tag{4.2 - 19h}$$

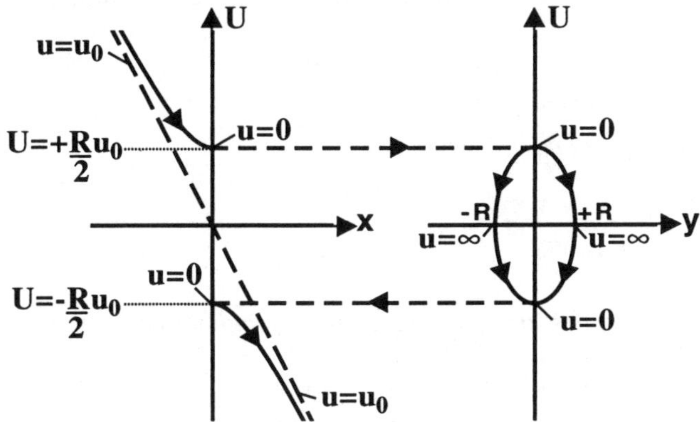

Fig. 4.2 - 7. Potential U of the flow around a plate at right angles for $H = 0$ according to (4.2 - 19f)

i) Borda mouth

The flow of water through the mouth of a straight canal which enters a lake or sea can be simulated with the Borda model mouth [Kober 1957 B, Prandtl & Tietjens 1957a&b B] illustrated in Fig. 4.2 - 8. If b is the width of the canal and u_0 the velocity of the flow in the canal far from the mouth, the Borda model mouth is described by the analytic complex function $w = f(z)$ defined by the relation

$$z = \frac{2}{u_0} w + \frac{b}{2\pi} exp\left[\frac{4\pi}{bu_0} w\right] \tag{4.2 - 20a}$$

with $z = x + iy$ and $w = f(z) = -U + iH$.

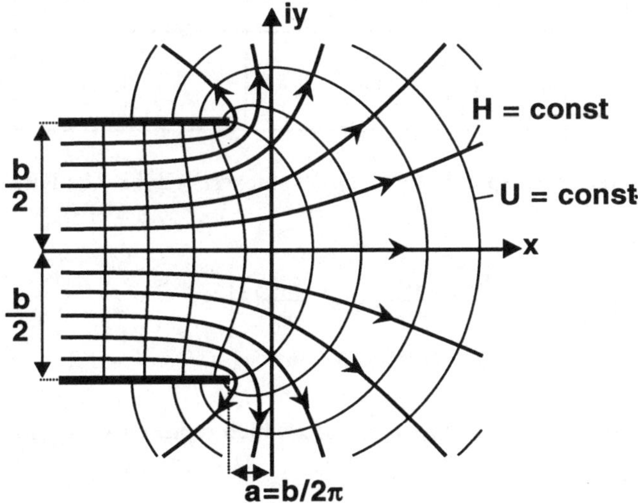

Fig. 4.2 - 8. Flow through the Borda mouth (4.2 - 20a-g)

This equation relates the coordinates x and y to the potential U and the Hamiltonian H as follows

$$x = -\frac{2}{u_0}U + \frac{b}{2\pi}exp\left[-\frac{4\pi}{bu_0}\right]cos\frac{4\pi}{bu_0}H \quad \text{and}$$

$$y = \frac{2}{u_0}H + \frac{b}{2\pi}exp\left[-\frac{4\pi}{bu_0}U\right]sin\frac{4\pi}{bu_0}H \quad . \tag{4.2 - 20b}$$

In order to understand this flow one has to consider specific *streamlines* characterized by a constant Hamiltonian H:

α) The *center line of the canal* corresponds to the zero Hamiltonian

$$H(x,y) = H = 0 \quad \begin{cases} x(U) = -\frac{2}{u_0}U + \frac{b}{2\pi}exp\left[-\frac{4\pi}{bu_0}U\right] \\ y(U) = 0 \end{cases} \quad . \tag{4.2 - 20c}$$

Partial differentiation with respect to x gives the velocity u on the center line $y = 0$ as a function of the potential U

$$u(U) = -2U_x = u_0\left[1 + exp(-\frac{4\pi}{bu_0}U)\right]^{-1} \quad . \tag{4.2 - 20d}$$

These equations yield the following velocities on the center line $z = x$ of the canal:

y	x	u	U
0	$-\infty$	u_0	$+\infty$
0	b/2	$u_0/2$	0
0	$+\infty$	0	$-\infty$

β) The two *dams of the canal* correspond to the two values $H = \pm bu_0/4$ of the Hamiltonian according to

$$H(x,y) = H = \pm\frac{bu_0}{4} \quad \begin{cases} x(U) = -\frac{2}{u_0}U - \frac{b}{2\pi}exp\left[-\frac{4\pi}{bu_0}U\right] \\ y(U) = \pm b/2 \end{cases} \quad . \tag{4.2 - 20e}$$

Differentiation with respect to U yields the maximum of x which represents the ends of the two dams

$$x = x_{max} = x(U = 0) = -b/2\pi = -a \quad . \tag{4.2 - 20f}$$

The partial differentiation of $x(U)$ of (4.2 - 20e) with respect to x gives the velocities $u(z = x \pm ib/2)$ at the two dams as functions of U

$$u(U) = -2U_x = u_0 \left[1 - exp\left(-\frac{4\pi}{bu_0}U\right)\right]^{-1} \quad . \qquad (4.2 - 20g)$$

Thus, one finds the following potentials U and velocities u inside and outside of the dams at $z = x \pm ib/2$.

	y	x	u	U
inside	$\pm$ b/2	$-\infty$	u_0	$+\infty$
inside	$\pm$ b/2	$-(1+e^{-1})a$	$(1-e^{-1})^{-1}u_0$	$+u_0a/2$
ends	$\pm$ b/2	$x_{max} = -a$	∞	0
outside	$\pm$ b/2	$-(e-1)a$	$-(e^{-1})^{-1}u_0$	$-u_0a/2$
outside	$\pm$ b/2	$-\infty$	0	$-\infty$

The potential U and the velocity u are positive on the inner sides of the dams, whereas they are negative on the outer sides. At the ends of the dams, which form sharp edges, the velocity u reaches infinity.

Finally, it should be mentioned that the Borda model mouth corresponds to the *Rogowski profile of high-voltage electrical engineering.* W. Rogowski considered a vacuum capacitor made of two electrodes whose profiles correspond to the streamlines of the Borda model mouth defined by a constant Hamiltonian H. In this case the lines with a constant potential U represent the field lines of the static electric field $\vec{E}$ between the two electrodes.

4.2.5 Hamiltonian Systems

By definition Hamiltonian systems are characterized by a *zero potential*

$$U = U(x, y) = U(r, \varphi) = 0 \quad . \qquad (4.2 - 11a)$$

Therefore, they possess neither sources nor sinks according to

$$div\, \vec{\upsilon} = q = u_x + \upsilon_y = \upsilon_{r,r} + r^{-1}\upsilon_r + r^{-1}\upsilon_{\varphi,\varphi} = -\Delta U = 0 \quad . \qquad (4.2 - 10a)$$

The zero potential $U = 0$ implies the following *standard representation* of two-dimensional Hamiltonian systems in Cartesian coordinates

$$\dot{x} = u(x, y) = +H_y(x, y)$$
$$\dot{y} = \upsilon(x, y) = -H_x(x, y) \qquad (4.2 - 21a)$$

and in polar coordinates

$$\begin{aligned} \dot{r} &= \upsilon_{\mathrm{r}}(r,\varphi) = r^{-1}H_{\varphi}(r,\varphi) \\ r\dot{\varphi} &= \upsilon_{\varphi}(r,\varphi) = -H_{\mathrm{r}}(r,\varphi) \quad . \end{aligned} \tag{4.2 - 21b}$$

The *streamlines* and *trajectories* of Hamiltonian systems coincide with the lines defined by a constant Hamiltonian

$$H(x,y) = H = const \quad . \tag{4.2 - 11i}$$

Thus, the Hamiltonians act as *stream functions.*

The local rotation $\omega(\vec{r})$ at a position $\vec{r}$ of a Hamiltonian system is related to the corresponding Hamiltonian H by the *Poisson equation*

$$-[curl\,\vec{\upsilon}]_{\mathrm{z}} = -2\omega(\vec{r}) = +\Delta H(\vec{r}) \quad . \tag{4.2 - 3c}$$

The *Stokes law* in two dimensions relates the local rotation $\omega(\vec{r})$ with the *circulation* $\Gamma(C)$ on a closed curve C that encloses the area A of the xy plane as follows

$$\begin{aligned} \Gamma(C) &= \oint_{\mathrm{C}} \vec{\upsilon}\cdot d\vec{r} = \oint_{\mathrm{C}} (u dx + \upsilon dy) = \int_{\mathrm{A(C)}} [curl\,\vec{\upsilon}]_{\mathrm{z}}\, dA \\ &= 2 \int_{\mathrm{A(C)}} \omega(x,y) dx\, dy \quad . \end{aligned} \tag{4.2 - 22}$$

The fixed *singular points* $\vec{r}_{\mathrm{S}}$ of a two-dimensional Hamiltonian system (4.2 - 21a) are either *vortex or saddle points*. These singular points are discussed in Section 4.3, listed in Table 4.3 - 1 and illustrated in Fig. 4.3 - 3. This statement can be proved by the linearization of the Hamiltonian systems (4.2 - 21a) at their singular points $\vec{r}_{\mathrm{S}}$ as described in Section 4.1.4. The relevant Jacobi matrix (4.1 - 10c) has the form

$$\boldsymbol{J} = \begin{bmatrix} u_{\mathrm{x}} & u_{\mathrm{y}} \\ \upsilon_{\mathrm{x}} & \upsilon_{\mathrm{y}} \end{bmatrix} = \begin{bmatrix} H_{\mathrm{xy}} & H_{\mathrm{yy}} \\ -H_{\mathrm{xx}} & -H_{\mathrm{xy}} \end{bmatrix} \quad \text{for} \quad x = x_{\mathrm{S}}, y = y_{\mathrm{S}} \quad . \tag{4.2 - 23a}$$

Its eigenvalues are

$$\Lambda(\boldsymbol{J}) = \lambda_{1,2} = \pm[-det\,\boldsymbol{J}]^{1/2} = \pm\left[H_{\mathrm{xy}}^2 - H_{\mathrm{xx}}H_{\mathrm{yy}}\right]^{1/2} \quad . \tag{4.2 - 23b}$$

They indicate a

$$\begin{aligned} &\textit{vortex point}, \quad \text{if } det\,\boldsymbol{J} > 0 \quad , \\ &\textit{saddle point}, \quad \text{if } det\,\boldsymbol{J} < 0 \quad . \end{aligned} \tag{4.2 - 23c}$$

The existence of vortex points implies *periodic solutions* of the two-dimensional autonomous Hamiltonian systems (4.2 - 21a) with closed trajectories and streamlines defined by the condition (4.2 - 11i) that requires $H = const.$

Since the periodic solutions of a two-dimensional autonomous Hamiltonian system form closed trajectories in the phase space (x, y) with $H = const$, one can calculate the area I in the phase space enclosed by a trajectory as a function of H as follows

$$I = I(H) = \oint_{H} y dx \quad . \tag{4.2 - 24}$$

This area I determines the period $T(H)$ of the motion corresponding to the trajectory defined by $H = const$ by the relation [Plaschko & Brod 1995 B]

$$T = T(H) = \frac{d}{dH} I(H) \quad . \tag{4.2 - 24b}$$

In the following section these equations are applied to the pure rotation for illustration.

a) Pure rotation

The motion of a merry-go-round represents a pure rotation around a fixed vertical axis. The pure rotation with a constant circular frequency ω around the z axis is illustrated in Fig. 4.2 - 9. It obeys the *vector equation*

$$\vec{\upsilon} = \vec{\omega} \times \vec{r} \quad \text{with} \quad \vec{\upsilon} = [u, \upsilon, 0]; \quad \vec{\omega} = [0, 0, \omega]; \quad \vec{r} = [x, y, z] \quad . \tag{4.2 - 25a}$$

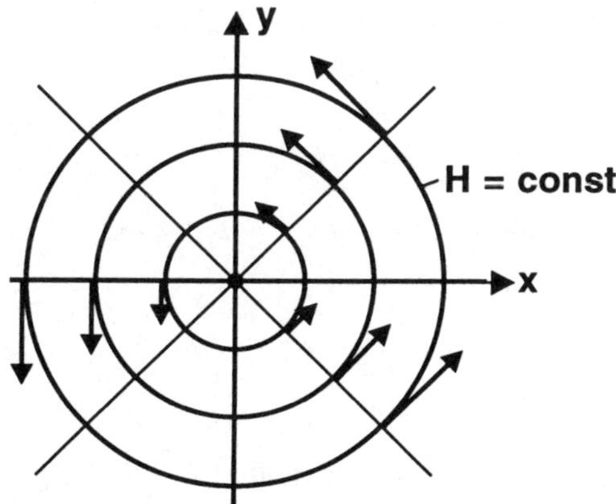

Fig. 4.2 - 9. Pure rotation (4.2 - 25a-i)

Therefore, it constitutes a system that can be written in Cartesian coordinates as

$$\dot{x} = u(x,y) = +H_y(x,y) = -\omega y$$
$$\dot{y} = \upsilon(x,y) = -H_x(x,y) = +\omega x \qquad (4.2 - 25b)$$

and in polar coordinates as

$$\dot{r} \ = \upsilon_r(r,\varphi) = 0 \quad \text{or} \quad r = r_0 = const$$
$$r\dot{\varphi} = \upsilon_\varphi(r,\varphi) = \omega r \quad . \qquad (4.2 - 25c)$$

Therefore, the pure rotation around the z axis can be defined by the *Hamiltonian*

$$H = -\frac{\omega}{2} r^2 = -\frac{\omega}{2}\left(x^2 + y^2\right) \qquad (4.2 - 25d)$$

and the *zero potential* $U = 0$.

The *streamlines* and *trajectories* of the pure rotation are the lines determined by a constant Hamiltonian. This implies

$$r^2 = -2H / \omega = r_0^2 = const \quad . \qquad (4.2 - 25e)$$

Thus, the streamlines form concentric circles in agreement with the first equation of (4.2 - 25c).

Consequently, the area *I*(*H*) of the trajectory with $H = const$ in the phase space (*x*, *y*) is according to (4.2 - 24a)

$$I(H) = -\pi r_0^2 = 2\pi H / \omega \quad . \qquad (4.2 - 25f)$$

The period *T*(*H*) of this trajectory can be deduced from (4.2 - 24b). The result is

$$T(H) = \frac{d}{dH} I(H) = 2\pi / \omega = const \qquad (4.2 - 25g)$$

in agreement with the fact that the *local rotation* $\omega(\vec{r})$ *does not depend on the position* $\vec{r}$

$$\omega(\vec{r}) = -\frac{1}{2}\Delta H = \omega = const \quad . \qquad (4.2 - 25h)$$

Furthermore, it should be mentioned that due to the zero potential $U = 0$ the pure rotation has *neither sources nor sinks* according to

$$div\ \vec{\upsilon} = q = -\Delta U = 0 \quad . \qquad (4.2 - 10a)$$

The solution of the system (4.2 - 25a&b) is

$$x(t) = r_0 \cos(\omega t - \alpha) \quad \text{and} \quad y(t) = r_0 \sin(\omega t - \alpha) \qquad (4.2\text{ - }25i)$$

with r_0 and α arbitrary. These equations demonstrate that the *projection* of a pure rotation on a straight line through the origin represents a *harmonic oscillation.*

b) Shear flow

The shear flow on a wall illustrated in Fig. 4.2 - 10 is relevant in *fluid dynamics* with respect to *Prandtl's boundary layer* [Anderson 1988 B, Prandtl & Tietjens 1957b B]. The simple shear flow in the x-direction can be represented by the following system in Cartesian coordinates

$$\begin{aligned} \dot{x} &= u(x,y) = +H_y(x,y) = u_0 + \alpha y \\ \dot{y} &= \upsilon(x,y) = -H_x(x,y) = 0 \quad . \end{aligned} \qquad (4.2\text{ - }26a)$$

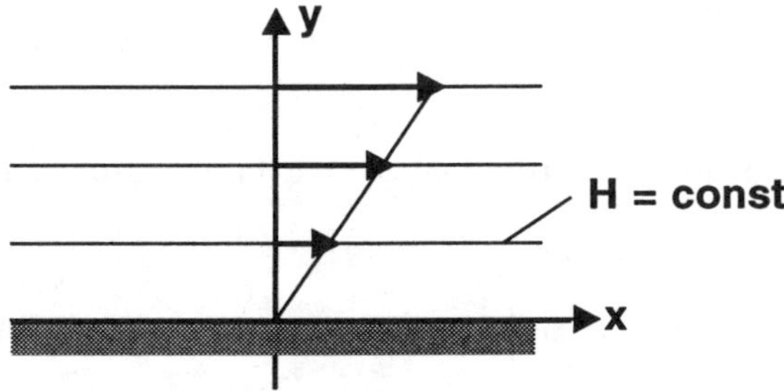

Fig. 4.2 - 10. Shear flow (4.2 - 26a-d)

In Fig. 4.2 - 10 the velocity u_0 on the wall is zero. As the pure rotation the shear flow implies a *local rotation* $\omega(\vec{r})$ *independent of the position* $\vec{r}$

$$\omega(\vec{r}) = \frac{1}{2}\left(\upsilon_x - u_y\right) = -\alpha/2 = \omega = const \quad . \qquad (4.2\text{ - }26b)$$

The shear flow corresponds to a Hamiltonian system and thus possesses *neither sources nor sinks* as manifested by

$$div\,\vec{\upsilon} = q = u_x + \upsilon_y = 0 \quad . \qquad (4.2\text{ - }10a)$$

According to the (4.2 - 26a&b) the shear flow can be described by the zero potential $U = 0$ and by the Hamiltonian

$$H = H(y) = u_0 y + \frac{1}{2}\alpha y^2 = u_0 y - \omega y^2 \quad . \qquad (4.2\text{ - }26c)$$

The solution of the system (4.2 - 26a) that represents the shear flow is

$$x(t) = (u_0 + \alpha y_0)t \quad \text{and} \quad y(t) = y_0 \tag{4.2 - 26d}$$

with arbitrary y_0 and u_0.

c) Circular vortex flows and related oscillations

Two-dimensional circular vortex flows can be described by Hamiltonian systems with a zero potential and an axially symmetric Hamiltonian H as follows:

$$U = 0 \quad \text{and} \quad H = H(r, \varphi) = H(r) \tag{4.2 - 27a}$$

where r and φ indicate the polar coordinates. By taking account of equations (4.2 - 21b) one finds the following *standard representation* of the corresponding systems of equations

$$\begin{aligned} \dot{r} &= \upsilon_r(r) = 0 \quad \text{or} \quad r = r_0 = const \\ r\dot{\varphi} &= \upsilon_\varphi(r) = H_r(r) \quad . \end{aligned} \tag{4.2 - 27b}$$

The corresponding local variation of the *local rotation* $\omega(\vec{r})$ exhibits also axial symmetry

$$\omega(\vec{r}) = \omega(r) = -\frac{1}{2}\Delta H = -\frac{1}{2}\left[H_{rr}(r) + r^{-1}H_r(r)\right] \quad . \tag{4.2 - 27c}$$

The *local rotation* $\omega(r)$ is related to the circulation $\Gamma(C)$ on a circle of radius r with the center at the origin $r = 0$ according to *Stokes' law* (4.2 - 22)

$$\Gamma(C) = 2\pi\, r\, \upsilon_\varphi(r) = 4\pi \int_0^r \omega(r) r dr \quad . \tag{4.2 - 27d}$$

This equation implies the following relations between $\upsilon_\varphi(r)$ and $\omega(r)$

$$r\dot{\varphi} = \upsilon_\varphi(r) = -H_r(r) = 2\, r^{-1} \int_0^r \omega(r) r dr \quad \text{with} \tag{4.2 - 27e}$$

$$\omega(r) = \frac{1}{2r} \frac{d}{dr}\left[r\, \upsilon_\varphi(r)\right] \quad . \tag{4.2 - 27f}$$

Thus, one can evaluate the transverse velocity $\upsilon_\varphi(r)$ for a given local variation $\omega(r)$ of the local rotation and vice versa.

The *oscillations* related to the circular vortex flows correspond to the periodic solutions of (4.2 - 27b). They can be derived from (4.2 - 27e) and written as

$$\left.\begin{matrix} x(t) \\ y(t) \end{matrix}\right\} = r \left\{\begin{matrix} cos \\ sin \end{matrix}\right. \left[2r^{-2} \int_0^r \omega(r) r\, dr \cdot t - \alpha \right] \tag{4.2 - 27g}$$

where α is an arbitrary phase. Thus, the *period* $T(r)$ of the oscillation depends on its amplitude r according to

$$T(r) = \pi r^2 \left[\int_0^r \omega(r) r\, dr \right]^{-1} = 2\pi r / \upsilon_\varphi(r) \quad . \tag{4.2 - 27h}$$

The inverse $r(H)$ of the Hamiltonian function $H(r)$ of (4.2 - 27a) permits to calculate the period T of a circular vortex flow as a function of H with the aid of (4.2 - 24a&b) as follows

$$T(H) = -2\pi r(H) \frac{d}{dH} r(H) \quad . \tag{4.2 - 27i}$$

Typical examples of Hamiltonian systems with axially symmetric Hamiltonians are discussed in the following.

d) Vortex filament of finite radius

The first example is the vortex flow around a *rectilinear vortex filament of finite radius* r_0. This filament is defined by the following radial variation $\omega(r)$ of the local rotation

$$\omega(r \le r_0) = \omega = const \quad \text{and} \quad \omega(r > r_0) = 0 \quad . \tag{4.2 - 28a}$$

The application of (4.2 - 27e) results in the following radial variation of the transverse velocity

$$\upsilon_\varphi(r \le r_0) = \omega r \quad \text{and} \quad \upsilon_\varphi(r \ge r_0) = \omega r_0^2 \cdot r^{-1} \quad . \tag{4.2 - 28b}$$

The corresponding axially symmetric *Hamiltonian is* described by

$$\begin{aligned} &H(r \le r_0) = -\frac{1}{2}\omega r^2 \quad \text{and} \\ &H(r > r_0) = -\frac{1}{2}\omega r_0^2 [1 + 2\ell n(r / r_0)] \quad . \end{aligned} \tag{4.2 - 28c}$$

The vortex filament (4.2 - 28a) implies the oscillations

$$\left.\begin{matrix} x(t) \\ y(t) \end{matrix}\right\} = r \left\{\begin{matrix} cos(\omega t - \alpha) \\ sin(\omega t - \alpha) \end{matrix}\right. \qquad \text{for } r \leq r_0 \quad ,$$

$$\left.\begin{matrix} x(t) \\ y(t) \end{matrix}\right\} = r \left\{\begin{matrix} cos \\ sin \end{matrix}\right. \left[\omega(r_0 / r)^2 t - \alpha\right] \text{ for } r \geq r_0 \quad . \tag{4.2 - 28d}$$

e) Kepler vortex

The *Kepler vortex* as second example is defined by the zero potential $U = 0$ and the axially symmetric *Hamiltonian*

$$H = H(r) = -C(r / r_0)^{1/2} \quad . \tag{4.2 - 29a}$$

The corresponding transverse velocity is according to (4.2 - 27e)

$$r\dot{\varphi} = \upsilon_\varphi(r) = (C / 2r_0)(r / r_0)^{-1/2} \quad . \tag{4.2 - 29b}$$

This equation in combination with (4.2 - 27f) yields the *local rotation*

$$\omega(r) = (C / 8r_0^2)(r / r_0)^{-3/2} \quad . \tag{4.2 - 29c}$$

The period $T(r)$ of the oscillation given by (4.2 - 27h) fulfills *Kepler's law*

$$T^2(r) = 16\pi^2 r_0 C^{-2} r^3 \quad . \tag{4.2 - 29d}$$

f) Rectilinear vortex as Hamiltonian system

The third example is the *rectilinear vortex* already introduced in Section 4.2.4f. This vortex can also be represented as a *pure Hamiltonian system* with a zero potential $U = 0$ and the axially symmetric *Hamiltonian*

$$H = H(r) = \frac{\Gamma}{2\pi} \ell n\, r \quad . \tag{4.2 - 30a}$$

Its *transverse velocity* can be derived with the help of (4.2 - 21b)

$$r\dot{\varphi} = \upsilon_\varphi(r) = \frac{\Gamma}{2\pi r} \quad . \tag{4.2 - 30b}$$

This equation corresponds to (4.2 - 17b). The rectilinear vortex represents the *oscillation*

$$\left.\begin{matrix} x(t) \\ y(t) \end{matrix}\right\} = r \left\{\begin{matrix} cos \\ sin \end{matrix}\right. \left[(\Gamma / 2\pi r^2)t - \alpha\right] \tag{4.2 - 30c}$$

with an arbitrary phase α and a *period* T determined by (4.2 - 27h) and (4.2 - 24a&b)

$$T(r) = (2\pi r)^2 \Gamma^{-1} = (2\pi)^2 \Gamma^{-1} exp(4\pi H / \Gamma) \quad . \tag{4.2 - 30d}$$

The application of (4.2 - 27f) to (4.2 - 30b) yields a zero angular velocity

$$\omega(r > 0) = 0 \quad . \tag{4.2 - 30e}$$

In summary, the rectilinear vortex has a *complex representation* (4.2 - 17a) with a non-zero potential U and a Hamiltonian H which fulfill the Riemann-Cauchy conditions (4.2 - 12b) as well as a *real representation* with a zero potential $U = 0$ and the non-zero Hamiltonian (4.2 - 30a).

4.2.6 Gradient Systems and Flows

Gradient systems and flows are by definition characterized by a *zero Hamiltonian H*

$$H = H(x, y) = H(r, \varphi) = 0 \quad . \tag{4.2 - 9a}$$

As a consequence these systems and flows are *irrotational*

$$\left[curl\, \vec{\upsilon}\right]_z = 2\omega = -\Delta H = 0 \quad . \tag{4.2 - 8a}$$

Therefore, they can be represented in the vector notation by

$$\vec{\upsilon}(\vec{r}) = -grad\, U(\vec{r}) \quad , \tag{4.2 - 9b}$$

in Cartesian coordinates by

$$\begin{aligned} \dot{x} &= u = -U_x(x, y) \\ \dot{y} &= \upsilon = -U_y(x, y) \end{aligned} \tag{4.2 - 9c}$$

and in polar coordinates by

$$\begin{aligned} \dot{r} \;\; &= \;\; \upsilon_r = -U_r(r, \varphi) \\ r\dot{\varphi} &= \upsilon_\varphi = -r^{-1} U_\varphi(r, \varphi) \quad . \end{aligned} \tag{4.2 - 9d}$$

The *local source and sink strength* q of a gradient field is given by

$$div\, \vec{\upsilon} = q = -\left(U_{xx} + U_{yy}\right) = -U_{rr} - r^{-1} U_r - r^{-2} U_{\varphi\varphi} = -\Delta U \quad . \tag{4.2 -3a}$$

Thus q and U are related by a *Poisson equation.*

According to (4.2 - 9f) the streamlines $\vec{r}(t)=[x(t),y(t),0]$ are *perpendicular to the equipotential lines* which are defined by

$$U(x,y)=U=const \qquad (4.2 - 9h)$$

because

$$d\vec{r}\parallel grad\,U \quad \text{and} \quad d\vec{r}\perp(U=const) \quad . \qquad (4.2 - 9g)$$

a) Singular points

The stationary *singular points* $\vec{r}_S$ of the two-dimensional autonomous gradient systems (4.2 - 9c) are either *nodes or saddle points.* These are described in Section 4.3, listed in Table 4.3 - 1 and plotted in Fig. 4.3 - 3. These systems have neither spirals nor vortex points. Thus, they show *no periodic solutions.* These statements can be proved by the linearization of the gradient systems at their singular points $\vec{r}_S$ as described in Section 4.1.4. The relevant Jacobi matrix ***J*** (4.1 - 10c) is

$$\boldsymbol{J}=\begin{bmatrix} u_x & u_y \\ \upsilon_x & \upsilon_y \end{bmatrix}=-\begin{bmatrix} U_{xx} & U_{xy} \\ U_{xy} & U_{yy} \end{bmatrix} \quad \text{with} \quad x=x_S,\, y=y_S \quad . \qquad (4.2 - 31a)$$

This Jacobi matrix ***J*** has *only real eigenvalues*

$$\Lambda(\boldsymbol{J})=\lambda_{1,2}(\boldsymbol{J})=-\frac{1}{2}\Delta U\pm\frac{1}{2}\left[\left(U_{xx}-U_{yy}\right)^2+4U_{xy}{}^2\right]^{1/2} \qquad (4.2 - 31b)$$

because the root in this equation contains the sum of two squares. As a consequence the singular points are either nodes or saddle points.

The Jacobi matrix ***J*** (4.2 - 31a) of the gradient system (4.2 - 9c) is the *negative Hessian matrix* ***H*** [Tu 1982 B] of the potential *U*(*x*) defined as

$$\boldsymbol{H}=\begin{bmatrix} U_{xx} & U_{xy} \\ U_{xy} & U_{yy} \end{bmatrix} \quad . \qquad (4.2 - 32a)$$

This *symmetric* matrix has the following characteristics

$$tr\,\boldsymbol{H}=U_{xx}+U_{yy}=\Delta U \quad , \quad det\,\boldsymbol{H}=U_{xx}U_{yy}-U_{xy}{}^2 \quad , \qquad (4.2 - 32b)$$

$$\Lambda(\boldsymbol{H})=\lambda_{1,2}(\boldsymbol{H})=+\frac{1}{2}\Delta U\pm\frac{1}{2}\left[\left(U_{xx}-U_{yy}\right)^2+4U_{xy}{}^2\right]^{1/2} \qquad (4.2 - 32c)$$

where "*tr*" and "*det*" indicate the trace and the determinant, whilst $\lambda_{1,2}$ represent the two real eigenvalues.

During the motion $\vec{r}(t)$ of a gradient system its *potential* $U(\vec{r})$ *decreases* except when the system is at rest in equilibrium

$$\frac{d}{dt}U(\vec{r}) = +grad\,U(\vec{r})\cdot\vec{\upsilon}(\vec{r}) = -|grad\,U(\vec{r})|^2 \leq 0 \quad . \tag{4.2 - 33}$$

Thus, this potential U represents a Lyapunov function according to Section 4.6.

The *acceleration* $\vec{a}(\vec{r})$ of a gradient system is determined by the *Hessian matrix* $\boldsymbol{H}$ (4.2 - 32a) as follows

$$\begin{aligned}\vec{a}(\vec{r}) = \ddot{\vec{r}} = \dot{\vec{\upsilon}}(\vec{r}) = -grad\left[\frac{d}{dt}U(\vec{r})\right] = +grad\left\{|grad\,U(\vec{r})|^2\right\} = \\ = \boldsymbol{H}\;grad\,U(\vec{r}) = \boldsymbol{H}\;\vec{\upsilon}(\vec{r})\end{aligned} \tag{4.2 - 34a}$$

or in Cartesian coordinates

$$\begin{pmatrix} a_x \\ a_y \end{pmatrix} = \begin{pmatrix} \ddot{x} \\ \ddot{y} \end{pmatrix} = \begin{pmatrix} \dot{u} \\ \dot{\upsilon} \end{pmatrix} = \boldsymbol{H}\begin{pmatrix} U_x \\ U_y \end{pmatrix} = \boldsymbol{H}\begin{pmatrix} u \\ \upsilon \end{pmatrix} \quad \text{with} \quad \boldsymbol{H} = \begin{pmatrix} U_{xx} & U_{xy} \\ U_{xy} & U_{yy} \end{pmatrix} \quad . \tag{4.2 - 34b}$$

The Hessian matrix $\boldsymbol{H}$ as well as the Jacobi matrix $\boldsymbol{J} = -\boldsymbol{H}$ determines the *behavior of the gradient system near equilibrium states*. The equilibrium states k are defined by position vectors $\vec{r}_k$ which fulfill the condition

$$grad\,U(\vec{r}_k) = \vec{0} \quad \text{or} \quad U_x(x_k, y_k) = U_y(x_k, y_k) = 0 \quad . \tag{4.2 - 35a}$$

The application of this condition in (4.2 - 33) leads to the conclusion that the potential $U(\vec{r}_k)$ remains constant

$$\frac{d}{dt}U(\vec{r}_k) = 0 \quad . \tag{4.2 - 35b}$$

In the *neighborhood* $\vec{r}_k + \delta\vec{r}$ of an *equilibrium state k* corresponding to the position vector $\vec{r}_k$ the potential $U(\vec{r}_k + \delta\vec{r})$ can be approximated by the vector equation

$$U(\vec{r}_k + \delta\vec{r}_k) \approx U(\vec{r}_k) + \delta\vec{r}\,\boldsymbol{H}(U(r_k))\,\delta\vec{r} \tag{4.2 - 35c}$$

or in Cartesian coordinates

$$\begin{aligned}U(x_k + \delta x, y_k + \delta y) \approx U(x_k, y_k) + \\ + \frac{1}{2}\left[U_{xx}(x_k, y_k)\delta x^2 + 2U_{xy}(x_k, y_k)\delta x\delta y + U_{yy}(x_k, y_k)\delta y^2\right] \quad .\end{aligned} \tag{4.2 - 35d}$$

The corresponding linear system of equations in Cartesian coordinates is

$$\begin{aligned}\frac{d}{dt}(\delta x) &= -\frac{\partial}{\partial \delta x} U\left(x_{\mathrm{k}} + \delta x, y_{\mathrm{k}} + \delta y\right) \\ &= -U_{\mathrm{xx}}\left(x_{\mathrm{k}}, y_{\mathrm{k}}\right)\delta x - U_{\mathrm{xy}}\left(x_{\mathrm{k}}, y_{\mathrm{k}}\right)\delta y \quad .\end{aligned} \tag{4.2 - 35e}$$

It corresponds to the vector equation

$$\frac{d}{dt}\delta\vec{r} = -\boldsymbol{H}\big(U(\vec{r}_{\mathrm{k}})\big)\,\delta\vec{r} \quad . \tag{4.2 - 35f}$$

Equation (4.2 - 32c) implies that the two eigenvalues $\lambda_{1,2}(\boldsymbol{H})$ of the Hessian matrix $\boldsymbol{H}$ are real. Hence *no periodic solutions* are possible. This *excludes spiraling nodes and limit cycles* [Tu 1982 B] near the equilibrium positions. The character of the equilibrium positions is determined by the real eigenvalues $\lambda_{1,2}(\boldsymbol{H})$ as follows

$$\begin{array}{lcl} \lambda_1(\boldsymbol{H}) \le \lambda_2(\boldsymbol{H}) < 0 & : & \text{instable equilibrium} \\ \lambda_1(\boldsymbol{H}) < 0 < \lambda_2(\boldsymbol{H}) & : & \text{saddle point} \\ 0 < \lambda_1(\boldsymbol{H}) \le \lambda_2(\boldsymbol{H}) & : & \text{stable equilibrium} \end{array} \quad . \tag{4.2 - 35g}$$

An *example* is a gradient system with the following potential U

$$U(x,y) = x^4 - 2x^2 + 2y^2 \quad . \tag{4.2 - 36a}$$

The equilibrium states with $U_{\mathrm{x}} = U_{\mathrm{y}} = 0$ are at the positions

$$(x_{\mathrm{k}}, y_{\mathrm{k}}) = (-1,0)\,;(0,0)\,;(1,0) \quad . \tag{4.2 - 36b}$$

The corresponding Hessian matrices have the form

$$\boldsymbol{H}_{\mathrm{k}} = \begin{pmatrix} 8 & 0 \\ 0 & 4 \end{pmatrix};\begin{pmatrix} -4 & 0 \\ 0 & 4 \end{pmatrix};\begin{pmatrix} 8 & 0 \\ 0 & 4 \end{pmatrix} \quad . \tag{4.2 - 36c}$$

Their eigenvalues are

$$\lambda_{1,2}(\boldsymbol{H}_{\mathrm{k}}) = (8,4)\,;(-4,4)\,;(8,4) \quad . \tag{4.2 - 36d}$$

By taking into account these eigenvalues one can determine the character of the equilibrium positions:

$$\begin{aligned} &\alpha)\ \text{stable equilibrium at } (\pm 1, 0) \quad , \\ &\beta)\ \text{saddle point at } (0,0) \quad . \end{aligned} \tag{4.2 - 36e}$$

b) Radial gradient flows

Radial gradient flows are characterized by *potentials* of *axial symmetry*

$$U(r,\varphi) = U(r) \quad . \tag{4.2 - 37a}$$

These flows are represented by systems of the form

$$\begin{aligned} \dot{r} &= \upsilon_r(r) = -U_r(r) \\ r\dot{\varphi} &= \upsilon_\varphi(r) = 0 \end{aligned} \quad . \tag{4.2 - 37b}$$

Therefore, the *equipotential curves* $U(r) = const$ represent *circles* whilst the *streamlines* correspond to *radial beams.*

The local source and sink strength q shows also axial symmetry because of the *Poisson equation*

$$q = q(r) = -\Delta U(r) = -U_{rr} - r^{-1}U_r \quad . \tag{4.2 - 37c}$$

In this case the *Gauss law* in two dimensions takes the simple form

$$\int_0^{2\pi} \upsilon_r \, r d\varphi = 2\pi r \upsilon_r = \int_0^r q 2\pi r dr \quad . \tag{4.2 - 37d}$$

This law relates the source or sink strength $q(r)$ to the radial velocity $\upsilon_r(r)$ and vice versa by

$$\upsilon_r(r) = r^{-1} \int_0^r q(r) r \, dr \tag{4.2 - 37e}$$

and

$$q(r) = r^{-1} \frac{d}{dr}\left[r\,\upsilon_r(r)\right] \quad . \tag{4.2 - 37f}$$

c) Rectilinear source of finite radius

As first *example* of a radial gradient flow serves the *radial flow from a rectilinear source of finite radius* r_0. The corresponding radial variation of the local source and sink strength is

$$q(r) = \begin{cases} q(r \le r_0) = q_0 \\ q(r > r_0) = 0 \end{cases} \quad . \tag{4.2 - 38a}$$

The radial flow $\upsilon_r(r)$ can be evaluated with (4.2 - 37d). The result is

$$v_r(r) = \begin{cases} \frac{1}{2} q_0 r & \text{for} \quad r \le r_0 \\ \frac{1}{2} q_0 r_0^2 / r & \text{for} \quad r \ge r_0 \end{cases} \quad . \tag{4.2 - 38b}$$

The first equation (4.2 - 37b) permits to evaluate the corresponding the potential $U(r)$

$$U(r) = \begin{cases} -\frac{1}{4} q_0 r^2 & \text{for} \quad r \le r_0 \\ -\frac{1}{4} q_0 r_0^2 [1 + 2\ell n(r / r_0)] & \text{for} \quad r > r_0 \end{cases} \quad . \tag{4.2 - 38c}$$

d) Rectilinear Bessel source

A second *example* is the radial flow that originates in a *source and sink strength* $q(r)$ whose variation with the distance r from the z axis is described by the *Bessel function of zero order* $J_0(s)$ [Abramowitz & Stegun 1965 B] according to

$$q = q(r) = q_0 \, J_0(r / r_0) \quad . \tag{4.2 - 39a}$$

This strength $q(r)$ determines the potential $U(r)$ by the *Poisson equation*

$$\Delta U(r) + q(r) = U_{rr} + r^{-1} U_r + q = 0 \quad . \tag{4.2 - 37b}$$

The solution of this equation yields the *potential*

$$U(r) = U_0 + q_0 \, r_0^2 \, J_0(r / r_0) \tag{4.2 - 39b}$$

and the radial velocity

$$v_r(r) = -U_r(r) = q_0 \, r_0 \, J_1(r / r_0) \quad . \tag{4.2 - 39c}$$

In the last equation $J_1(s)$ indicates the *Bessel function of first order* [Abramowitz & Stegun 1965 B].

e) Rectilinear source or sink of zero diameter

The last *example* concerns the *rectilinear source or sink* whose *complex representation* as an ideal flow was reported in Section 4.2.4e. This source or sink has also a *real representation* as *gradient flow* with a non-zero *potential $U(r)$ and a zero Hamiltonian*

$$U(r) = -\frac{Q}{2\pi} \ell n \, r \quad \text{and} \quad H(r) = 0 \quad . \tag{4.2 - 40a}$$

The corresponding *radial velocity* is

$$\dot{r} = \upsilon_r(r) = -U_r(r) = \frac{Q}{2\pi r} \qquad (4.2 - 16c)$$

in agreement with Section 4.2.4e.

4.3 Two-dimensional Linear Autonomous Systems

According to the survey on systems of differential equations in Section 4.1 the discussion of linear autonomous or time-independent systems, which are also called *d'Alembert systems* [Kamke 1957 B], can be restricted to *homogeneous systems*. The d'Alembert systems are basic and relatively simple. Among these the two-dimensional homogenous d'Alembert systems are the simplest because they concern only two time-dependent variables.

4.3.1 Representations

In Cartesian coordinates the *standard representation* of the systems considered is of the form

$$\begin{aligned} dx / dt &= \dot{x} = u(x, y) = a_{11}\, x + a_{12} y \\ dy / dt &= \dot{y} = \upsilon(x, y) = a_{21}\, x + a_{22} y \end{aligned} \qquad (4.3 - 1a)$$

while the vector representation is written as

$$\dot{\vec{r}} = \vec{\upsilon}(\vec{r}) = \boldsymbol{A}\, \vec{r} \quad \text{with} \quad \boldsymbol{A} = \begin{bmatrix} a_{11} & a_{12} \\ a_{21} & a_{22} \end{bmatrix} \quad . \qquad (4.3 - 1b)$$

The elements a_{jk} of the *characteristic matrix* $\boldsymbol{A}$ are assumed to be real. The relevant parameters of the matrix $\boldsymbol{A}$ are

$$S = tr\, \boldsymbol{A} = 2\overline{\alpha} = a_{11} + a_{22} \quad , \qquad (4.3 - 2a)$$

$$D = det\, \boldsymbol{A} = a_{11}\, a_{22} - a_{12}\, a_{21} \quad , \qquad (4.3 - 2b)$$

$$\Delta = 4\omega_0^2 sign\, \Delta = S^2 - 4D \qquad (4.3 - 2c)$$

where S indicates the *trace*, D the *determinant* and Δ the *discriminant* of the matrix $\boldsymbol{A}$. The *eigenvalues* $\Lambda(\boldsymbol{A})$ of the matrix $\boldsymbol{A}$ are determined by the secular equation

$$|\boldsymbol{A} - \Lambda(\boldsymbol{A})\boldsymbol{I}| = 0 \qquad (4.3 - 3a)$$

that corresponds to the quadratic equation

$$\alpha^2 - S\alpha + D = 0 \quad \text{with} \quad \Lambda(\boldsymbol{A}) = \alpha_{1,2} \quad . \tag{4.3 - 3b}$$

This equation is also fulfilled by the matrix $\boldsymbol{A}$ itself

$$\boldsymbol{A}^2 - S\boldsymbol{A} + D\boldsymbol{I} = 0 \quad , \quad \text{where} \quad \boldsymbol{I} = \begin{bmatrix} 1 & 0 \\ 0 & 1 \end{bmatrix} \tag{4.3 - 3c}$$

represents the two-dimensional *unit matrix*.

The solution of (4.3 - 3b) yields the eigenvalues

$$\Lambda(\boldsymbol{A}) = \alpha_{1,2} = \frac{1}{2}\left(S \pm \Delta^{1/2}\right) \quad . \tag{4.3 - 3d}$$

As a consequence the discriminant Δ classifies the eigenvalues $\alpha_{1,2}$ of the matrix $\boldsymbol{A}$ as follows

$$\begin{aligned} &\Delta > 0 : \alpha_{1,2} = \overline{\alpha} \pm \omega_0 \quad \text{real different} \quad , \\ &\Delta = 0 : \alpha_{1,2} = \overline{\alpha} \quad \text{real equal} \quad , \\ &\Delta < 0 : \alpha_{1,2} = \overline{\alpha} \pm i\,\omega_0 \quad \text{complex conjugate} \quad , \\ &\text{where} \quad 2\overline{\alpha} = S \quad \text{and} \quad 4\omega_0^2 = |\Delta| \quad . \end{aligned} \tag{4.3 - 3e}$$

The two-dimensional d'Alembert systems represent special *two-dimensional autonomous systems*. These have been reviewed in Section 4.2. Their standard representation is

$$\begin{aligned} \dot{x} &= u(x,y) = -U_x(x,y) + H_y(x,y) \\ \dot{y} &= \upsilon(x,y) = -U_y(x,y) - H_x(x,y) \quad . \end{aligned} \tag{4.2 - 1}$$

The *potentials* U and the *Hamiltonians* H of the two-dimensional d'Alembert systems (4.2 - 1a&b) have the form

$$U = U(x,y) = -\frac{1}{2}\overline{\alpha}\left(x^2 + y^2\right) + \beta\, xy \quad , \tag{4.3 - 4a}$$

$$H = H(x,y) = -\frac{1}{2}\omega\left(x^2 + y^2\right) + \gamma\, xy \quad . \tag{4.3 - 4b}$$

The application of (4.2 -1) to these functions yields the following representation of the d'Alembert systems

$$\dot{x} = u = -U_x + H_y = (\overline{\alpha} + \gamma)x - (\omega + \beta)y$$
$$\dot{y} = \upsilon = -U_x - H_x = (\omega - \beta)x + (\overline{\alpha} - \gamma)y \quad . \tag{4.3 - 4c}$$

In this representation the *characteristic parameters* of the matrix **A** correspond to

$$S = 2\overline{\alpha} \quad , \tag{4.3 - 5a}$$

$$D = \left(\overline{\alpha}^2 + \omega^2\right) - \left(\beta^2 + \gamma^2\right) \quad , \tag{4.3 - 5b}$$

$$\Delta = sign\,\Delta \cdot 4\omega_0^2 = 4\left(\beta^2 + \gamma^2 - \omega^2\right) \quad . \tag{4.3 - 5c}$$

The related kinematic characteristics are

$$div\,\vec{\upsilon} = 2\overline{\alpha} = S \quad , \tag{4.3 - 6a}$$

$$\left(curl\ \vec{\upsilon}\right)_z = 2\omega \quad . \tag{4.3 - 6b}$$

The representation of the two-dimensional d'Alembert systems in *polar coordinates* (4.2 - 4a&b) makes use of representation of the potential U (4.3 - 4a) and the Hamiltonian H (4.3 - 4b) in polar coordinates

$$U = U(r,\varphi) = -\frac{1}{2}r^2(\overline{\alpha} - \beta\ sin\ 2\varphi) \quad , \tag{4.3 - 7a}$$

$$H = H(r,\varphi) = -\frac{1}{2}r^2(\omega - \gamma\ sin\ 2\varphi) \quad . \tag{4.3 - 7b}$$

If one applies these two functions to the representation (4.2 - 6) of two-dimensional autonomous systems in polar coordinates one finds

$$\frac{d}{dt}\ell n\ r = \dot{r}/r = \overline{\alpha} - \beta\ sin 2\varphi + \gamma\ cos 2\varphi \quad ,$$
$$\dot{\varphi} = \omega - \beta\ cos 2\varphi - \gamma\ sin 2\varphi \quad . \tag{4.3 - 7c}$$

These equations yield easy solutions of many two-dimensional d'Alembert systems.

4.3.2 Associated Differential Equations

Each two-dimensional d'Alembert system is associated by a *linear differential equation of second order*. This is verified by the elimination of y in the d'Alembert system (4.3 - 1a). The result is

$$\ddot{x} - S\dot{x} + Dx = 0 \quad \text{with} \quad S = tr\,\boldsymbol{A} \quad \text{and} \quad D = \det \boldsymbol{A} \quad . \tag{4.3 - 8a}$$

If the solution $x = x(t)$ of this equation is known, the second variable $y = y(t)$ can be determined with the equation

$$y = -(a_{11} / a_{12})x + (1 / a_{12})\dot{x} \quad . \qquad (4.3 - 9a)$$

It is also possible to eliminate first the variable x instead of y in the d'Alembert system (4.3 - 1a). The result is the same differential equation as before

$$\ddot{y} - S\,\dot{y} + Dy = 0 \quad \text{with} \quad S = tr\,\boldsymbol{A} \quad \text{and} \quad D = det\,\boldsymbol{A} \quad . \qquad (4.3 - 8b)$$

The solution $y = y(t)$ of this equation determines also the variable $x = x(t)$ by the relation

$$x = -(a_{22} / a_{21})y + (1 / a_{21})\dot{y} \quad . \qquad (4.3 - 9b)$$

a) Equivalent d'Alembert systems

D'Alembert systems with the same associated differential equations (4.3 - 8a&b) are called *equivalent*. These equations (4.3 - 8a&b) have the same coefficients D and S as the eigenvalue equation (4.3 - 3a) of the matrix $\boldsymbol{A}$. As a consequence, two d'Alembert systems with the characteristic matrices $\boldsymbol{A}_1$ and $\boldsymbol{A}_2$ are equivalent if these matrices are related by the *similarity transformation*

$$\boldsymbol{A}_2 = \boldsymbol{T}\,\boldsymbol{A}_1\,\boldsymbol{T}^{-1} \quad \text{with} \quad det\,T \neq 0 \qquad (4.3 - 10)$$

where $\boldsymbol{T}$ indicates a *non-singular matrix.*

The linear differential equation of second order (4.3 - 8a) that is associated to the d'Alembert system (4.3 - 1a) can be transformed into an equivalent d'Alembert system named *companion system* [Birkhoff & Rota 1989 B]

$$\begin{aligned} \dot{x} &= u = y \\ \dot{y} &= v = -Dx + Sy \end{aligned} \quad \text{corresponding to} \quad \boldsymbol{A} = \begin{bmatrix} 0 & 1 \\ -D & S \end{bmatrix} \quad . \qquad (4.3- 11)$$

b) Canonical matrices

Among the equivalent matrices $\boldsymbol{A}$ of equivalent d'Alembert systems there exists a set of especially simple basic matrices called *canonical matrices* $\boldsymbol{A}_{\text{can}}$. The solutions $x = x(t)$ and $y = y(t)$ of the corresponding d'Alembert systems are also simple. These canonical matrices are classified according to the sign of their discriminant Δ:

α) *positive discriminant:* $\Delta > 0$, *sign* $\Delta = +1$

$$\boldsymbol{A}_{\text{can}} = \begin{bmatrix} \alpha_1 & 0 \\ 0 & \alpha_2 \end{bmatrix} \qquad (4.3 - 12a)$$

with the corresponding solution

$$x(t) = x(0)\, exp\, \alpha_1 t \quad \text{and} \quad y(t) = y(0)\, exp\, \alpha_2 t \quad . \tag{4.3 - 12b}$$

β) *zero discriminant:* $\Delta = 0$, *sign* $\Delta = 0$

$$\boldsymbol{A}_{\text{can}} = \boldsymbol{J} = \begin{bmatrix} \overline{\alpha} & 1 \\ 0 & \overline{\alpha} \end{bmatrix} = \text{Jordan Matrix} \tag{4.3 - 13a}$$

with the corresponding solution

$$x(t) = x(0)\, exp\, \overline{\alpha}t + y(0)\, t\, exp\, \overline{\alpha}t \quad \text{and} \quad y(t) = y(0)\, exp\, \overline{\alpha}t \quad . \tag{4.3 - 13b}$$

γ) *negative discriminant*: $\Delta = -4\omega^2$, *sign* $\Delta = -1$

$$\boldsymbol{A}_{\text{can}} = \begin{bmatrix} \overline{\alpha} & -\omega \\ +\omega & \overline{\alpha} \end{bmatrix} \tag{4.3 - 14a}$$

with the corresponding system of differential equations in polar coordinates (4.3 - 7c)

$$\dot{r} = \overline{\alpha}\, r \quad \text{and} \quad \dot{\varphi} = \omega \quad .$$

This system has the solution

$$r(t) = r(0)\, exp\, \overline{\alpha}t \quad \text{and} \quad \varphi(t) = \omega\, t + \varphi(0) \quad . \tag{4.3 - 14b}$$

c) Solutions

In the following we discuss the *solutions of the associated differential equations* (4.3 - 8a). The ansatz

$$x = x(t) = A\, exp(\alpha\, t) + B\, t\, exp(\alpha\, t) \tag{4.3 - 15a}$$

yields the following simultâneous conditions for α:

$$\begin{aligned} A\left(\alpha^2 - S\alpha + D\right) + B(2\alpha - S) &= 0 \quad , \\ B\left(\alpha^2 - S\alpha + D\right) \qquad\qquad &= 0 \quad . \end{aligned} \tag{4.3 - 15b}$$

These conditions can be fulfilled for $B = 0$ as well as for $B \neq 0$:

α) for $\quad B = 0: \quad \alpha^2 - S\alpha + D = 0 \quad .$ (4.3 - 15c)

This equation corresponds to the eigenvalue equation (4.3 - 3a) of the matrix $\boldsymbol{A}$ characteristic for the two-dimensional d'Alembert system.

β) for $B \neq 0$: $\Delta = 0, \alpha_{1,2} = \overline{\alpha} = S/2$. (4.3 - 15d)

For $B \neq 0$ the discriminant Δ is zero. This implies that α_1 and α_2 are degenerate.

On the basis of these results and (4.3 - 9a&b) one can distinguish between *three types of solutions of the associated differential equations and of the d'Alembert systems.* This distinction can be made with the aid of the classification (4.3 - 3d) by the eigenvalues $\alpha_{1,2}$ of the characteristic matrix $\boldsymbol{A}$:

α) exponential solutions: $\Delta = 4\omega_0^2 > 0$

$$\Delta > 0: \quad x(t) = \left[A \cosh \omega_0 t + B \sinh \omega_0 t\right] \exp \overline{\alpha}\, t \quad . \tag{4.3 - 16a}$$

β) critical solutions: $\Delta = 0$

$$\Delta = 0: \quad x(t) = [A + Bt] \exp \overline{\alpha}\, t \quad . \tag{4.3 - 16b}$$

γ) oscillatory solutions: $\Delta = -4\omega_0^2 < 0$

$$\Delta < 0: \quad x(t) = \left[A \cos \omega_0 t + B \sin \omega_0 t\right] \exp \overline{\alpha}\, t \quad . \tag{4.3 - 16c}$$

These solutions $x = x(t)$ determine and classify also the *solutions of the corresponding d'Alembert systems,* since the solutions $y = y(t)$ can be derived directly from (4.3 - 9a).

4.3.3 Stability

Stability is an important feature of systems of differential equations. In homogeneous d'Alembert systems (4.1 - 7a&b) stability concerns the solutions at and in the neighborhood of the singular or critical point (4.1 - 8) in the origin $\vec{r}_S = \vec{0}$. In these circumstances the various types of stability are defined with respect to the distance r from the critical point in the origin as a function of time t

$$r(t) = +\left[\sum_{j=1}^{n} x_j^2(t)\right]^{1/2} \geq 0 \tag{4.3 - 17a}$$

where n is the dimension of the system.

a) Classification

The following consideration deals with a system of differential equations that has a singular or critical point in the origin $\vec{r}_S = \vec{0}$. In this critical point the system is α) *strictly or asymptotically stable,* if for each solution $\vec{r}(t)$ of the system with $r(t_0) < r_0$

$$\lim_{t \to \infty} r(t) = 0 \quad . \tag{4.3 - 17b}$$

On this condition the critical point in the origin $\vec{r}_S = \vec{0}$ is called *attractor.*

β) *exponentially stable,* if for $C > 0$, $\tau > 0$ and $r(t_0) \leq r_0$ [Slotine & Li 1991 B]

$$r(t) \leq C\, r(t_0)\, exp\left[-\frac{1}{\tau}(t - t_0)\right] \quad . \tag{4.3 - 17c}$$

The exponential stability represents a special type of *strict stability.*

γ) *stable,* if for each solution $\vec{r}(t)$ with $r(t_0) < r_0$ and $C > 0$

$$r(t) \leq C \quad \text{for} \quad t > t_0 \quad . \tag{4.3 - 17d}$$

δ) *instable,* if the solutions $\vec{r}(t)$ fulfill neither the condition α) for strict stability nor the condition γ) for stability.

b) Stability diagram of d'Alembert systems

The *stability of two-dimensional d'Alembert systems* is determined by the trace S (4.3 - 2a) and by the determinant D of (4.3 - 2b) of the characteristic matrix $\boldsymbol{A}$ (4.3 - 1b). S and D are coefficients of the associated differential equations (4.3 - 8a&b).

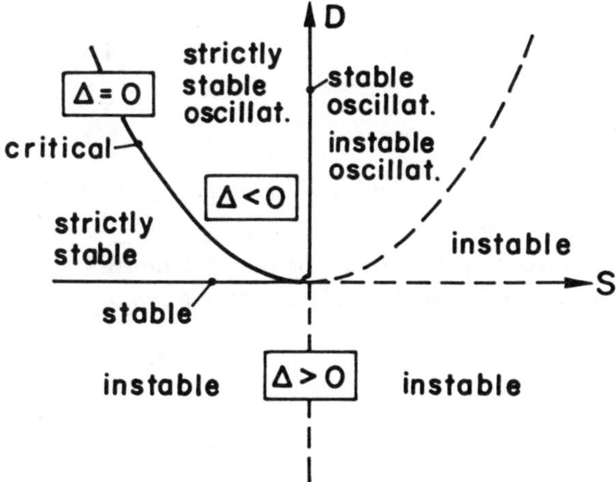

Fig. 4.3 - 1. Stability diagram of the two-dimensional homogeneous d'Alembert systems and the associated differential equations in the S-D plane

Fig. 4.3 - 1 represents the stability diagram [Birkhoff & Rota 1989 B] of the two-dimensional homogeneous d'Alembert systems in the *S*-*D* plane. Essential boundaries are the straight lines $S = 0$ and $D = 0$ as well as the parabola $S^2 - 4D = \Delta = 0$. The decision on the type of stability in any range within these boundaries can be made with the aid of the eigenvalues (4.3 - 3d) of the matrix **A** or by taking into account the solutions (4.3 - 16a-c) of the associated differential equation.

4.3.4 Analysis of the Critical Point

Homogeneous linear autonomous systems (4.1 - 7a&b) possess a single critical point (4.1 - 8) in the origin $\vec{r}_S = \vec{0}$. Apart from the stability of these systems at the critical point, the characteristic behavior of trajectories and streamlines at and in the close neighborhood of this critical point is of fundamental interest. Trajectories and streamlines are identical because the systems (4.1 - 7a&b) are autonomous. In the following the two-dimensional homogeneous d'Alembert systems are analysed and classified according to the behavior of their streamlines at the origin.

a) Representation of trajectories

The trajectories and streamlines as solutions of the *two-dimensional* homogeneous d'Alembert systems can be described by

α) a parameter representation where x and y are given as functions of the time t as parameter

$$\begin{aligned} x &= x(t) \quad \text{with} \quad x_0 = x(t_0) \\ y &= y(t) \quad \text{with} \quad y_0 = y(t_0) \quad . \end{aligned} \tag{4.3 - 18a}$$

β) an explicit representation where

$$y = y(x, x_0, y_0) \quad . \tag{4.3 - 18b}$$

γ) an implicit representation in the form

$$\Phi(x, y, x_0, y_0) = 0 \quad . \tag{4.3 - 18c}$$

This implicit representation can be derived by division of the two equations (4.3 - 1a) of the system. This eliminates the time t as variable. The result is a *homogeneous differential equation of first order*

$$\frac{dy}{dx} = \frac{a_{21}\ x + a_{22}\ y}{a_{12}\ x + a_{12}\ y} \tag{4.3 - 19a}$$

that can be solved by the introduction of the new variable [Kamke 1956 B]

$$g = y / x \quad . \tag{4.3 - 19b}$$

This procedure yields the implicit solution

$$\int_{g_0=y_0/x_0}^{g=y/x} \frac{a_{12}\,g + a_{11}}{a_{12}\,g^2 + (a_{11} - a_{22})g - a_{21}}\,dg + \ell n(x/x_0)$$

$$= \Phi(x, y, x_0, y_0) = 0 \qquad (4.3 - 19c)$$

This implicit solution is rarely used for linear two-dimensional d'Alembert systems. Yet it is well suited for nonlinear two-dimensional systems of differential equations, which are described in the following Section 4.4.

b) Classification

The behavior of the two-dimensional homogeneous d'Alembert systems and their trajectories at the critical point in the origin $\vec{r}_S = \vec{0}$ is determined essentially by the trace S and the determinant D of the characteristic matrix $\boldsymbol{A}$. As for the stability classification in the preceding Section 4.3.3 one can divide the S-D plane into regions where the d'Alembert systems and their streamlines behave identically at the critical point in the origin $\vec{r}_S = \vec{0}$. These regions, which contain different *classes of equivalent d'Alembert* systems, are shown in Fig. 4.3 - 2.

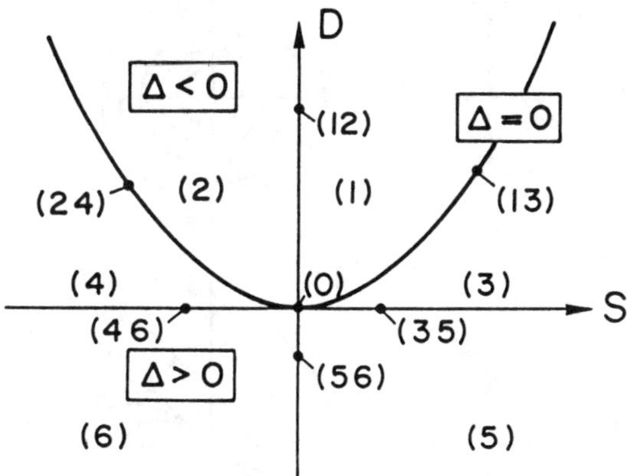

Fig. 4.3 - 2. Classification of the two-dimensional homogeneous d'Alembert systems at the critical point in the origin $\vec{r}_S = \vec{0}$

The different types or classes of solutions of the two-dimensional homogeneous d'Alembert systems in the neighborhood of the critical point in the origin [Birkhoff & Rota 1989 B, Percival & Richards 1982 B, Slotine & Li 1991 B, Tu 1992 B, Verhulst 1985 B] are listed in Table 4.3 - 1 and illustrated by their trajectories and streamlines in Fig. 4.3 - 3.

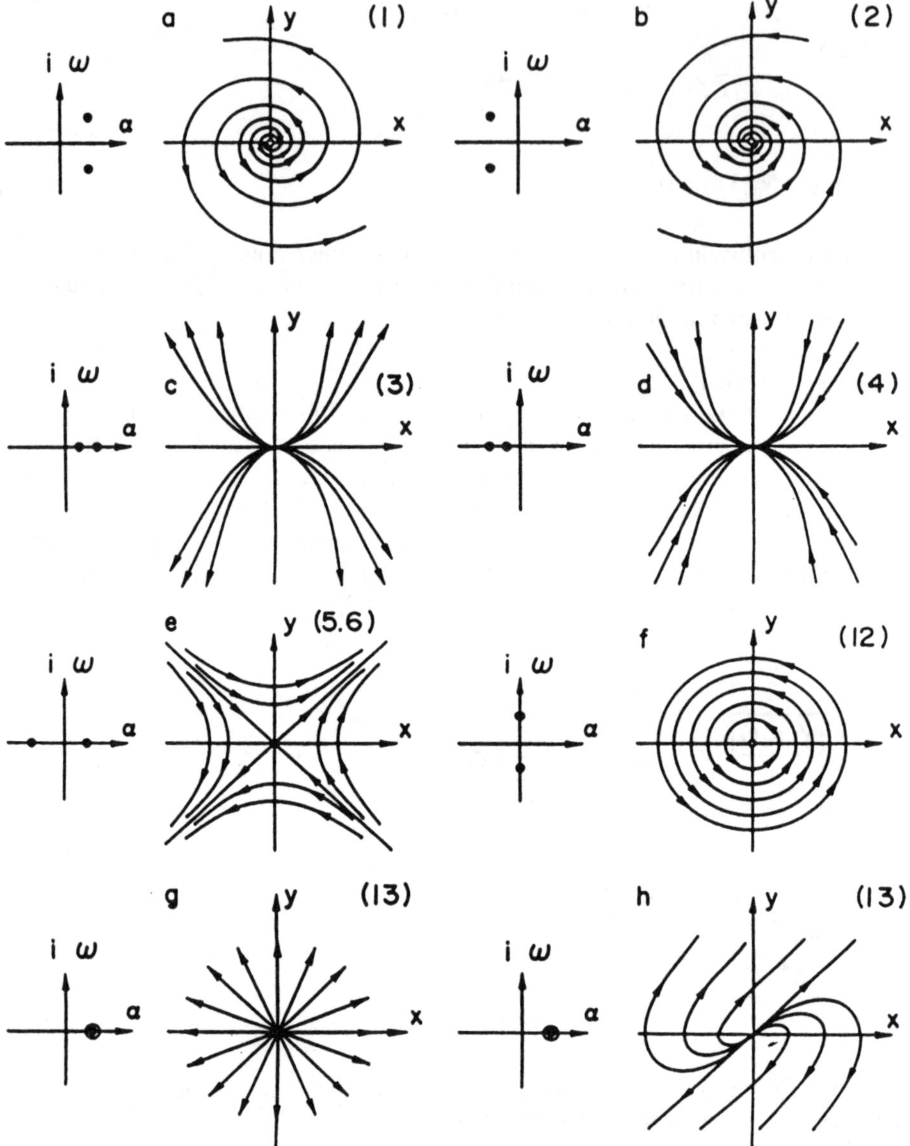

Fig. 4.3 - 3. Typical trajectories of the two-dimensional homogeneous d'Alembert systems at the critical point in the origin $\vec{r}_S = \vec{0}$. All these figures are classified and characterized in Table 4.3 - 1

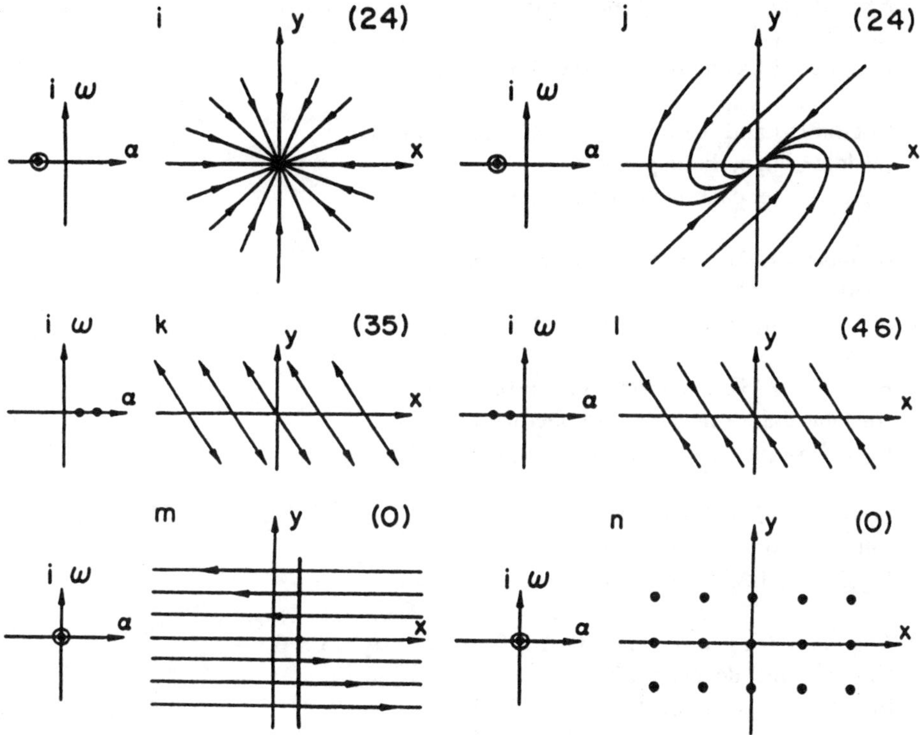

Fig. 4.3 - 3. continued

Table 4.3 - 1:
Classification of the solutions of the two-dimensional homogeneous d'Alembert systems (4.3 - 1a&b) at and in the neighborhood of the critical point in the origin $\vec{r}_S = \vec{0}$. The regions and boundaries of equivalent solutions in the *S-D* plane are indicated in Fig. 4.3 - 2 whilst the corresponding trajectories are shown in Fig. 4.3 - 3.

- *Region (1):* $S > 0, D > 0, \Delta < 0, \ \overline{\alpha} > 0$
  "instable spiral or focus"
  Fig. 4.3 - 3a, canonical matrix (4.3 - 13a)

- *Region (2):* $S < 0, D > 0, \Delta < 0, \ \overline{\alpha} < 0$
  "spiral attractor, stable spiral or focus"
  Fig. 4.3 - 3b, canonical matrix (4.3 - 13a)

- *Region (3):* $S > 0, D > 0, \Delta > 0, \ \alpha_1 > 0, \ \alpha_2 > 0, \ \alpha_1 \neq \alpha_2$
  "instable node"
  Fig. 4.3 - 3c, canonical matrix (4.3 - 11a)

- *Region (4):* $S < 0, D > 0, \Delta > 0, \alpha_1 < 0, \alpha_2 < 0, \alpha_1 \neq \alpha_2$
  "stable node"
  Fig. 4.3 - 3d, canonical matrix (4.3 - 11a)

- Region (5) and (6): D < 0, $\Delta > 0$, sign $\alpha_1 \neq$ sign α_2
  "saddle or hyperbolic point"
  Fig. 4.3 - 3e, canonical matrix (4.3 - 11a)

- *Boundary (12):* $S = 0, D > 0, \Delta < 0, \overline{\alpha} = 0$
  "vortex point"
  Fig. 4.3 - 3f, canonical matrix (4.3 - 13a)

- *Boundary (13):* $S > 0, D > 0, \Delta = 0, \overline{\alpha} > 0$
  "instable star or instable improper node"
  Figs. 4.3 - 3g&h, canonical matrix (4.3 - 12a)

- *Boundary (24):* $S < 0, D > 0, \Delta = 0, \overline{\alpha} < 0$
  "attractor: star or inproper node"
  Figs. 4.3 - 3i&j, canonical matrix (4.3 - 12a)

- *Boundary (35):* $S > 0, D = 0, \Delta > 0, \alpha_1 = 0, \alpha_2 = S > 0$
  "instable line, degenerate"
  Fig. 4.3 - 3k, canonical matrix (4.3 - 11a)
  example: $x(t) = x(0)$, $y(t) = y(0)\ exp\ \ S\ t$

- *Boundary (46):* $S < 0, D = 0, \Delta > 0, \alpha_1 = 0, \alpha_2 = S < 0$
  "stable line, degenerate"
  Fig. 4.3 - 3 ℓ, canonical matrix (4.3 - 11a)
  example: $x(t) = x(0)$, $y(t) = y(0)\ exp - |S|t$

- *Boundary (56):* $S = 0, D < 0, \Delta > 0, \alpha_1 = -\alpha_2$
  "saddle or hyperbolic point"
  Fig. 4.3 - 3e, canonical matrix (4.3 - 11a)

- *Origin (0):* $S = D = \Delta = \alpha_1 = \alpha_2 = 0$

  α) "shear point, degenerate": $\omega \neq 0$
  Fig. 4.3 - 3m
  example: $\dot{x} = -2\omega y, \dot{y} = 0$: $\quad x(t) = x(0) - 2\omega y(0)t, y(t) = y(0)$

  β) "stationary state": $\omega = 0$
  Fig. 4.3 - 3n
  example: $\dot{x} = 0, \dot{y} = 0$: $\quad x(t) = x(0), y(t) = y(0)$

4.3.5 Propagators

In the following the propagators of homogeneous d'Alembert systems with finite dimensions are discussed with special emphasis on those of only two dimensions.

a) System propagator and characteristic matrix

Homogeneous d'Alembert systems of finite dimensions can be solved in general with the aid of system propagators. Accordingly, the vector equation of such a d'Alembert system in the standard form

$$\dot{\vec{r}} = \boldsymbol{A}\,\vec{r} \tag{4.1 - 1b}$$

has the solution

$$\vec{r}(t_0 + t) = \boldsymbol{P}_{\mathrm{S}}(t)\,\vec{r}(t_0) \tag{4.3 - 20a}$$

where $\boldsymbol{P}_{\mathrm{S}}(t)$ represents the *system propagator*

$$\boldsymbol{P}_{\mathrm{S}}(t) = exp(t\,\boldsymbol{A}) = \sum_{\mathrm{k}=1}^{\infty} \frac{t^{\mathrm{k}}}{k!}\boldsymbol{A}^{\mathrm{k}} \quad . \tag{4.3 - 20b}$$

In the solution (4.3 - 20a) the initial time t_0 can be chosen arbitrarily, because the d'Alembert system (4.1 - 1b) is autonomous.

The relation (4.3 - 20b) between the system propagator $\boldsymbol{P}_{\mathrm{S}}(t)$ and the characteristic matrix $\boldsymbol{A}$ of the d'Alembert system implies also relations between their characteristic functions. If

$$tr\,\boldsymbol{A} = S \;\text{ and }\; \Lambda(\boldsymbol{A}) = \alpha_{\mathrm{j}}, j = 1,\ldots,n \tag{4.3 - 21a}$$

represent trace and eigenvalues of $\boldsymbol{A}$, then the trace, the determinant and the eigenvalues of $\boldsymbol{P}_{\mathrm{S}}(t)$ can be described by

$$\begin{aligned} tr\,\boldsymbol{P}_{\mathrm{S}}(t) &= \sum_{\mathrm{j}=1}^{\mathrm{n}} exp(\alpha_{\mathrm{j}}\,t) \quad , \\ det\,\boldsymbol{P}_{\mathrm{S}}(t) &= exp(St) \quad , \\ \Lambda(\boldsymbol{P}_{\mathrm{S}}(t)) &= exp(\alpha_{\mathrm{j}}\,t) \quad . \end{aligned} \tag{4.3 - 21b}$$

If the characteristic matrix is a sum of two matrices $\boldsymbol{A}$ and $\boldsymbol{B}$ then there exist the relations [Bronson 1988 B]

$$exp(tA)\,exp(t\boldsymbol{B}) = \prod_{k=1}^{\infty} exp\left\{\frac{t^k}{k!}\boldsymbol{C}^k\right\}$$

$$= exp(t(\boldsymbol{A}+\boldsymbol{B}))\,exp\left(\frac{t^2}{2}(\boldsymbol{A}\,\boldsymbol{B}-\boldsymbol{B}\,\boldsymbol{A})\right) \dots \qquad (4.3 - 21c)$$

$$= exp(t(\boldsymbol{A}+\boldsymbol{B})) \quad \text{for} \quad \boldsymbol{A}\,\boldsymbol{B} = \boldsymbol{B}\,\boldsymbol{A} \quad .$$

These relations determine the system propagator $\boldsymbol{P}_S$ of the sum of $\boldsymbol{A}$ and $\boldsymbol{B}$

$$\boldsymbol{P}_S[\boldsymbol{A}+\boldsymbol{B}] = exp(t(\boldsymbol{A}+\boldsymbol{B})) =$$

$$= exp(t\boldsymbol{A})\cdot exp(t\boldsymbol{B})\cdot exp\left\{-\frac{t^2}{2}(\boldsymbol{AB}-\boldsymbol{BA})\right\}\dots \qquad (4.3 - 21d)$$

$$= exp(t\boldsymbol{A})\cdot exp(t\boldsymbol{B}) \quad \text{for} \quad \boldsymbol{AB} = \boldsymbol{BA} \quad .$$

With respect to the time t, the system operators obey the following rules [Bellman 1966 B]

$$\boldsymbol{P}_S(t_2)\,\boldsymbol{P}_S(t_1)\,\vec{r}(t_0) = \boldsymbol{P}_S(t_2)\,\vec{r}(t_0+t_1) = \vec{r}(t_0+t_1+t_2) \qquad (4.3 - 22a)$$

$$exp(t_1\,\boldsymbol{A})\,exp(t_2\boldsymbol{A}) = exp((t_1+t_2)\boldsymbol{A}) \quad . \qquad (4.3 - 22b)$$

b) Evaluation of the system propagator

The system propagator $\boldsymbol{P}_S(t)$ can be written as a linear combination of a finite number of powers of the characteristic matrix $\boldsymbol{A}$ [Bronson 1993 B]. This can be demonstrated with the aid of (4.3 - 20b) and the fact that $\boldsymbol{A}$ *fulfills its own eigenvalue or secular equation*

$$|\alpha\boldsymbol{I}-\boldsymbol{A}| = \alpha^n + a_{n-1}\,\alpha^{n-1} + + \quad + a_0 = \prod_{k=1}^{n}(\alpha-\alpha_k) \qquad (4.3 - 23a)$$

where n indicates the dimension of $\boldsymbol{A}$ and the α_k, $k = 1, 2, ..., n$ its eigenvalues. This yields the matrix equation

$$\boldsymbol{A}^n = -a_{n-1}\boldsymbol{A}^{n-1} - a_{n-2}\boldsymbol{A}^{n-2} - - \quad - a_1\boldsymbol{A} - a_0\boldsymbol{I} \qquad (4.3 - 23b)$$

where $\boldsymbol{I}$ is the n-dimensional unit matrix.

By taking into account (4.3 - 2a&b) and (4.3 - 23b) one can represent the system propagator $\boldsymbol{P}_S(t)$ by the finite sum

$$\boldsymbol{P}_S(t) = p_0\,\boldsymbol{I} + p_1\,t\,\boldsymbol{A} + p_2\,t^2\boldsymbol{A}^2 + + + p_{n-1}\,t^{n-1}\,\boldsymbol{A}^{n-1} \qquad (4.3 - 23c)$$

with the associated characteristic polynomial of the variable αt

$$P(\alpha t) = p_0 + p_1 \alpha t + p_2 (\alpha t)^2 + + p_{n-1} (\alpha t)^{n-1} \quad . \tag{4.3 - 23d}$$

In its diagonal form, the characteristic matrix $\boldsymbol{A}$ can be written as

$$\boldsymbol{A} = \{a_{jk}\} = \{\alpha_k \delta_{jk}\} \quad . \tag{4.3 - 23e}$$

Here, δ_{jk} indicates the Kronecker δ and α_k the $k-th$ eigenvalue. The introduction of the diagonal matrix $\boldsymbol{A}$ (4.3 - 23e) in (4.3 - 23c) results in

$$exp(\alpha_j t) = P(\alpha_j t) \quad \text{with} \quad j = 1, 2, \ldots, n \quad . \tag{4.3 - 23f}$$

These equations are valid, if the eigenvalues α_j are all different. If, however, the $(k+1)$ eigenvalues $\alpha_m = \alpha_{m+1} = \ldots = \alpha_{m+k}$ are equal, in other words degenerate, then one has to apply also the equations

$$exp(\alpha_m t) = P(\alpha_m t) = \frac{1}{t} \frac{dP}{d\alpha}(\alpha_m t) = \ldots = \frac{1}{t^k} \frac{d^k P}{d\alpha^k}(\alpha_m t) \quad . \tag{4.3 - 23g}$$

The total of n independent equations (4.3 - 23f&g) permits the evaluation of the coefficients p_m, $m = 0, 1, 2 \ldots, n-1$ of the polynomial (4.3 - 23c) that determines the system propagator $\boldsymbol{P}_S(t)$.

The restriction to the *two-dimensional homogeneous d'Alembert systems* (4.3 - 1a&b) reduces (4.3 - 23c&d) to

$$\boldsymbol{P}_S(t) = p_0 \boldsymbol{I} + t \, p_1 \boldsymbol{A} \quad , \tag{4.3 - 24a}$$

$$P(\alpha t) = p_0 + p_1 \alpha t \quad . \tag{4.3 - 24b}$$

If α_1 and α_2 are the real or complex eigenvalues of the two-dimensional characteristic matrix $\boldsymbol{A}$, then the corresponding system propagator $\boldsymbol{P}_S(t)$ exhibits the three following forms

$$(\alpha_1 - \alpha_2) \boldsymbol{P}_S(t) = \left(e^{\alpha_1 t} - e^{\alpha_2 t}\right) \boldsymbol{A} - \left(\alpha_2 e^{\alpha_1 t} - \alpha_1 e^{\alpha_2 t}\right) \boldsymbol{I} \quad \text{for} \quad \alpha_1 \neq \alpha_2 \quad , \tag{4.3 - 25a}$$

$$\boldsymbol{P}_S(t) = t \, e^{\alpha t} \boldsymbol{A} + (1 - \alpha t) e^{\alpha t} \boldsymbol{I} \quad \text{for} \quad \alpha_1 = \alpha_2 = \alpha \quad , \tag{4.3 - 25b}$$

$$\boldsymbol{P}_S(t) = \alpha^{-1} \left(e^{\alpha t} - 1\right) \boldsymbol{A} + \boldsymbol{I} \qquad \text{for} \quad \alpha_1 = \alpha \neq 0, \ \alpha_2 = 0 \quad . \tag{4.3 - 25c}$$

c) Propagators of associated differential equations

The homogeneous d'Alembert systems (4.3 - 1a&b) as well as their associated differential equations (4.3 - 8a) can be solved with propagators. The corresponding solutions have the form

$$\begin{pmatrix} x(t) \\ y(t) \end{pmatrix} = \boldsymbol{P}_{\mathrm{S}}(t) \begin{pmatrix} x(0) \\ y(0) \end{pmatrix} \quad \text{for the d'Alembert system} \quad , \tag{4.3 - 26a}$$

$$\begin{pmatrix} x(t) \\ \dot{x}(t) \end{pmatrix} = \boldsymbol{P}(t) \begin{pmatrix} x(0) \\ \dot{x}(0) \end{pmatrix} \quad \text{for the associated equation} \quad . \tag{4.3 - 26b}$$

The propagator $\boldsymbol{P}(t)$ can be used for the solution of all types of linear differential equations of second order as demonstrated by (2.2 - 15a-c) and (2.3 - 18a&b).

According to (4.3 - 9a) the propagator $\boldsymbol{P}_{\mathrm{S}}(t)$ of the d'Alembert system (4.3 - 1a) is related to the propagator $\boldsymbol{P}(t)$ of the associated differential equation (4.3 - 8a) by the *time-independent similarity transformation*

$$\boldsymbol{P}_{\mathrm{S}}(t) = \boldsymbol{T}\,\boldsymbol{P}(t)\,\boldsymbol{T}^{-1} \tag{4.3 - 27a}$$

with the transformation matrices

$$\boldsymbol{T}^{-1} = \begin{pmatrix} 1 & 0 \\ a_{11} & a_{12} \end{pmatrix} \quad \text{and} \quad \boldsymbol{T} = \begin{pmatrix} 1 & 0 \\ \frac{a_{11}}{a_{12}} & \frac{1}{a_{12}} \end{pmatrix} \quad . \tag{4.3 - 27b}$$

4.4 Two-dimensional Quadratic Autonomous Systems

In many cases the behavior of a two-dimensional autonomous or time-independent system at a singular point $\vec{r}_{\mathrm{S}}$ can be determined with the aid of the first approximation in the form of a two-dimensional linear autonomous system described in Section 4.3. If, however, this linear system is degenerate refuge has to be taken to the second approximation that constitutes a quadratic system. For this reason the following gives a systematic survey on two-dimensional quadratic autonomous systems. Additional information on nonlinear systems and their critical points is found in the following sections, Chapter 6, and in the literature [Guckenheimer & Holmes 1983 B, Verhulst 1990 B].

4.4.1 Classification

A two-dimensional autonomous system with a singular point in its origin $\vec{r}_{\mathrm{S}} = \vec{0}$ can be approximated in its standard form by the *quadratic system*

$$dx/dt = \dot{x} = u(x,y) \approx a_{11}x + a_{12}y + c_{11}x^2 + c_{12}xy + c_{13}y^2$$
$$dy/dt = \dot{y} = \upsilon(x,y) \approx a_{21}x + a_{22}y + c_{21}x^2 + c_{22}xy + c_{23}y^2 \qquad (4.4 - 1)$$
$$\text{with} \quad u(0,0) = \upsilon(0,0) = 0 \quad \text{and} \quad \boldsymbol{A} = \begin{Bmatrix} a_{11} & a_{12} \\ a_{21} & a_{22} \end{Bmatrix} \quad ,$$

if $u(x, y)$ and $\upsilon(x,v)$ are analytical in the origin. ***A*** represents the characteristic matrix of the two-dimensional linear autonomous system discussed in the previous Section 4.3.

The most important aspect of quadratic autonomous systems is their behavior at the singular point in the origin $\vec{r}_S = \vec{0}$. This behavior can be *classified on the basis of the characteristic matrix* ***A*** of the corresponding linear system:

a) Corresponding linear system non-degenerate
The behavior of the quadratic system (4.4 - 1) at its singular point in the origin $\vec{r}_S = \vec{0}$ is determined by the corresponding linear system (4.3 - 1a&b) if the determinant of the characteristic matrix ***A*** differs from zero

$$det\,\boldsymbol{A} = a_{11}a_{22} - a_{12}a_{21} = \alpha_1\alpha_2 \neq 0$$
$$\text{with} \quad \Lambda(\boldsymbol{A}) = \alpha_1, \alpha_2 \quad . \qquad (4.4 - 2a)$$

This implies that the two eigenvalues α_1, α_2 of the matrix ***A*** are different from zero and that thc *corresponding linear system* (4.3 - 1a&b) *is non-degenerate.* With respect to the behavior of the system (4.4 - 1) the condition (4.4 - 2a) permits therefore to neglect the quadratic terms and to set $c_{ij} = 0$. On this condition the behavior of the quadratic system (4.4 - 1) corresponds therefore to those of the linear systems (4.3 - 1a&b) of Section 4.3.

b) Corresponding linear system degenerate
If the determinant of the characteristic matrix ***A*** vanishes

$$det\,\boldsymbol{A} = a_{11}a_{22} - a_{12}a_{21} = \alpha_1\,\alpha_2 = 0$$
$$\text{with} \quad \Lambda(\boldsymbol{A}) = \alpha_1, \alpha_2 \quad . \qquad (4.4 - 2b)$$

either one or both of the eigenvalues $\Lambda(\boldsymbol{A}) = \alpha_1, \alpha_2$ of the matrix ***A*** are zero. The *corresponding linear system* (4.3 - 1a&b) *is degenerate.* As a consequence, the quadratic terms and coefficients c_{ij} have to be taken into account with regard to the behavior of the quadratic system (4.4 - 1) at the singular point in its origin $\vec{r}_S = \vec{0}$.

Three different types of quadratic systems with a zero determinant (4.4 - 2b) of the matrix ***A*** can be distinguished. This distinction is made with respect to the *eigenvalues* $\Lambda(\boldsymbol{A}) = \alpha_1, \alpha_2$ and the *canonical forms* $\boldsymbol{A}_{can}$ of this matrix ***A***.

α) The *type I quadratic systems* (4.4 - 1) are characterized by the zero matrix

$$A_{\text{can}} = \begin{pmatrix} 0 & 0 \\ 0 & 0 \end{pmatrix} \quad \text{with} \quad \Lambda(A) = 0,0 \quad . \tag{4.4 - 3a}$$

These systems of type I are discussed for the *standard representation*

$$\begin{aligned} \dot{x} &= u(x,y) = D\,x^2 + E\,xy + F\,y^2 \\ \dot{y} &= \upsilon(x,y) = A\,x^2 + B\,xy + C\,y^2 \quad . \end{aligned} \tag{4.4 - 3b}$$

β) The *type II quadratic systems* (4.4 - 1) are characterized by the matrix

$$A_{\text{can}} = \begin{pmatrix} 0 & 0 \\ \eta & 0 \end{pmatrix} \quad \text{with} \quad \Lambda(A) = 0,0 \quad \text{and} \quad \eta \neq 0 \quad . \tag{4.4 - 4a}$$

They have the *standard representation* for $\eta = 1$:

$$\begin{aligned} \dot{x} &= u(x,y) = \qquad D\,x^2 + E\,xy + F\,y^2 \\ \dot{y} &= \upsilon(x,y) = x + A\,x^2 + B\,xy + C\,y^2 \quad . \end{aligned} \tag{4.4 - 4b}$$

γ) The *type III quadratic systems* (4.4 - 1) are characterized by the following canonical form and eigenvalues $\Lambda(\boldsymbol{A})$ of the matrix $\boldsymbol{A}$

$$A_{\text{can}} = \begin{pmatrix} 0 & 0 \\ 0 & \mu \end{pmatrix} \quad \text{with} \quad \Lambda(A) = 0, \mu \quad \text{where} \quad \mu \neq 0 \quad . \tag{4.4 - 5a}$$

The *standard representation* for $\mu = 1$ of these type III systems is

$$\begin{aligned} \dot{x} &= u(x,y) = \qquad Dx^2 + E\,xy + F\,y^2 \\ \dot{y} &= \upsilon(x,y) = y + A\,x^2 + B\,xy + C\,y^2 \quad . \end{aligned} \tag{4.4 - 5b}$$

The solutions of the three types I, II and III of quadratic systems (4.4 - 1) are discussed and classified in the following sections. The main emphasis is put on the behavior of these solutions at the singular point in the origin with $x_S = y_S = 0$. On this occasion the types I to III of systems are further separated into *subtypes* with respect to the parameters *A* to *F*.

4.4.2 Type I Quadratic Systems

The trajectories and streamlines of the type I systems defined by equations(4.4 - 3ab) are determined by the *homogeneous equation*

$$\frac{dy}{dx} = \frac{Ax^2 + Bxy + Cy^2}{Dx^2 + Exy + Fy^2} \tag{4.6 - 6}$$

that can be derived by division of the two equations (4.4 - 3b). This equation is usually solved by the introduction of the new variable

$$u = y / x = \tan\varphi \quad . \tag{4.4 - 7a}$$

This results in the new equation

$$x\frac{d}{dx}\left(K + Du + \frac{E}{2}u^2 + \frac{F}{3}u^3\right) = -P(u) \tag{4.4 - 7b}$$

with the arbitrary constant K. $P(u)$ represents the *characteristic polynomial*

$$P(u) = Fu^3 + (E - C)u^2 + (D - B)u - A \tag{4.4 - 8}$$

with the three roots u_k, $k = 1, 2, 3$.

Equation (4.4 - 7b) requires that there exist solutions of the form

$$x / y = u_k = \tan\varphi_k \quad \text{with} \quad P(u_k) = 0 \quad \text{and} \quad k = 1,2,3 \quad . \tag{4.4 - 9a}$$

Since $P(u)$ is a real cubic polynomial at least one of the roots u_k and one of the solutions (4.4 - 9a) are real. For this reason the type I systems show *no proper oscillations*.

For a non-zero polynomial $P(u)$ equation (4.4 - 7b) can be transformed into

$$-d(\ell n\, x) = P^{-1}(u)\left(D + Eu + Fu^2\right)du \quad . \tag{4.4 - 9b}$$

In general this equation can be solved by integration. The resulting solution supplements the linear solution(s) (4.4 - 9a).

a) Hamiltonian systems

The simplest type I systems are Hamiltonian systems defined by the following Hamiltonian

$$H(x,y) = \frac{1}{3}\left(Fy^3 - 3Cy^2x + 3Dyx^2 - Ax^3\right) \quad . \tag{4.4 - 10a}$$

They have the form

$$\begin{aligned} \dot{x} &= u(x,y) = \quad H_y(x,y) = Dx^2 - 2Cxy + Fy^2 \\ \dot{y} &= \upsilon(x,y) = -H_x(x,y) = Ax^2 - 2Dxy + Cy^2 \end{aligned} \tag{4.4 - 10b}$$

and fulfill the condition

$$div\ \vec{\upsilon} = u_x(x,y) + \upsilon_y(x,y) = 0 \quad . \tag{4.4 - 10c}$$

Their characteristic polynomial $P(u)$ is related to the Hamiltonian as follows

$$P(u) = 3x^{-3}H(x,y) = Fu^3 - 3Cu^2 + 3Du - A \quad . \qquad (4.4 - 10d)$$

The trajectories and streamlines of the Hamiltonian systems are determined by the condition that the Hamiltonian $H(x,y)$ is a constant

$$H(x,y) = H = const \quad . \qquad (4.2 - 11i)$$

b) Classification into subtypes

The type I systems and their solutions can be classified according to the roots u_k of the characteristic polynomial $P(u)$. This real cubic polynomial has at least one real root u_1 which can be reduced to zero by the following linear orthogonal transformation

$$\begin{aligned} x' &= x \; cos \; \varphi_1 + y \; sin \; \varphi_1 \\ y' &= x \; sin \; \varphi_1 + y \; cos \; \varphi_1 \qquad (4.4 - 11) \\ &\text{with} \quad u_1 = tan \; \varphi_1 = y_1 / x_1 \quad \text{and} \quad u_1' = tan \; \varphi_1' = y_1' / x_1' = 0 \quad . \end{aligned}$$

Therefore, one can choose $u_1 = 0$ for the general classification of the type I systems.

c) Type I α systems

The type I α systems are characterized by the degeneracy of all three roots u_k of the characteristic polynomial $P(u)$ defined by (4.4 -8). Because of the assumption $u_1 = 0$, these systems fulfill the condition

$$u_k = 0, 0, 0 \quad . \qquad (4.4 - 12a)$$

The corresponding parameter relations are

$$A = 0, B = D, C = E \quad \text{and} \quad F \neq 0 \quad . \qquad (4.4 - 12b)$$

Thus, the type I α systems have the form

$$\begin{aligned} \dot{x} &= Dx^2 + Exy + Fy^2 \\ \dot{y} &= \qquad Dxy + Ey^2 \quad . \end{aligned} \qquad (4.4 - 12c)$$

Accordingly (4.4 - 9b) is reduced to

$$-d(\ell n\, x) = \left\{ u^{-1} + (E/F)u^{-2} + (D/F)u^{-3} \right\} du \quad . \qquad (4.4 - 13a)$$

The solution of this equation can be written as

$$y = y_\infty \, exp\left\{(E/F)u^{-1} + (D/2F)u^{-2}\right\} \quad . \tag{4.4 - 13b}$$

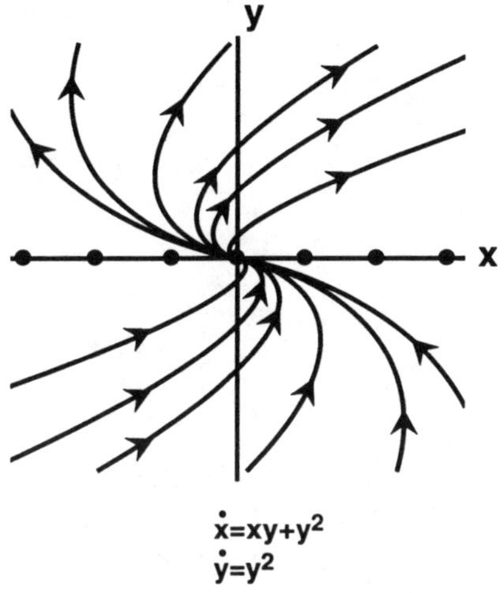

Fig. 4.4 - 1a. Type I α system (4.4 - 14a-d) with singular points on the x axis

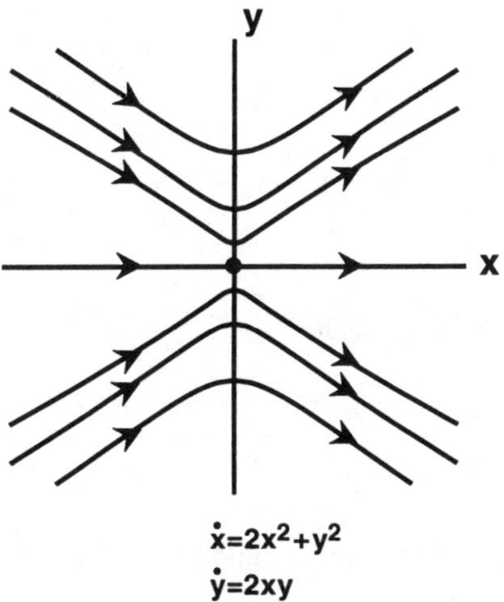

Fig. 4.4 - 1b. Type I α system (4.4 - 15a-c) with a singular point in the origin

Fig. 4.4 - 1a &b show the two typical examples determined by the following equations

$$A = B = D = 0; C = E = F = 1 \quad , \tag{4.4 - 14a}$$

$$\dot{x} = xy + y^2 \quad \text{and} \quad \dot{y} = y^2 \quad , \tag{4.4 - 14b}$$

$$x = x_0, y = 0 \quad , \tag{4.4 - 14c}$$

$$x = y \cdot \ell n(y / y_\infty) \quad , \tag{4.4 - 14d}$$

and

$$A = C = E = 0; B = D = 2; F = 1 \quad , \tag{4.4 - 15a}$$

$$\dot{x} = 2x^2 + y^2 \quad \text{and} \quad \dot{y} = 2xy \quad , \tag{4.4 - 15b}$$

$$x^2 = y^2 \, \ell n(y / y_\infty) \quad . \tag{4.4 - 15c}$$

d) Type I β systems

The type I β systems are characterized by three real roots u_k of the characteristic polynomial $P(u)$ with the additional condition that two of them are equal. In accordance with this condition and the assumption $u_1 = 0$ one can presume

$$u_1 = u_2 = 0, u_3 \neq 0 \; real \quad . \tag{4.4 - 15a}$$

These roots imply the relations

$$A = 0, B = D, C \neq E \quad . \tag{4.4 - 15b}$$

In order to survey the type I β systems one also assumes

$$u_3 = \infty \quad \text{and} \quad \varphi_3 = arctan\, u_3 = \pi / 2 \quad . \tag{4.4 - 16a}$$

This implies $F = 0$ and the following form of the type I β systems

$$\begin{aligned} \dot{x} &= Dx^2 + Exy \\ \dot{y} &= \qquad\quad Dxy + Cy^2 \end{aligned} \quad . \tag{4.4 - 16b}$$

These systems can be solved with the aid of (4.4 - 9b). This equation is reduced by the conditions (4.4 - 15b) and (4.4 - 16a) to the form

$$(E - C) d(\ell n\, x) + \left(E u^{-1} + D u^{-2}\right) du = 0 \quad . \tag{4.4 - 17}$$

The two equations (4.4 - 16b) and (4.4 - 17) yield the solutions

$$x = 0 \quad \text{and} \quad 1 / y = C(t - t_\infty) \quad , \tag{4.4 - 18a}$$

$$y = 0 \quad \text{and} \quad 1 / x = D(t - t_\infty) \quad , \tag{4.4 - 18b}$$

$$y^{E}x^{-D} = K\,exp(Dx / y) \quad \text{with} \quad E \neq C \qquad (4.4 - 18c)$$

where K indicates an arbitrary constant.

The *first example* to be considered is the type I β system with

$$A = C = 0;\, B = D = E = 1 \quad , \qquad (4.4 - 19a)$$

$$\dot{x} = x^2 + xy \quad \text{and} \quad \dot{y} = xy \quad , \qquad (4.4 - 19b)$$

$$x = 0,\, y = y_0 \quad , \qquad (4.4 - 19c)$$

$$y = 0 \quad \text{and} \quad 1 / x = (t - t_\infty) \quad , \qquad (4.4 - 19d)$$

$$x = y\,\ell n(y / y_\infty) \quad . \qquad (4.4 - 19e)$$

Equation (4.4 - 19e) demonstrates that the type I β system (4.4 - 19b) illustrated in Fig. 4.4 - 2a has the same streamlines (4.4 - 14d) as the type I α system (4.4 - 14b) shown in Fig. 4.4 - 1a. The differences between these systems are their temporal behavior on one hand and the different sets of critical points on the other hand.

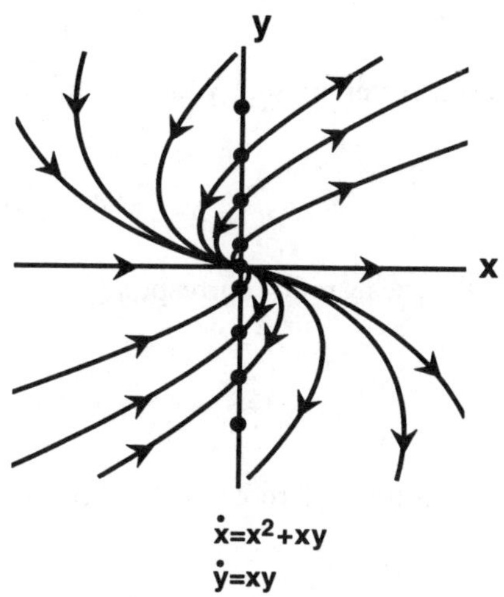

Fig. 4.4 - 2a. Type I β system (4.4 - 19a-c) with singular points on the y axis. Its streamlines are identical to those of Fig. 4.4 - 1a, yet its trajectories travel differently.

The *second example* of a type I β system is the Hamiltonian system with

$$A = B = D = F = 0;\, E = -2C = 2 \quad , \qquad (4.4 - 20a)$$

$$\dot{x} = 2xy \quad \text{and} \quad \dot{y} = -y^2 \quad . \qquad (4.4 - 20b)$$

Its trajectories and streamlines in Fig. 4.4 - 2b are determined by the Hamiltonian

$$H(x, y) = xy^2 = H = const \quad . \tag{4.4 - 20c}$$

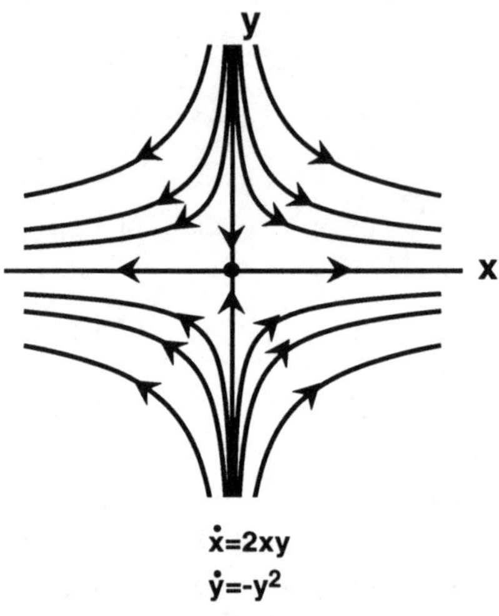

Fig. 4.4 - 2b. Hamiltonian type I β system (4.4 - 20a-c)

e) Type I γ systems

The type I γ systems are characterized by three different real roots u_k of the characteristic polynomial $P(u)$. Thus, one

$$u_1 = 0, u_2 \neq u_3 \; real \tag{4.4 - 21a}$$

by taking into account the assumption $u_1 = 0$. These roots correspond to the conditions

$$A = 0 \quad \text{and} \quad (C - E)^2 + 4(B - D)F > 0 \quad . \tag{4.4 - 21b}$$

For a characterisation of the type I γ systens the following roots (4.4 - 21a) are adequate:

$$u_k = 0, -\sqrt{3}, +\sqrt{3} \quad . \tag{4.4 - 22a}$$

These imply the relations

$$A = 0; B = D + 3F; C = E \quad . \tag{4.4 - 22b}$$

A simple type I γ system that fulfills conditions (4.4 - 23a&b) is the ideal system or flow

$$\begin{aligned} \dot{x} &= H_y(x,y) = -U_x(x,y) = x^2 - y^2 \\ \dot{y} &= -H_x(x,y) = -U_y(x,y) = -2xy \end{aligned} \qquad (4.4 - 23a)$$

with the parameters

$$A = C = E = 0;\ B = -2;\ D = +1;\ F = -1 \quad . \qquad (4.4 - 23b)$$

The solutions illustrated in Fig. 4.4 - 3 are determined by the fact that the Hamiltonian is a constant

$$H(x,y) = x^2 y - \frac{1}{3} y^3 = H = const \quad . \qquad (4.4 - 23c)$$

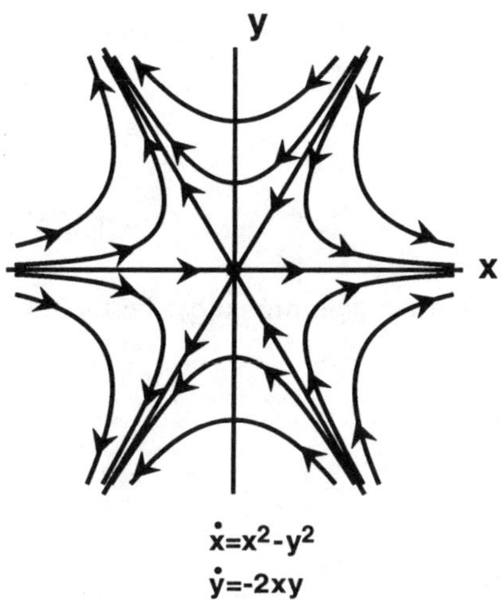

Fig. 4.4 - 3. Ideal type I γ system (4.4 - 23a-d)

The potential of this system is

$$U(x,y) = xy^2 - \frac{1}{3} x^3 \quad . \qquad (4.4 - 23d)$$

f) Type I δ systems

The type I δ systems are characterized by one real and two complex conjugate roots u_k of the characteristic polynomial $P(u)$. Thus one chooses

$$u_k = 0,\ a - ib,\ a + ib \quad \text{with} \quad a, b \ real \qquad (4.4 - 24a)$$

in agreement with the assumption $u_1 = 0$. These roots u_k imply the parameter relations

$$A = 0; (C - E)^2 + 4(B - D)F < 0 \quad . \qquad (4.4 - 24b)$$

For a comparison with the type I γ systems characterized by the roots u_k (4.4 - 22a) the following roots u_k are chosen for the type I δ systems

$$u_k = 0, -\sqrt{3}\,i, +\sqrt{3}\,i \quad . \qquad (4.4 - 25a)$$

They imply the parameter relations

$$A = 0;\ B = D - 3F;\ C = E \quad . \qquad (4.4 - 25b)$$

A simple Hamiltonian system of type I γ is

$$\begin{aligned} \dot{x} &= x^2 + y^2 = \ H_y(x, y) \\ \dot{y} &= -2xy \ \ = -H_x(x, y) \end{aligned} \qquad (4.4 - 26a)$$

with the parameters

$$A = 0,\ B = -2,\ C = 0,\ D = +1,\ E = 0,\ F = +1 \quad . \qquad (4.4 - 26b)$$

The solutions of this system illustrated in Fig. 4.4 - 4 are determined by the fact that the Hamiltonian is a constant

$$H(x, y) = x^2 y + \frac{1}{3} y^3 = H = const \quad . \qquad (4.4 - 26c)$$

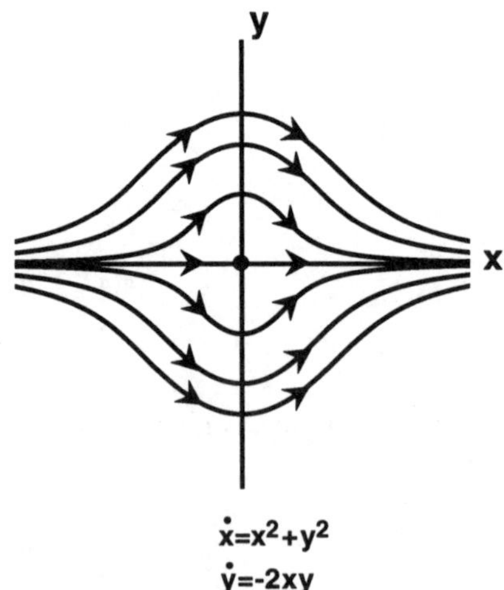

Fig. 4.4 - 4. Hamiltonian type I δ system (4.4 - 26a-c)

4.4.3 Type II Quadratic Systems

Trajectories and streamlines of the type II systems defined by equations (4.4 - 4a&b) are determined by the differential equation

$$\frac{dy}{dx} = \frac{Ax^2 + Bxy + Cy^2 + x}{Dx^2 + Exy + Fy^2} \tag{4.4 - 27}$$

that can be derived from (4.4 - 4b).

a) Hamiltonian systems

Part of the type II systems is *Hamiltonian*. These systems are defined by the Hamiltonian

$$H(x,y) = -\frac{1}{2}x^2 - \frac{1}{3}\left[Ax^3 - 3Dx^2y + 3Cxy^2 - Fy^3\right] \quad . \tag{4.4 - 28a}$$

The type II systems have the form

$$\begin{aligned} \dot{x} &= u(x,y) = +H_y(x,y) = Dx^2 - 2Cxy + Fy^2 \\ \dot{y} &= \upsilon(x,y) = -H_x(x,y) = Ax^2 - 2Dxy + Cy^2 + x \\ &\text{with} \quad B = -2D \quad \text{and} \quad E = -2C \end{aligned} \tag{4.4 - 28b}$$

and fulfill the condition

$$div\,\vec{\upsilon} = u_x(x,y) + \upsilon_y(x,y) = 0 \quad . \tag{4.4 - 28c}$$

Trajectories and streamlines of these Hamiltonian systems are determined by constant values of the Hamiltonian

$$H(x,y) = H = const \quad . \tag{4.2 - 11i}$$

The type II systems are classified in the following. This classification is based on their behavior exactly at and in the immediate neighborhood of the singularity in the origin $x_S = y_S = 0$. It differs from the classification of the type I systems that takes into account the asymptotic behavior of the trajectories at infinity.

b) Type II α systems

The type II α systems are characterized by a non-zero parameter F

$$F \neq 0 \quad . \tag{4.4 - 29a}$$

This condition demands that a trajectory that reaches of passes through the singularity in the origin $x_S = y_S = 0$ can be represented by a series of the form

$$y(x) = \sum_{k=2}^{\infty} a_k x^{k/3} = a_2 x^{2/3} + a_3 x^1 + a_4 x^{4/3} + \ldots$$
$$= (3/2\ F)x^{2/3} + \frac{3C-2E}{F} x + \ldots \qquad (4.4 - 29b)$$

with $x^{2/3}$ as leading term.

A *first example* is the system

$$\dot{x} = x^2 + y^2 \quad \text{and} \quad \dot{y} = x \qquad (4.4 - 30a)$$

with the parameters

$$A = B = C = E = 0;\ D = F = 1 \quad . \qquad (4.4 - 30b)$$

Its trajectories plotted in Fig. 4.4 - 5a, can be represented analytically as follows

$$x^2 = x_0^2 exp(2y) + \frac{1}{2}\left[exp(2y) - 1 - 2y - 2y^2\right]$$
$$= x_0^2 exp(2y) + \frac{1}{3}\left[2y^3 + y^4 + \ldots\right] \quad . \qquad (4.4 - 30c)$$

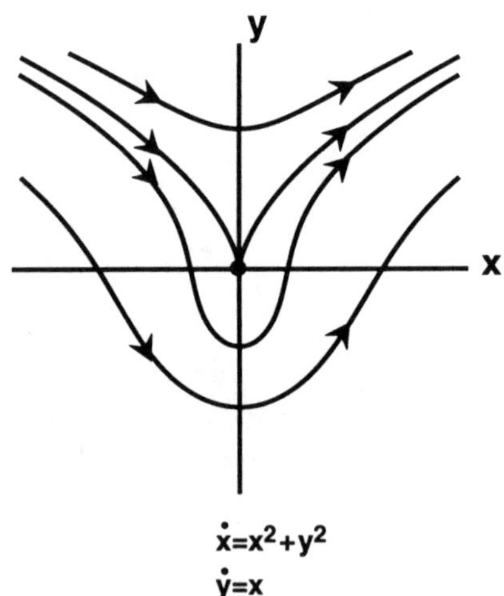

Fig. 4.4 - 5a. Type II α system (4.4 - 30a-c)

As *second example* serves the Hamiltonian type II α system

$$\dot{x} = (y - x)^2 \quad \text{and} \quad \dot{y} = (y - x)^2 + x \qquad (4.4 - 31a)$$

with the parameters

$$A = C = D = F = 1; B = E = -2 \quad . \tag{4.4 - 31b}$$

Its trajectories shown in Fig. 4.4 - 5b are defined by constant values of the Hamiltonian

$$H(x,y) = -\frac{1}{2}x^2 + \frac{1}{3}(y-x)^3 = H = const \quad . \tag{4.4 - 31c}$$

Therefore, they can be written as

$$y = \left[\frac{3}{2}\left(x^2 - x_0^2\right)\right]^{1/3} + x \quad . \tag{4.4 - 31d}$$

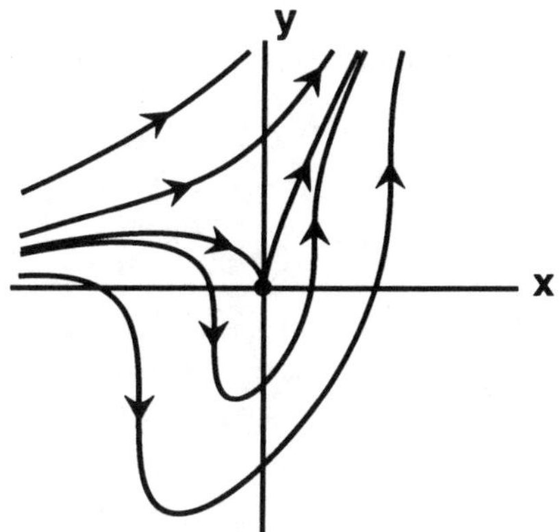

Fig. 4.4 - 5b. Hamiltonian type II α system (4.4 - 31a-d)

c) Type II β systems

The type II β systems fulfill the following parameter conditions

$$F = 0 \quad \text{and} \quad E \neq \frac{4C}{m}; \quad m = 2,3,4, \ldots \quad . \tag{4.4 - 32a}$$

These conditions imply two trajectories through the singularity in the origin $x_S = y_S = 0$. The first is the straight line

$$x = 0 \quad \text{with} \quad \dot{y} = Cy^2 \quad , \tag{4.4 - 32b}$$

whilst the second can be represented by the following series

$$x(y) = \sum_{k=2}^{\infty} b_k y^k = b_2 y^2 + b_3 y^3 + \dots$$
$$= \frac{1}{2}(E-2C)y^2 + \frac{1}{2}\frac{(E-2C)}{(3E-4C)}(D-2B)y^3 + \dots \qquad (4.4 - 32c)$$

with y^2 in the leading term.

The condition $F = 0$ implies an *additional singularity* of a type II β system besides the singularity in the origin $x_S = y_S = 0$. Its position is

$$x_2 = -\left[A - B(D/E) + C/D/E)^2\right]^{-1} \quad ,$$
$$y_2 = +(D/E)\left[A - B(D/E) + C(D/E)^2\right]^{-1} \quad . \qquad (4.4 - 33)$$

A *first example* of a type II β system is Scherrer's Hamiltonian system [Scherrer 1996 J]

$$\dot{x} = +H_y(x,y) = -xy$$
$$\dot{y} = -H_x(x,y) = x + \frac{3}{2}x^2 + \frac{1}{2}y^2 \qquad (4.4 - 34a)$$

with the Hamiltonian

$$H(x,y) = -\frac{1}{2}\left(x^2 + x^3 + xy^2\right) \qquad (4.4 - 34b)$$

and the additional singular point

$$x_2 = -2/3, \; y_2 = 0 \quad . \qquad (4.4 - 34c)$$

The first of the two trajectories passing the origin is the straight line (4.4 - 32b). The second forms the circle

$$\left(x + \frac{1}{2}\right)^2 + y^2 = \frac{1}{4} \quad . \qquad (4.4 - 34d)$$

These and other trajectories are illustrated in Fig. 4.4 - 6a.

The singularity of this system in the origin $x_S = y_S = 0$ is peculiar because for an initial point inside the circle (4.4 - 34d) the solution of the system is periodic, whereas for an initial point outside the circle the solution is instable. In addition, the solution is strictly stable for an initial point on this circle because it ends in the origin with $x_S = y_S = 0$.

A second example of a type II β system is the simple non-Hermitian system

$$\dot{x} = x^2 + xy \quad \text{and} \quad \dot{y} = x \tag{4.4 - 35a}$$

that has the straight line of singular points

$$x = 0, y = y_0 \tag{4.4 - 35b}$$

as solution and no second singularity (4.4 - 33).

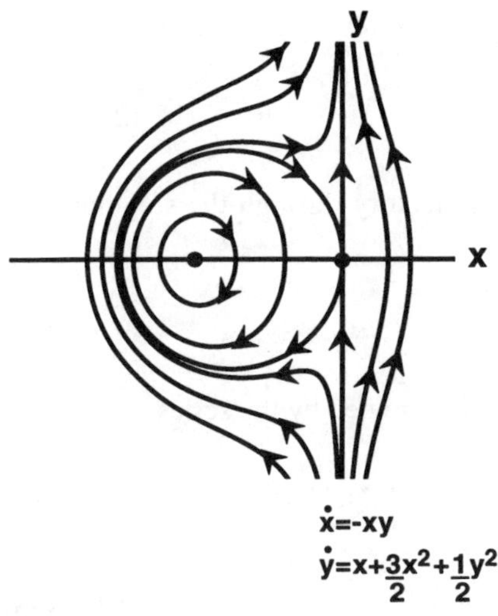

Fig. 4.4 - 6a. Scherrer's Hamiltonian type II β system (4.4 - 34a-d) with periodic solutions

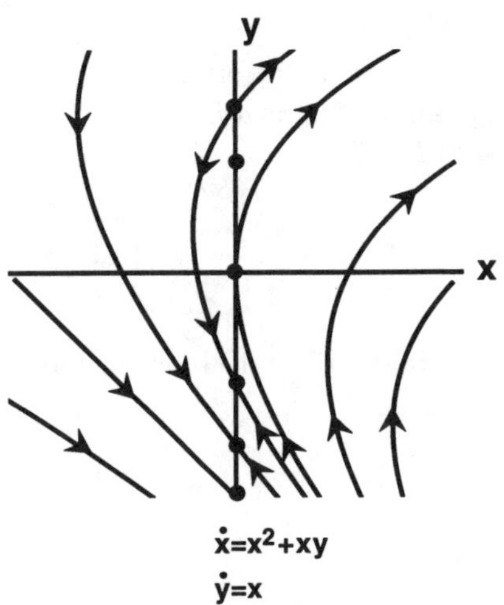

Fig. 4.4 - 6b. Type II β system (4.4 - 35a-d) with singular points on the y axis

Its trajectories illustrated in Fig. 4.4 - 6b have the form

$$x(y) = (1 + x_0) exp(y) - 1 - y \quad \text{with} \quad x_0 = x(0) \quad . \tag{4.4 - 35c}$$

The second trajectory through the origin is accordingly

$$x(y) = exp(y) - 1 - y = \frac{1}{2} y^2 + \frac{1}{6} y^3 + \ldots \quad . \tag{4.4 - 35d}$$

d) Type II γ systems

The type II γ systems are defined by the parameter conditions

$$F = 0 \quad \text{and} \quad E = 2C \quad . \tag{4.4 - 36a}$$

These systems are characterized by the straight trajectory through the singularity in the origin

$$x = 0 \quad \text{with} \quad \dot{y} = Cy^2 \quad . \tag{4.4 - 32b}$$

The other trajectories that cut the x axis can be approximated by the series

$$x(y) = \sum_0^\infty c_k y^k = x_0 + c_1 y + c_2 y^2 + \ldots$$
$$= x_0 + x_0 \frac{d}{1 + Ax_0} y + \ldots \tag{4.4 - 36b}$$
$$\text{with} \quad x_0 = x(0) \neq 0 \quad .$$

Also these systems exhibit a second singularity besides that in the origin $x_S = y_S = 0$. Its coordinates are given by (4.4 - 33) when taking into account that $E = 2C$.

As *first example* serves the type II γ system with $D = 0$

$$\begin{aligned} \dot{x} &= xy \\ \dot{y} &= x + x^2 + y^2 \quad . \end{aligned} \tag{4.4 - 37a}$$

Its second singular point is at $x = 1$, $y = 0$, whilst its trajectories fulfill the equation

$$\frac{d(y^2)}{d(\ell n(\pm x))} - y^2 = x + x^2 \tag{4.4 - 37b}$$

with the solution

$$y^2 = \pm[y^2(\pm 1) - 1]x + x^2 + x\, \ell n(\pm x) \quad . \tag{4.4 - 37c}$$

These trajectories, the two critical points and the relevant straight trajectory through the origin are illustrated in Fig. 4.4 - 7a. This figure reveals that this system has two ranges of stability. Initial points in the range of the closed trajectories imply strictly stable solutions that end in the origin, while initial points outside this range result in instable solutions.

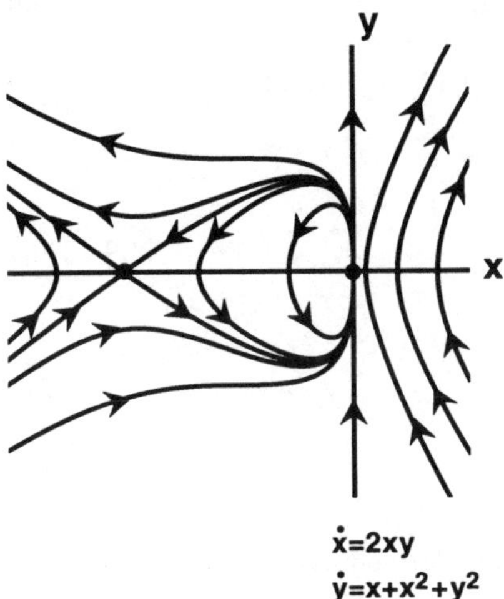

Fig. 4.4 - 7a. Type II γ system (4.4 - 37a-c) with a region of strict stability

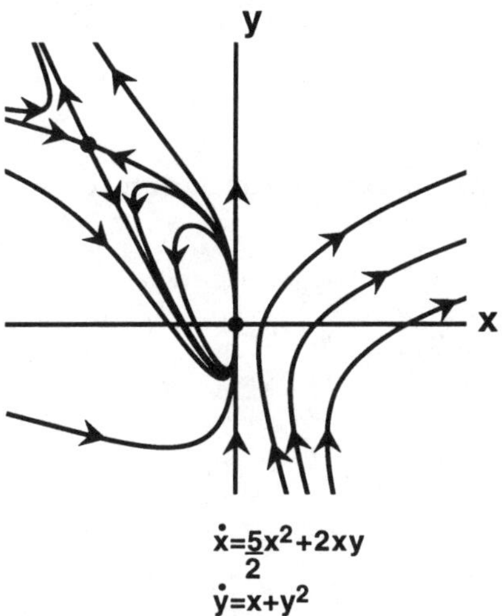

Fig. 4.4 - 7b. Type II γ system (4.4 - 38a&b). This figure represents a deformation of Fig. 4.4 - 7a.

The *second example* of a type II γ system to be considered is the following system with $A = 0$

$$\dot{x} = \frac{5}{2}x^2 + 2xy$$
$$\dot{y} = x \quad + y^2 \quad . \qquad (4.4 - 38a)$$

It has a second singularity at x = -16/25, y = 4/5. The trajectories of this system shown in Fig. 4.4 - 7b are similar to those of the system (4.4 - 37a), apart from deformation.

The trajectories that cut the x axis in a point outside the origin $x_S = y_S = 0$ can be approximated by

$$x(y) = x_0 + \frac{5}{2}x_0 y + \left(1 + \frac{25}{8}x_0\right)y^2 + \ldots \qquad (4.4 - 38b)$$
$$\text{with} \quad x_0 = x(0) \neq 0$$

in agreement with (4.4 - 36b).

4.4.4 Type III Quadratic Systems

Trajectories and streamlines of the type III systems defined by equations (4.4 - 5ab) are determined by the differential equation

$$\frac{dy}{dx} = \frac{Ax^2 + Bxy + Cy^2 + y}{Dx^2 + Exy + Fy^2} \quad . \qquad (4.4 - 39)$$

a) Gradient systems

There exists a variety of type III *gradient systems* that are characterized by the potential

$$U(x,y) = -\frac{1}{2}y^2 - \frac{1}{3}\left[Dx^3 + 3Ax^2y + 3Fxy^2 + Cy^3\right] \quad . \qquad (4.4 - 40a)$$

These systems have the form

$$\dot{x} = u(x,y) = -U_x(x,y) = Dx^2 + 2Axy + Fy^2$$
$$\dot{y} = \upsilon(x,y) = -U_y(x,y) = Ax^2 + 2Fxy + Cy^2 + y \qquad (4.4 - 40b)$$
$$\text{with} \quad B = 2F \quad \text{and} \quad E = 2A \quad .$$

They obey the condition

$$\left[curl\,\vec{\upsilon}\right]_z = \upsilon_x(x,y) - u_y(x,y) = 0 \quad . \qquad (4.4 - 40c)$$

In these systems the trajectories and streamlines are perpendicular to the equipotential lines with

$$U(x, y) = U = const \quad . \tag{4.2 - 9h}$$

They can be characterized by the vector equation

$$\vec{v} \times d\vec{r}(U = const) = -grad\, U \times d\vec{r}(U = const) = 0 \quad . \tag{4.2 - 9f}$$

On the contrary, there exist no type III Hamiltonian systems.

In the following the type III systems are classified according to the behavior of their trajectories at and near the singularity in the origin $x_S = y_S = 0$ similar to the classification of the type II systems.

b) Type III α systems

The type III α systems are characterized by the parameter condition

$$F \neq 0 \quad . \tag{4.4 - 41a}$$

This condition implies that the trajectory passing the singular point at the origin $x_S = y_S = 0$ can be represented by the series

$$x(y) = \sum_{k=2}^{\infty} d_k y^k = \frac{F}{2} y^2 \left[1 + \frac{1}{3}(E - C)y + \ldots \right] \tag{4.4 - 41b}$$

with y^2 in the leading term.

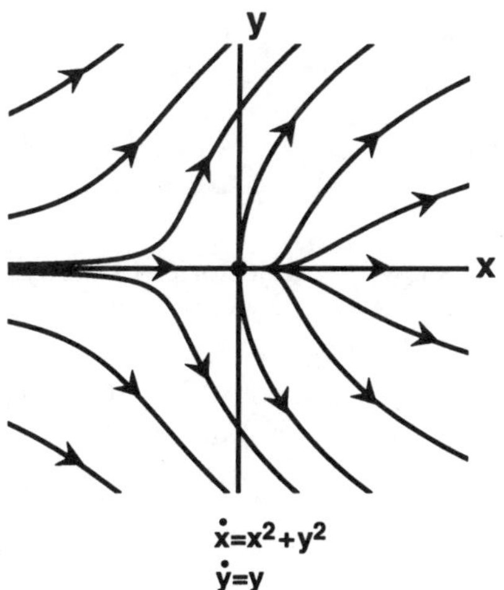

Fig. 4.4 - 8. Type III α system (4.4 - 42a-c)

As *example* Fig. 4.4 - 8 shows the trajectories of the type III α system

$$\dot{x} = x^2 + y^2 \quad \text{and} \quad \dot{y} = y \tag{4.4 - 42a}$$

with the parameters

$$A = B = C = E = 0; D = F = 1 \quad . \tag{4.4 - 42b}$$

The trajectory that passes the singular point in the origin $x_S = y_S = 0$ can be approximated by

$$x(y) = \frac{1}{2} y^2 + \frac{1}{16} y^4 + \ldots \quad . \tag{4.4 - 42c}$$

c) Type III β systems

The type III β systems fulfill the parameter conditions

$$F = 0 \quad \text{and} \quad A \neq 0 \quad . \tag{4.4 - 43a}$$

As a consequence the trajectories and streamlines of these systems that pass through the singularity in the origin $x_S = y_S = 0$ can be represented by the series

$$y(x) = \sum_{k=2}^{\infty} c_k x^k = -Ax^2 [1 + (2D - B)x + \ldots] \quad . \tag{4.4 - 43b}$$

The condition $F = 0$ may imply a second singular point besides that in the origin $x_S = y_S = 0$. Its coordinates are

$$\begin{aligned} x_2 &= -(E/D)\left[A(E/D)^2 - B(E/D) + C\right]^{-1} \quad , \\ y_2 &= \left[A(E/D)^2 - B(E/D) + C\right]^{-1} \quad . \end{aligned} \tag{4.4 - 43c}$$

A simple *example* is illustrated in Fig. 4.4 - 9. It shows the trajectories of the type III β system

$$\dot{x} = x^2 \quad \text{and} \quad \dot{y} = x^2 + y \tag{4.4 - 44a}$$

with the parameters

$$B = C = E = F = 0; A = D = 1 \quad . \tag{4.4 - 44b}$$

This system does not exhibit the second singularity (4.4 -41c) mentioned above.

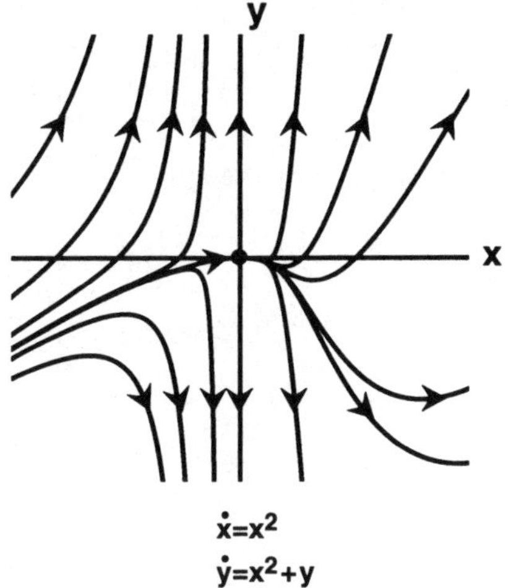

Fig. 4.4 - 9. Type III β system (4.4 - 44a-e)

The trajectory passing through the singularity in the origin $x_S = y_S = 0$ of this system can be approximated by

$$y(x) = -x^2 - 2x^3 + \dots \qquad (4.4 - 44c)$$

in accordance with (4.4 - 41b). The trajectories of the system (4.4 - 44a) fulfill the differential equation

$$\frac{dy}{dx} - x^{-2} y = 1 \quad , \qquad (4.4 - 44d)$$

which can be solved analytically. The result is

$$y(x) = \left[y(\infty) - \int_{x}^{\infty} exp\left(\frac{1}{x}\right) dx \right] exp\left(-\frac{1}{x}\right) \quad . \qquad (4.4 - 44e)$$

d) Type III γ systems

The III γ systems are characterized by the parameter conditions

$$A = F = 0 \quad \text{and} \quad D \neq 0 \quad . \qquad (4.4 - 45a)$$

Therefore, they can be written as

$$\begin{aligned} \dot{x} &= Dx^2 + Exy = x(Dx + Ey) \\ \dot{y} &= Bxy + Cy^2 + y = y(1 + Bx + Cy) \end{aligned} \quad . \qquad (4.4 - 45b)$$

They usually show a second singular point with the coordinates

$$x_2 = E[CD - BE]^{-1} \quad \text{and} \quad y_2 = -D[CD - BE]^{-1} \quad . \tag{4.4 - 45c}$$

For small $|y| << |x|$ the type III γ systems can be approximated by the system

$$\dot{x} = Dx^2 \quad \text{and} \quad \dot{y} = y(1 + Bx) \tag{4.4 - 46a}$$

whose trajectories obey the relation

$$\frac{dy}{dx} = y\left(\frac{1}{Dx^2} + \frac{B}{Dx}\right) \quad . \tag{4.4 - 46b}$$

This differential equation has the solution

$$y = y_\infty x^{B/D} \exp(-1/Dx) \tag{4.4 - 46c}$$

that characterizes the irregular singularities of the type III γ systems in the origin $x_S = y_S = 0$.

For *illustration* Fig. 4.4 - 10a shows the trajectories of the system

$$\dot{x} = x^2 + xy \quad \text{and} \quad \dot{y} = y \tag{4.4 - 47a}$$

with the parameters

$$A - B = C = F = 0; D = E = 1 \quad . \tag{4.4 - 47b}$$

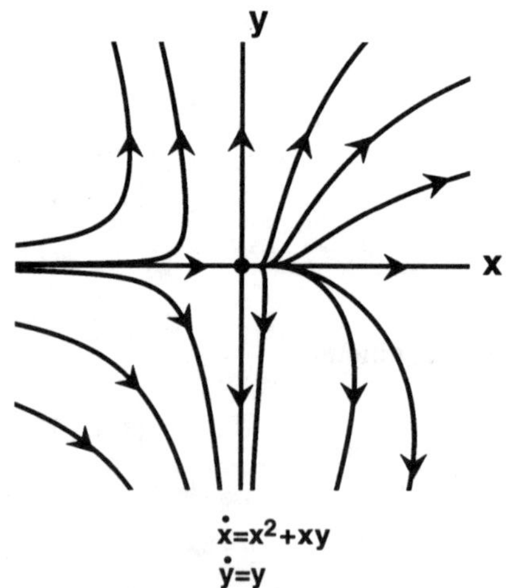

Fig. 4.4 - 10a. Type III γ system (4.4 - 47a&b) with a single singular point in the origin

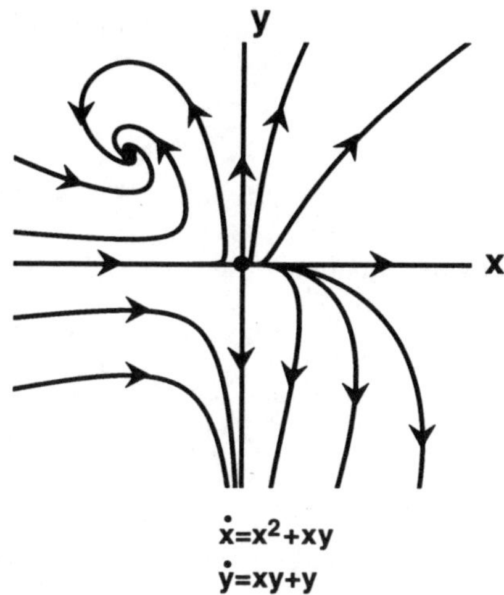

Fig. 4.4 - 10b. Type III γ system (4.4 - 48a-c) with a second singular point outside the origin

This system does not exhibit a second singularity.

Fig. 4.4 - 10b presents as further *example* the trajectories of the system

$$\dot{x} = x^2 + xy \quad \text{and} \quad \dot{y} = y + xy \tag{4.4 - 48a}$$

with the parameters

$$A = C = F = 0; B = D = E = 1 \quad . \tag{4.4 - 48b}$$

On the contrary to the system (4.4 - 47a) the system (4.4 - 48a) includes a second singularity at the coordinates

$$x_2 = -1 \quad \text{and} \quad y_2 = +1 \quad . \tag{4.4 - 48c}$$

e) Type III δ systems

The type III δ systems obey the parameter conditions

$$A = D = F = 0 \quad \text{and} \quad E \neq 0 \quad . \tag{4.4 - 49a}$$

and can therefore be represented by

$$\begin{aligned} \dot{x} &= Exy \\ \dot{y} &= Bxy + Cy^2 + y = y(1 + Bx + Cy) \quad . \end{aligned} \tag{4.4 - 49b}$$

Consequently, these systems exhibit the straight line of singularities

$$x = x_0, \, y = 0 \qquad (4.4 - 49c)$$

plus a singularity at

$$x = 0, \, y = -1/C \quad . \qquad (4.4 - 49d)$$

The trajectories of these systems are determined by the equation

$$\frac{dy}{dx} - \frac{C}{Ex} \cdot y = \frac{B}{E} + \frac{1}{Ex} \qquad (4.4 - 49e)$$

with the solution

$$y(x) = \frac{1}{E} \ell n \, x + \frac{B}{E} x + K \cdot x^{C/E} \qquad (4.4 - 49f)$$

where K is an arbitrary constant.

As *example* Fig. 4.4 - 11 shows the trajectories of the type III δ system

$$\dot{x} = xy \quad \text{and} \quad \dot{y} = y + xy + 2y^2 \qquad (4.4 - 50a)$$

with the parameters

$$A = D = E = 0; 2B = C = 2E = 2 \quad . \qquad (4.4 - 50b)$$

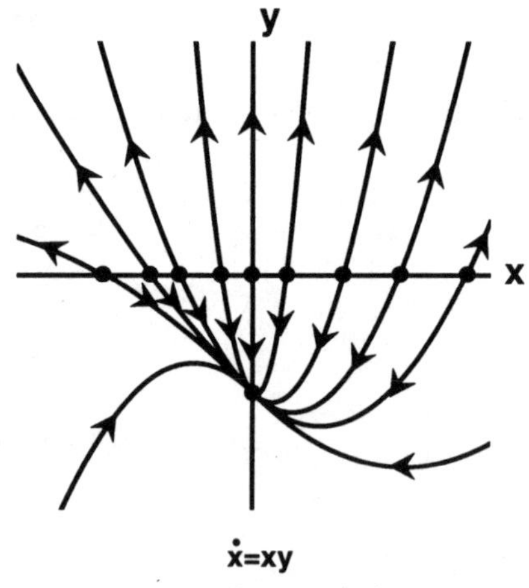

Fig. 4.4 - 11. Type III δ system (4.4 - 50a-e) with singular points on the x axis and one singular point on the y axis

These trajectories are defined by the equation

$$y(x) = \ell n\, x + x + K \cdot x^2 \qquad (4.4 - 50c)$$

with the constant K.

This system contains singularities on the straight line

$$x = x_0;\ y = 0 \qquad (4.4 - 50d)$$

and one at the point

$$x = 0,\ y = -\frac{1}{2} \quad . \qquad (4.4 - 50e)$$

4.5 Limit Cycles of Two-dimensional Autonomous Nonlinear Systems

In the following we consider an isolated closed curve K in the two-dimensional space (x, y) of a two-dimensional *autonomous* system of differential equations or in the phase space $(x, \dot{x})$ of a differential equation of second order. This curve K represents a *limit cycle* if on one hand it corresponds to a solution of the system of the differential equation, and if on the other hand the trajectories and streamlines in the phase space spiral towards or away from it [Bogoljubow & Mitropolsky 1965 B, Delamotte 1993 Z, Pontrjagin 1985 B, Slotine & Li 1991 B, Verhulst 1990 B, Zwillinger 1989 B]. If the neighboring trajectories and streamlines spiral towards the curve K, it is called a *stable limit cycle*. If, on the contrary, they spiral away from it, it is named an *instable limit cycle*. A *semistable limit cycle* is a curve K where on one side the streamlines or the trajectories spiral towards it and on the other side they spiral away from it.

Stable limit cycles constitute a special type of *attractors*. They correspond to *stationary oscillations* of oscillators and systems. The trajectories and streamlines which spiral towards the stable limit cycles represent the *transients* of these oscillations.

If two or more limit cycles form a concentric system of closed curves K in the (x, y) space or in the $(x, \dot{x})$ phase space, then the stable and the instable limit cycles alternate. In this situation a critical point inside this concentric system of limit cycles can be considered as *a degenerate limit cycle*.

4.5.1 Limit Cycles of Axially Symmetric Systems

Axially symmetric autonomous systems are best suited for the discussion of limit cycles because in these systems the *limit cycles form concentric circles*. The adequate description of axially symmetric two-dimensional systems makes use of the *polar coordinates*

$$r = +\left(x^2 + y^2\right)^{1/2} \quad \text{and} \quad \varphi = arc\,tg\left(y / x\right) \quad . \qquad (4.2 - 4a)$$

In these coordinates the two-dimensional autonomous systems have the standard form

$$\begin{aligned} \dot{r} &= \upsilon_r = -U_r + \frac{1}{r} H_\varphi \quad \text{with} \quad U = U(r, \varphi) \\ r\dot{\varphi} &= \upsilon_\varphi = -\frac{1}{r} U_\varphi - H_r \quad \text{with} \quad H = H(r, \varphi) \end{aligned} \qquad (4.2 - 6)$$

where U indicates the potential and H the Hamiltonian.

An *axially symmetric system* is characterized by a potential U and a Hamiltonian H which are *independent of the angle* φ

$$U = U(r) \quad \text{and} \quad H = H(r) \quad . \qquad (4.5 - 1a)$$

On this condition, the system (4.2 - 6) is reduced to the form

$$\begin{aligned} \dot{r} &= \upsilon_r(r) = -U_r \quad \text{with} \quad U = U(r) \quad , \\ r\dot{\varphi} &= \upsilon_\varphi(r) = -H_r \quad \text{with} \quad H = H(r) \quad . \end{aligned} \qquad (4.5 - 1b)$$

The simplest system of this type is the *pure rotation* with constant circular frequency ω_0. It is defined by

$$U(r) = const \quad \text{and} \quad H(r) = -\frac{1}{2} \omega_0 \, r^2 \quad . \qquad (4.5 - 2a)$$

The application of these functions in the system (4.5 - 1b) results in

$$r = const \quad \text{and} \quad \dot{\varphi} = \omega_0 = const \quad . \qquad (4.5 - 2b)$$

The *existence of limit cycles* in axially symmetric systems is determined by the potential $U(r)$. This requires at least one radius r_k that fulfills the condition

$$-\dot{r} = U_r\left(r_k\right) = 0 \quad \text{with} \quad k = 0, 1, 2, \ldots \quad . \qquad (4.5 - 3a)$$

As a consequence, isolated limit cycles in the form of circles exist only for nonlinear potentials $U(r)$ with at least one extremum as illustrated in Fig. 4.5 - 1.

The stability of a limit cycle or circle with a radius $r = r_k$, $k = 0, 1, 2, \ldots$ is determined by the sign of $\dot{r}(r)$ for small deviations Δr from r_k as shown in Fig. 4.5 - 1. The following situations occur

$$\dot{r}(r_k + \Delta r)\Delta r \begin{cases} < 0 & \text{stable} \\ = 0 & \text{indifferent, semistable} \\ > 0 & \text{instable} \end{cases} \quad . \tag{4.5 - 3b}$$

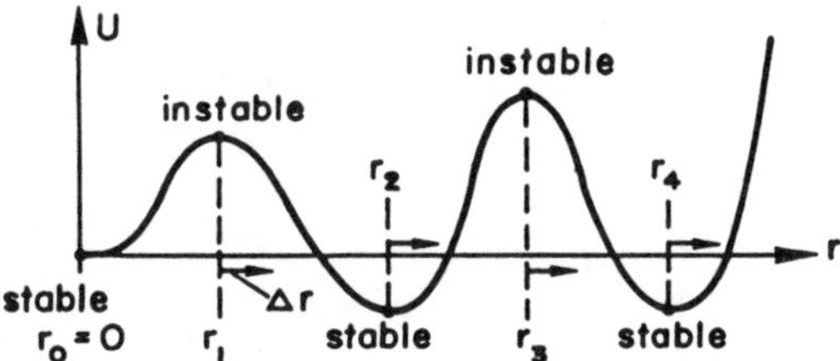

Fig. 4.5 - 1. Potential $U(r)$ of an axially symmetric two-dimensional autonomous system of differential equations with $r^2 = x^2 + y^2$

If the potential $U(r)$ is analytic at $r = r_k$, the product (4.5 - 3b) can be written as

$$\dot{r}(r_k + \Delta r)\Delta r = -\sum_{m=2}^{\infty} \Delta r^m \frac{1}{(m-1)!} \frac{d^m U}{dr^m}(r_k) \quad . \tag{4.5 - 3c}$$

The combination of the relations (4.5 - 3b) and the series (4.5 - 3c) yields the following classification of the limit cycles with respect to stability

$$U_{rr}(r_k) \begin{cases} > 0 & \text{stable} \\ = 0 & \text{indifferent, semistable} \\ < 0 & \text{instable} \end{cases} \quad . \tag{4.5 - 3d}$$

Three typical examples of limit cycles in axially symmetric autonomous systems are presented in the following.

a) System with a stable limit cycle

The *standard example* [Verhulst 1990 B, Zwillinger 1989 B] of a stable limit cycle in an axially symmetric system and the corresponding instable critical point in the origin is defined by the following potential U and Hamiltonian H:

$$U(r) = \frac{1}{2} r^2 \left[\frac{1}{2} r^2 - 1 \right] \quad \text{and} \quad H(r) = -\frac{1}{2} r^2$$
$$\text{with} \quad U_r(r) = r(r^2 - 1); \quad U_{rr}(r) = 3r^2 - 1 \quad . \tag{4.5 - 4a}$$

The potential $U(r)$ is illustrated in Fig. 4.5 - 2a. The system with $U(r)$ and $H(r)$ of (4.5 - 4a) has the following form in polar coordinates

$$\dot{r} = -U_{\mathrm{r}} = r\left(1-r^2\right)$$
$$\dot{\varphi} = -\frac{1}{r} H_{\mathrm{r}} = 1 \qquad \qquad (4.5 - 4b)$$

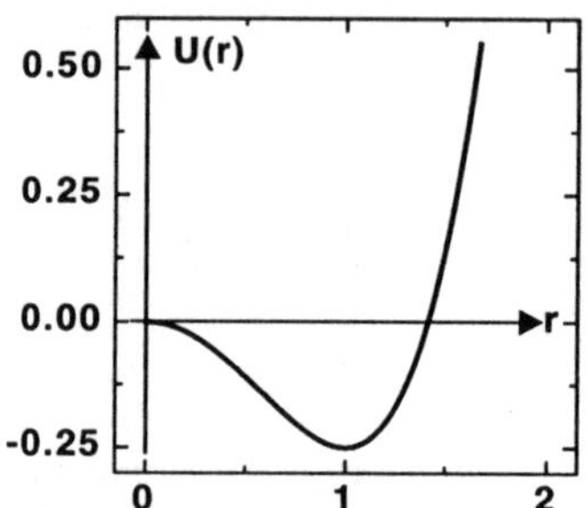

Fig. 4.5 - 2a. Potential $U(r)$ of the axially symmetric system (4.5 - 4b&c) with a stable limit cycle with $r = r_1 = 1$ and an instable critical point in the origin with $r = r_0 = 0$

In most books [e.g. Verhulst 1990 B, Zwillinger 1989 B] this system is described in Cartesian coordinates

$$\dot{x} = -y + x\left[1-x^2-y^2\right]$$
$$\dot{y} = +x + y\left[1-x^2-y^2\right] \qquad \qquad (4.5 - 4c)$$

According to condition (4.5 - 3a) and classification (4.5 - 3d) this system includes *an instable critical point* at the origin $r = r_0 = 0$ and a *stable limit cycle* with the radius $r = r_1 = 1$. This is demonstrated by the following data concerning the potential $U(r)$ defined by (4.5 - 4a)

$$U_{\mathrm{r}}(0) = 0, \quad U_{\mathrm{rr}}(0) = -1 \quad ,$$
$$U_{\mathrm{r}}(1) = 0, \quad U_{\mathrm{rr}}(1) = +2 \quad . \qquad \qquad (4.5 - 4d)$$

The system described by (4.5 - 4b&c) has the *general solution* in polar coordinates [Zwillinger 1989 B]

$$\varphi(t) = t \quad ,$$
$$r(t) = +\left[1 - e^{-2\mathrm{t}} + r(0)^{-2} e^{-2\mathrm{t}}\right]^{-1/2} = r(0)e^{\mathrm{t}}\left[1 + r(0)^2\left(e^{2\mathrm{t}} - 1\right)\right]^{-1/2} \quad . \qquad \qquad (4.5 - 4e)$$

This solution can be derived with the ansatz

$$r = +1/\sqrt{y} \quad \text{and} \quad y = 1/r^2 \quad .$$

For long times $t \to \infty$ the solution converges versus the limit cycle with $r = 1$ independent of the initial condition $r = r(0)$

$$r(t \to \infty) = 1 \quad . \tag{4.5 - 4f}$$

This is illustrated in Fig. 4.5 - 2b.

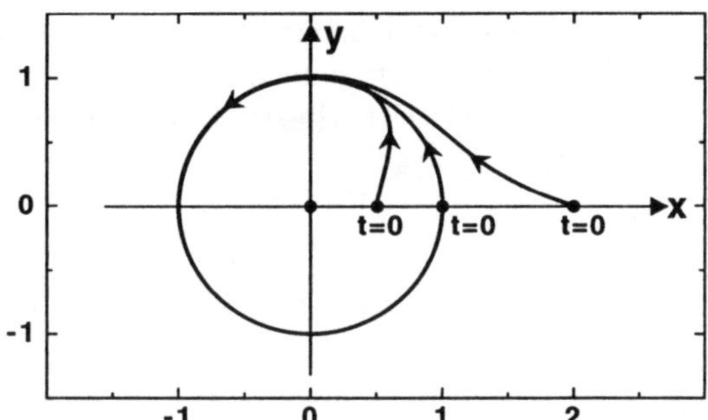

Fig. 4.5 - 2b. Trajectories of the axially symmetric system (4.5 - 4b&c) with a stable limit cycle

b) System with an instable and a stable limit cycle

An axially symmetric system with a stable critical point in the origin, an instable and a stable limit cycle is defined by the following potential U and Hamiltonian H

$$U(r) = \frac{1}{4} r^2 (r-2)^2 \quad \text{and} \quad H(r) = -\frac{1}{2} r^2$$
$$\text{with} \quad U_{\mathrm{r}}(r) = r(r-1)(r-2) \; ; \; U_{\mathrm{rr}}(r) = 3r^2 - 6r + 2 \quad . \tag{4.5 - 5a}$$

This potential $U(r)$ is plotted in Fig. 4.5 - 3a.

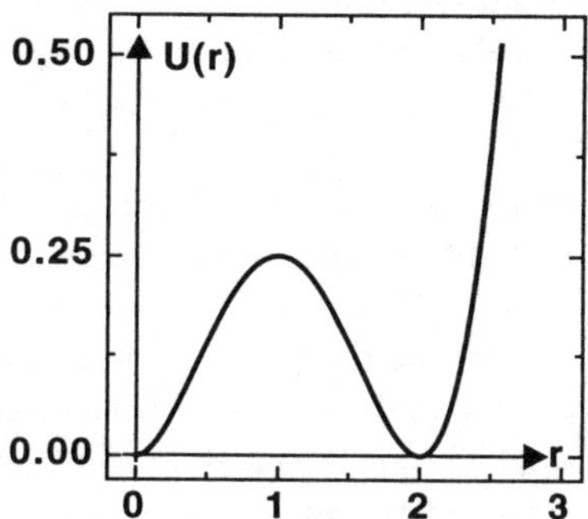

Fig. 4.5 - 3a. Potential $U(r)$ of the axially symmetric system (4.5 - 5b) including a stable limit cycle with $r = r_2 = 2$, an instable limit cycle with $r = r_1 = 1$ and a stable critical point in the origin with $r = r_0 = 0$

The system described by the potential $U(r)$ and the Hamiltonian $H(r)$ of (4.5 - 5a) can be represented in polar coordinates by

$$\begin{aligned} \dot{r} &= -U_{\mathrm{r}} = -r(r-1)(r-2) \\ \dot{\varphi} &= -\frac{1}{r} H_{\mathrm{r}} = 1 \quad . \end{aligned} \qquad (4.5 - 5b)$$

This system is characterized by a stable critical point in the origin with $r = r_0 = 0$, an instable limit cycle with $r = r_1 = 1$ and a stable limit cycle with $r = r_2 = 2$. Critical point and limit cycles can be derived from its potential $U(r)$. This potential, which is illustrated in Fig. 4.5 - 3a, exhibits the following derivates

$$\begin{aligned} U_{\mathrm{r}}(0) &= 0\,, \quad U_{\mathrm{rr}}(0) = +2 \quad , \\ U_{\mathrm{r}}(1) &= 0\,, \quad U_{\mathrm{rr}}(0) = -1 \quad , \\ U_{\mathrm{r}}(2) &= 0\,, \quad U_{\mathrm{rr}}(0) = +2 \quad . \end{aligned} \qquad (4.5 - 5c)$$

The system (4.5 - 5b) has the following general solution in polar coordinates

$$\begin{aligned} \varphi(t) &= t \quad , \\ r(t) &= 1 + sign\big(r(0)-1\big)\cdot\Big[1 - e^{-2\mathrm{t}} + \big(r(0)-1\big)^{-2} e^{-2\mathrm{t}}\Big]^{-1/2} \\ &= 1 + \big(r(0)-1\big)\, e^{\mathrm{t}} \cdot \Big[1 + \big(r(0)-1\big)^2 \big(e^{2\mathrm{t}} - 1\big)\Big]^{-1/2} \quad . \end{aligned} \qquad (4.5 - 5d)$$

This solution can be derived with the ansatz

$$r = 1 + sign\big(r(0)-1\big)\cdot\big(1/\sqrt{y}\big) \quad \text{and} \quad y = (r-1)^{-2} \quad .$$

Fig. 4.5 - 3b shows the solution (4.5 - 5d) for the initial conditions $r(0) = 2/3, 3/2, 3$. The solution for $r(0) = 2/3 < 1$ spirals towards the stable critical point in the origin with $r = 0$, whilst those for $r(0) = 3/2, 3 > 1$ approach the stable limit cycle with $r = 2$.

The system in consideration represents a model for the starting of the piston engines used in motor cars and airplanes as well as of femtosecond-KLM (Kerr-lens modelocking) solid-state lasers [Ippen 1994 J, Keller et al. 1991 J, Salin et al. 1991 J]. Both, piston engines and KLM lasers, do not start by their own. They need an impulse for starting. In our model this type of start corresponds to a rapid transition from rest in the stable critical point in the origin with $r = r_0 = 0$ to a state near the stable limit cycle with $r = r_2 = 2$. This transition caused by an impulse at time $t = -\tau < 0$ can be incorporated in our model system (4.5 - 5b) by introduction of a Dirac delta function (2.3 - 46c):

$$\dot{\varphi} = 1$$
$$\dot{r} = R\,\delta(t+\tau) - r(r-1)(r-2) \qquad (4.5 - 6a)$$
$$\text{with} \quad 0 < \tau << 1 \quad .$$

Thus, one assumes that the system is at rest in the stable critical point in the origin $r = r_0 = 0$ at times $t < -\tau$. The impulse to start the system is applied between times $t = -2\tau$ and $t = 0$. Since $0 < \tau < < 1$ it is permitted to write

$$r(0) \cong \int_{-2\tau}^{0} R\,\delta(t+\tau)dt = R \quad . \qquad (4.5 - 6b)$$

For times $t > 0$ both, the systems (4.5 - 5b) and (4.5 - 6a) have the solution (4.5 - 5d) with the initial condition $r(0) = R$. Thus, the *condition for successfully starting the system to oscillate permanently* is

$$R = r(0) > 1 \quad . \qquad (4.5 - 6c)$$

On this condition the system approaches the stable limit cycle with $r = r_2 = 2$ and begins a stationary oscillation. For $R = r(0) < 1$ the system returns to rest in the stable critical point in the origin with $r = r_0 = 0$.

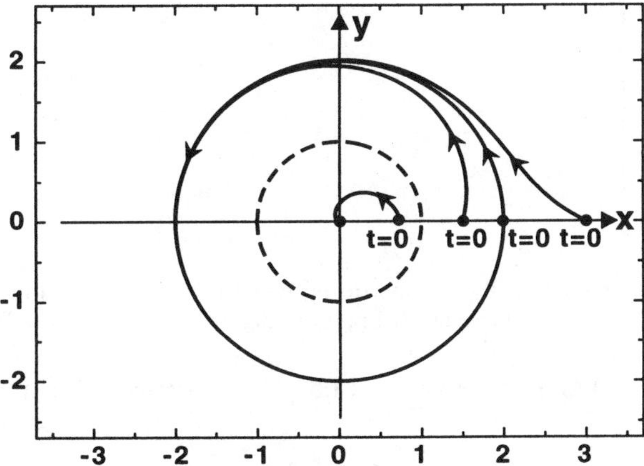

Fig. 4.5 - 3b. Trajectories of the axially symmetric systems (4.5 - 5b) with a stable and an instable limit cycle

c) System with semistable limit cycle

The last system to be discussed exhibits a semistable limit cycle defined by the following potential $U(r)$ and Hamiltonian $H(r)$

$$U(r) = \frac{1}{6} r^2 \left(r^4 - 3r^2 + 3\right) \quad \text{and} \qquad H(r) = -\frac{1}{2} r^2$$

$$\text{with} \quad U_{\mathrm{r}}(r) = r\left(r^2 - 1\right)^2 \quad , \quad U_{\mathrm{rr}}(r) = \left(5r^2 - 1\right)\left(r^2 - 1\right) \quad , \qquad (4.5 - 7a)$$

$$U_{\mathrm{rrr}}(r) = 4r\left(5r^2 - 3\right) \quad .$$

The potential $U(r)$ is shown in Fig. 4.5 - 4 a. The system defined by $U(r)$ and $H(r)$ of (4.5 - 7a) can be represented in polar coordinates by

$$\dot{\varphi} = -\frac{1}{r} H_{\mathrm{r}} = 1 \quad \text{and} \quad \dot{r} = -r\left(1 - r^2\right)^2 \quad . \qquad (4.5 - 7b)$$

This system includes a *stable critical point* in the origin with $r = r_0 = 0$ and a *semistable limit cycle* with $r = r_1 = 1$. This can be verified by consideration of Fig. 4.5 - 4a and by taking account of

$$\begin{aligned} &U_{\mathrm{r}}(0) = 0 \,, \quad U_{\mathrm{rr}}(0) = 1 \,, \quad U_{\mathrm{rrr}}(0) = 0 \quad , \\ &U_{\mathrm{r}}(1) = 0 \,, \quad U_{\mathrm{rr}}(1) = 0 \,, \quad U_{\mathrm{rrr}}(1) = 8 \quad . \end{aligned} \qquad (4.5 - 7c)$$

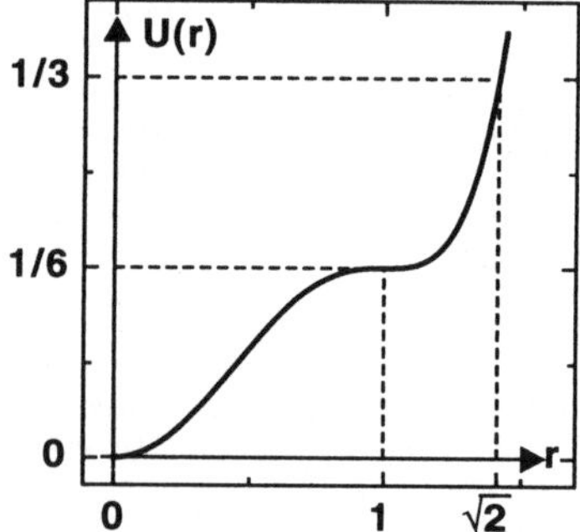

Fig. 4.5 - 4a. Potential $U(r)$ of the axially symmetric system (4.5 - 7b) including a semistable limit cycle with $r = r_1 = 1$ and a stable critical point in the origin with $r = r_0 = 0$

In polar coordinates the general solution of the system (4.5 - 7b) for $r(0) \neq 0$, 1 can be written as

$$\varphi(t) = t \quad ,$$

$$\left[1 - \frac{1}{r^2}\right] exp\left[\frac{1}{r^2 - 1}\right] = \left[1 - \frac{1}{r(0)^2}\right] exp\left[\frac{1}{r(0)^2 - 1}\right] exp\,(2t) \quad . \qquad (4.5 - 7d)$$

This solution can be derived with the aid of the ansatz

$$r = cosh\, u \quad \text{for} \quad r(0) > 1 \quad \text{and} \quad r = cos\, u \quad \text{for} \quad r(0) < 1 \quad .$$

The limits of this solution are

$$\begin{aligned} r(+\infty) &= 1 \quad \text{for} \quad r(0) > 1 \quad , \\ r(+\infty) &= 0 \quad \text{for} \quad r(0) < 1 \quad , \\ r(-\infty) &= 1 \quad \text{for} \quad r(0) < 1 \quad . \end{aligned} \tag{4.5 - 7e}$$

These data demonstrate that there exists a semistable limit cycle with $r = r_1 = 1$. The general solutions (4.5 - 7d) for the initial conditions $r(0)^2 = 1/2$, 2 are illustrated in Fig. 4.5 - 4b.

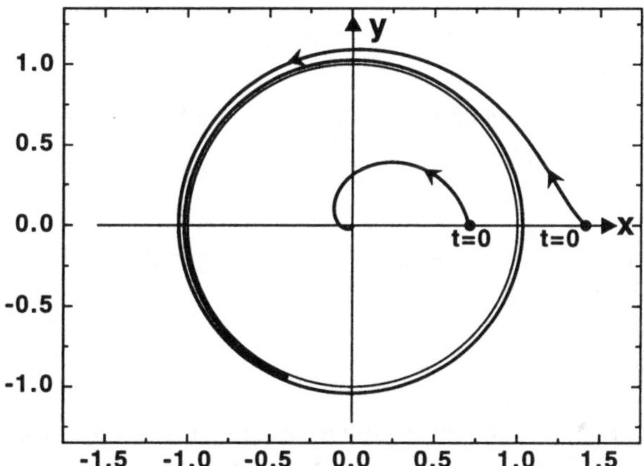

Fig. 4.5 - 4b. Trajectories of the axially symmetric system (4.5 - 7b) with a semistable limit cycle

4.5.2 Existence of Limit Cycles

Three *theorems on the existence of limit cycles* are relevant for the two-dimensional autonomous systems of explicit differential equations of first order

$$\begin{aligned} \dot{x} &= u(x,y) = -U_x(x,y) + H_y(x,y) \\ \dot{y} &= \upsilon(x,y) = -U_y(x,y) - H_x(x,y) \end{aligned} \tag{4.2 - 1}$$

with the vector representation

$$\dot{\vec{r}} = \vec{\upsilon}(\vec{r}) \quad . \tag{4.2 - 2}$$

a) Theorem of Poincaré and Bendixon

If a trajectory or streamline of a two-dimensional autonomous system or a differential equation of second order is situated within a closed, singly connected region G of the

(x, y) plane or the $(x, \dot{x})$ phase plane, then one of the three following statements are valid:

α) The trajectory or streamline leads to a stable critical point.
β) The trajectory or streamline approaches asymptotically a limit cycle.
γ) The trajectory or streamline is a limit cycle.

b) Theorem of Bendixon on the nonexistence of limit cycles
If the *divergence* of a system (4.2 - 1) or (4.2 - 2), which is defined by

$$\begin{aligned} div\, \vec{\upsilon}(\vec{r}) &= u_x(x,y) + \upsilon_y(x,y) \\ &= -\Delta U(x,y) = -U_{x,x}(x,y) - U_{yy}(x,y) \quad , \end{aligned} \tag{4.5 - 8}$$

is either positive or negative everywhere in a closed, singly connected region G, then there exists *no limit cycle.*

c) Index theorem of Poincaré
If there exists a limit cycle in a closed, singly connected region G, then the following relation is valid

$$N - S = 1 \tag{4.5 - 9}$$

where S indicates the number of saddle points and N the total number of nodes, focii or spirals, and centers. These types of singular points are described in the analysis of the critical point of homogeneous d'Alembert systems in Section 4.3.4.

The relation (4.5 - 9) is called index theorem because Poincaré introduced specific indices j for the different types of critical points and limit cycles when he proved this theorem [Bogoljubow & Mitropolsky 1965 B, Verhulst 1990 B]. These indices are $j = 0$ for regular points, $j = -1$ for saddle points and $j = +1$ for nodes, focii or spirals, centers and limit cycles.

The index theorem (4.5 - 9) demonstrates that the existence of a limit cycle in a closed, singly connected region G requires at least one node, center, focus or spiral in G.

Limit cycles of two-dimensional autonomous *systems without axial symmetry* and the related differential equations are discussed in Section 2.5.4 for *Smith* oscillators and in Section 2.5.5 for *van der Pol* oscillators.

4.6 Stability Criteria of Lyapunov

The *qualitative theory* of a single differential equation or of a system of differential equations was introduced by H. Poincaré in the 1880's [Poincaré 1892a B, 1885 J]. This theory does not require exact or approximative solutions of differential equations. It gives information on the properties of the solutions by taking account of the structure of the single differential equation or of the system of differential equations. A relevant type of information concerns the stability of solutions. The first significant qualitative theory on stability was formulated by A.M. Lyapunov in the 1890's [Lyapunov 1892/1992 B] who introduced the concept of *Lyapunov functions* [Beltrami 1987 B, Guckenheimer & Holmes 1983 B, Hahn 1959 B, Hairer et al. 1980 B, LaSalle & Lefschetz 1961 B, 1967 B, Lyapunov 1892/1992 B, Slotine & Li 1991 B, Verhulst 1990 B, Zwillinger 1989 B]. These functions are related to the potential $U(\vec{r})$ of the two-dimensional autonomous gradient systems discussed in Section 4.2.6. During its motion $\vec{r}(t)$ the gradient system obeys the condition

$$\frac{d}{dt}U(\vec{r}) = -|grad\ U(\vec{r})|^2 \le 0 \quad . \tag{4.2 - 33}$$

This inequality is characteristic for Lyapunov functions.

4.6.1 Conservative Field of Force

To a physicist the concept of the Lyapunov functions can be elucidated with the aid of the dynamics of a point of mass in a conservative static field of force

$$\vec{F}(\vec{r}) = -grad\ E_{\text{pot}}(\vec{r}) = -grad\ V(\vec{r}) \tag{4.6 - 1a}$$

with the potential energy

$$E_{\text{pot}}(\vec{r}) = V(\vec{r}) = -\int_{\vec{0}}^{\vec{r}} \vec{F}(r) \cdot d\vec{r} + V(\vec{0}) \quad . \tag{4.6 - 1b}$$

The work δW performed on a point of mass during a displacement from $\vec{r}$ to $\vec{r} + d\vec{r}$ amounts to

$$\delta W = -\vec{F}(\vec{r}) \cdot d\vec{r} = +grad\ V(\vec{r}) \cdot d\vec{r} = dV \quad . \tag{4.6 - 2a}$$

Consequently, the momentary power P applied at the position $\vec{r}$ is

$$P = \frac{\delta W}{dt} = +grad\ V(\vec{r}) \cdot \vec{v}(\vec{r}) = \dot{V}(\vec{r}) \quad . \tag{4.6 - 2b}$$

The power P is positive when the point of mass gains potential energy, and negative when it looses potential energy.

Equation (4.6 - 2b) is based on the assumption that the point of mass moves according to a stationary velocity field, which corresponds to an autonomous system of explicit differential equations described by

$$\dot{\vec{r}} = \vec{\upsilon}(\vec{r}) \quad \text{with} \quad \frac{\partial}{\partial t}\vec{\upsilon} = \vec{0} \quad . \tag{4.2 - 2}$$

For this system of differential equations the loss of potential energy means stabilization, whilst the gain of potential energy signifies destabilization near a critical point $\vec{r}_S$ with

$$\vec{\upsilon}(\vec{r}_S) = \vec{0} \quad . \tag{4.1 - 4b}$$

In conclusion one finds for

$$\dot{V}(\vec{r}) \begin{cases} < 0 & \text{an instable solution} \quad . \\ = 0 & \text{a stable solution } (4.3-17\,\text{d}) \quad . \\ > 0 & \text{an asymptotically stable solution } (4.3-17\,\text{b}) \quad . \end{cases} \tag{4.6 - 3}$$

Thus $\dot{V}(\vec{r})$ gives information on the stability of the solutions of the system (4.2 - 2) near a critical point before they are known precisely or approximately. This is the purpose of a Lyapunov function.

4.6.2 Lyapunov Functions and Stability

A *Lyapunov function* is by definition a scalar positive definite function $V(\vec{r})$ in a space or in a phase space, which fulfills the following conditions in an open region B around a critical point in the origin with $\vec{r} = \vec{r}_S = \vec{0}$

$$\textit{grad}\, V(\vec{r}) \quad \text{exists} \quad , \tag{4.6 - 4a}$$

$$V(\vec{r}) \quad \text{and} \quad \textit{grad}\, V(\vec{r}) \quad \text{are continous} \quad , \tag{4.6 - 4b}$$

$$V\left(\vec{r} = \vec{0}\right) = 0 \quad , \tag{4.6 - 4c}$$

$$V\left(\vec{r} \neq \vec{0}\right) > 0 \quad , \tag{4.6 - 4d}$$

$$\dot{V}(\vec{r}) \leq 0 \quad . \tag{4.6 - 4e}$$

With the aid of this Lyapunov function $V(\vec{r})$ the following theorems on the stability of autonomous systems and related differential equations can be formulated:

a) Stability theorem of Lyapunov
If a Lyapunov function defined by (4.6 - 4a-e) exists in an open region B around the critical point in the origin $\vec{r} = \vec{r}_S = \vec{0}$, then the solutions of the autonomous system

(4.2 - 2) of differential equations and the associated differential equations are *stable* according to (4.3 - 17d) [LaSalle & Lefschetz 1961 B, 1967 B].

b) Theorem on the asymptotic stability

A more specific statement than the stability theorem of Lyapunov is possible with the aid of the theorem on the asymptotic stability [LaSalle & Lefschetz 1961 B, 1967 B]:

If the Lyapunov function $V(\vec{r})$ as well as its negative time derivative $-\dot{V}(\vec{r})$ are positive definite, then the solutions of the autonomous system (4.2 - 2) and the associated differential equations are *asymptotically stable* according to (4.3 - 17b).

A positive definite $-\dot{V}(\vec{r})$ means that the condition (4.6 - 4e) is *replaced* by the two conditions

$$\dot{V}\left(\vec{r}=\vec{0}\right)=0 \tag{4.6 - 4f}$$

$$\dot{V}\left(\vec{r}\neq\vec{0}\right)<0 \quad . \tag{4.6 - 4g}$$

c) Theorem on the attractor in the origin

The previous theorems b) and c) on stability can be formulated even more specifically, if the conditions (4.6 - 4 a-d) and (4.6 - 4f&g) are supplemented by the following condition:

The Lyapunov function $V(\vec{r})$ fulfills the inequalities

$$0<V\left(\vec{r}\right)<E \tag{4.6 - 4h}$$

in a region B_{E} around the critical point in the origin with $\vec{r}=\vec{r}_{\mathrm{S}}=\vec{0}$ that is situated within the region B defined in relation to (4.6 - 4a-g). The regions B and B_{E} are illustrated in Fig. 4.6 - 1.

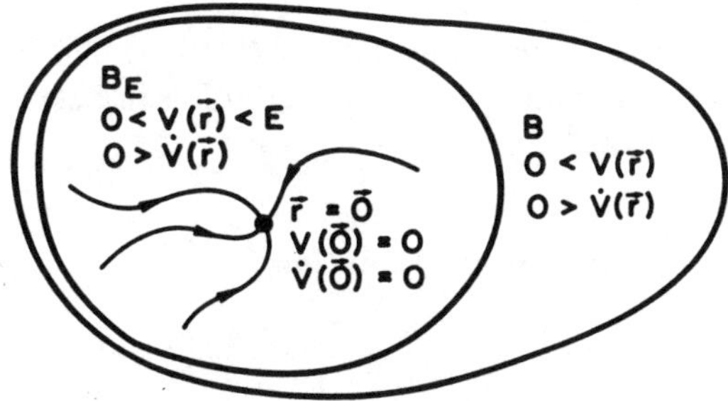

Fig. 4.6 - 1. Theorem on the attractor in the origin

If this condition as well as (4.6 - 4a-d) and (4.6 - 4 f&g) are obeyed, then each solution of the autonomous system (4.2 - 2) and the associated differential equations is *asymptotically stable* according to (4.3 - 17b) and approaches the origin $\vec{r} = \vec{r}_S = \vec{0}$ for $t \to \infty$.

d) Instability theorem of Tchetayev

The instability of the solutions of an autonomous system (4.2 - 2) and the associated differential equations at the critical point in the origin with $\vec{r} = \vec{r}_S = \vec{0}$ is determined by the instability theorem of Tchetayev [LaSalle & Lefschetz 1961 B, 1967 B].

As prerequisite of this theorem it is first assumed that the scalar function $V(\vec{r})$ fulfills the conditions (4.6 - 4a-c) in the open region B. Secondly, it is also assumed that in a region B_I within B also contains the singular point $\vec{r} = \vec{r}_S = \vec{0}$ as an edge point as illustrated in Fig. 4.6 - 2, the function $V(\vec{r})$ obeys three conditions

$$\dot{V}(\vec{r}) > 0 \quad \text{for } \vec{r} \text{ in } B \quad , \tag{4.6 - 4i}$$

$$V(\vec{r}) = 0 \quad \text{for } \vec{r} \text{ on the edge of } B_I \quad , \tag{4.6 - 4j}$$

$$V(\vec{r}) > 0 \quad \text{for } \vec{r} \text{ in } B_I \quad . \tag{4.6 - 4k}$$

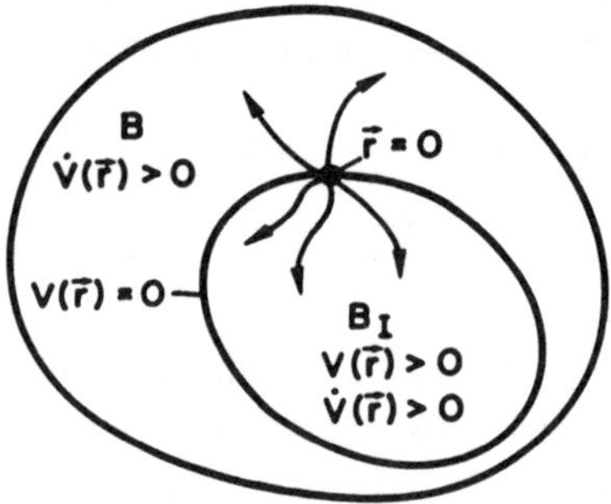

Fig. 4.6 - 2. Instability theorem of Tchetayev

On these assumptions, the solutions of the autonomous system (4.2 - 2) and the associated differential equations are *instable* according to the definition in Section 4.3.3 on the stability at the critical point in the origin $\vec{r} = \vec{r}_S = 0$.

4.6.3 The Hamilton Function as Lyapunov Function

As mentioned before the potential $U(\vec{r})$ of a *gradient field* described in Section 4.2.6 can act as Lyapunov function and thus fulfills the inequality (4.2 - 30).

Since the construction of a Lyapunov function for a given system of differential equations is difficult in general, one has to discuss the possibility of applying the *Hamiltonian function* $H(\vec{r})$ *as Lyapunov function* $V(\vec{r})$ in two-dimensional systems of the form

$$\dot{x} = u(x,y) = -U_x(x,y) + H_y(x,y)$$
$$\dot{y} = \upsilon(x,y) = -U_y(x,y) - H_x(x,y) \quad . \qquad (4.2 - 1)$$

In these systems, the Hamiltonian *H(x,y)* can be applied as Lyapunov function *V(x,y)* *if it is positive definite* in *x* and *y*

$$V = V(x,y) = H(x,y) > 0 \quad \text{if} \quad [x,y] \neq [0,0] \quad . \qquad (4.6 - 5a)$$

The temporal variation of this Lyapunov function is determined by the system (4.2 - 1) as follows

$$\dot{V} = \dot{H} = -\left[H_x(x,y)U_x(x,y) + H_y(x,y)\,U_y(x,y)\right] \quad . \qquad (4.6 - 5b)$$

a) Lyapunov function of Liénard oscillators

The Hamiltonian *H(x,y)* can be applied as Lyapunov function *V(x,y)* for the Liénard oscillators described in Section 2.5. These are defined by the differential equation

$$\ddot{x} - S(x)\cdot\dot{x} + D(x)\cdot x = 0 \qquad (2.5 - 1a)$$

corresponding to the autonomous system

$$\dot{x} = -U_x(x) + y$$
$$\dot{y} = -H_x(x) \qquad (2.5 - 1b)$$

with the following potential *U* and Hamiltonian *H*

$$U = U(x) \quad \text{and} \quad H = H(x,y) = \frac{1}{2}F\left(x^2\right) + \frac{1}{2}y^2 \quad . \qquad (2.5 - 1c)$$

The functions of the differential equation (2.5 - 1a) and of the system (2.5 - 1b) are related by

$$S(x) = -U_{xx}(x) \quad \text{and} \quad D(x) = d\,F(x^2)\,/\,dx^2 \quad . \qquad (2.5 - 1d)$$

The introduction of the Hamiltonian of (2.5 - 1c) as *Lyapunov function*

$$V = V(x,y) = H(x,y) = \frac{1}{2}F(x^2) + \frac{1}{2}y^2 \qquad (4.6 - 6a)$$

and the subsequent combination of this function with (4.6 - 5b) result in

$$\dot{V} = -H_x(x)U_x(x) = +x\,D(x)\int_0^x S(x)\,dx \qquad (4.6 - 6b)$$

in agreement with [LaSalle & Lefschetz 1961 B, 1967 B].

b) Lyapunov function of harmonic oscillators
According to definition (2.2 - 20) a harmonic oscillator with or without damping or amplification represents a linear Liénard oscillator with the parameters

$$S(x) = -2/\tau \quad \text{and} \quad D(x) = \Omega^2 \quad . \qquad (4.6 - 7a)$$

With these parameters (2.5 - 1c&d) and (4.6 - 6a) characterize the corresponding Hamiltonian as *Lyapunov function*

$$V = V(x,y) = H(x,y) = \frac{1}{2}\Omega^2 x^2 + \frac{1}{2}y^2 \qquad (4.6 - 7b)$$

whose temporal variation is given by (4.6 - 5b)

$$\dot{V} = \dot{H} = -(2/\tau)\,\Omega^2 x^2 = -Q^{-1}\Omega^3 x^2 \quad . \qquad (4.6 - 7c)$$

In this equation Q represents the quality factor defined by (2.2 - 21).

The theorems of Section 4.6.2 yield the expected *stability criteria* for harmonic oscillations

$$Q \begin{cases} > 0 & \text{asymptotically stable } (4.3-17\,\text{b}) \quad , \\ = 0 & \text{stable } (4.3-17\,\text{d}) \quad , \\ < 0 & \text{instable} \quad . \end{cases} \qquad (4.6 - 7d)$$

4.7 Population Dynamics

Growth and decay of a single population or a number of interacting populations of self-reproducing organisms are nonlinear in general [Beltrami 1987 B]. As a consequence, the models of population dynamics imply complicated nonlinear single differential equations or systems. Models whose equations can be solved analytically are rare. Among these the best known are the *Malthus model* [Tu 1992 B] and the *logistic model* [Beltrami 1987 B, Percival & Richards 1982 B, Tu 1992 B, Verhulst 1838 J] for single populations and the *Lotka-Volterra model* [Beltrami 1987 B, Goel et al 1971 J, Lotka 1920 J, 1925 B, Tu 1992 B, Volterra 1931 B, 1937 J] for the interacting populations of predator and prey. The logistic model has gained additional interest because its discrete mathematical formulation, which is presented in Section 6.2, yields a standard example of *deterministic chaos* [Beltrami 1987 B, Froyland

1992 B, Percival & Richards 1982 B, Schuster 1984 B, Verhulst 1990 B]. The generalization of the Lotka-Volterra model yields the *quadratic model* [Beltrami 1987 B] whose characteristic system of differential equations has not been solved analytically.

In the following considerations a *population* x (or y) is defined as the number of identical individuals at a given time t.

4.7.1 Single Populations

The dynamics of single populations include two relatively simple standard models [Beltrami 1987 B, Tu 1992 B]:

a) Malthus Model

The Malthus model [Tu 1992 B] is based on the assumption that the population $x(t) \geq 0$ increases per individual (per capita) with a constant growth rate $r = 1/\tau\,[s^{-1}]$, which corresponds to the difference between the average birth rate and the average death rate. This assumption yields the relation

$$\dot{x} = r x = (1/\tau)x \quad . \tag{4.7 - 1a}$$

This differential equation has the solution

$$x(t) = x(0)\,exp(rt) = x(0)\,exp(t/\tau) \quad , \tag{4.7 - 1b}$$

which predicts an infinite growth of the population $x(t)$. In most cases, this is impossible because of the finite resources of the vital space or habitat of the population.

b) Continuous logistic model

In the continuous logistic model the infinite growth of the population $x(t)$ is avoided by assuming a growth rate r per capita that decreases for an increasing population as follows

$$r = (1/\tau)\left[1 - \left(x(t)/x_{\mathrm{m}}\right)\right] \quad . \tag{4.7 - 2a}$$

In this equation, x_{m} indicates the *carrying capacity* of the vital space or habitat defined as the maximum population that can be sustained by its resources. As a consequence of the modified growth rate r, the variation $\dot{x}(t)$ of the population is now determined by the differential equation

$$\dot{x} = (1/\tau)x\left[1 - \left(x/x_{\mathrm{m}}\right)\right] = f(x) \quad . \tag{4.7 - 2b}$$

This equation has *two stationary solutions*: an instable solution with $x = 0$ and a stable solution with $x = x_m$. The *general solution* for an arbitrary initial condition $x(0) \geq 0$ can be written as

$$x(t) = 0 \quad \text{for} \quad x(0) = 0 \quad ,$$

$$x(t) = \frac{x_m}{1 + \left[\{x_m / x(0)\} - 1\right] \cdot exp(-t/\tau)} \quad \text{for} \quad x(0) > 0 \quad . \tag{4.7 - 2c}$$

It is illustrated in Fig. 4.7 - 1. For positive $x(0) > 0$ this solution $x(t)$ tends to x_m with increasing time t.

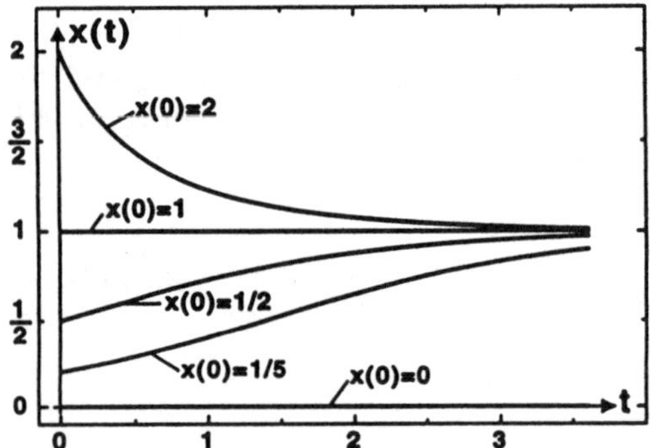

Fig. 4.7 - 1. Variation of the population $x(t)$ according to the continuous logistic model with $\tau = 1$ and $x_m = 1$

4.7.2 The Lotka-Volterra Model

The Lotka-Volterra model [Beltrami 1987 B, Goel et al. 1971 J, Lotka 1920 J, 1925 B, Tu 1992 B, Volterra 1931 B, 1937 J] describes the relation between the populations of predator and prey, e.g. fox and hare, shark and prey fish.

In a vital space or habitat with unlimited resources and without predator the population $x(t)$ of the prey grows to infinity according to the Malthus model (4.7 - 1b)

$$\dot{x} = (1/\tau_B)x \quad . \tag{4.7 - 3a}$$

On the contrary, the population $y(t)$ of the predator decreases in the absence of prey according to the relation

$$\dot{y} = -(1/\tau_R)y \quad . \tag{4.7 - 3b}$$

In the presence of both, prey and predator, there exists an interaction W between their populations $x(t)$ and $y(t)$. In a first approximation this interaction is assumed to be *bilinear*

$$W = C\,x\,y \quad \text{with} \quad C = const \quad . \tag{4.7 - 3c}$$

All together, the three relations (4.7 - 3a-c) yield the following *Lotka-Volterra system of equations*

$$\begin{aligned} \tau_B\,\dot{x} &= \tau_B\,u = +x - x\left(y / y_0\right) \quad \text{with} \quad x \geq 0 \\ \tau_R\,\dot{y} &= \tau_R\,\upsilon = -y + \left(x / x_0\right)y \quad \text{with} \quad y \geq 0 \end{aligned} \tag{4.7 - 4a}$$

$$\text{where} \quad 2\omega = \left(curl\,\vec{\upsilon}\right)_z = \upsilon_x - u_y = \left(1/\,\tau_B y_0\right)x + \left(1/\,\tau_R x_0\right)y \geq 0 \quad . \tag{4.7 - 4b}$$

This system has two *stationary solutions:* an instable solution with $x = y = 0$ and a stable solution with $x = x_0$, $y = y_0$.

The Lotka-Volterra system (4.7 - 4a) can be *normalized* as follows

$$\begin{aligned} \tau_B\,\dot{X} &= +X - XY \quad \text{with} \quad X = x / x_0 \geq 0 \\ \tau_R\,\dot{Y} &= -Y + XY \quad \text{with} \quad Y = y / y_0 \geq 0 \end{aligned} \tag{4.7 - 4b}$$

$$\text{where} \quad 2\omega = \left(curl\,\vec{\upsilon}\right)_z = \upsilon_x - u_y = \left(1/\,\tau_B\right)X + \left(1/\,\tau_R\right)Y \geq 0 \quad . \tag{4.8 - 5b}$$

The instable *stationary solution* of (4.7 - 5a) is $X = Y = 0$, whereas the stable is situated at $X = Y = 1$.

The Lotka-Volterra system (4.7 - 5a) is autonomous. Consequently, the time t can be eliminated as variable. This results in the equation

$$\frac{dY}{dX} = \left(\tau_B / \tau_R\right)\frac{-Y + XY}{+X - XY} \tag{4.7 - 6a}$$

$$\text{or} \quad \tau_B\left(1 - X^{-1}\right)dX + \tau_R\left(1 - Y^{-1}\right)dY = 0 \quad . \tag{4.7 - 6b}$$

The integration of (4.7 - 6b) yields the *equation of the trajectories* in the X-Y-plane

$$\tau_B\left(X - \ell n\,X\right) + \tau_R\left(Y - \ell n\,Y\right) = const \geq \tau_B + \tau_R \tag{4.7 - 7}$$

These trajectories are illustrated in Fig. 4.7 - 2. According to the positive $\omega > 0$ of (4.7 - 5b) they move counter-clockwise around the stationary point $X = Y = 1$.

For small deviations $\delta X = X - 1$ and $\delta Y = Y - 1$ from the stable stationary solution $X = Y = 1$, the trajectories form *ellipses* according to the approximation

$$\tau_B \ \delta X^2 + \tau_R \ \delta Y^2 \approx const \geq \tau_B + \tau_R \quad . \tag{4.7 - 8}$$

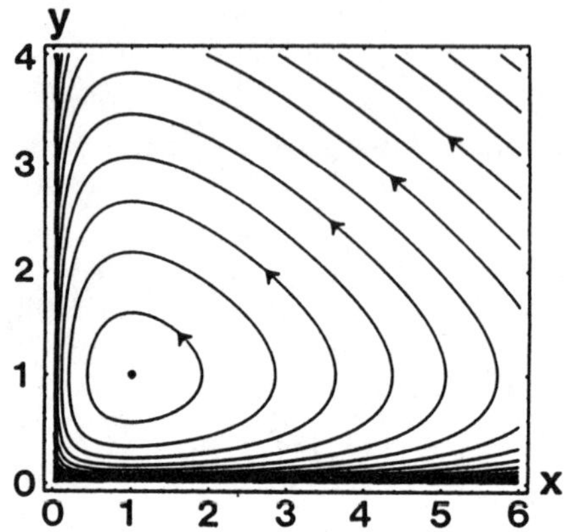

Fig. 4.7 - 2. *X-Y*-diagram of the normalized Lotka-Volterra model with $\tau_B = 5$ and $\tau_R = 10$

The corresponding approximated Lotka-Volterra system

$$\begin{aligned} &\tau_B \ \delta\dot{X} = -\delta Y \\ &\tau_R \ \delta\dot{Y} = +\delta X \\ &\text{with} \quad \delta\ddot{X} + \omega_{LV}^2 \ \delta X = \delta\ddot{Y} + \omega_{LV}^2 \delta Y = 0 \end{aligned} \tag{4.7 - 9}$$

describes a vortex which rotates counter-clockwise with the circular frequency

$$\omega_{LV} \approx +(\tau_B \tau_R)^{-1/2} \quad . \tag{4.7 - 10a}$$

The temperal variation of the normalized populations $X(t)$ and $Y(t)$ of the prey and the predator are shown in Fig. 4.7 - 3.

In conclusion, the Lotka-Volterra model of the populations of prey and predator exhibits the following *phenomena*

α) The stationary population ($X = Y = 1$ or $x = x_0$, $y = y_0$) is singular.

β) Instationary populations never become stationary.

γ) Instationary populations are periodic with a period T, which depends on the initial conditions. Almost stationary populations vary with the period

$$T \approx 2\pi(\tau_B \tau_R)^{1/2} \quad . \tag{4.8 - 10b}$$

The Lotka-Volterra model does not suffice to explain two interacting populations with one suffering extinction or those populations who start instationary and become stationary. These phenomena require *more sophisticated models* [Beltrami 1987 B, Tu 1992 B].

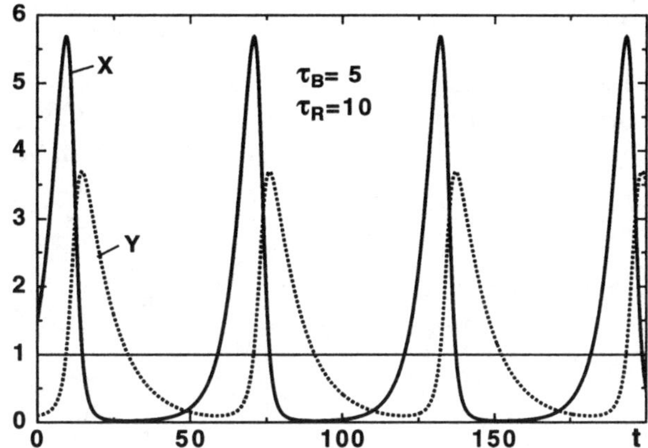

Fig. 4.7 - 3. Temporal variation of the prey population $X(t)$ and the predator population $Y(t)$ according to the normalized Lotka-Volterra model with $\tau_B = 5$ and $\tau_R = 10$

4.8 Conservative Linear Mechanical Systems

A mechanical system is *conservative*, if it can be described with a potential energy V explicitly independent of time t. In a conservative *linear* mechanical system with a stable equilibrium in the origin $\vec{r}_0 = \vec{0}$ the potential energy V represents a *positive definite quadratic form* of the coordinates x_j, $j = 1, 2, ..., n$

$$V = \frac{1}{2}\vec{r}\,\boldsymbol{F}\,\vec{r} = \frac{1}{2}\sum_{j,k=1}^{n} f_{jk}\,x_j\,x_k \tag{4.8 - 1a}$$

where $\boldsymbol{F}$ indicates a symmetric matrix

$$\boldsymbol{F} = \left\{f_{jk}\right\} = \left\{f_{kj}\right\} = \boldsymbol{F}^{\mathrm{T}} \quad . \tag{4.8 - 1b}$$

$\boldsymbol{F}^{\mathrm{T}}$ indicates the transposed matrix of $\boldsymbol{F}$. A quadratic form is called positive definite if it is positive for all x_j, except when all x_j are zero. In that case it is also zero.

The kinetic energy T of a mechanical system is a *positive definite quadratic form* of the components $\dot{x}_j$, $j = 1, 2, .., n$ of the velocities

$$T = \frac{1}{2}\dot{\vec{r}}\,\boldsymbol{M}\,\dot{\vec{r}} = \frac{1}{2}\sum_{j,k=1}^{n} \mu_{jk}\,\dot{x}_j\,\dot{x}_k \tag{4.8 - 2a}$$

with the symmetric matrix

$$M = \{\mu_{jk}\} = \{\mu_{kj}\} = M^T \quad . \tag{4.8 - 2b}$$

4.8.1 Lagrange Mechanics

For given matrices $\boldsymbol{F}$ and $\boldsymbol{M}$ the oscillation equations of a conservative mechanical system can be derived on the basis of *Lagrange mechanics* [Goldstein 1978 B, Kuypers 1982 B]. This implies the introduction of the *Lagrange function* defined by

$$L = T - V \quad . \tag{4.8 - 3a}$$

Then, the oscillation equations can be determined with the aid of the Lagrange equations

$$\frac{d}{dt} \left(\partial L / \partial \dot{x}_j\right) - \left(\partial L / \partial x_j\right) = 0 \quad . \tag{4.8 - 3b}$$

The result is

$$\sum_{k=1}^{n} \left(\mu_{jk}\, \ddot{x}_k + f_{jk}\, x_k\right) = 0, \quad j = 1, 2, \ldots n \tag{4.8 - 4a}$$

corresponding to the vector equation

$$M\, \ddot{\vec{r}} + F\, \vec{r} = \vec{0} \quad . \tag{4.8 - 4b}$$

The multiplication with M^{-1} transforms this equation into the *acceleration field*

$$\ddot{\vec{r}} = -\Omega^2\, \vec{r} = -M^{-1} F\, \vec{r} \quad \text{with} \quad \Omega^2 = \{\Omega^2_{jk}\} \quad , \tag{4.8 - 5a}$$

where Ω^2 is the characteristic matrix. This vector equation can be represented by the following system of differential equations

$$\ddot{x}_j = -\sum_{k=1}^{n} \Omega^2_{jk}\, x_k \quad . \tag{4.8 - 5b}$$

4.8.2 Oscillations

The oscillation equations (4.8 - 5a&b) are solved with the ansatz

$$\vec{r}(t) = \vec{a}\, cos(\omega t - \varphi) \quad \text{with} \quad \vec{a} = [a_1, a_2 \ldots a_n] \quad . \tag{4.8- 6a}$$

In combination with (4.8 - 4b) and (4.8 - 5a) this ansatz yields the homogeneous time-independent equations for the determination of the vector $\vec{a}$

$$(\boldsymbol{F} - \omega^2 \boldsymbol{M})\,\vec{a} = \vec{0} \tag{4.8 - 6b}$$

and

$$(\Omega^2 - \omega^2 \boldsymbol{I})\vec{a} = \vec{0} \quad \text{with} \quad \boldsymbol{I} = \left\{\delta_{jk}\right\} \quad . \tag{4.8 - 6c}$$

In (4.8 - 6c) $\boldsymbol{I}$ indicates the *n*-dimensional unity matrix.

Nontrivial solutions $\vec{a} \neq \vec{0}$ of (4.8 - 6b&c) require that the determinants of their matrices vanish:

$$det\left[\boldsymbol{F} - \omega^2 \boldsymbol{M}\right] = 0 \quad , \quad \text{respectively} \tag{4.8 - 7a}$$

$$det\left[\Omega^2 - \omega^2 \boldsymbol{I}\right] = 0 \quad . \tag{4.8 - 7b}$$

These are two forms of the *secular equation of the system*. It has *n* roots or eigenvalues $\omega_m^2 \geq 0$, $m = 1, ..., n$. Each circular eigenfrequency $\omega_m \geq 0$ corresponds to one eigenvector $\vec{a}_m$ that can be determined by solving one of the equations (4.8 - 6a&b) after replacing ω by ω_m. If a number k of the ω_m are *degenerate*, i.e. equal, they still yield k different eigenvectors $\vec{a}_m$.

Because of the homogeneity of (4.8 - 6a&b) each eigenvector $\vec{a}_m$ can be multiplied with an arbitrary factor C_m without effect. In conclusion the *general solution* of a system described by (4.8 - 4a&b) can be written as

$$\vec{r}(t) = \sum_{m=1}^{n} C_m \cos\left(\omega_m t - \varphi_m\right)\vec{a}_m \qquad \text{with} \quad \vec{a}_m = \left[a_{m1}, a_{m2}, \ldots a_{mn}\right] \quad . \tag{4.8 - 8}$$

In this equation, the C_m signify arbitrary amplitudes and the φ_m indicate arbitrary phases. Thus, it can be adapted to all pairs of *initial conditions* $\vec{r}(0)$ and $\dot{\vec{r}}(0)$. For this purpose the C_m and the φ_m are evaluated with the aid of the equations

$$\vec{r}(0) = \sum_{m=1}^{n} C_m \cos\varphi_m \cdot \vec{a}_m \quad , \quad \text{and} \tag{4.8 - 9a}$$

$$\dot{\vec{r}}(0) = \sum_{m=1}^{m} C_m \sin\varphi_m \cdot \omega_m\,\vec{a}_m \quad . \tag{4.8 - 9b}$$

The functions

$$q_{\mathrm{m}}(t) = C_{\mathrm{m}}\, cos(\omega_{\mathrm{m}} t - \varphi_{\mathrm{m}}) \qquad (4.8 - 10a)$$

are frequently called normal coordinates. They determine the general solution (4.8 - 8) in combination with the eigenvectors a_{m}

$$\vec{r}(t) = \sum_{\mathrm{m}=1}^{\mathrm{n}} q_{\mathrm{m}}(t)\vec{a}_{\mathrm{m}} \quad . \qquad (4.8 - 10b)$$

The application of normal coordinates yields a simple representation of the potential and the kinetic energy by diagonal matrices ***F*** and ***M***

$$V = \frac{1}{2}\sum_{\mathrm{m}=1}^{\mathrm{n}} f_{\mathrm{m}} q_{\mathrm{m}}^2 \quad \text{with} \quad \boldsymbol{F} = \{f_{\mathrm{k}}\, \delta_{\mathrm{jk}}\} \quad , \text{ and} \qquad (4.8 - 10c)$$

$$T = \frac{1}{2}\sum_{\mathrm{m}=1}^{\mathrm{n}} \mu_{\mathrm{m}} \dot{q}_{\mathrm{m}}^2 \quad \text{with} \quad \boldsymbol{M} = \{\mu_{\mathrm{k}}\, \delta_{\mathrm{jk}}\} \quad . \qquad (4.8 - 10d)$$

4.8.3 Molecular Vibrations

The conservative linear mechanical systems form a basis of the theory of molecular vibrations [Steele 1971 B, Wilson et al. 1955 B].

A molecule consisting of r atoms possesses $f = 3r$ *degrees of freedom* with respect to the motion of its nuclei in the three-dimensional space. These degrees of freedom include those of the *translation* and the *rotation* of the entire molecule plus those of the *vibration*. With respect to the rotation two types of molecules behave differently.

α) For a *nonlinear molecule*, the translation of the entire molecule and its center of mass as well as the rotation of the entire molecule around its center of mass claim three degrees of freedom. If no external force or torque acts on the molecule, than the eigenfrequencies ω_{m} of translation and rotation are zero. On this condition, the following statement on the circular eigenfrequencies ω_{m} of a nonlinear molecule is valid

$$\begin{aligned} \omega_{\mathrm{m}} &= 0 \quad \text{for} \quad 1 \le m \le 6 \qquad \text{(translation \& rotation)} \quad , \\ \omega_{\mathrm{m}} &> 0 \quad \text{for} \quad 7 \le m \le n = 3r \quad \text{(vibration)} \quad . \end{aligned} \qquad (4.8 - 11a)$$

Thus, the proper vibrations of a nonlinear molecule have

$$f(vibr.) = 3r - 6 = f - 6 \qquad (4.8 - 11b)$$

degrees of freedom.

β) For a *linear molecule* the translation of the entire molecule and its center of mass also claims three degrees of freedom, whereas its rotation around its center of mass has only two degrees of freedom. Thus, circular eigenfrequencies ω_m of a linear molecule fulfill the relations

$$\begin{aligned} &\omega_m = 0 \quad \text{for} \quad 1 \le m \le 5 \qquad \text{(translation \& rotation)} \quad , \\ &\omega_m > 0 \quad \text{for} \quad 6 \le m \le n = 3r \quad \text{(vibration)} \quad . \end{aligned} \tag{4.8 - 12a}$$

Consequently, the proper vibrations of a linear molecule have

$$f(vibr.) = 3r - 5 = f - 5 \tag{4.8 - 12b}$$

degrees of freedom.

The *degeneracy of the molecular vibrations* and their circular eigenfrequencies $\omega_m > 0$ are to a large extent determined by the *symmetry of the molecule* [Herzberg 1945 B, Steele 1971 B, Wilson et al. 1955 B]. Therefore, *group theory* plays an essential role in the calculation of the vibrations of small molecules [Herzberg 1945 B, Steele 1971 B, Wilson et al. 1955 B]. Apart from the determination of the inherent degeneracies of the circular eigenfrequencies, group theory also permits the reduction of the matrices $\boldsymbol{F}$ and $\boldsymbol{M}$ of the oscillation equations (4.8 - 4b) as well as the factorization of the secular equation (4.8 - 7a&b).

As an *example of the inherent degeneracies* of the vibrations of a small nonlinear molecule one should mention the circular eigenfrequencies ω_m of *methane* CH_4 with $f = 3r = 15$ degrees of freedom of the nuclei

$$\begin{aligned} &\omega_{1-6} = 0, \ \omega_7 = 5,75 \cdot 10^{14} \ s^{-1}, \ \omega_{8-9} = 2,61 \cdot 10^{14} s^{-1} \quad , \\ &\omega_{10-12} = 5,95 \cdot 10^{14} \ s^{-1}, \ \omega_{13-15} = 2,58 \cdot 10^{14} s^{-1} \quad . \end{aligned}$$

4.9 Time-dependent Linear Systems

In this section the focus is on the solution of homogeneous and inhomogeneous time-dependent or *non-autonomous* linear systems. Difficulties are encountered because the homogeneous time-dependent systems do usually not exhibit fixed singular points in contrary to the corresponding time-invariant or autonomous systems. After a general discussion of the solutions and their stability, the considerations are restricted to two-dimensional systems with emphasis on *self-adjoint systems*.

4.9.1 Homogenous Systems of Arbitrary Dimensions

Homogenous time-dependent linear systems can be represented according to (4.1 - 5a&b) and (4.1 - 6a) either by

$$dx_j / dt = \dot{x}_j = v_j = \sum_{k=1}^{n} a_{jk}(t)\, x_k \tag{4.9 - 1a}$$

or by the vector equation

$$\frac{d}{dt}\vec{r} = \dot{\vec{r}} = \vec{v}(t) = \boldsymbol{A}(t)\vec{r} \quad \text{with} \quad \boldsymbol{A}(t) = \{a_{jk}(t)\} \tag{4.9 - 1b}$$

with the time-dependent characteristic matrix $\boldsymbol{A}(t)$.

a) Propagator solution

The linear dependence of the solution $\vec{r}(t)$ of (4.9 - 1b) on the initial state $\vec{r}(0)$ implies a *propagator solution* in the form [Hainer et al. 1987 B]

$$\vec{r}(t) = \boldsymbol{P}_S(t, t_0)\ \vec{r}(0) \quad . \tag{4.9 - 2}$$

The matrix $\boldsymbol{P}_S(t, t_0)$ is called *system propagator* or *resolvent*. It can be represented in various ways:

α) The systems propagator $\boldsymbol{P}_S(t, t_0)$ can be written as an *exponential function* if and only if the characteristic matrix $\boldsymbol{A}(t)$ and its integral $\boldsymbol{B}(t, t_0)$ over time t commute [Zwillinger 1989 B]

$$\boldsymbol{B}(t, t_0)\, \boldsymbol{A}(t) = \boldsymbol{A}(t)\, \boldsymbol{B}(t, t_0) \quad \text{with} \tag{4.9 - 3a}$$

$$\boldsymbol{B}(t, t_0) = \{b_{jk}(t, t_0)\} = \int_{t_0}^{t} dt\, \boldsymbol{A}(t) = \left\{ \int_{t_0}^{t} dt\, a_{jk}(t) \right\} \quad . \tag{4.9 - 3b}$$

On this condition, the system propagator can be represented as follows

$$\boldsymbol{P}_S(t, t_0) = exp[+\boldsymbol{B}(t, t_0)] = exp\left[\int_{t_0}^{t} dt\, \boldsymbol{A}(t) \right] = \sum_{m=1}^{\infty} \frac{1}{m!} \boldsymbol{B}^m(t, t_0) \quad . \tag{4.9 - 3c}$$

Its *inverse* has the corresponding form

$$\boldsymbol{P}_S^{-1}(t, t_0) = \boldsymbol{P}_S(t_0, t) = exp[-\boldsymbol{B}(t, t_0)] \quad . \tag{4.9 - 3d}$$

These equations demonstrate that the propagator $\boldsymbol{P}_S(t, t_0)$ commutes with the characteristic matrix $\boldsymbol{A}(t)$

$$\boldsymbol{P}_S(t, t_0)\ \boldsymbol{A}(t) = \boldsymbol{A}(t)\ \boldsymbol{P}_S(t, t_0) \quad . \tag{4.9 - 3e}$$

β) The propagator solution described corresponds to a special *similarity transformation* of the system (4.9 - 1b)

$$\vec{r}(t) = \boldsymbol{T}(t,t_0)\,\vec{r}(t_0) \quad \text{or} \tag{4.9 - 4a}$$

$$\vec{r}(t_0) = \boldsymbol{T}(t_0,t)\,\vec{r}(t) = \boldsymbol{T}^{-1}(t,t_0)\,\vec{r}(t) \quad . \tag{4.9 - 4b}$$

The matrix $\boldsymbol{T}(t, t_0)$ can be determined by the following calculation

$$\frac{d}{dt}\left[\vec{r}(t_0)\right] = \dot{\boldsymbol{T}}(t_0,t)\,\vec{r}(t) + \boldsymbol{T}(t_0,t)\,\dot{\vec{r}}(t) = \left[\dot{\boldsymbol{T}}(t_0,t) + \boldsymbol{T}(t_0,t)\,\boldsymbol{A}(t)\right]\vec{r}(t) = \vec{0} \quad .$$

It yields for $\boldsymbol{T}(t_0, t)$ the equation

$$\dot{\boldsymbol{T}}(t_0,t) = -\boldsymbol{T}(t_0,t)\,\boldsymbol{A}(t) \quad , \tag{4.9 - 4c}$$

which has the solution

$$\boldsymbol{T}(t_0,t) = \boldsymbol{P}_S(t_0,t) = exp\left[-\boldsymbol{B}(t,t_0)\right] \quad \text{or} \tag{4.9 - 4d}$$

$$\boldsymbol{T}(t,t_0) = \boldsymbol{P}_S(t,t_0) = exp\left[+\boldsymbol{B}(t,t_0)\right] \quad . \tag{4.9 - 4e}$$

γ) The system propagator can also be described with the aid of the *Wronski matrix* [Hairer et al. 1987 B, Kamke 1956 B]. For n solutions $\vec{r}_m(t)$, $m = 1,2, \ldots n$ of the n-dimensional homogeneous time-dependent linear system (4.9 - 1a&b) the Wronski matrix is defined as

$$\boldsymbol{W}(t) = \{W_{km}(t)\} = \{x_{km}(t)\} = \begin{bmatrix} x_{11}(t) & .. & x_{1n}(t) \\ : & & : \\ x_{n1}(t) & .. & x_{nn}(t) \end{bmatrix} \tag{4.9 - 5a}$$

$$\text{with} \quad \vec{r}_m(t) = \left[x_{1m}(t) \ldots x_{km}(t) \ldots x_{nm}(t)\right] \quad .$$

This Wronski matrix fulfills the vector equation (4.9 - 1b) of the homogeneous linear system [Hairer et al. 1987 B] since it is composed of the individual solutions $\vec{r}_m(t)$

$$\frac{d}{dt}\boldsymbol{W}(t) = \boldsymbol{A}(t)\,\boldsymbol{W}(t) \quad . \tag{4.9 - 5b}$$

In the following it is assumed that the n solutions $\vec{r}_m(t)$, $m = 1,2, \ldots n$ of the system are *linearly independent* and thus form a *basis of the solutions*. Consequently, each solution of the system (4.9 - 1a&b) can be represented as a linear combination of these n linearly independent solutions $\vec{r}_{\rm m}$. This implies

$$\begin{aligned} &\vec{r}(t) = \boldsymbol{W}(t)\ \vec{c} \quad \text{and} \quad \vec{r}(0) = \boldsymbol{W}(0)\ \vec{c} \\ &\text{with} \quad \vec{c} = \left[c_1, \ldots c_{\rm k}, \ldots c_{\rm n}\right] \quad . \end{aligned} \tag{4.9 - 5c}$$

The vector $\vec{c}$ in this equation is constant. Since it is assumed that the solutions $\vec{r}_{\rm m}(t)$ of the system (4.9 - 1a&b) are linearly independent, one can form the inverse of the Wronski matrices $W(0)$ and $W(t)$ and thus eliminate the vector $\vec{c}$. The result is the system propagator

$$\begin{aligned} &\boldsymbol{P}_{\rm S}(t, t_0) = \boldsymbol{W}(t)\ \boldsymbol{W}^{-1}(0) \\ &\text{with} \quad \vec{r}(t) = \boldsymbol{P}_{\rm S}(t, t_0)\ \vec{r}(0) \quad . \end{aligned} \tag{4.9 - 5d}$$

In conclusion, one can formulate the solution $\vec{r}(t)$ of a system (4.9 - 1a&b) for any initial condition $\vec{r}(0)$, if n linearly independent solutions $\vec{r}_{\rm m}(t)$, m = 1,2, ... n are known. Unfortunately, little is known on the construction of these solutions for time-dependent $\boldsymbol{A}(t)$ [Hairer et al. 1987 B].

If the Wronski matrix $\boldsymbol{W}(t)$ is formed by n linearly independent solutions $\vec{r}_{\rm m}(t)$ of the system (4.9 - 1a&b), then its determinant *det* $\boldsymbol{W}(t)$ differs from zero. Its temporal variation is determined by the *Abel-Liouville-Jakobi-Ostrogradskii identity* [Hairer et al. 1987 B]

$$det\{\boldsymbol{W}(t)\} = det\{\boldsymbol{W}(t_0)\}\ exp \int_{t_0}^{t} dt'\ tr\ \boldsymbol{A}(t') \tag{4.9 - 5e}$$

where $tr\,\boldsymbol{A}(t)$ indicates the trace of $\boldsymbol{A}(t)$.

b) Solutions by series

A homogeneous time-dependent linear system (4.9 - 1a&b) can be solved for times t close to the time point t_0 with the aid of series. These are determined by the behavior of the characteristic matrix $\boldsymbol{A}(t)$ at this time point t_0. For simplicity one performs the shift $t \rightarrow t - t_0$ of the time scale and sets $t_0 = 0$. The following is devoted to the two types of series that can be applied if the characteristic matrix $\boldsymbol{A}(t)$ is either regular or regularly singular at the time point $t_0 = 0$.

α) By definition the characteristic matrix $\boldsymbol{A}(t)$ is *regular* at the time point $t_0 = 0$, if it can be written as a Taylor matrix series

$$\begin{aligned} &\boldsymbol{A}(t) = \sum_{\rm m=0}^{\infty} t^{\rm m}\ \boldsymbol{A}_{\rm m} \\ &\text{with} \quad \boldsymbol{A}_{\rm m} = 0 \quad \text{for} \quad m < 0 \quad \text{and} \quad \frac{d}{dt} \boldsymbol{A}_{\rm m} = \boldsymbol{0} \quad . \end{aligned} \tag{4.9 - 6a}$$

If $A(t)$ is regular, the solution $\vec{r}(t)$ of the system (4.9 - 1a&b) for a given $\vec{r}(0)$ can be represented by the series

$$\vec{r}(t)=\begin{bmatrix} I+\frac{t}{1!}A_0+\frac{t^2}{2!}\left(A_0^2+A_1\right)+ \\ +\frac{t^3}{3!}\left(A_0^3+A_0A_1+2A_1A_0+2A_2\right)+\dots \end{bmatrix}\vec{r}(0) \qquad (4.9 - 6b)$$

where I indicates the n-dimensional unity matrix.

β) The characteristic matrix $A(t)$ is called *regularly or weakly singular* at the time point $t_0 = 0$, if it can be represented by the Laurent matrix series [Hairer et al. 1987 B]

$$A(t)=\sum_{m=-1}^{\infty} t^{m} A_{m}$$
$$\text{with} \quad A_m = \mathbf{0} \quad \text{for} \quad m<-1 \quad \text{and} \quad \frac{d}{dt}A_m=\mathbf{0} \quad . \qquad (4.9 - 7a)$$

If $A(t)$ is regularly singular, the solution of the system (4.9 - 1a&b) is performed with the ansatz of the following vector series

$$\vec{r}(t)=t^{\alpha}\sum_{m=0}^{\infty} t^{m}\,\vec{r}_m \quad \text{with} \quad \frac{d}{dt}\,\vec{r}_m=\vec{0} \quad . \qquad (4.9 - 7b)$$

Application of equations (4.9 - 7a&b) in the vector representation (4.9 - 1b) of the system and subsequent comparison of the coefficients of identical powers of t yields

$$\begin{aligned} &[\alpha I - A_{-1}]\vec{r}_0=\vec{0} \quad , \\ &[(\alpha+1)I - A_{-1}]\vec{r}_1 = A_0\vec{r}_0 \quad , \\ &[(\alpha+2)I - A_{-1}]\vec{r}_2 = A_0\,\vec{r}_1 + A_1\,\vec{r}_0 \quad , \\ &[(\alpha+3)I - A_{-1}]r_3 = A_0\,\vec{r}_2 + A_1\,\vec{r}_1 + A_2\,\vec{r}_2 \quad , \quad \text{etc.} \end{aligned} \qquad (4.9 - 7c)$$

The first equation of (4.9 - 7c) is fulfilled if α is one of the n eigenvalues α_p of A_{-1} and $\vec{r}_0$ equals the corresponding eigenvector $\vec{r}_{0p}$. This implies the *secular equation* for $\alpha = \alpha_p$

$$det\left[A_{-1}-\alpha_p I\right]=0, \quad p=1,2,\dots n \quad . \qquad (4.9 - 7d)$$

If α_p and $\vec{r}_{op}$ are determined, then the corresponding $\vec{r}_{mp}$ can be evaluated with the aid of the remaining equations of (4.9 - 7c).

4.9.2 Stability of Homogeneous Systems

Relatively little is known on the stability of homogeneous time-dependent linear systems [Slotine & Li 1991 B]. In the following two standard stability criteria are discussed:

a) Strict stability

A homogeneous time-dependent linear system (4.9 - 1a&b) is *strictly stable* at the singular point in the origin with $\vec{r}_S = \vec{0}$, if all eigenvalues $\alpha_{sp}(t)$ of the *symmetric part* $\boldsymbol{A}_S(t)$ of the characteristic matrix $\boldsymbol{A}(t)$ have *negative real parts* [Slotine & Li 1991 B]

$$\boldsymbol{A}_S(t) = \frac{1}{2}\left[\boldsymbol{A}(t) + \boldsymbol{A}^T(t)\right] \tag{4.9 - 8a}$$

$$Re\ \alpha_{sp}(t) = Re\ \Lambda\left\{\boldsymbol{A}_S(t)\right\} < 0, \quad p = 1,2,\ldots n \quad . \tag{4.9 - 8b}$$

In (4.5 - 8a) $\boldsymbol{A}^T(t)$ indicates the transposed matrix of $\boldsymbol{A}(t)$.

As a consequence of this criteria, it *does not suffice* for strict stability that the real parts $Re\ \alpha_p(t)$ of the n eigenvalues $\alpha_p(t)$, p = 1, 2, ..., n of the characteristic matrix $\boldsymbol{A}(t)$ are negative.

This statement is elucidated by the system with the characteristic matrix

$$\boldsymbol{A}(t) = \begin{bmatrix} -1 & 0 \\ 2exp(2t) & -1 \end{bmatrix} \quad . \tag{4.9 - 9a}$$

Its eigenvalues $\Lambda\{\boldsymbol{A}(t)\}$ are all negative

$$\alpha_{1,2}(t) = \alpha_{1,2} = -1 \quad . \tag{4.9 - 9b}$$

Nevertheless, the general solution of this system is instable

$$\begin{aligned} x(t) &= x(0)\ exp(-t) \quad , \\ y(t) &= \left[y(0) - x(0)\right] exp(-t) + x(0)\ exp(+t) \quad . \end{aligned} \tag{4.9 - 9c}$$

This reflects the fact that not all eigenvalues $\alpha_{sp}(t)$ of the symmetric part $\boldsymbol{A}_S(t)$ of the matrix $\boldsymbol{A}(t)$ are negative

$$\alpha_{S1,2}(t) = -1 \pm exp(2t) \quad . \tag{4.9 - 9d}$$

b) Exponential stability

An additional stability criteria exists for n-dimensional *d'Alembert systems with a time-dependent perturbation* of the form

$$\dot{\vec{r}} = [A + S(t)]\,\vec{r} \quad \text{with} \quad \frac{d}{dt}A = \mathbf{0} \quad . \tag{4.9 - 10a}$$

The constant matrix A represents the d'Alembert system, while the time-dependent matrix $S(t)$ forms the perturbation. In addition it is presumed that the d'Alembert system exhibits *Hurwitz stability*. This stability signifies that the characteristic matrix A of the d'Alembert system has exclusively eigenvalues α_p with negative real parts

$$Re\,\alpha_p = Re\,\Lambda\{A\} < 0, \quad p = 1,2,\ldots n \quad . \tag{4.9 - 10b}$$

On this condition, the d'Alembert system with a time-dependent perturbation is *exponentially stable* (4.3 - 17c) at the singular point in the origin with $\vec{r}_S = \vec{0}$, if the perturbation matrix S fulfills the following requirements

$$\lim_{t\to\infty} S(t) = \mathbf{0} \quad \text{and} \quad \int_0^\infty det\,\{S(t)\}dt = C \neq \pm\infty \quad . \tag{4.9 - 10c}$$

In the first equation $\mathbf{0}$ indicates the zero matrix.

4.9.3 Two-dimensional Homogeneous Systems

A two-dimensional homogeneous time-dependent linear system can be described by a potential U and a Hamiltonian H the same way as the two-dimensional d'Alembert systems (4.3 - 4a-c). In this case, however, the potential U as well as the Hamiltonian H and the related parameters vary with time t

$$U(t,x,y) = -\frac{1}{2}\overline{\alpha}(t)\left(x^2 + y^2\right) + \beta(t)\,x\,y \quad , \tag{4.9 - 11a}$$

$$H(t,x,y) = -\frac{1}{2}\,\omega(t)\left(x^2 + y^2\right) + \gamma(t)\,x\,y \quad . \tag{4.9 - 11b}$$

The corresponding system of differential equations in *Cartesian coordinates* can be derived by application of (4.2 - 1). The result is

$$\begin{aligned} u = \dot{x} &= a_{11}(t)\,x + a_{12}(t)\,y = \left(\overline{\alpha}(t) + \gamma(t)\right)x - \left(\omega(t) + \beta(t)\right)y \\ \upsilon = \dot{y} &= a_{21}(t)\,x + a_{22}(t)\,y = \left(\omega(t) - \beta(t)\right)x + \left(\overline{\alpha}(t) - \gamma(t)\right)y \end{aligned} \quad . \tag{4.9 - 11c}$$

In polar coordinates, the potential U of (4.9 - 11a) and the Hamiltonian H of (4.9 - 11b) have the form

$$U(t,r,\varphi) = -\frac{1}{2} r^2 \left[\overline{\alpha}(t) - \beta(t) \sin 2\varphi\right] \quad , \tag{4.9 - 12a}$$

$$H(t,r,\varphi) = -\frac{1}{2} r^2 \left[\omega(t) - \gamma(t) \sin 2\varphi\right] \quad . \tag{4.9 - 12b}$$

The corresponding representation of the system (4.9 - 11c) in *polar coordinates* can be evaluated with the aid of (4.2 - 6)

$$\begin{aligned} &\frac{d}{dt} \ell n r = \dot{r} / r = \overline{\alpha}(t) - \beta(t) \sin 2\varphi + \gamma(t) \cos 2\varphi \quad , \\ &\dot{\varphi} = \omega(t) - \beta(t) \cos 2\varphi - \gamma(t) \sin 2\varphi \quad . \end{aligned} \tag{4.9 - 12c}$$

According to (4.9 - 11c) and (4.9 - 12c) the two-dimensional homogeneous time-dependent linear systems are determined by the time-dependent parameters $\overline{\alpha}, \beta, \gamma, \omega$.

a) Simple systems

In order to gain a first insight into the behavior of time-dependent systems one should consider the most simple of the two-dimensional homogeneous time-dependent systems. They can be characterized by the form of their trajectories:

α) *Star:* $\beta \equiv \gamma \equiv \omega \equiv 0$

$$r(t) = r(0) \cdot exp \int_0^t \overline{\alpha}(t')\, dt' \quad , \quad \varphi(t) = \varphi(0) \tag{4.9 - 13}$$

The trajectories form radial rays.

β) *Saddle I:* $\overline{\alpha} \equiv \gamma \equiv \omega \equiv 0$

$$x^2(t) - y^2(t) = x^2(0) - y^2(0) \tag{4.9 - 14}$$

The trajectories form hyperbolas.

γ) *Saddle II:* $\overline{\alpha} \equiv \beta \equiv \omega \equiv 0$

$$x(t)\, y(t) = x(0)\, y(0) \tag{4.9 - 15}$$

The trajectories form hyperbolas.

δ) *Rotation or vortex:* $\overline{\alpha} \equiv \beta \equiv \gamma \equiv 0$

$$r(t) = r(0) \quad , \quad \varphi(t) = \varphi(0) + \int_0^t \omega(t')dt' \tag{4.9 - 16}$$

The trajectories form circles.

ε) *Rotating system:* $\beta \equiv \gamma \equiv 0$
In polar coordinates (4.2 - 4a&b) the solution has the form

$$\begin{aligned} \gamma(t) &= \gamma(0) + \int_0^t \omega(t')dt' \\ r(t) &= r(0) \cdot exp \int_0^t \overline{\alpha}(t')dt' \end{aligned} \tag{4.9 - 17a}$$

whilst in *Cartesian coordinates* the solution can be written as [Kamke 1956 B]

$$\begin{aligned} x(t) &= [x(0)\cos\varphi - y(0)\sin\varphi]\rho \quad , \\ y(t) &= [x(0)\sin\varphi + y(0)\cos\varphi]\rho \quad , \\ \text{with} \quad \varphi &= \int_0^t \omega(t')\,dt' \ , \quad \rho = exp\int_0^t \overline{\alpha}(t')dt' \quad . \end{aligned} \tag{4.9 - 17b}$$

b) Selfadjoint systems
Selfadjoint systems play a dominant role in the theory of linear differential equations of second order [Birkhoff & Rota 1989 B]. They are defined by the following parameter conditions

$$\overline{\alpha} \equiv \overline{\gamma} \equiv 0 \quad , \quad \omega(t) + \beta(t) = R(t) \quad , \quad \omega(t) - \beta(t) = Q(t) \tag{4.9 - 18a}$$

which imply the following form of these systems

$$\begin{aligned} \dot{x} &= -R(t)\,y \\ \dot{y} &= +Q(t)\,x \end{aligned} \quad . \tag{4.9 - 18b}$$

The associated second-order differential equations for x and y are selfadjoint in the following sense

$$\frac{d}{dt}\left[\frac{\dot{x}}{R(t)}\right] + Q(t)\,x = 0 \quad , \tag{4.9 - 18c}$$

$$\frac{d}{dt}\left[\frac{\dot{y}}{Q(t)}\right]+R(t)\,y=0 \quad . \tag{4.9 - 18d}$$

Examples of well-known selfadjoint differential equations [Kamke 1956 B] are the

α) *Laguerre* differential equation

$$\frac{d}{dt}[t\,\dot{x}]+\left[\lambda-\frac{1}{4}(t+2)\right]x=0 \quad . \tag{4.9 - 19a}$$

β) *Euler* differential equation

$$\frac{d}{dt}[t\,\dot{x}]-\left[\nu^2 t^{-1}\right]x=0 \quad . \tag{4.9 - 19b}$$

γ) *Bessel* differential equation

$$\frac{d}{dt}[t\,\dot{x}]+\left[t-\left(\nu^2 t^{-1}\right)\right]x=0 \quad . \tag{4.9 - 19c}$$

δ) *Legendre* differential equation

$$\frac{d}{dt}\left[\left(1-t^2\right)\dot{x}\right]-\nu(\nu+1)x=0 \quad . \tag{4.9 - 19d}$$

The analysis of selfadjoint second-order differential equations often makes use of the *Prüfer substitution* [Birkhoff & Rota 1989 B, Zwillinger 1989 B]. With respect to the selfadjoint differential equation (4.9 - 18c) the Prüfer substitution corresponds to the transformation

$$\begin{aligned} x(t) &= r(t)\cos\varphi(t) \\ \dot{x}(t) &= -R(t)\,r(t)\sin\varphi(t) \quad . \end{aligned} \tag{4.9 - 20a}$$

It transforms (4.9 - 18c) into the system

$$\begin{aligned} \frac{d}{dt}\ell n\, r &= \dot{r}/r = -\beta(t)\sin 2\varphi \\ \dot{\varphi} &= -\beta(t)\cos 2\varphi + \omega(t) \quad . \end{aligned} \tag{4.9 - 20b}$$

Thus, the Prüfer substitution applied to a selfadjoint second-order differential equation (4.9 - 18c) generates the *corresponding selfadjoint system* (4.9 - 18b) with $\overline{\alpha}\equiv\gamma\equiv 0$ *in the representation* (4.9 - 12c) *with polar coordinates.*

4.9.4 Inhomogeneous Systems

According to the survey in Section 4.1 the standard form of a n-dimensional inhomogeneous time-dependent linear system is

$$dx_{\mathrm{j}} / dt = \dot{x}_{\mathrm{j}} = v_{\mathrm{j}} = \sum_{\mathrm{k}=1}^{\mathrm{n}} a_{\mathrm{jk}}(t)\, x_{\mathrm{k}} + b_{\mathrm{j}}(t) \quad . \tag{4.1 - 5a}$$

The entity of these equations corresponds to the vector equation

$$\frac{d}{dt}\vec{r} = \dot{\vec{r}} = \vec{v} = \mathbf{A}(t)\vec{r} + \vec{b}(t) \quad . \tag{4.1 - 5b}$$

In the following two different methods to solve this inhomogeneous system are discussed:

a) Variation of the constant
If one assumes that n linearly independent solutions of the corresponding homogeneous n-dimensional system with $\vec{b} = \vec{0}$ (4.1 - 6a) are known, then one can solve the inhomogeneous system (4.1 - 5a&b) by variation of the constant [Hairer et al. 1987 B]. The n linearly independent solutions of the homogeneous system with $\vec{b} = \vec{0}$ determine the *Wronski matrix* $\mathbf{W}(t)$ according to (4.9 - 5a). Since the n linearly dependent solutions form a basis of the general solution, the latter can be represented with the aid of the Wronski matrix as follows

$$\vec{r}(t) = \mathbf{W}(t)\, \vec{c} \tag{4.9 - 5c}$$

where $\vec{c}$ indicates a constant vector.

In order to solve the inhomogeneous system (4.1 - 5a&b) by variation of the constant , one presumes that this vector $\vec{c}$ depends on time t

$$\vec{c} = \vec{c}(t) \quad . \tag{4.9 - 21a}$$

By taking into account (4.1 - 5 a&b) and (4.9 - 5a-c) one finds

$$\dot{\vec{r}} = \mathbf{A}\,\vec{r} + \vec{b} = \frac{d}{dt}(\mathbf{W}\,\vec{c}) = \dot{\mathbf{W}}\,\vec{c} + \mathbf{W}\,\dot{\vec{c}} = \mathbf{A}\mathbf{W}\,\vec{c} + \mathbf{W}\,\vec{c} = \mathbf{A}\vec{r} + \mathbf{W}\,\dot{\vec{c}} \quad .$$

This results in the relation

$$\dot{\vec{c}}(t) = \mathbf{W}^{-1}(t)\, \vec{b}(t) \quad . \tag{4.9 - 21b}$$

The integration of this equation and the subsequent multiplication with the Wronski matrix $\mathbf{W}(t)$ yields the solution of the inhomogeneous system (4.1 - 5a&b).

$$\vec{r}(t) = \mathbf{W}(t)\vec{c}(t) = \mathbf{W}(t)\left[\mathbf{W}^{-1}(t_0)\vec{r}(t_0) + \int_{t_0}^{t} \mathbf{W}^{-1}(s)\vec{b}(s)\,ds\right] \quad . \tag{4.9 - 21c}$$

The introduction of the system propagator $\boldsymbol{P}_S(t,t_0)$ defined by (4.9 - 5d) modifies this solution as follows

$$\vec{r}(t) = \boldsymbol{P}_S(t,t_0)\,\vec{r}(t_0) + \int_{t_0}^{t} \mathbf{P}_S(t,s)\,\vec{b}(s)\,ds \quad . \tag{4.9 - 21d}$$

b) Modification of the Wronski determinant

Another scheme to solve the inhomogeneous system (4.1 - 5a&b) is based on a modification of the Wronski determinant [Kamke 1956 B]. In this scheme the Wronski determinant defined by (4.9 - 5a) is modified by replacing its $m-th$ column by the components $b_k(t)$ of the perturbation vector $\vec{b}(t)$. This procedure results in the modified Wronski matrix $\boldsymbol{W}_m(t)$

$$\mathbf{W}_m(t) = \begin{bmatrix} x_{11}(t) & .. & x_{1,m-1}(t) & b_1(t) & .. & x_{1n}(t) \\ : & & : & : & & : \\ x_{n1}(t) & .. & x_{n,m-1}(t) & b_n(t) & .. & x_{nn}(t) \end{bmatrix} \quad . \tag{4.9 - 22a}$$

The determinants $det\boldsymbol{W}_m(t)$, $m = 1, 2, ..., n$ determine the general solution of the inhomogeneous system as follows [Kamke 1956 B]

$$\vec{r}(t) = \sum_{m=1}^{n}\left[C_m + \int_{t_0}^{t}\{det\,\mathbf{W}_m(0)\,/\,det\,\mathbf{W}(s)\}\,ds\right]\vec{r}_m(t) \tag{4.9 - 22b}$$

where $\vec{r}_m(t)$, $m = 1, 2, ..., n$, represent n linearly independent solutions of the corresponding homogeneous system with $\vec{b}(t) \equiv \vec{0}$.

4.10 Three-dimensional Systems and Flows

Three-dimensional systems of explicit differential equations of first order describe velocity fields of fluids. Consequently the characteristics of these systems correspond to those of fluid kinematics. A three-dimensional flow or system can be defined by a scalar potential U and a vector potential $\vec{A}$. In plane flows the vertical component of the vector potential $\vec{A}$ is the Hamiltonian H used for the description of two-dimensional systems in Sections 4.2 to 4.6 and 4.9.

4.10.1 Concepts of Fluid Kinematics

In order to understand the behavior of three-dimensional systems it is necessary to know the basic concepts of fluid kinematics. The most important are introduced in the following sections.

a) Velocity fields

Three-dimensional flows of fluids can be represented by *velocity fields* [Anderson 1988 B, Chung 1988 B, Kellogg 1953 B, Lüst 1978 B, Prandtl & Tietjens 1957a B, Spurk 1997 B] that correspond to the vector equation of systems of explicit differential equations of first order

$$\dot{\vec{r}} = \vec{\upsilon} = \vec{\upsilon}(t, \vec{r}) \quad . \tag{4.1 - 1b}$$

In this equation t indicates the time, $\vec{r} = [x, y, z]$ the position vector and $\vec{\upsilon} = [u, \upsilon, w]$ the velocity vector. In Cartesian coordinates this velocity field has the form

$$\begin{aligned} \dot{x} &= u = u(t,x,y,z) \\ \dot{y} &= \upsilon = \upsilon(t,x,y,z) \\ \dot{z} &= w = w(t,x,y,z) \end{aligned} \quad . \tag{4.10 - 1}$$

b) Trajectories and streamlines

The velocity field (4.1 - 1b) determines the paths or *trajectories* of the fluid particles in the flow. They represent the solutions of the three-dimensional system (4.10 - 1). If each particle is labeled by its position vector $\vec{r}_0 = [x_0, y_0, z_0]$ at time $t = 0$, its trajectory can be described by its time-dependent position vector [Lüst 1978 B]

$$\vec{r} = \vec{r}(t, \vec{r}_0) \quad \text{with} \quad \vec{r}_0 = \vec{r}(0, \vec{r}_0) \quad . \tag{4.10 - 2a}$$

As solutions of the system (4.10 - 1) the trajectories can be written in Cartesian coordinates as follows

$$\begin{aligned} x &= x(t, x_0, y_0, z_0) & & x_0 = x(0, x_0, y_0, z_0) \\ y &= y(t, x_0, y_0, z_0) & \text{with} \quad & y_0 = y(0, x_0, y_0, z_0) \\ z &= z(t, x_0, y_0, z_0) & & z_0 = z(0, x_0, y_0, z_0) \end{aligned} \quad . \tag{4.10 - 2b}$$

An *example* represent the trajectories of the time-dependent vector field

$$\begin{aligned} \dot{x} &= u(t,x,y,z) = (1+t)^{-1} x \\ \dot{y} &= \upsilon(t,x,y,z) = y \\ \dot{z} &= w(t,x,y,z) = 0 \end{aligned} \quad . \tag{4.10 - 3a}$$

These trajectories are described by the solutions of this system

$$\left.\begin{aligned} x &= x(t, x_0, y_0, z_0) = x_0(1+t) \\ y &= y(t, x_0, y_0, z_0) = y_0 e^{t} \end{aligned}\right\} \quad \text{or} \quad \frac{y}{y_0} = exp\left\{\frac{x}{x_0} - 1\right\} \tag{4.10 - 3b}$$

$$z = y(t, x_0, y_0, z_0) = z_0 \quad .$$

The trajectories of the fluid particles give no information on the momentary state of a flow. This information is given by the *streamlines* also called field lines or lines of flow. By definition a streamline is parallel to the momentary velocity field in each point and therefore obeys the vector equation

$$d\vec{r}_S \times \vec{v}(\vec{r}, t) = \vec{0} \quad . \tag{4.10 - 4a}$$

As a consequence the streamlines of time-dependent flows vary with time. They can be represented as parameter curves

$$\vec{r}_S = \vec{r}_S(\sigma; t, \vec{r}_0) \quad \text{with} \quad \vec{r}_0 = \vec{r}_S(0; t, \vec{r}_0) \tag{4.10 - 4b}$$

where σ is designates the running parameter. Because of relation (4.10 - 4a) the streamline vector $\vec{r}_S$ and its parameter σ are determined by the streamline equation

$$\frac{d}{d\sigma} \vec{r}_S\,(\sigma; t, \vec{r}_0) = \vec{v}(t, \vec{r}_S(\sigma; t, \vec{r}_0)) \tag{4.10 - 4c}$$

and its corresponding system

$$\begin{aligned} \frac{d}{d\sigma} x_S &= u(t, x_S, y_S, z_S) \\ \frac{d}{d\sigma} y_S &= v(t, x_S, y_S, z_S) \qquad \text{with } t \text{ fixed} \quad . \\ \frac{d}{d\sigma} z_S &= w(t, x_S, y_S, z_S) \end{aligned} \tag{4.10 - 4d}$$

As *example* serves the system (4.10 - 3a) considered before. This system yields the following equations (4.10 - 4d) for the determination of the streamlines

$$\begin{aligned} \frac{d}{d\sigma} x_S &= (1+t)^{-1} x_S \\ \frac{d}{d\sigma} y_S &= y_S \qquad . \\ \frac{d}{d\sigma} z_S &= 0 \end{aligned} \tag{4.10 - 3c}$$

Their solution is

$$\left.\begin{aligned} x_S &= x_0 \, exp\{(1+t)^{-1}\sigma\} \\ y_S &= y_0 \, exp\{\sigma\} \end{aligned}\right\} \text{ or } \frac{y}{y_0} = \left[\frac{x}{x_0}\right]^{(1+t)} \qquad (4.10 - 3d)$$

$$z_S = z_0 \quad .$$

By comparison of the systems (4.10 - 1) and (4.10 - 4d) as well as of (4.10 - 3a) and (4.10 - 3c) one finds that *the trajectories* of the fluid particles *and the streamlines differ for time-dependent flows*. On the contrary, they are *identical for stationary flows*. For these flows the parameter σ can be identified with the time t and both, the system (4.10 - 1) that determines the trajectories and the system (4.10 - 4d) relevant for the streamlines, have the common form

$$\begin{aligned} \frac{d}{dt} x &= \dot{x} = u(x,y,z) \\ \frac{d}{dt} y &= \dot{y} = \upsilon(x,y,z) \\ \frac{d}{dt} z &= \dot{z} = w(x,y,z) \end{aligned} \quad . \qquad (4.10 - 5)$$

An *example* is the system

$$\begin{aligned} \frac{dx}{dt} &= \dot{x} = -\omega y \\ \frac{dy}{dt} &= \dot{y} = +\omega x \\ \frac{dz}{dt} &= \dot{z} = \upsilon_a \end{aligned} \qquad (4.10 - 6a)$$

with the solution

$$\begin{aligned} x &= r_0 \, cos(\omega t - \varphi) \\ y &= r_0 \, sin(\omega t - \varphi) \\ z &= z_0 + \upsilon_a t \end{aligned} \quad . \qquad (4.10 - 6b)$$

The corresponding trajectories form spirals around the z-axis with the pitch $p = 2\pi \, \upsilon_a / \omega$ and coincide with the streamlines.

c) Divergence and continuity equation

The divergence of a velocity field $\vec{\upsilon}(t,\vec{r})$ that is defined as

$$div\ \vec{\upsilon} = div[u, \upsilon, w] = u_x + \upsilon_y + w_z = \frac{\partial}{\partial x}u + \frac{\partial}{\partial y}\upsilon + \frac{\partial}{\partial z}w \tag{4.10 - 7}$$

is a relevant characteristic of a flow. In this equation, the indices x, y, z indicate the partial derivatives $\partial / \partial x$, $\partial / \partial y$, $\partial / \partial z$. The significance of the divergence of a velocity field in fluid dynamics originates in the *continuity equation* [Kneubühl 1994 B, Lüst 1978 B, Prandtl & Tietjens 1957a B] that can be written in the form

$$\rho\, q = \frac{\partial}{\partial t}\rho + div(\rho\vec{\upsilon}) = \frac{d}{dt}\rho + \rho\ div\ \vec{\upsilon} \quad . \tag{4.10 - 8a}$$

This equation relates the local velocity $\vec{\upsilon}[\mathrm{m/s}]$ and the local *density* $\rho[\mathrm{kg/m^3}]$ with the local *source or sink density* $q[1/\mathrm{s} = \mathrm{m^3/m^3 s}]$ of the flow [Lüst 1978 B]. The density q is positive for a source and negative for a sink. $\partial / \partial t$ designates the partial or *local time derivative* whilst d/dt indicates the total or *convective time derivative.* These derivatives fulfill relation concerning the density ρ

$$\frac{d}{dt}\rho = \frac{\partial}{\partial t}\rho + \vec{\upsilon}\cdot grad\ \rho \quad , \tag{4.10 - 9a}$$

or in Cartesian coordinates

$$\frac{d}{dt}\rho(t,x,y,z) = \frac{d}{dt}\rho = \frac{\partial}{\partial t}\rho + \frac{\partial}{\partial x}\rho\frac{dx}{dt} + \frac{\partial}{\partial y}\rho\frac{dy}{dt} + \frac{\partial}{\partial z}\rho\frac{dz}{dt} \quad . \tag{4.10 - 9b}$$

Here, $\partial\rho / \partial t$ corresponds to the local density variation of the fluid while $d\rho / dt$ describes the density variation of a fluid particle moving with the flow.

The divergence $div\ \vec{\upsilon}$ (4.10 - 7) of a velocity field can be interpreted differently. First, one assumes that the flow of a *compressible fluid is free from sources and sinks.* Then one finds with the aid of (4.10 - 8a)

$$div\ \vec{\upsilon} = -\rho^{-1}\frac{d}{dt}\rho = -\frac{d}{dt}\ell n\, \rho \quad \text{for} \quad q = 0 \quad . \tag{4.10 - 8b}$$

In this case the divergence $div\ \vec{\upsilon}$ corresponds to the relative density decrease of a fluid particle moving in the flow.

Secondly, one presumes a flow of an *incompressible fluid* with a constant density ρ. This assumption permits to modify the continuity equation (4.10 - 8a) as follows

$$div\ \vec{\upsilon} = q \quad \text{for} \quad \rho = const \quad . \tag{4.10 - 8c}$$

As a consequence, the divergence of the velocity field $\vec{\upsilon}(\vec{r},t)$ of an incompressible fluid represents the local source or sink density $q(\vec{r},t)$.

Equation (4.10 - 8c) can be transformed into an integral equation with the aid of the *Gauss theorem* [Kellogg 1953 B]

$$\int_{A} \vec{v} \cdot \vec{n} \, dA = \int_{V(A)} div \; \vec{v} \, dV \quad \text{with} \quad |\vec{n}| = 1 \quad . \tag{4.10 - 10}$$

The resulting integral equation is

$$\int_{A} \vec{v}(t,\vec{r}) \cdot \vec{n} \, dA = Q(V(A)) \quad . \tag{4.10 - 8d}$$

As illustrated in Fig. 4.10 - 1 *A* represents a closed simply connected surface which encloses the volume *V*(*A*). The normal vector $\vec{n}$ on the surface element *dA* points outside. *Q*(*V*(*A*)) indicates the *source or sink strength* [Lüst 1978 B] of the volume *V*(*A*). Its unit is [*Q*] = m³/s.

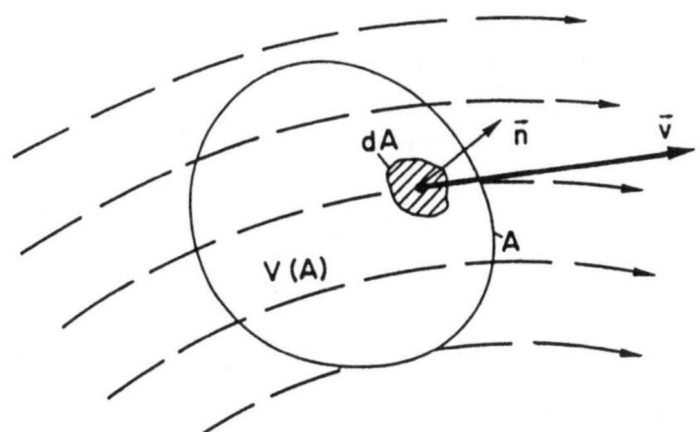

Fig. 4.10 - 1. Continuity equation and Gauss' theorem

Equations (4.10 - 8c&d) imply that a *flow of an incompressible fluid without sources or sinks* is characterized by the following *continuity equation* in differential and integral form

$$div \; \vec{v}(t,\vec{r}) = 0 \quad \text{and} \tag{4.10 - 11a}$$

$$\int_{A} \vec{v}(t,\vec{r}) \cdot \vec{n} \, dA = 0 \quad . \tag{4.10 - 11b}$$

The following discussions of this section *are restricted to flows of incompressible fluids* since they show the optimum similarity to the three-dimensional systems (4.10 - 1) of explicit differential equations of first order.

d) Local rotation and vortices

Local rotation and *vortices* are frequent and important phenomena in fluid dynamics. On the contrary the *proper rotation* of a fluid around a fixed axis as on a merry-go-round is exceptional. It may occur in a cylindrical vessel that is filled with a liquid and rotates at a constant angular velocity ω around its axis.

In order to understand local rotation one may consider the *momentary rotation of a rigid body* around a time-dependent axis with a time-dependent circular frequency $\omega(t) = 2\pi\, \nu(t)$. The axis is usually represented by a time-dependent unity vector $\vec{e}(t)$ with $|\vec{e}(t)| = 1$. With these prerequisites the *momentary rotation* of a rigid body can be described by the vector equation

$$\dot{\vec{r}} = \vec{\upsilon}(t,\vec{r}) = [\vec{\omega}(t) \times \vec{r}] \quad \text{with} \quad \vec{\omega}(t) = \omega(t)\cdot\vec{e}(t) \quad \text{and} \quad |\vec{e}(t)| = 1 \qquad (4.10 - 12a)$$

where $\vec{r}$ indicates the position vector. This motion is illustrated in Fig. 4.10 - 2. If the rotation axis is fixed, i.e. if $\vec{e}(t) = \vec{e} = const$, then this equation describes a *proper rotation* around a fixed axis. If, however, $\vec{e}(t)$ varies with time *t*, then this equation represents the *motion of a solid top* with one point fixed.

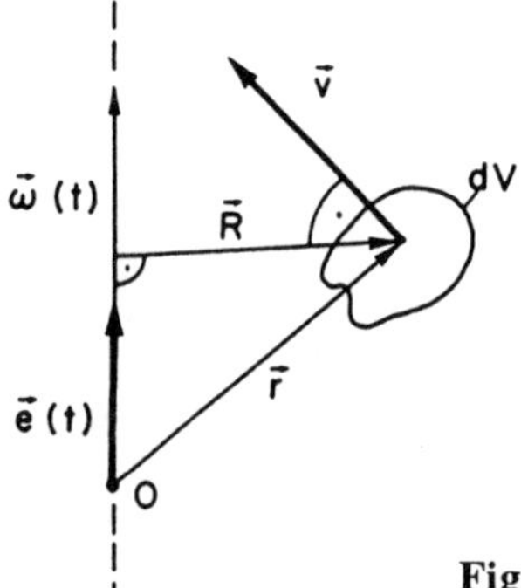

Fig. 4.10 - 2. Proper rotation

The vector equation (4.10 - 12a) can be written differently. If the position vector $\vec{r}$ is decomposed into a vector $\vec{r}_p$ parallel and a vector $\vec{R}$ perpendicular the momentary rotation axis $\vec{e}(t)$, then (4.10 - 10a) can be reduced to

$$\dot{\vec{r}} = \vec{\upsilon} = \vec{\upsilon}(t,\vec{R}) = [\vec{\omega}(t)\ x\ \vec{R}] \quad . \qquad (4.10 - 12b)$$

More important is the following *matrix representation* of the momentary rotation

$$\begin{pmatrix}\dot{x}\\ \dot{y}\\ \dot{z}\end{pmatrix} = \begin{pmatrix}u\\ \upsilon\\ w\end{pmatrix} = \Omega \begin{pmatrix}x\\ y\\ z\end{pmatrix} = \begin{bmatrix}0 & -\omega_3(t) & +\omega_2(t)\\ +\omega_3(t) & 0 & -\omega_1(t)\\ -\omega_2(t) & +\omega_1(t) & 0\end{bmatrix}\begin{pmatrix}x\\ y\\ z\end{pmatrix} \qquad (4.10 - 12c)$$

with the *skew-symmetric matrix* Ω whose elements are the components of the vector $\vec{\omega}(t) = [\omega_1(t), \omega_2(t), \omega_3(t)]$. This vector is related to the vector field (4.10 - 12a) by the equation

$$\vec{\omega}(t) = \frac{1}{2}\ curl\ \vec{\upsilon}(t,\vec{r}) = \frac{1}{2}\ curl\left[\vec{\omega}(t)\ \times \vec{r}\right] \tag{4.10 - 12d}$$

where the differential operator *"curl"* or *"rot"* is defined as

$$curl\ \vec{\upsilon} = curl\left[u, \upsilon, w\right] = \left[w_{\mathrm{y}} - \upsilon_{\mathrm{z}}, u_{\mathrm{z}} - w_{\mathrm{x}}, \upsilon_{\mathrm{x}} - u_{\mathrm{y}}\right] \quad . \tag{4.10 - 13}$$

The indices *x,y,z* indicate the partial differentiations $\partial / \partial x, \partial / \partial y, \partial / \partial z$.

The *local rotation* of a fluid particle at the position $\vec{r}$ at time t is determined by the vector field $\vec{\upsilon}(t,\vec{r})$. The *motion and deformation of an initially spherical fluid* particle in a velocity field $\vec{\upsilon}(t,\vec{r})$ can in a first approximation be decomposed into [Lüst 1978 B, Prandtl & Tietjens 1957a B]

α) a translation,
β) a rotation,
γ) an elliptic deformation in three orthogonal axes.

This results in the following approximation

$$\begin{aligned} &\vec{\upsilon}(t,\vec{r} + \Delta\vec{r}) \cong \vec{\upsilon}(t,\vec{r}) + \left[\vec{\omega}(t,\vec{r}) \times \Delta\vec{r}\right] + \boldsymbol{D}(t,\vec{r})\Delta\vec{r} \\ &\text{with} \quad \boldsymbol{D}(\vec{r},t) = \left\{D_{\mathrm{jk}}(\vec{r},t)\right\} \quad . \end{aligned} \tag{4.10 - 14a}$$

This approximation is illustrated in Fig. 4.10 - 3.

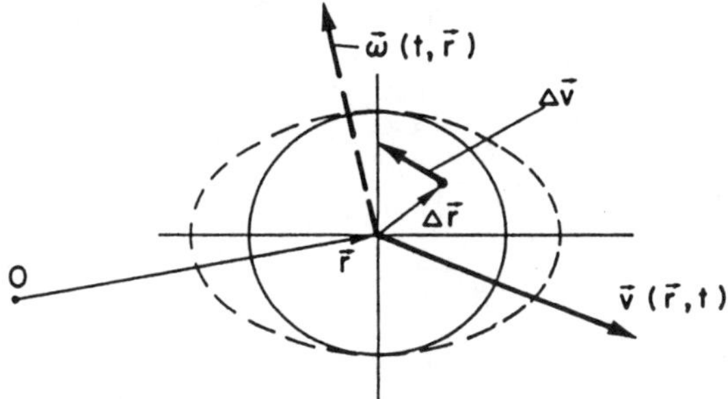

Fig. 4.10 - 3. Motion and deformation of an initially spherical fluid particle

The *local rotation* $\vec{\omega}(t,\vec{r})$ obeys the same relation to the velocity field $\vec{\upsilon}(t,\vec{r})$ as the momentary rotation (4.10 - 12d)

$$\vec{\omega}(t,\vec{r}) = \frac{1}{2}\ curl\ \vec{\upsilon}(\vec{r},t) \quad . \tag{4.10 - 14b}$$

In (4.10 - 14a) $\boldsymbol{D}(t, \vec{r})$ designates the symmetric deformation matrix. It contains information on the divergence and on the source or sink density $q(t,\vec{r})$ of the velocity field $\vec{\upsilon}(t,\vec{r})$ of the flow of an incompressible fluid

$$tr\ \boldsymbol{D}(t,\vec{r}) = div\ \ \vec{\upsilon}(t,\vec{r}) = q(t,\vec{r}) \quad . \tag{4.10 - 14c}$$

In this equation "*tr*" means the "trace" or "Spur" of the matrix $\boldsymbol{D}$. It is defined as

$$tr\ \boldsymbol{D} = tr\ \left\{D_{\mathrm{jk}}\right\} = \sum_{\mathrm{j=1}}^{3} D_{\mathrm{jj}} \quad . \tag{4.10 - 15}$$

The concept of the local rotation $\vec{\upsilon}(t,\vec{r})$ characterized by (4.10 - 14b) is also relevant for the description of *vortices*. A characteristic of vortices is the *circulation*, *vorticity* or *vortex strength* Γ [Kneubühl 1994 B, Lüst 1976 B, Prandtl & Tietjens 1957a B] defined by

$$\Gamma = \oint_{\mathrm{S}} \vec{\upsilon}(t,\vec{r}) \cdot d\vec{r} \tag{4.10 - 16a}$$

where S represents a closed curve as illustrated in Fig. 4.2 - 4. This integral can be transformed with the aid of *Stokes' theorem* [Kellogg 1953 B]

$$\oint_{\mathrm{S}} \vec{\upsilon} \cdot d\vec{r} = \int_{\mathrm{A(s)}} \left(curl\ \vec{\upsilon}\right) \cdot \vec{n}\, d\,A \quad \text{with} \quad |\vec{n}| = 1 \quad . \tag{4.10 - 17}$$

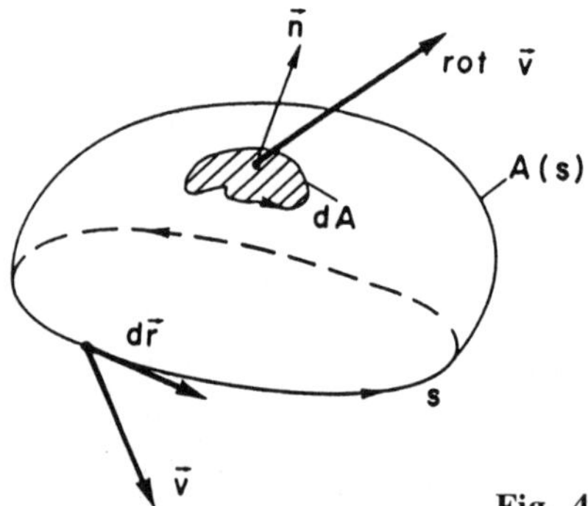

Fig. 4.10 - 4. Stokes' theorem

As demonstrated in Fig. 4.10 - 4 A(s) represents an arbitrary simple surface spanned by the closed curve s. The curve s runs in the clockwise sense if one looks in the direction of the normal vector $\vec{n}$ with $|\vec{n}| = 1$.

The application of Stokes' theorem (4.10 - 17) to (4.10 - 16a) results in a relation between the local rotation $\vec{\upsilon}(t,\vec{r})$ and the circulation Γ

$$\Gamma = \int_{A(s)} \left(rot\ \vec{v}(t,\vec{r})\right)\cdot\vec{n}\, dA = 2\int_{A(s)} \vec{\omega}(t,\vec{r})\cdot\vec{n}\, dA \tag{4.10 - 16b}$$

4.10.2 Gradient Systems as Potential Flows

Potential flows are characterized by a *scalar flow potential* $U(t,\vec{r})$ that determines the velocity field $\vec{v}(t,\vec{r})$ by the relation

$$\dot{\vec{r}}(t,\vec{r}) = \vec{v}(t,\vec{r}) = -grad\ U(t,\vec{r}) \quad . \tag{4.10 - 18}$$

In fluid dynamics the negative sign in this equation is often replaced by the positive sign. From the mathematical point of view the relation (4.10 - 18) represents a *gradient system* of explicit differential equations of first order. The *gradient "grad"* is defined by the operator equation

$$grad\ U = \left[U_x, U_y, U_z\right] = \left[\frac{\partial U}{\partial x}, \frac{\partial U}{\partial y}, \frac{\partial U}{\partial z}\right] \tag{4.10 - 19}$$

This gradient obeys the relation

$$curl\ (grad\ U) = \vec{0} \quad . \tag{4.10 - 20}$$

As a consequence, a potential flow contains *no vortices*

$$curl\ \vec{v}(t,\vec{r}) = 2\vec{\omega}(t,\vec{r}) = \vec{0} \quad . \tag{4.10 - 21}$$

Equation (4.10 - 18) permits the calculation of the velocity field $\vec{v}(t,\vec{r})$ for a given flow potential $U(t,\vec{r})$. The flow potential $U(t,\vec{r})$ of a given velocity field $\vec{v}(t,\vec{r})$, which contains no vortices in a closed singly connected region G according to (4.10 - 21), can be determined with the integral

$$U(t,\vec{r}) = U(t,\vec{r}_0) - \int_{\vec{r}_0,s}^{\vec{r}} \vec{v}(t,\vec{r})\cdot d\vec{r} \quad \text{with} \quad t = const \tag{4.10 - 22}$$

where s represents a path from $\vec{r}_0$ to $\vec{r}_1$ within the region G.

The combination of the field equation (4.10 - 18) of the potential flow with the continuity equation (4.10 - 8c) of the incompressible fluids yields the *Poisson equation*

$$div\left(grad\ U(t,\vec{r})\right) = \Delta\, U(t,\vec{r}) = -q(t,\vec{r}) \tag{4.10 - 23}$$

where Δ indicates the *Laplace operator* defined as

$$\Delta U = div(grad\,U) = U_{xx} + U_{yy} + U_{zz} = \frac{\partial^2}{\partial x^2}U + \frac{\partial^2}{\partial y^2}U + \frac{\partial^2}{\partial z^2}U \quad . \qquad (4.10 - 24)$$

Equation (4.10 - 23) permits the calculation of the source or sink density $q(t,\vec{r})$ of an incompressible fluid for a given potential $U(t,\vec{r})$. The inverse problem, i.e. the determination of the potential $U(t,\vec{r})$ for a given source density $q(t,\vec{r})$ of an incompressible fluid, can be solved with the aid of the *Poisson integral* [Madelung 1943 B, Lüst 1978 B]

$$U(t,\vec{r}) = \frac{1}{4\pi}\int_{V^*} \frac{q(t,\vec{r}^*)dV^*}{|\vec{r}^* - \vec{r}|} \quad \text{with } t = const \qquad (4.10 - 25)$$

where V^* indicates the entire three-dimensional space of $\vec{r}^*$.

a) Point source or sink

A basic *example* of a potential flow of an incompressible fluid is a point source or sink in the origin $\vec{r} = \vec{0}$ illustrated in Fig. 4.10 - 5. Its source or sink density is discontinuous

$$q(t,\vec{r}) = Q(t)\,\delta(\vec{r}) \quad . \qquad (4.10 - 26a)$$

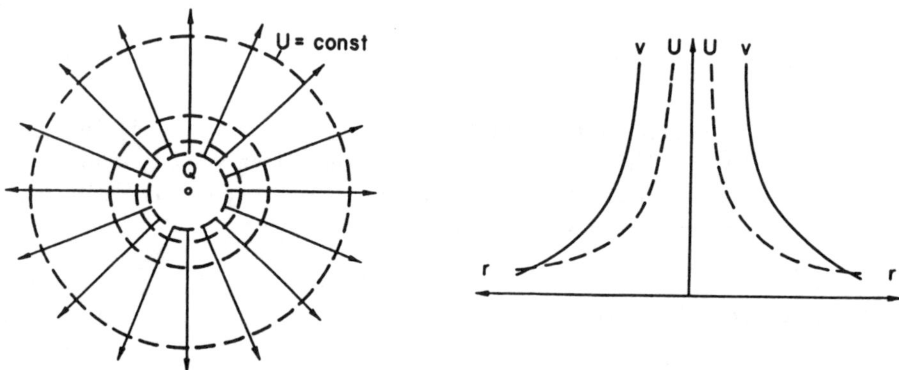

Fig. 4.10 - 5. Velocity υ, flow potential U, equipotential surfaces and streamlines of a three-dimensional point source

In this equation $Q(t)$ designates the source or sink strength and $\delta(\vec{r})$ the three-dimensional Dirac delta function. $Q(t)$ is positive for a source and negative for a sink. The application of (4.10 - 25) to (4.10 - 26a) results in the following flow potential [Madelung 1943 B]

$$U(t,\vec{r}) = \frac{Q(t)}{4\pi\, r} \quad . \tag{4.10 - 26b}$$

The corresponding velocity field $\vec{v}(t,\vec{r})$ can be evaluated with (4.10 - 18). The result is

$$\vec{v}(t,\vec{r}) = -grad\, U(t,\vec{r}) = \frac{Q(t)}{4\pi}\frac{\vec{r}}{r^3} \quad . \tag{4.10 - 26c}$$

b) First representation of the dipole source

Another relevant *example* of a potential flow of an incompressible fluid is the dipole source illustrated in Fig. 4.10 - 6. In principle it consists of a point source with the source strength + $Q(t)$ at the position $+\Delta\vec{r}\,/\,2$ and another point source of strength $-Q(t)$ at the position $-\Delta\vec{r}\,/\,2$. The corresponding source density can be represented by

$$q(t,\vec{r}) = +Q(t)\,\delta(\vec{r} - \frac{1}{2}\Delta\vec{r}) - Q(t)\,\delta(\vec{r} + \frac{1}{2}\Delta\vec{r}) \quad . \tag{4.10 - 27a}$$

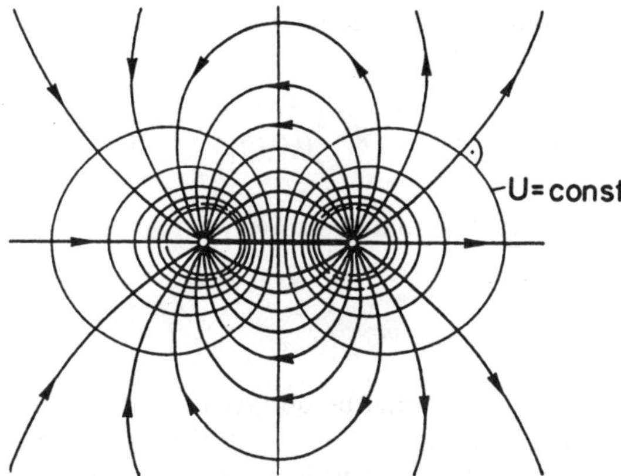

Fig. 4.10 - 6. Equipotential surfaces and streamlines of a three-dimensional dipole source

The ideal dipole source is defined as the limit where the distance Δr becomes zero on the condition that the product $\Delta r \cdot Q(t)$ does not vanish identically. This implies

$$\vec{P}(t) = \lim_{\Delta r \to 0} \; Q(t)\,\Delta\vec{r} \neq \vec{0} \quad . \tag{4.10 - 27b}$$

The vector $\vec{P}(t)$ represents the *dipole moment* of the dipole source.

Flow potential and velocity field of the dipole source can be calculated from (4.10 - 27a&b) by making use of (4.10 - 25) and (4.10 - 18). The result is [Madelung 1943 B]

$$U(t,\vec{r}) = -\frac{r^{-3}}{4\pi}\left(\vec{P}(t)\cdot\vec{r}\right) \quad \text{and} \tag{4.10 - 27c}$$

$$\vec{v}(t,\vec{r}) = -grad\, U(t,\vec{r}) = \frac{r^{-3}}{4\pi}\vec{P}(t) - \frac{3r^{-5}}{4\pi}\left(\vec{P}(t)\cdot\vec{r}\right)\vec{r} \quad . \tag{4.10 - 27d}$$

Another representation of the dipole source will be discussed in Section 4.10.3 d.

4.10.3 Flows without Sources and Sinks

A flow of an incompressible fluid exhibits neither sources nor sinks if the source density $q(t,\vec{r})$ is zero

$$q(t,\vec{r}) = div\ \vec{v}(t,\vec{r}) = 0 \quad . \tag{4.10 - 28}$$

This condition is fulfilled, if the velocity field $\vec{v}(t,\vec{r})$ can be represented as rotation of a *vector potential* $\vec{A}(t,\vec{r})$

$$\vec{v}(t,\vec{r}) = curl\ \vec{A}(t,\vec{r}) \quad \text{with} \quad \vec{A}(t,\vec{r}) = \left[F(t,\vec{r}), G(t,\vec{r}), H(t,\vec{r})\right] \quad . \tag{4.10 - 29}$$

This implies a zero source density $q(t,\vec{r})$ according to the relation

$$q(t,\vec{r}) = div\left(curl\ \vec{A}(t,\vec{r})\right) = 0 \quad . \tag{4.10 - 30}$$

With the aid of (4.10 - 29) the vector field $\vec{v}(t,\vec{r})$ can be derived from the vector potential $\vec{A}(t,\vec{r})$ by partial differentiations. Consequently, each velocity field $\vec{v}(t,\vec{r})$ is thus related to manifold of vector potentials $\vec{A}(t,\vec{r})$. This manifold can be reduced by an additional condition for $\vec{A}(t,\vec{r})$ without lack of generality. A frequent condition is the *gauge equation*

$$div\ \vec{A}(t,\vec{r}) = 0 \quad . \tag{4.10 - 31}$$

This equation and the relation

$$curl\left(curl\ \vec{A}\right) = grad\left(div\ \vec{A}\right) - \Delta\vec{A} \tag{4.10 - 32}$$

permit to relate the vector potential $\vec{A}(t,\vec{r})$ to the local rotation $\vec{\omega}(t,\vec{r})$ by the *Poisson equation*

$$\vec{\omega}(t,\vec{r})=\frac{1}{2}\,curl\;\vec{\upsilon}(t,\vec{r})=-\frac{1}{2}\,\Delta\vec{A}(t,\vec{r})\quad\text{or}\qquad(4.10-33a)$$

$$\Delta\vec{A}(t,\vec{r})=-2\,\vec{\omega}(t,\vec{r})\qquad(4.10-33b)$$

In Cartesian coordinates $\Delta\vec{A}$ has the form

$$\Delta\vec{A}=[\Delta F,\Delta G,\Delta H]\qquad(4.10-34)$$

where Δ represents the Laplace operator defined by (4.10 - 24).

Equations (4.10 - 33a&b) permit to calculate the local rotation $\vec{\omega}(t,\vec{r})$ of a flow determined by a vector potential $\vec{A}(t,\vec{r})$. The inverse problem, i.e. the determination of the vector potential $\vec{A}(t,\vec{r})$ of a flow with a given local rotation $\vec{\omega}(t,\vec{r})$, can be solved by the following *Poisson integral* [Lüst 1978 B, Madelung 1943 B]

$$\vec{A}(t,\vec{r})=\frac{1}{2\pi}\int_{V^*}\frac{\vec{\omega}(t,\vec{r}\,^*)\,dV^*}{|\vec{r}\,^*-\vec{r}|}\quad\text{with}\quad t=const\quad,\qquad(4.10-35)$$

where V^* indicates the entire three-dimensional space of $\vec{r}\,^*$.

a) Vortex filaments

An important flow of an incompressible fluid without source or sink is a vortex around a filament W. A filament W is a curve described by

$$\vec{r}_W=\vec{r}_W(s)\qquad(4.10-36a)$$

where s indicates the curve parameter. It represents a vortex filament if two requirements are fulfilled. First the local rotation $\vec{\omega}(\vec{r})$ must vanish outside the filament

$$\vec{\omega}(\vec{r}\neq\vec{r}_w(s))=\frac{1}{2}\,curl\;\vec{\upsilon}(\vec{r}\neq\vec{r}_w(s))=\vec{0}\quad.\qquad(4.10-36b)$$

Secondly the circulation Γ (4.10 - 16 a&b) on a closed curve s^* that travels once around the filament W does not vanish and differs for different curves s^*

$$\Gamma(s^*\;around\;W)=\int_{s^*}\vec{\upsilon}\cdot d\vec{r}=const\neq0\quad.\qquad(4.10-36c)$$

The vector potential $\vec{A}(t,\vec{r})$ of a flow around a vortex filament W can be determined by [Madelung 1943 B]

$$\vec{A}(\vec{r}) = -\frac{\Gamma}{4\pi}\int_{s}\frac{d\vec{r}_{w}}{|\vec{r}_{w}(s)-\vec{r}|} \quad , \tag{4.10 - 37a}$$

whilst the corresponding velocity field $\vec{v}(\vec{r})$ is given by [Madelung 1943 B]

$$\vec{v}(\vec{r}) = \frac{\Gamma}{4\pi}\int_{s}\frac{[(\vec{r}_{w}(s)-\vec{r})\times d\vec{r}_{w}]}{|\vec{r}_{w}(s)-\vec{r}|^{3}} \quad . \tag{4.10 - 37b}$$

The vortex filament W as well as the vectors of this equation are illustrated in Fig. 4.10 - 7.

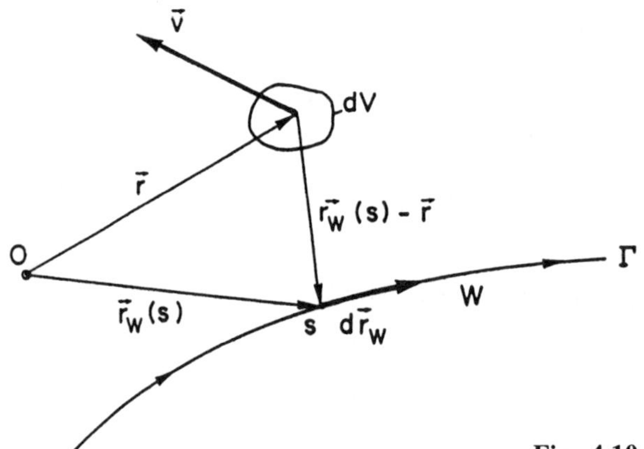

Fig. 4.10 - 7. Vortex filament W

c) Theorems on vortex motion

Vortex motion of a non-viscous homogeneous, i.e. incompressible fluid under the influence of irrotational, i.e. gradient fields of force are essentially determined by the following theorems [Lüst 1979 B, Prandtl & Tietjens 1957a B]:

Thomson's theorem on the permanence of circulation states that the circulation Γ (4.10 - 16a&b) on a closed curve in the fluid is constant for all time.

This statement was later supplemented by *Helmholtz's vortex theorems*, which are valid on the same conditions:

α) The vortex filaments W must be either closed curves in the fluid or end on the boundaries of the fluid.

β) The circulations Γ on closed curves around the vortex filament W in different vortex cross-sections are identical.

γ) Fluid particles, which at any time are part of a vortex line, always belong to this vortex line. As a consequence a vortex always contains the same fluid particles.

c) Second representation of the dipole source
In Section 4.10.2c it was demonstrated that a dipole source creates an irrotational or gradient flow with the potential $U(t,\vec{r})$ described by (4.10 - 27c) and with the velocity field $\vec{\upsilon}(t,\vec{r})$ determined by (4.10 - 27d). Since the source and the sink that form the dipole source compensate each other with respect to the source or sink density in the origin, the dipole source can also be characterized as free of sources and sinks. This permits to relate the velocity field $\vec{\upsilon}(t,\vec{r})$ to a vector potential $\vec{A}(t,\vec{r})$. This has the form [Madelung 1943 B]

$$\vec{A}(t,\vec{r}) = \frac{r^{-3}}{4\pi}\left[\vec{P}(t)\times\vec{r}\right]$$
$$\text{where} \quad div\,\vec{A}(t,\vec{r}) = 0 \quad \text{and} \quad \vec{\upsilon}(t,\vec{r}) = curl\,\vec{A}(t,\vec{r}) \quad . \qquad (4.10 - 38)$$

4.10.4 Decomposition of Velocity Fields

An arbitrary three-dimensional flow of an incompressible fluid can be decomposed into an irrotational flow free of vortices and a flow free of sources and sinks according to the equation

$$\vec{\upsilon}(t,\vec{r}) = \vec{\upsilon}_{\upsilon f}(t,\vec{r}) + \vec{\upsilon}_{sf}(t,\vec{r})$$
$$\text{with } curl\,\vec{\upsilon}_{\upsilon f}(t,\vec{r}) = 0 \quad \text{and} \quad div\,\vec{\upsilon}_{sf}(t,\vec{r}) = 0 \quad . \qquad (4.10 - 39)$$

For this reason, a general three-dimensional flow requires both, a scalar potential $U(t,\vec{r})$ and a vector potential $\vec{A}(t,\vec{r})$ for its description

$$\vec{\upsilon}(t,\vec{r}) = -grad\;U(t,\vec{r}) + curl\;\vec{A}(t,\vec{r}) = \vec{\upsilon}_{\upsilon f}(t,\vec{r}) + \vec{\upsilon}_{sf}(t,\vec{r})$$
$$\text{with} \quad div\,\vec{A}(t,\vec{r}) = 0 \quad . \qquad (4.10 - 40)$$

In Cartesian coordinates equation (4.10 - 40) corresponds to the general three-dimensional system of differential equations

$$\begin{aligned} \dot{x} &= u = -U_x + H_y - G_z \\ \dot{y} &= \upsilon = -U_y + F_z - H_x \\ \dot{z} &= w = -U_z + G_x - F_y \end{aligned} \quad \text{with} \begin{cases} \vec{\upsilon}(t,\vec{r}) = \left[u(t,\vec{r}),\,\upsilon(t,\vec{r}),\,w(t,\vec{r})\right] \;, \\ \vec{A}(t,\vec{r}) = \left[F(t,\vec{r}),\,G(t,\vec{r}),\,H(t,\vec{r})\right] \;, \\ div\,\vec{A}(t,\vec{r}) = F_x + G_y + H_z = 0 \;. \end{cases} \qquad (4.10 - 41)$$

If a *two-dimensional plane flow* in a three-dimensional medium is parallel to the xy plane of the Cartesian coordinates xyz, then its characteristic quantities can be written as

$$\begin{aligned} \vec{r} &= [x, y, z] \quad , \quad \vec{\upsilon} = [u(t,x,y), \upsilon(t,x,y), 0] \quad , \\ U &= U(t,x,y) \quad , \quad \vec{A} = [0, 0, H(t,x,y)] \quad . \end{aligned} \tag{4.10 - 42}$$

Hence, neither velocity field nor scalar and vector potential depend on the coordinate z. Consequently, the three-dimensional system (4.10 - 41) can be reduced to two dimensions

$$\begin{aligned} \dot{x} &= u = -U_{\mathrm{x}} + H_{\mathrm{y}} \\ \dot{y} &= \upsilon = -U_{\mathrm{y}} - H_{\mathrm{x}} \end{aligned} \quad . \tag{4.10 - 43}$$

Plane flows free of sources and sinks are characterized by $\Delta U = 0$, whilst irrotational plane flows imply $\Delta H = 0$. Plane Hamiltonian systems obey the relations

$$\begin{aligned} &\dot{x} = +H_{\mathrm{y}} \quad \text{and} \quad \dot{y} = -H_{\mathrm{x}} \\ &\text{with} \quad U(t,x,y) = 0 \quad \text{and} \quad \vec{A}(t,x,y) = [0,0,H(t,x,y)] \quad . \end{aligned} \tag{4.10 - 44}$$

These relations demonstrate that the *z-component of the vector potential* $\vec{A}(t,\vec{r})$ *corresponds to the Hamiltonian* $H(t, x, y)$ of the two-dimensional system in the xy plane.

Finally it should be noticed that the two-dimensional systems of differential equations (4.10 - 43) are discussed in detail in Sections 4.2 to 4.5, 4.7 and 4.9.

4.10.5 Three and Higher Dimensional d'Alembert Systems

In principle, d'Alembert systems (4.1 - a&b) with dimensions $n > 2$ can be solved as the two-dimensional d'Alembert systems discussed in Section 4.3.

An important aspect of all homogeneous d'Alembert systems

$$\dot{\vec{r}} = \vec{\upsilon} = \boldsymbol{A}\,\vec{r} \quad \text{with} \quad \boldsymbol{A} = \{a_{\mathrm{ij}}\} \tag{4.1 - 7b}$$

is the *stability* of the solutions in the critical or singular point in the origin with $\vec{r}_{\mathrm{S}} = \vec{0}$. This stability is determined by the eigenvalues $\Lambda(\boldsymbol{A}) = \alpha_{\mathrm{i}}$ of the characteristic matrix $\boldsymbol{A}$ in (4.1 - 7b). These are solutions of the eigenvalue or secular equation

$$|\alpha \boldsymbol{I} - \boldsymbol{A}| = \alpha^{\mathrm{n}} + a_{\mathrm{n-1}}\,\alpha^{\mathrm{n-1}} +++ a_0 = \prod_{\mathrm{k=1}}^{\mathrm{n}} (\alpha - \alpha_{\mathrm{k}}) \quad . \tag{4.3 - 23a}$$

The coefficients a_{m-1}, m = 1, 2, ... n are real. $\boldsymbol{I}$ indicates the n-dimensional unity matrix.

The following *stability criteria of Hurwitz* are valid for d'Alembert systems [Birkhoff & Rota 1989 B]: The solutions of a homogeneous d'Alembert system at the singular point in the origin with $\vec{r}_S = \vec{0}$ are

α) *strictly stable*
if and only if each eigenvalue α_k of the characteristic matrix $\boldsymbol{A}$ has a negative real part

$$Re\ \alpha_k < 0 \quad \text{for all} \quad k = 1, 2, \ldots n \quad . \tag{4.10 - 45a}$$

β) *stable*
if and only if each multiple degenerate eigenvalue $\alpha_k = \alpha_{k+1} = = = \alpha_{k+r}$ has a negative real part and, in addition, no single non-degenerate eigenvalue α_j has a positive real part

$$Re\ \alpha_j \text{ (non-degenerate)} \leq 0 \quad \text{and} \quad Re\ \alpha_k \text{ (degenerate)} < 0 \quad . \tag{4.10 - 45b}$$

The real polynomial (4.3 - 23a) that represents the eigenvalue equation of the characteristic matrix $\boldsymbol{A}$ is called stable, if all its roots α_k have negative real parts. This assumption implies special restrictions concerning the real coefficients of this polynomial. These are known as *Routh-Hurwitz conditions* [Gantmacher 1959 B, Hairer et al. 1987 B].

The *behavior of the solutions* of two-dimensional d'Alembert systems *at the singular point in the origin* with $\vec{r}_S = \vec{0}$ is analyzed in Section 4.3.4. A similar analysis has been performed for the *homogeneous three-dimensional d'Alembert systems* [Reitmann 1996 B, Verhulst 1990 B].

5. Transfer Systems

Transfer systems play a significant role in electrical engineering and cybernetics. This chapter offers a survey on their basic properties with emphasis on time-invariant linear and delay systems plus their application in feedback-control systems and resonant oscillating circuits.

5.1 Linear Time-invariant Systems

Linear time-invariant (LTI) systems represent the simplest basic transfer systems. Consequently they are suited for explaining the basic concepts of transfer systems and for the discussion of their relevant properties.

5.1.1 Basic Concepts

A transfer system can be defined in a mathematical sense by a rule by which a time-varying *input* or *excitation* $x(t)$ is mapped into a time-varying *output* $y(t)$. This rule can be expressed symbolically by [Schetzen 1989 B]

$$\boldsymbol{T}\{x(t)\} = y(t) \qquad (5.1 - 1)$$

where $\boldsymbol{T}\{\ \}$ indicates the *system operator.*

It is possible that two or more inputs $x_1(t), x_2(t), \cdots$ can be mapped by the system operator into the same output $y_1(t)$. Yet one input $x_1(t)$ cannot be mapped into more than one output $y_1(t)$. Thus, the system operator constitutes either a *one-to-one or a many-to-one mapping.*

a) Time-invariant systems

A *time-invariant* or *Volterra system* is characterized by a system operator which does not vary with time. Under this condition a time translation of the input $x(t)$ results in the same time translation of the output $y(t)$. Consequently the time invariance is defined by

$$\boldsymbol{T}\{x(t+\tau)\} = y(t+\tau) \quad \text{for any } \tau \quad . \qquad (5.1 - 2)$$

b) Linear systems

A *linear system* maps a linear combination of two inputs $x_1(t)$ and $x_2(t)$ generates the same linear combination of the two corresponding outputs $y_1(t)$ and $y_2(t)$ as follows

$$T\{a\,x_1(t)+b\,x_2(t)\}=a\,T\{x_1(t)\}+b\,T\{x_2(t)\}=a\,y_1(t)+b\,y_2(t) \quad . \qquad (5.1 - 3a)$$

This implies the mapping of the zero input into the zero output

$$T\{x(t)=0\}=y(t)=0 \quad . \qquad (5.2 - 3b)$$

c) LTI systems

For a *linear time-invariant* (LTI) *system* also called a *first-order Volterra system* the system operator also fulfills the condition of time invariance (5.1 - 2). Therefore one can write

$$T\left\{\sum_{k} a_k\, x(t-\tau_k)\right\}=\sum_{k} a_k\, T\{x(t-\tau_k)\}=\sum_{k} a_k\, y(t-\tau_k) \quad . \qquad (5.1 - 4)$$

5.1.2 System Operators of LTI Systems

The system operator of a LTI system acts as convolution. Consequently, system operator and convolution show identical properties as demonstrated in the following.

a) Convolution as system operator

The response $y(t)$ of a linear time-independent (LTI) system is determined by the *convolution* of the input $x(t)$ with the system *impulse response function* $\phi(t)$ as follows

$$y(t)=T\{x(t)\}=\int_{-\infty}^{-\infty} d\,\vartheta\,\phi(t-\vartheta)\,x(\vartheta)=\phi(t)*x(t) \quad . \qquad (5.1 - 5a)$$

In this equation the asterisk * designates the convolution defined by (3.2 - 24a). Examples of convolutions are listed in Appendix A.4. The definition (5.1 - 5) of the system operator as convolution characterizes the LTI systems as first-order Volterra systems with the impulse response $\phi(t)$ as *first-order Volterra kernel* [Schetzen 1989 B].

According to (3.2 - 24b) the convolution is commutative. Therefore, the system operators of LTI systems can also be represented by

$$y(t)=T\{x(t)\}=\int_{-\infty}^{\infty} d\,\tau\,\phi(\tau)\,x(t-\tau)=x(t)*\phi(t) \quad . \qquad (5.1 - 5b)$$

b) Differentiation and integration of signals
From (5.1 - 5b) one can conclude that the system operator of LTI systems *commutes* with the differentiation and integration of the signals

$$\begin{aligned} &\boldsymbol{T}\left\{\frac{d}{dt}x(t)\right\}=\frac{d}{dt}\boldsymbol{T}\{x(t)\}=\frac{d}{dt}y(t) \quad , \quad \text{or} \\ &\phi(t)*\frac{d}{dt}x(t)=\frac{d}{dt}\phi(t)*x(t)=\frac{d}{dt}\{\phi(t)*x(t)\} \quad , \end{aligned} \tag{5.1 - 6a}$$

$$\begin{aligned} &\boldsymbol{T}\left\{\int^{t} x(\vartheta)d\vartheta\right\}=\int^{t} \boldsymbol{T}\{x(\vartheta)\}d\vartheta=\int^{t} y(\vartheta)d\vartheta \quad , \quad \text{or} \\ &\phi(t)*\int^{t} x(\vartheta)d\vartheta=\int^{t} \phi(\vartheta)d\vartheta*x(t)=\int^{t}\{\phi(\vartheta)*x(\vartheta)\}d\vartheta \end{aligned} \tag{5.1 - 6b}$$

c) Tandem connection
The tandem or series connection of two LTI systems Nos. 1 & 2 is described by the two equations

$$\begin{aligned} &y(t)=\boldsymbol{T}_1\{x(t)\}=\phi_1(t)*x(t) \quad , \\ &z(t)=\boldsymbol{T}_2\{y(t)\}=\phi_2(t)*y(t) \end{aligned} \tag{5.1 - 7a}$$

or by

$$z(t)=\boldsymbol{T}_2\{\boldsymbol{T}_1\{x(t)\}\}=\phi_2(t)*\{\phi_1(t)*x(t)\} \quad . \tag{5.1 - 7b}$$

Because the convolution is *associative* the tandem connection of the two LTI systems Nos. 1 & 2 represents also an LTI system with its own system operator

$$\begin{aligned} z(t)=\boldsymbol{T}\{x(t)\}=\phi(t)*x(t)&=\{\phi_1(t)*\phi_2(t)\}*x(t) \\ &=\phi_1(t)*\phi_2(t)*x(t) \quad . \end{aligned} \tag{5.1 - 7c}$$

5.1.3 Impulse Response Functions

The impulse response functions $\phi(t)$ relate the inputs $x(t)$ and outputs $y(t)$ of LTI systems by convolution according to (5.1 - 5a&b). As a rule they have to satisfy physical restraints. The first restraint to be discussed is the causality.

a) Causality
A transfer system is *causal*, if for any input $x(t)$ the response $y(t)$ at any instant of time t does not depend on the future $x(\tau)$ of the input at times $\tau > t$. In order to fulfill this condition the convolutions (5.1 - 5a&b) have to be modified as follows

$$y(t) = \boldsymbol{T}\{x(t)\} = \int_{-\infty}^{t} d\vartheta\, \phi(t-\vartheta)x(\vartheta) = \int_{0}^{\infty} d\vartheta\, \phi(\vartheta)x(t-\vartheta) \quad . \qquad (5.1 - 8a)$$

This modification can be taken into account in the basic convolutions

$$y(t) = \boldsymbol{T}\{x(t)\} = \int_{-\infty}^{+\infty} d\vartheta\, \phi(t-\vartheta)\, x(\vartheta) = \int_{-\infty}^{+\infty} d\tau\, \phi(\tau)\, x(t-\tau) \qquad (5.1 - 5a\&b)$$

by the assumption

$$\phi(t<0) = 0 \quad \text{or} \quad \phi(t) = H(t)f(t) \qquad (5.1 - 8b)$$

where $H(t)$ represents the Heaviside step function (2.3 - 46a-c) and $f(t)$ an arbitrary function.

b) Stability

Another important requirement to transfer systems is stability. Frequently a transfer system is called stable, if every bounded input gives rise to a bounded output [Schetzen 1989 B]. This is referred to as BIBO or bounded-input bounded-output stability.

For LTI systems it has been demonstrated [Schetzen 1989 B] that for a bounded input $x(t)$, which obeys the condition

$$|x(t)| < M \quad , \qquad (5.1 - 9a)$$

the output $y(t)$ fulfills the condition

$$|y(t)| > M \int_{-\infty}^{+\infty} |\phi(t)|\, dt \quad . \qquad (5.1 - 9b)$$

If one requires that $y(t)$ is bounded, this inequality implies the condition

$$\int_{-\infty}^{+\infty} |\phi(t)|\, dt = C < \infty \quad . \qquad (5.1 - 9d)$$

This condition yields also

$$\left| \int_{-\infty}^{+\infty} \phi(t)\, dt \right| = N \leq C < \infty \quad . \qquad (5.1 - 9d)$$

This relation does not guarantee a bounded $y(t)$ by its own.

c) Impulse response function as response to a δ-function input
The impulse response function $\phi(t)$ corresponds to the transient response $y(t)$ of a LTI system to an input $x(t)$ in form of the unit impulse or δ-function impulse

$$x(t) = \delta(t) \qquad (5.1 - 10a)$$

where $\delta(t)$ represents the Dirac delta function defined by (2.3 - 46c). This is demonstrated by the convolution

$$y(t) = \boldsymbol{T}\{\delta(t)\} = \int_{-\infty}^{\infty} d\tau\, \phi(\tau)\, \delta(t-\tau) = \phi(t) \qquad (5.1 - 10b)$$

or

$$y(t) = \delta(t) * \phi(t) = \phi(t) \quad . \qquad (5.1 - 10c)$$

Thus, the Dirac delta function represents the identity function of convolution.

5.1.4 Transients

According to (5.1 - 10b) the impulse response function $\phi(t)$ represents the response of the corresponding LTI system to the simplest transient $x(t)$ in form of a Dirac delta function. There exist, however, two other simple transients of common interest:

a) Step response
A transient of similar importance is the transient $x(t)$ in form of a *unit step* defined by the Heaviside step function (2.3 - 46a-c)

$$x(t) = H(t) = \begin{cases} 0 & \text{for} \quad t < 0 \\ 1/2 & \text{for} \quad t = 0 \\ 1 & \text{for} \quad t > 0 \end{cases} \quad . \qquad (5.1 - 11a)$$

The *unit step response* $W(t)$ represents the integral over the transfer function

$$y(t) = \boldsymbol{T}\{H(t)\} = W(t) = \int_{-\infty}^{t} \phi(\tau)\, d\tau \quad . \qquad (5.1 - 11b)$$

Because of (5.1 - 8b) *causality* implies that the *unit step* response $W(t)$ is zero for negative times $t < 0$ and for time zero $t = 0$

$$W(t \leq 0) = 0 \quad . \qquad (5.1 - 11c)$$

A LTI system is *normalized* if

$$W(\infty) = \int_{-\infty}^{+\infty} \phi(\tau)\,d\tau = \int_{0}^{+\infty} \phi(\tau)\,d\tau = 1 \quad . \tag{5.1 - 11d}$$

Because of the time invariance of a LTI system, the shape of a unit step response is not changed by a shift of the time scale

$$y(t) = \boldsymbol{T}\{H(t-t_0\} = W(t-t_0) \quad \text{with} \quad t > t_0 \quad . \tag{5.1 - 11e}$$

b) Relaxation

The relaxation of a LTI system manifests itself in its response $y(t)$ to an input $x(t)$ that is constant for negative times $t < 0$ and varies for positive times

$$x(t) = x_0[1 - H(t)] + H(t)x_1(t) = \begin{cases} x_0 & \text{for} \quad t \leq 0 \\ x_1(t) & \text{for} \quad t \geq 0 \end{cases} \tag{5.1 - 12a}$$

$$\text{with} \quad x_1(0) = x_0 \quad .$$

This response $y(t)$ consists of two components

$$x(t) = x_0\, \psi(t) + \int_{-\infty}^{t} d\tau\, \phi(\tau) x(t-\tau) \tag{5.1 - 12b}$$

with [Schötzau & Kneubühl 1994 J]

$$\psi(t) = \int_{t}^{\infty} \phi(\tau)\,d\tau = W(\infty) - W(t) \quad \text{with} \quad \psi(t<0) = W(\infty) \quad . \tag{5.1 - 12c}$$

The *relaxation function* $\psi(t)$ describes the relaxation of the LTI system from the constant response $y(t<0) = W(\infty)\,x_0$ to the constant input $x(t<0) = x_0$ at negative times t < 0.

An *example* is the LTI system with *exponential relaxation*. This system is of interest in physics and technology because it represents the simplest approach to a relaxation problem. It is characterized by the following functions

$$\phi(t>0) = \Theta^{-1}\, exp(-t/\Theta) \quad \text{with} \quad \Theta > 0 \quad , \tag{5.1 - 13a}$$

$$W(t>0) = 1 - exp(-t/\Theta) \quad \text{with} \quad W(\infty) = 1 \quad , \tag{5.1 - 13b}$$

$$\psi(t>0) = 1 - W(t>0) = exp(-t/\Theta) \quad . \tag{5.1 - 13c}$$

Characteristic functions of more complicated relaxing systems are listed in reference [Schötzau & Kneubühl 1994 J].

5.1.5 Response Evaluation by Laplace Transformation

The Laplace transformation defined by (3.2 - 23a&b) is a valuable tool for the evaluation of the response of LTI systems to transients. As an example we formulate the Laplace transform of the response $y(t)$ of (5.1 - 12b) to the transient input $x(t)$ of (5.1 - 12a)

$$\begin{aligned} & y(p) = x_0\,\psi(p) + \phi(p)\,x(p) \\ & \text{with} \quad x(p) = \boldsymbol{L}\{x(t)\}; \quad y(p) = \boldsymbol{L}\{y(t)\}; \quad \phi(p) = \boldsymbol{L}\{\phi(t)\} \\ & \text{and} \quad \psi(p) = \boldsymbol{L}\{\psi(p)\} = p^{-1}\left[W(\infty) - \phi(p)\right] \quad . \end{aligned} \tag{5.1 - 14a}$$

For a *causal system* defined by

$$x(t \le 0) = x_0 = 0 \quad \text{and} \quad \dot{x}(t \le 0) = \dot{x}_0 = 0 \tag{3.1 - 6}$$

the first term of (5.1 - 12b) that represents the relaxation is zero. Consequently, (5.1 - 14a) is reduced to

$$y(p) = \phi(p)\,x(p) \quad . \tag{5.1 - 14b}$$

5.1.6 Fourier Analysis

The Fourier analysis is a clue to the comprehension of LTI systems since it yields the eigenvalue spectra of LTI system operators used for the determination of outputs for harmonic, periodic and transient inputs.

a) Response to a harmonic input

The response $y(t)$ to a harmonic input $x(t)$ is the basis of the Fourier analysis of a LTI system. The linearity of such a system permits the complex representation of the harmonic input $x(t)$ and its corresponding output $y(t)$.

The basic harmonic input $x(t)$ to be considered is the *phasor* [Schetzen 1989 B]

$$x(t) = exp(-i\omega t) \quad . \tag{5.1 - 15a}$$

The phasor is an *eigenfunction* of the system operators of LTI systems. Therefore the response $y(t)$ of an LTI system is just a multiple of the phasor itself

$$y(t) = \boldsymbol{T}\{exp(-i\omega t)\} = e(\omega)\,exp(-i\omega t) = e(\omega)\,x(t) \quad . \tag{5.1 - 15b}$$

The factor $e(\omega)$ represents the *eigenvalue* of the LTI system operator. Because the circular frequency ω is a continuous parameter the *eigenvalue spectrum* of a LTI system operator is also *continuous*. The eigenvalue $e(\omega)$ can be evaluated by the convolution (5.1 - 5b). The result is

$$e(\omega) = \phi(\omega) \quad \text{with}$$
$$\phi(\omega) = F\{\phi(t)\} = \int_{-\infty}^{\infty} \phi(t)\,exp(+i\omega t)\,dt \tag{5.1 - 15c}$$

Consequently, the eigenvalue $e(\omega)$ equals the transfer function $\phi(\omega)$ defined as Fourier transform (2.2 - 46a) of the impulse response $\phi(t)$. As a consequence the relation between the amplitudes $x(\omega)$ and $y(\omega)$ of stationary complex harmonic inputs $x(t)$ and outputs $y(t)$

$$x(t) = x(\omega)\,exp(-i\omega t) \quad ,$$
$$y(t) = y(\omega)\,exp(-i\omega t) \tag{5.1 - 16a}$$

is determined by the eigenvalue $e(\omega)$ of equation (5.1 - 15b)

$$y(\omega) = e(\omega)\,x(\omega) = \phi(\omega)\,x(\omega) \quad . \tag{5.1 - 16b}$$

A classical *example* represent the eigenvalues $e(\omega)$ of the normalized LTI system with exponential relaxation defined by (5.1 - 13a-c)

$$e(\omega) = \phi(\omega) = [1 - i\omega\Theta]^{-1} = \left[1 + \omega^2\Theta^2\right]^{-1/2} exp(i\varphi)$$
$$\text{with} \quad \varphi = arctan(\omega\Theta) \tag{5.1 - 17}$$

The first factor represents the amplitude damping, while the second determines the phase shift.

b) Response to a periodic input

A *periodic input* $x(t)$ with the period T can be written as *Fourier series*

$$x(t) = x(t+T) = \sum_{m=-\infty}^{+\infty} F_m exp(-im\omega_1 t)$$
$$\text{with} \quad \omega_1 = 2\pi / T \quad . \tag{5.1 - 18a}$$

Properties and examples of Fourier series are listed in Appendix A.1.

The response $y(t)$ of LTI system to such a periodic input $x(t)$ is also periodic

$$y(t) = \sum_{m=-\infty}^{+\infty} F_m \phi(m\omega_1) exp(-im\omega_1 t) \quad . \tag{5.1 - 18b}$$

Here $\phi(\omega)$ indicates the Fourier transform (2.2 - 46a) of the impulse response $\phi(t)$. This equation can be deduced from the linearity of LTI systems and (5.1 - 16b).

c) Response to a transient
A transient input $x(t)$ can be described by a *Fourier integral*

$$x(t) = \frac{1}{2\pi} \int_{-\infty}^{+\infty} F(\omega) exp(-i\omega t) d\omega \tag{2.2 - 46b}$$

where $F(\omega)$ is the Fourier transform (2.2 - 46a) of $x(t)$. Properties and examples of Fourier integrals and Fourier transforms are listed in appendix A.2.

If $F(\omega)$ is known the response $y(t)$ of a LTI system to this input $x(t)$ can be evaluated by the Fourier integral

$$y(t) = \frac{1}{2\pi} \int_{-\infty}^{+\infty} \phi(\omega) F(\omega) exp(-i\omega t) d\omega \quad . \tag{5.1 - 19}$$

Again $\phi(\omega)$ corresponds to the Fourier transform of the impulse response $\phi(t)$. This equation can also be deduced from the linearity of LTI systems and (5.1 - 16b).

5.1.7 Averages and Correlations

In practice, the input $x(t)$ of a transfer system exhibits random fluctuations or noise that is transferred to the output $y(t)$. Therefore, the transfer of random inputs by LTI systems is relevant. This transfer is described by statistical measures such as average, mean square, auto- and cross-correlation and power spectra [Schetzen 1989 B].

a) Statistical measures
The *average* of a time function is defined as

$$\overline{f(t)} = \lim_{\tau \to \infty} \frac{1}{2\tau} \int_{-\tau}^{+\tau} f(t) dt \quad . \tag{5.1 - 20}$$

Accordingly, the *mean square* of a time function $f(t)$ is the average of the square of this function

$$\overline{f^2(t)} = \lim_{\tau \to \infty} \frac{1}{2\tau} \int_{-\tau}^{+\tau} f^2(t) dt \quad . \tag{5.1 - 21}$$

The *autocorrelation* $\gamma(s)$ of a real time function $f(t)$ is defined by

$$\gamma(s) = \lim_{\tau\to\infty} \frac{1}{2\tau} \int_{-\tau}^{+\tau} f(t)\, f(t+s)\, dt \quad . \tag{5.1 - 22a}$$

The autocorrelation $\gamma(s)$ of a real time function $f(t)$ is symmetric and exhibits a maximum at the origin $s = 0$ [Bracewell 1986]

$$\gamma(-s) = \gamma(s) \quad , \tag{5.1 - 22b}$$

$$\gamma(0) \geq \gamma(s) \quad . \tag{5.1 - 22c}$$

The definition of the *crosscorrelation* $\gamma_{fg}(s)$ of two real time functions $f(t)$ and $g(t)$ is similar to that of the autocorrelation $\gamma(s)$

$$\gamma_{fg}(s) = \lim_{\tau\to\infty} \frac{1}{2\tau} \int_{-\tau}^{+\tau} f(t)\, g(t+s)\, dt \quad . \tag{5.1 - 23a}$$

The autocorrelation $\gamma(s)$ is *antisymmetric*

$$\gamma_{fg}(-s) = \gamma_{gf}(s) = \lim_{\tau\to\infty} \frac{1}{2\tau} \int_{-\tau}^{+\tau} g(t)\, f(t+s)\, ds \quad . \tag{5.1 - 23b}$$

b) Wiener-Khintchine theorem

In the physical interpretation of a real time function $f(t)$ as a signal, the *mean square* represents its total *average power* [Bracewell 1986 B, Schetzen 1989 B]

$$P_{av} = \overline{y^2(t)} \tag{5.1 - 24a}$$

The average power of this signal between the circular frequencies zero and ω can be written as integral

$$P_{av}(\omega) = \int_0^{\omega} \sigma(\omega')\, d\omega'$$
$$\text{with} \quad \sigma(\omega) = \frac{d}{d\omega} P_{av}(\omega) \tag{5.1 - 24b}$$

where $\sigma(\omega)$ designates the *power spectral density* or *power spectrum.*

The *Wiener-Khintchine theorem* postulates that the power spectral density $\sigma(\omega)$ corresponds to the Fourier transform (2.2 - 46a) of the autocorrelation function $\gamma(\tau)$ defined by (5.1 - 22a)

$$\sigma(\omega) = \int_{-\infty}^{+\infty} \gamma(\tau) exp(+i\omega\tau) d\tau \qquad (5.1 - 25a)$$

$$\gamma(\tau) = \frac{1}{2\pi} \int_{-\infty}^{+\infty} \sigma(\omega) exp(-i\omega\tau) d\omega \qquad (5.1 - 25b)$$

A relevant example represents the *Johnson* or *Nyquist noise* [Duley 1976 B, Pöschl 1956 B, van der Ziel 1976 B] of a resistor with the resistance $R[\Omega]$ and the absolute temperature T[K]. It constitutes a *white noise* [Schetzen 1989] since it has a constant spectral power density

$$\sigma(\omega) = \frac{d}{d\omega} P_{\text{av}}(\omega) = \frac{2}{\pi} RkT = \sigma_0 = const$$
$$\text{with} \quad k = 1.3807 \cdot 10^{-23}\ \text{JK}^{-1} \qquad (5.1 - 26a)$$

In this equation k indicates Boltzmann's constant. According to (5.1 - 25a&b) the autocorrelation function $\gamma(\tau)$ corresponding to this constant power spectral density σ_0 is a multiple of the Dirac delta function

$$\gamma(\tau) = \sigma_0\, \delta(\tau) \qquad (5.1 - 26b)$$

In this equation $\gamma(\tau)$ represents the autocorrelation of the voltage $U(t)$ or of the current $i(t)$. The Dirac delta autocorrelation (5.1 - 26b) signifies that a signal $U(t)$ or $i(t)$ at time t has *no relation* to a signal $U(t)$ or $i(t')$ at a different time $t' \neq t$. This is *characteristic for white noise.*

c) Relations between input and output

According to (5.1 - 5a) a LTI system is determined by its impulse response function $\phi(t)$. The transfer function $\phi(\omega)$ as its Fourier transform provides the basis of the relations between the statistical measures of input and output [Schetzen 1989 B].

The averages $\overline{x(t)}$ and $\overline{y(t)}$ of input x(t) and y(t) are related by

$$\overline{y(t)} = \phi(\omega = 0)\, \overline{x(t)}$$
$$\text{with} \quad \phi(\omega = 0) = \int_{-\infty}^{+\infty} \phi(t) dt \quad . \qquad (5.1 - 27)$$

The *mean square of the output* $y(t)$ is determined by the Fourier transform $\phi(\omega)$ of the impulse response function $\phi(t)$ and by the power spectrum $\sigma(\omega)$ of the input $x(t)$

$$\overline{y^2(t)} = \int_{-\infty}^{+\infty} |\phi(\omega)|^2\, \sigma(\omega) d\omega \quad . \qquad (5.1 - 28)$$

The *crosscorrelation between input* $x(t)$ *and output* $y(t)$ equals the convolution between the impulse response function $\phi(t)$ and the autocorrelation $\gamma(t)$ of the input

$$\gamma_{xy}(\tau) = \phi(\tau) * \gamma(\tau) = \int_{-\infty}^{+\infty} \phi(s)\,\gamma(\tau - s)\,ds \quad . \tag{5.1 -29a}$$

This equation is relevant because it permits the *determination of the impulse response function* $\phi(t)$ with white or Nyquist noise as input $x(t)$. The introduction of the autocorrelation function (5.1 - 26b) of white noise in (5.1 - 29a) results in the following crosscorrelation between input and output

$$\gamma_{xy}(\tau) = \sigma_0\,\phi(\tau) \quad . \tag{5.1 - 29b}$$

Thus, the impulse response function $\phi(t)$ is proportional to the crosscorrelation of input and output for white noise.

5.1.8 Transfer Differential Equations

The transfer of the input $x(t)$ to the output $y(t)$ by the system operator $\boldsymbol{T}\{...\}$ and the corresponding convolution (5.1 - 5a) can be determined by a differential equation if the impulse response function $\phi(t)$ fulfills a homogeneous differential equation

$$\sum_{n=0}^{N} a_n\,\phi^{(n)}(t) = 0 \quad . \tag{5.1 - 30a}$$

This can be demonstrated by the following representation of the system operator $\boldsymbol{T}\{...\}$ and its time derivatives

$$\begin{aligned} y(t) = \quad & \boldsymbol{T}\{x(t)\} = \int_{-\infty}^{t} d\vartheta\,\phi(t-\vartheta)\,x(\vartheta) \quad , \\ \dot{y}(t) = \quad & \dot{\boldsymbol{T}}\{x(t)\} = \int_{-\infty}^{t} d\vartheta\,\dot{\phi}(t-\vartheta)\,x(\vartheta) + \phi(+0)\,x(t) \quad , \\ y^{(n)}(t) = \; & \boldsymbol{T}^{(n)}\{x(t)\} = \int_{-\infty}^{t} d\vartheta\,\phi^{(n)}(t-\vartheta)\,x(\vartheta) + \\ & + \sum_{r=0}^{n-1} \phi^{(r)}(+0)\,x^{(n-r-1)}(t) \quad . \end{aligned} \tag{5.1 - 30b}$$

The linear combination of the $y^{(n)}(t > 0)$; $n = 1, 1, 2...$ according to (5.1 - 30a) yields the *transfer differential equation* for the transfer of $x(t)$ to $y(t > 0)$

$$\sum_{n=0}^{N} a_n\, y^{(n)}(t) = \sum_{m=1}^{N} x^{(m-1)}(t) \sum_{k=m}^{N} a_k\, \phi^{(k-m)}(+0) \tag{5.1 - 30c}$$

a) Exponential relaxation

The normalized LTI system with exponential relaxation, which is defined by the normalized characteristic functions (5.1 - 13a-c), can be described by a type (5.1 - 30c) equation. The relevant exponential transfer function (5.1 - 13a) fulfills the linear differential equation

$$\phi(t) + \Theta \frac{d}{dt}\phi(t) = 0 \tag{5.1 - 31a}$$

with the initial condition

$$\phi(+0) = 1/\Theta \quad . \tag{5.1 - 31b}$$

According to (5.1 - 30c) the input $x(t)$ and the output $y(t)$ of this LTI system are related by the transfer differential equation

$$y(t) + \Theta \frac{d}{dt} y(t) = x(t) \quad . \tag{5.1 - 31c}$$

The solution of this equation corresponds to the convolution (5.1 - 8a).

$$y(t) = \int_{-\infty}^{t} d\vartheta \frac{1}{\Theta} exp\left(-\frac{t-\vartheta}{\Theta}\right) x(\vartheta) = \int_{-\infty}^{t} d\vartheta\; \phi(t-\vartheta)\, x(\vartheta) \tag{5.1 - 31d}$$

b) Harmonic input and output

For harmonic input $x(t)$ and output $y(t)$ according to (5.1 - 16a), equations (5.1 - 16b) and (5.1 - 30c) yield the following relation

$$y(\omega) = \boldsymbol{T}\{x(\omega)\} = \frac{\sum_{m=1}^{N} (i\omega)^{m-1} \sum_{k=m}^{N} a_k \phi^{k-m}(+0)}{\sum_{n=0}^{N} a_n (-i\omega)^{n}}\, x(\omega) \quad . \tag{5.1 - 32}$$

Again the normalized LTI system with exponential relaxation described by the normalized characteristic functions (5.1 - 13a-c) is taken as an *example*. From (5.1 - 33) it can be deduced that $N = 1$, $a_0 = 1$, and $a_1 = \Theta$ while (5.1 - b) presumes $\phi(+0) = 1/\Theta$. These data if used in (5.1 - 32) provide a relation derived before, i.e.

$$\phi(\omega) = [1 - i\omega\Theta]^{-1} \quad . \tag{5.1 - 17}$$

5.1.9 Nonlinear Time-invariant Systems

Nonlinear time-invariant systems with nonlinear transfer equations of the form

$$y(t) = \boldsymbol{T}_{\mathrm{NL}}(x(t)) = \int_0^{\infty} d\tau\, \phi(\tau)\, f(x(t-\tau)) \qquad (5.1 - 33)$$

are rather complicated. $\phi(\tau)$ is the original impulse response function and $f(x)$ indicates a nonlinear function. These systems can be analyzed with Volterra and Wiener theories [Schetzen 1989 B]. In general, the output $y(t)$ is related to the input $x(t)$ of a nonlinear time-invariant system by the Volterra series [Volterra 1959 B]

$$\begin{aligned} y(t) = & \int_{-\infty}^{+\infty} d\tau_1\, \phi_{\mathrm{a}}(\tau_1)\, x(t-\tau_1) \\ & + \int_{-\infty}^{+\infty}\int_{-\infty}^{+\infty} d\tau_1\, d\tau_2\, \phi_2(\tau_1, \tau_2)\, x(t-\tau_1)\, x(t-\tau_2) \\ & + \int_{-\infty}^{+\infty}\int_{-\infty}^{+\infty}\int_{-\infty}^{+\infty} d\tau_1\, d\tau_2\, d\tau_3\, \phi_3(\tau_1, \tau_2, \tau_3)\, x(t-\tau_1)\, x(t-\tau_2)\, x(t-\tau_3) \qquad (5.1 - 34) \\ & +++ \end{aligned}$$

with $\quad \phi_{\mathrm{n}}(\tau_1, \ldots \tau_{\mathrm{n}}) = 0 \quad$ for any $\tau_{\mathrm{j}} < 0$.

In Section 5.3.3 nonlinear time-invariant delay systems will be presented as *examples*.

Nonlinear time-invariant transfer systems are of basic interest in electrical system engineering. Recently, a nonlinear time-invariant transfer operator was introduced in a new comprehensive theory on dynamics of high-power arcs [Schötzau & Kneubühl 1994 J].

5.2 Circuits

Transfer systems form components of feedback-control systems as well as of oscillating resonant circuits. Examples of these circuits are shown in Figs. 5.2 - 1&2. Each of these two circuits consists of a *linear transfer system* and a *linear feedback.* The linear transfer system is characterized by the linear transfer operator

$$y(t) = \boldsymbol{T}\{x(t)\} = \int_{-\infty}^{t} d\,\vartheta\, \phi(t-\vartheta)\, x(\vartheta) = \int_0^{\infty} d\,\tau\, \phi(\tau)\, x(t-\tau) \qquad (5.1 - 5a)$$

while the feedback is described by the linear differential operator

$$z(t) = \Phi\{y(t)\} = \sum_{k=0}^{K} f_k \frac{d^{(k)}}{dt^k} y(t) \quad . \tag{5.2 - 1}$$

5.2.1 Feedback-control Systems

The behavior of the feedback-control system or control circuit illustrated in Fig. 5.2 - 1 is determined by the two coupled equations

$$y(t) = \boldsymbol{T}\{x(t) + z(t)\} \tag{5.2 - 2a}$$

$$z(t) = \Phi\{y(t)\} \tag{5.2 - 2b}$$

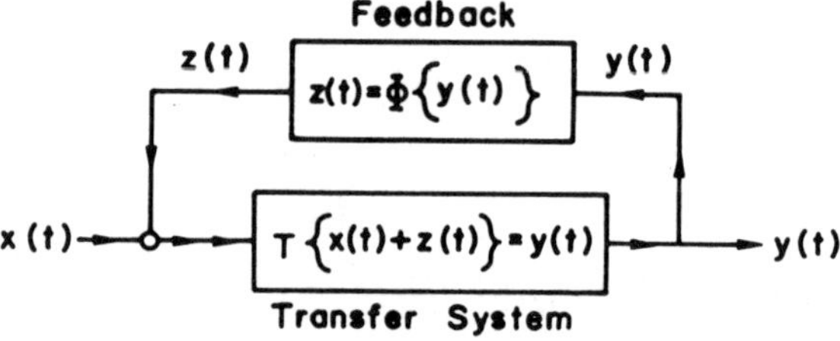

Fig. 5.2 - 1. Feedback-control system formed by transfer system and feedback in parallel

a) Harmonic excitations

The response of a feedback-control system or control circuit characterized by (5.2 - 2a&b) to continuous harmonic excitations can be investigated with the aid of complex harmonic inputs $x(t)$ and outputs $y(t)$ of (5.1 - 16a) to (5.2 - 2a&b). Taking into account (5.1 - 16b) one can thus transform (5.2 - 2a&b) into

$$y(\omega) = \phi(\omega)\left[x(\omega) + z(\omega)\right] \quad \text{with} \quad \phi(\omega) = \boldsymbol{F}\{\phi(t)\} \quad , \tag{5.2 - 3a}$$

$$z(\omega) = y(\omega) \cdot \sum_{k=0}^{K} f_k (-i\,\omega)^k \quad . \tag{5.2 - 3b}$$

The combination of these two equations permits the relation of the complex amplitudes $x(\omega)$ and $y(\omega)$ of input and output

$$y(\omega) = \phi(\omega)\left[1 - \phi(\omega)\sum_{k=0}^{K} f_k (-i\,\omega)^k\right]^{-1} x(\omega) \quad . \tag{5.2 - 3c}$$

An *example* is the control circuit that consists of a normalized transfer system with exponential relaxation described by (5.1 - 13a-c) and a feedback proportional to the velocity defined as

$$z(t) = \Phi_1\{y(t)\} = f_1 \frac{dy(t)}{dt} \quad . \tag{5.2 - 4a}$$

From (5.1 - 17), (5.2 - 2a&b) and (5.2 - 3c) one can deduce the following relation between the complex amplitudes $x(\omega)$ and $y(\omega)$

$$\begin{aligned} & y(\omega) = x(\omega)\left[1 - i\,\omega(\Theta - f_1)\right]^{-1} \\ & \text{with} \quad y(\omega) = x(\omega) \quad \text{if} \quad f_1 = \Theta \quad . \end{aligned} \tag{5.2 - 4b}$$

Consequently, the feedback compensates the damping of the normalized transfer system if $f_1 = \Theta$.

b) Transients

Transient processes in a control circuit can be described in a simple way for the initial conditions

$$x^{(n)}(t \le +0) = y^{(n)}(t \le +0) = 0 \quad \text{for} \quad n = 0, 1, 2, \ldots \quad . \tag{5.2 - 5a}$$

According to (3.1 - 6) these conditions imply a causal system. If the *Laplace transformation* (3.2 - 23a) is applied to the coupled equations (5.2 - 2a&b) on conditions (5.2 - 5a), it yields the following relation between the Laplace transforms $x(p)$ and $y(p)$ of input and output

$$y(p) = R(p)\,x(p) = \left[\phi(p)^{-1} - \sum_{k=0}^{K} f_k\, p^k\right]^{-1} x(p) \quad . \tag{5.2 - 5b}$$

$R(p)$ designates the transfer function [Pöschl 1956 B] of the entire control circuit, whilst $\phi(p)$ represents the Laplace transform of the impulse response function $\phi(t)$ of the transfer system defined by (5.1 - 5a). The control circuit is *stable*, if no pole p_r of the transfer function $R(p)$ exhibits a real part larger than zero. Thus, the *condition for stability* of a control circuit can be written as [Pöschl 1956 B]

$$Re\, p_r \le 0 \quad \text{for} \quad R(p_r) = \pm\infty, \quad r = 1, 2, \ldots \quad . \tag{5.2 - 5c}$$

5.2.2 Resonant Oscillating Circuits

The oscillating resonant circuit illustrated in Fig. 5.2 - 2 corresponds to the control circuit of Fig. 5.2 - 1 with zero input $x(t) = 0$. Therefore the characteristic equations (5.2 - 2a&b) of the control circuit are reduced to the *tandem equations*

$$y(t) = \boldsymbol{T}\{z(t)\} \quad , \tag{5.2 - 6a}$$

$$z(t) = \Phi\{y(t)\} \quad . \tag{5.2 - 6b}$$

The characteristic equations of oscillating resonant circuits as illustrated in Fig. 5.2 - 2 are found by elimination of either $z(t)$ or $y(t)$

$$y(t) = T\{\Phi\{y(t)\}\} \quad , \tag{5.2 - 6c}$$

$$z(t) = \Phi\{T\{z(t)\}\} \quad . \tag{5.2 - 6d}$$

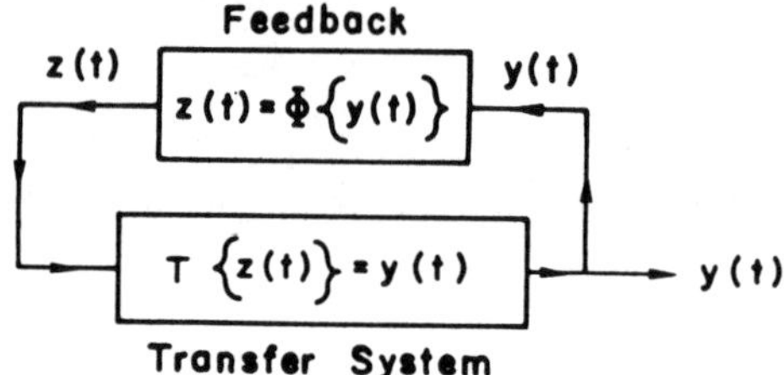

Fig. 5.2 - 2. Oscillating resonant circuit formed by transfer system and feedback

a) Oscillations

The oscillation condition of an oscillating resonant circuit composed of a linear transfer system defined by (5.1 - 5a) and a linear feedback characterized by (5.2 - 1) can be determined by assuming complex harmonic oscillations according to (5.1 - 16a). Thus (5.2 - 6c) is transformed into

$$\left[1 - \phi(\omega) \sum_{k=0}^{K} f_k(-i\,\omega)^k\right] y(\omega) = 0 \quad . \tag{5.2 - 7a}$$

By this procedure a similar equation is found for $z(\omega)$. A non-zero amplitude $y(\omega)$ requires the vanishing of the other factor in (5.2 - 7a). This requirement yields the *oscillation condition* of the resonant circuit

$$\phi(\omega) \sum_{k=0}^{K} f_k(-i\,\omega)^k = 1 \quad . \tag{5.2 - 7b}$$

This equation has solutions of the form

$$\omega = \omega_m + i\,\alpha_m \quad \text{with} \quad m = 1, 2, 3, \ldots \quad . \tag{5.2 - 7c}$$

For $\omega_m > 0$ these solutions correspond

$$\text{for} \begin{Bmatrix} \alpha_m > 0 & \text{to} & \text{increasing} \\ \alpha_m = 0 & \text{to} & \text{stationary} \\ \alpha_m < 0 & \text{to} & \text{decreasing} \end{Bmatrix} \text{oscillations} \quad . \tag{5.2 - 7d}$$

With respect to the oscillation condition (5.2 - 7b) it should be noticed that it gives no information on the amplitudes $y_m(\omega)$ and $z_m(\omega)$ of a stationary oscillation with a real resonant circular frequency ω_m. This disadvantage originates in the linearity of the resonant circuit considered. The amplitude of an oscillation of a resonant circuit is generally determined by nonlinear effects.

As *example* one may consider the oscillation condition of a resonant circuit formed by the normalized transfer system with exponential relaxation as defined by (5.1 - 13a-c) and a feedback characterized by

$$\Phi\{z(t)\} = \left\{ f_0 + f_1 \frac{d}{dt} + f_2 \frac{d^2}{dt^2} \right\} z(t) \quad . \tag{5.2 - 8a}$$

By making use of (5.1 - 17), (5.2 - 7b) and (5.2 - 8a) one finds the oscillation condition

$$f_2\, \omega^2 + i\, \omega (f_1 - \Theta) - (f_0 - 1) = 0 \quad . \tag{5.2 - 8b}$$

According to this condition, the resonant circuit can perform a stationary harmonic oscillation if

$$f_0 > 1 \; ; \; f_1 = \Theta \quad \text{and} \quad \omega = \omega_1 = +\sqrt{(f_0 - 1) / f_2} \quad . \tag{5.2 - 8c}$$

b) Transients

Transients in oscillating circuits are determined by the initial conditions, e.g. by the variable $y(t)$ and its derivatives $y^{(k)}(t)$, $k = 1, 2, \ldots$ at time $t = +0$. On these initial conditions the transient $y(t)$ can be evaluated with the Laplace transformation. For this purpose the characteristic equation (5.2 - 6c) of the oscillating circuit is transformed into

$$\boldsymbol{T}^{-1}\{y(t)\} = \Phi\{y(t)\} \quad . \tag{5.2 - 9a}$$

The right side of this equation represents the feedback (5.2 - 1) whilst the left side corresponds to the inverse transfer. In case of a *causal system* (3.1 - 6) the Laplace transformation of the inverse transfer is governed by (5.1 - 14b). For a causal system and the initial conditions mentioned the Laplace transformation of (5.2 - 9a) yields

$$y(p)\left[\sum_{k=0}^{K} f_k p^k - \phi^{-1}(p)\right] = \sum_{m=0}^{K-1} p^m \sum_{n=0}^{K-1} f_{n+1} y^{(n-1)}(t=+0) \qquad (5.2-9b)$$
$$= -y(p)/R(p) \quad .$$

This equation determines the Laplace transform $y(p)$ of the transient $y(t)$ for times $t > 0$. The corresponding transient $z(t)$ can be evaluated by making use of (5.2 - 1). $R(p)$ is the transfer factor defined by (5.2 - 5b). Its poles p_r determine on one hand the stability of control circuits or servoloops and on the other hand the oscillation of circuits. A *necessary condition for oscillation* is the existence of a pole p_n of $R(p)$ with a non-negative real part

$$Re\, p_n \geq 0 \quad . \qquad (5.2-9c)$$

5.3 Delay Systems

There exist systems which cause a delay of the output $y(t)$ versus the input $x(t)$. A measure of this effect is the delay time $\tau > 0$ defined by the input-output condition that $y(t < \tau) = 0$ for $x(t < 0) = 0$. Delay systems can be rather complicated since in general the delay of the input $x(t)$ also involves distortion. The subsequent introduction of a normalized delay system that does not deform the input facilitates the following discussion of the more sophisticated delay systems and circuits.

5.3.1 Normalized Delay System

In the normalized linear delay system the input $x(t)$ is delayed by a constant delay time τ without any other change

$$y(t) = \boldsymbol{T}\{x(t)\} = x(t-\tau) \quad \text{with} \quad \tau > 0 \qquad (5.3-1a)$$

As a consequence the transfer operator $\boldsymbol{T}\{...\}$ of the normalized delay system can be represented as an *inverse propagator*

$$y(t) = \boldsymbol{T}\{x(t)\} = \boldsymbol{P}^{-1}(t-\tau, t)\, x(t) = \boldsymbol{P}(t, t-\tau) x(t)$$
$$= exp\left\{-\tau \frac{d}{dt}\right\} x(t) = x(t-\tau) \quad . \qquad (5.3-1b)$$

Furthermore, the normalized delay system is characterized by the following impulse response function $\phi(t)$, unit-step response $W(t)$ and relaxation function $\psi(t)$

$$\phi(t > 0) = \delta(t-\tau) \quad \text{with} \quad \tau > 0 \quad , \qquad (5.3-2a)$$

$$W(t>0)=H(t-\tau) \quad \text{with} \quad W(\infty)=1 \quad , \tag{5.3 - 2b}$$

$$\psi(t>0)=1-W(t>0)=1-H(t-\tau) \quad . \tag{5.3 - 2c}$$

For the complex *stationary oscillations* (5.1 - 16a) the complex amplitudes $x(\omega)$ and $y(\omega)$ of input and output of a normalized delay system are related by

$$y(\omega)=\boldsymbol{T}\{x(\omega)\}=\phi(\omega)\,x(\omega)=exp(+i\omega\tau)\,x(\omega) \quad . \tag{5.3 - 3}$$

Transient processes in causal (5.2 - 5a) and non-causal systems such as the feedback-control system (5.2 - 5b&c) are usually calculated with the aid of the *Laplace transformation* (3.2 - 23a&b). The Laplace transform of the system operator of a normalized delay system is [Zwillinger 1989 B]:

$$\begin{aligned} &\boldsymbol{L}\{y(t)\}=y(p)=\boldsymbol{L}\{\boldsymbol{T}\{x(t)\}\}=\boldsymbol{L}\{x(t-\tau)\} \\ &=x(p)\,exp(-\tau\,p)+x_0 p^{-1}\left[1-exp(-\tau\,p)\right] \quad \text{if} \quad x(t<0)=x_0 \quad . \end{aligned} \tag{5.3 - 4a}$$

The Laplace transform of the corresponding inverse system operator is

$$\begin{aligned} &\boldsymbol{L}\{x(t)\}=x(p)=\boldsymbol{L}\{\boldsymbol{T}^{-1}\{y(t)\}\}=\boldsymbol{L}\{y(t+\tau)\}= \\ &=y(p)\,exp(+\tau\,p)+y_0\,p^{-1}\left[1-exp(+\tau\,p)\right] \quad \text{if} \quad y(t<\tau)=y_0 \quad . \end{aligned} \tag{5.3 - 4b}$$

5.3.2 Delay Systems in Circuits

Circuits with integrated delay systems are usually described with *delay equations* [Bainov & Mishev 1991 B, Barr 1995 J, Myskis 1955 B, Penney 1959 B, Saathy 1981 B, Zwillinger 1989 B]. In the following two of these circuits are discussed for illustration.

a) Circuit with integrating feedback
The first circuit consists of a normalized linear delay system defined by

$$y(t)=\boldsymbol{T}\{x(t)+z(t)\}=x(t-\tau)+z(t-\tau) \tag{5.3 - 5a}$$

and an integrating feedback characterized by

$$z(t)=\boldsymbol{\Phi}_{-1}\{y(t)\}=f_{-1}\int_{-\infty}^{t} y(\vartheta)d\vartheta \quad . \tag{5.3 - 5b}$$

The corresponding delay equation of the *control circuit* has the form

$$\frac{d}{dt}y(t) = f_{-1}\,y(t-\tau) + \frac{d}{dt}x(t-\tau) \quad , \tag{5.3 - 5c}$$

while that of the corresponding *oscillating resonant circuit* can be written as

$$\frac{d}{dt}y(t) = f_{-1}\,y(t-\tau) \quad . \tag{5.3 - 5d}$$

For the complex *stationary harmonic oscillations* (5.1 - 16a) the relation between the complex amplitudes $x(\omega)$ and $y(\omega)$ of input and output of the control circuit is

$$y(\omega) = \left[exp(-i\,\omega\,\tau) + f_{-1}\,/\,i\,\omega\right]^{-1} x(\omega) \quad . \tag{5.3 - 6a}$$

The oscillation condition of the corresponding resonant circuit is accordingly

$$exp(+i\,\omega\,\tau) = -i\,\omega\,/\,f_1 \quad . \tag{5.3 - 6b}$$

Its solutions are

$$\begin{aligned} &\omega_n = (2n+1)\,\pi\,/\,2\tau \quad \text{with} \quad n = 0, +1, \pm 2, \ldots \quad , \\ &\text{and} \quad \left(f_{-1}\right)_n = (-1)^{n+1}\,\omega_n \quad . \end{aligned} \tag{5.3 - 6c}$$

As a consequence the resonant circuit oscillates with a resonant circular frequency ω_n if the feedback parameter f_{-1} is matched to this frequency.

The transients in this oscillating resonant circuit can be evaluated with the aid of the Laplace transformation of the delay equation (5.3 - 5d) by making use of (5.3 - 4a). The result is

$$y(p)\,/\,y(t=0) = \frac{p + f_{-1}\left[1 - exp(-\tau\,p)\right]}{p\left[p - f_{-1}\,exp(-\tau\,p)\right]} \quad . \tag{5.3 - 7}$$

b) Circuit with feedback proportional to the velocity

The second circuit to be considered consists of a normalized delay system defined by

$$y(t) = \boldsymbol{T}\{x(t) + z(t)\} = x(t-\tau) + z(t-\tau) \tag{5.3 - 5a}$$

and a feedback proportional to the velocity

$$z(t) = \Phi_1\{y(t)\} = f_1\,\frac{d}{dt}y(t) \quad . \tag{5.2 - 4a}$$

The corresponding *control circuit* obeys the delay equation

$$y(t) = x(t-\tau) + f_1 \frac{d}{dt} y(t-\tau) \quad \text{or}$$
$$y(t+\tau) = x(t) + f_1 \frac{d}{dt} y(t) \quad , \tag{5.3 - 8a}$$

while the *oscillating resonant circuit* is ruled by the delay equation

$$y(t) = f_1 \frac{d}{dt} y(t-\tau) \quad \text{or}$$
$$y(t+\tau) = f_1 \frac{d}{dt} y(t) \quad . \tag{5.3 - 8b}$$

For complex stationary harmonic oscillations (5.1 - 16a) the delay equation (5.3 - 8a) of the *control circuit* yields the following relation between the complex amplitudes $x(\omega)$ and $y(\omega)$ of input and output

$$y(\omega) = x(\omega)\left[exp(-i\,\omega\,\tau) + i\,\omega\,f_1\right]^{-1} \quad . \tag{5.3 - 9a}$$

The oscillation condition for the corresponding *oscillating resonant circuit* can be derived from (5.3 - 8b). It reads

$$exp(+i\,\omega\,\tau) = -1 / i\,\omega\,f_1 \tag{5.3 - 9b}$$

and its solutions are

$$\omega_{\mathrm{n}} = (2n+1)\pi / 2\tau \quad \text{with} \quad n = 0, \pm 1, \pm 2, \ldots \quad ,$$
$$\text{with} \quad (f_1)_{\mathrm{n}} = (-1)^{\mathrm{n}}\,\omega_{\mathrm{n}}^{-1} \quad . \tag{5.3 - 9c}$$

This resonant circuit oscillates at a resonant circular frequency ω_n only if the feedback parameter f_1 is matched according to the second equation (5.3 - 9c).

For short delay times τ the function $y(t-\tau)$ can be represented by a Taylor series. In this situation, the delay equation (5.3 - 8b) can be approximated by

$$y(t) \approx f_1 \frac{d}{dt} y(t) - \tau\, f_1 \frac{d^2}{d\,t^2} y(t) \quad . \tag{5.3 - 9d}$$

Thus, the delay equation (5.3 - 8b) is approximated by an ordinary differential equation.

Transients in the oscillating resonant circuit described can be determined by application of the Laplace transformation to the delay equation (5.2 - 8b). Taking into account (5.3 - 4b) one finds for its Laplace transform

$$y(p) = y(t=0)\frac{exp(\tau p) - f_1 p - 1}{p\left[exp(\tau p) - f_1 p\right]} \quad . \tag{5.3 - 10}$$

5.3.3 Nonlinear Delay Systems

A nonlinear time-invariant delay system can be described by a system operator of the form

$$y(t) = \boldsymbol{T}_{\mathrm{NL}}\{x(t)\} = F(x(t-\tau)) \tag{5.3 - 11}$$

where $F(x)$ represents a nonlinear function of x. The nonlinearity of this operator impedes the application of the Fourier and the Laplace transformation to the calculations on nonlinear delay systems.

Two oscillating resonant circuits with nonlinear delay are discussed as *examples*:

a) Nonlinear delay circuit with integrating feedback

The first oscillating resonant circuit consists of a nonlinear delay system characterized by the system operator

$$y(t) = \boldsymbol{T}_{\mathrm{NL}}\{z(t)\} = 1 + [z(t-\tau)]^{-2} \tag{5.3 - 12a}$$

and an integrating feedback described by

$$z(t) = \Phi_{-1}\{y(t)\} = f_{-1} \int_{-\infty}^{t} y(\vartheta)\, d\vartheta \quad . \tag{5.3 - 5b}$$

The corresponding nonlinear delay equation is

$$\frac{1}{f_{-1}} \frac{d}{dt} z(t) = 1 + [z(t-\tau)]^{-2} \quad , \tag{5.3 - 12b}$$

while its solutions are

$$\begin{aligned} & z(t) = tan(\omega_{\mathrm{n}} t - \varphi_{\mathrm{n}}), \varphi_{\mathrm{n}} \text{ arbitrary} \\ & \text{with} \quad \omega_{\mathrm{n}} = (2n+1)\pi / 2\tau, \quad n = 0, \pm 1, \pm 2, \ldots \\ & \text{and} \quad (f_{-1})_{\mathrm{n}} = \omega_1 \quad . \end{aligned} \tag{5.3 - 12c}$$

Thus, an oscillation of this system requires matching of the feedback.

b) Nonlinear delay circuit with feedback proportional to the velocity

The second oscillating resonant circuit contains the nonlinear delay system determined by the system operator

$$y(t) = \boldsymbol{T}_{\mathrm{NL}}\{z(t)\} = z(t-\tau) + \alpha\left[A^2 - z^2(t-\tau)\right]^{1/2} \tag{5.3 - 13a}$$

and the linear velocity-dependent feedback

$$z(t) = \boldsymbol{\Phi}_1\{y(t)\} = f_1 \frac{d}{dt} y(t) \quad . \tag{5.2 - 4a}$$

This resonant circuit is characterized by the nonlinear delay equation

$$\frac{1}{f_1} \int^{t} z(\vartheta)\, d\vartheta = z(t-\tau) + \alpha\left[A^2 - z^2(t-\tau)\right]^{1/2} \tag{5.2 - 13b}$$

that yields the solutions

$$z(t) = A \cos\left(\omega_{\mathrm{k}} t - \varphi_{\mathrm{k}}\right) \quad \text{with} \quad \varphi_{\mathrm{k}} \quad \text{arbitrary} \quad . \tag{5.2 - 13c}$$

These solutions are determined by the resonance and feedback-matching conditions

$$\cot \omega_{\mathrm{k}} \tau = \alpha \quad \text{and} \quad \left(f_1\right)_{\mathrm{k}} = \omega_{\mathrm{k}}^{-1}\left(1+\alpha^2\right)^{-1/2} \quad . \tag{5.2 - 13d}$$

6. Instabilities & Chaos

An oscillation represents a highly regular motion of a physical, chemical, biological or ecological system. This motion can be disturbed by structural changes of the system and become instable or chaotic. On certain conditions a structural change of a system can have a drastic effect on the motion in the form of a bifurcation. This can imply changes of stability, generation of subharmonics by period doubling, new kinds of motion or chaos. Chaos can be produced even by completely deterministic systems. In this chapter these scenarios are discussed with special emphasis on *bifurcations* and *deterministic chaos* and exemplified by the continuous *Lorenz model* and by the discrete *logistic map*.

6.1 Bifurcations

Systems of physical and technical interest typically have parameters which appear in the defining systems of equations. As these system parameters are varied, changes may occur in the qualitative structure of the solutions for certain parameter values. These changes are called *bifurcations* while the corresponding parameter values are called *bifurcation values* [Beltrami 1987 B, Birkhoff & Rota 1989 B, Chow & Hale 1982 B, Guckenheimer & Holmes 1983 B, Hale 1969 B, Iooss & Joseph 1980 B, Plaschko & Brod 1995 B, Poston & Steward 1978 B, Reitmann 1996 B, Ruelle 1989a B, Schuster 1984 B, Tu 1992 B, Verhulst 1985 B, Zwillinger 1989 B]. The first explicit study on bifurcation has probably been performed by H. Poincaré [Poincaré 1885 J, Tu 1992 B]. Since bifurcations represent structural changes of dynamical systems they are also topic of the *catastrophe theory* [Arnold 1984 B, Poston & Steward 1978 B, Thom 1972 B, 1975 B]. In most cases bifurcations involve *changes between stability and instability* of solutions. There exist various basic types of bifurcations, e.g. saddle-node, transcritical, pitchfork and Hopf [Hopf 1942 J]. These are described in Sections 6.1.3 - 7.

6.1.1 Bifurcations in Autonomous Systems

An autonomous or time-invariant system of differential equations (4.1 - 3a&b) that depends on N system parameters μ_1, ..., μ_N can be represented either by its components

$$dx_j / dt = \dot{x}_j = \upsilon_j(x_1 \cdot\cdot x_k \cdot\cdot x_n, \mu_1 \cdot\cdot \mu_r \cdot\cdot \mu_N)$$
$$\text{with} \quad j = 1,2,\cdot\cdot,n; \quad k = 1,2,\cdot\cdot,n; \quad r = 1,2,\cdot\cdot,N \quad , \tag{6.1 - 1a}$$

or by a vector equation

$$\dot{\vec{r}} = \vec{\upsilon}\left(\vec{r},\vec{\mu}\right)$$
$$\text{with} \quad \vec{r} = \left[x_1 \cdots x_n\right]; \; \vec{\upsilon} = \left[\upsilon_1 \cdots \upsilon_n\right]; \; \vec{\mu} = \left[\mu_1 \cdots \mu_N\right] \quad . \tag{6.1 - 1b}$$

A solution of this system is described by

$$x_k = x_k\left(t, \mu_1 \cdot\cdot \mu_r \cdot\cdot \mu_N\right) \quad \text{or} \quad \vec{r} = \vec{r}\left(t,\vec{\mu}\right) \quad . \tag{6.1 - 2}$$

A stationary solution in the form of a critical or singular point

$$x_{Sk} = x_{Sk}\left(\mu_1 \cdot\cdot \mu_r \cdot\cdot \mu_N\right) \quad \text{or} \quad \vec{r}_S = \vec{r}_S\left(\vec{\mu}\right) \tag{6.1 - 3}$$

fulfills the condition

$$\upsilon_j\left(x_{S1} \cdot\cdot x_{Sk} \cdot\cdot x_{Sn}, \quad \mu_1 \cdot\cdot \mu_r \cdot\cdot \mu_N\right) = 0 \quad \text{or} \quad \vec{\upsilon}\left(\vec{r}_S, \vec{\mu}\right) = \vec{0} \quad . \tag{6.1 - 4}$$

The *Jacobian matrix* for a singular point $\vec{r}_S\left(\vec{\mu}\right)$ of the system is defined by [Froyland 1992 B, Zwillinger 1989 B]

$$J\left(\vec{r}_S,\vec{\mu}\right) = \frac{d\vec{\upsilon}}{d\vec{r}}\left(\vec{r}_S,\vec{\mu}\right) \tag{6.1 - 5a}$$

$$\text{or} \quad J_{jk} = \left(\frac{\partial \upsilon_j}{\partial x_k}\left(x_{S1} \cdot\cdot x_{Sk} \cdot\cdot x_{Sn}, \quad \mu_1 \cdot\cdot \mu_r \cdot\cdot \mu_N\right)\right) \quad . \tag{6.1 - 5b}$$

This matrix gives information on the stability of the stationary singular points $\vec{r}_S\left(\vec{\mu}\right)$. With respect to its eigenvalues $\lambda_j\left(\vec{r}_S,\vec{\mu}\right)$ where j = 1, 2, ..., n, the following situations can be distinguished:

α) For a given $\vec{\mu}$ a singular point $\vec{r}_S\left(\vec{\mu}\right)$ represents a *stable* stationary solution if all real parts of the eigenvalues $\lambda_j\left(\vec{r}_S,\vec{\mu}\right)$ are *negative*.

β) For a given $\vec{\mu}$ a singular point $\vec{r}_S\left(\vec{\mu}\right)$ represents an *instable* stationary solution if none of the real parts of the eigenvalues $\lambda_j\left(\vec{r}_S,\vec{\mu}\right)$ vanishes and *one or more* are *positive*.

γ) If for a given $\vec{\mu}$ one or more of the eigenvalues $\lambda_j(\vec{r}_S, \vec{\mu})$ related to the singular point $\vec{r}_S(\vec{\mu})$ vanish, then this $\vec{\mu}$ represents a *bifurcation point* $\vec{\mu}_B$ where the stability and the number of singular points $\vec{r}_S(\vec{\mu})$ may change. This case is characterized by a zero determinant of the Jacobian matrix

$$det\ J(\vec{r}_S, \vec{\mu}_B) = 0 \quad . \tag{6.1 - 6}$$

δ) If for a certain $\vec{\mu}_{HB}$ the Jacobian matrix of the singular point $\vec{r}_S(\vec{\mu}_{HB})$ has two purely imaginary conjugated eigenvalues $\lambda_{1,2}(\vec{r}_S, \vec{\mu}_{HB})$, whilst all other eigenvalues $\lambda_{j>2}(\vec{r}_S, \vec{\mu}_{HB})$ have non-vanishing real parts

$$\begin{aligned} &\lambda_{1,2}(\vec{r}_S, \vec{\mu}_{HB}) = \pm i\omega \quad , \\ &Re\ \lambda_{j>2}(\vec{r}_S, \vec{\mu}_{HB}) \neq 0 \quad , \end{aligned} \tag{6.1 - 7}$$

then this $\vec{\mu}_{HB}$ indicates the presence of the *Hopf bifurcation* [Verhulst 1990 B] described in Section 6.1.7.

6.1.2 Bifurcations in One Dimension

The subsequent sections are dedicated to basic bifurcations in one- and two-dimensional systems depending on a single system parameter μ. The one-dimensional systems have the form [Guckenheimer & Holmes 1983 B, Reitmann 1996 B, Verhulst 1990 B]

$$\dot{x} = u(x,\mu) = -U_x(x,\mu) \quad , \tag{6.1 - 8a}$$

where U designates the *potential* and the index x indicates the partial differentiation with respect to x. In the subsequent survey of bifurcations it is assumed that $u(x, \mu)$ can be represented by the *Taylor series*

$$\begin{aligned} \dot{x} = u(x,\mu) &= u(0,0) + u_\mu(0,0)\mu \\ &+ u_x(0,0)x + \frac{1}{2!}u_{xx}(0,0)x^2 + \frac{1}{3!}u_{xxx}(0,0)x^3 \\ &+ u_{x\mu}(0,0)\mu\, x + \frac{1}{2!}u_{xx\mu}(0,0)\mu\, x^2 + \ldots \quad . \end{aligned} \tag{6.1 - 8b}$$

Again, the indices indicate partial differentiations.

A stationary solution in the form of a *singular point* $x_S(\mu)$ obeys the condition

$$u(x_S,\mu) = -U_x(x_S,\mu) = 0 \quad . \tag{6.1 - 9}$$

The corresponding *Jacobian matrix* is a scalar that equals its eigenvalue and its determinant

$$J(x_S,\mu) = det\, J(x_S,\mu) = \lambda_1(x_S,\mu) \\ = u_x(x_S,\mu) = -U_{xx}(x_S,\mu) \quad . \tag{6.1 - 10}$$

As a consequence of (6.1 - 6) and (6.1 - 10) a *bifurcation point* μ_B is determined by

$$J(x_S,\mu_B) = u_x(x_S,\mu_B) = -U_{xx}(x_S,\mu_B) = 0 \quad . \tag{6.1 - 11}$$

6.1.3 Saddle-Node Bifurcation

The saddle-node bifurcation [Guckenheimer & Holmes 1983 B, Plaschko & Brod 1995 B, Reitmann 1996 B, Verhulst 1990 B] occurs in one-dimensional autonomous systems of the form (6.1 - 8a&b) on the conditions

$$u(0,0) = u_x(0,0) = u_{x\mu}(0,0) = 0 \quad . \tag{6.1 - 12a}$$

Furthermore, it is assumed that these systems can be approximated at the singular point $x_S = 0$ by

$$\dot{x} = u(x,\mu) \approx u_\mu(0,0)\mu + \frac{1}{2} u_{xx}(0,0)x^2 \quad . \tag{6.1 - 12b}$$

Studies of the saddle-node bifurcation are usually based on the *standard system* [Guckenheimer & Holmes 1983 B, Platschko & Brod 1995 B]

$$\dot{x} = u(x,\mu) = -U_x(x,\mu) = \mu - x^2 \tag{6.1 - 13a}$$

with the *potential*

$$U(x,\mu) = -\mu\, x + \frac{1}{3} x^3 \quad . \tag{6.1 - 13b}$$

For $\mu < 0$ there exists no singular point $x_S(\mu < 0)$ as stationary solution of (6.1 - 13a). For $\mu > 0$ however, (6.1 - 9) and (6.1 - 13a) yield two singular points

$$x_{S\,1,2}(\mu > 0) = \pm\mu^{1/2} \tag{6.1 - 14a}$$

with the corresponding *Jacobian* (6.1 - 10)

$$J(x_{S\,1,2},\mu > 0) = \lambda_1(x_{S\,1,2},\mu > 0) = -x_{S\,1,2} = \mp\mu^{1/2} \quad . \tag{6.1 - 14b}$$

According to the rules α) and β) of Section 6.1.1 this equation implies that the singular point $x_{S1} = + \mu^{1/2}$ represents a *stable* solution, whereas the singular point $x_{S2} = - \mu^{1/2}$ is an *instable* solution.

The combination of (6.1 - 11) and (6.1 - 14b) yields the *bifurcation point* $\mu_B = 0$. Fig. 6.1 - 1 shows the *bifurcation diagram* with $x_S(\mu)$. The directions of $\dot{x}$ for $x \neq x_S(\mu)$ are indicated by arrows.

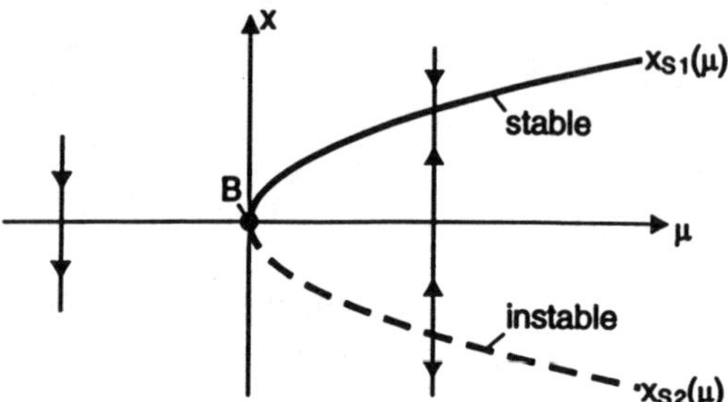

Fig. 6.1 - 1. Bifurcation diagram of the saddle-node bifurcation at *B*

More insight into the phenomenon of a saddle-node bifurcation is gained by the study of a *two-dimensional system*, e.g.

$$\begin{aligned} \dot{x} &= u(x, y, \mu) = -U_x(x, y, \mu) = \mu - x^2 \\ \dot{y} &= \upsilon(x, y, \mu) = -U_y(x, y, \mu) = -y \end{aligned} \qquad (6.1 - 15a)$$

with the *potential*

$$U(x, y, \mu) = -\mu\, x + \frac{1}{3} x^3 + \frac{1}{2} y^2 \quad . \qquad (6.1 - 15b)$$

Its trajectories are determined by

$$dy / y = \left(x^2 - \mu\right)^{-1} dx \quad . \qquad (6.1 - 15c)$$

The two singular points

$$\vec{r}_{S\,1,2}(\mu) = \left[x_{S\,1,2}(\mu), y_{S\,1,2}(\mu)\right] = \left[\pm \mu^{1/2}, 0\right] \qquad (6.1 - 16)$$

of the system exist for $\mu \geq 0$. They fulfill the condition

$$u(x_S, y_S, \mu) = \upsilon(x_S, y_S, \mu) = 0 \quad . \qquad (6.1 - 17)$$

The Jacobian matrices of these two singular points and their determinants are

$$J\left(\vec{r}_{S\,1,2},\mu\right)=\begin{bmatrix}\mp\mu^{1/2} & 0\\ 0 & -1\end{bmatrix} \quad , \tag{6.1 - 18a}$$

$$det\, J\left(\vec{r}_{S\,1,2},\mu\right)=\pm\mu^{1/2} \quad . \tag{6.1 - 18b}$$

The application of the rules α) and β) of Section 6.1.1 to (6.1 - 18a) demonstrates that the singular point [+ $\mu^{1/2}$, 0] is a *stable* solution, whereas the singular point [– $\mu^{1/2}$, 0] represents an *instable* solution. According to (6.1 - 6) and (6.1 - 18b) the *bifurcation point* of the system (6.1 - 15a) is $\mu_B = 0$.

Fig. 6.1 - 2 shows the *trajectories* of the system (6.1 - 15a) for $\mu = -1, 0, +1$, which have been derived from (6.1 - 15c). They represent the following functions

$$\mu=-1: \qquad y=y_0\; exp\left(arctan\; x\right) \quad , \tag{6.1 - 19a}$$

$$\mu=\mu_B=0: \quad y=y_\infty exp(-1/x) \quad , \tag{6.1 - 19b}$$

$$\mu=+1: \qquad \begin{aligned} &y=y_0\; exp\left(-arctanh\; x\right) \quad \text{for} \quad x^2\le 1 \quad ,\\ &y=y_\infty\; exp\left(-arccoth\; x\right) \quad \text{for} \quad x^2\ge 1 \quad . \end{aligned} \tag{6.1 - 19c}$$

The parameters y_0 and y_∞ label the different trajectories.

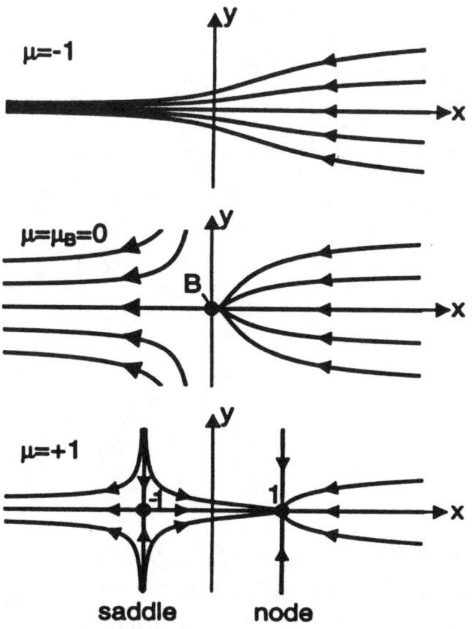

Fig. 6.1 - 2. Trajectories of the two-dimensional system (6.1 - 15a) for $\mu = -1, 0, +1$ with the saddle-node bifurcation at *B*

Fig. 6.1 - 2 demonstrates that the continuous flow of the system (6.1 - 15a) for $\mu < 0$ is split by the bifurcation into a saddle and a node for $\mu \geq 0$.

6.1.4 Transcritical Bifurcation

The transcritical bifurcation [Guckenheimer & Holmes 1983 B, Plaschko & Brod 1995 B, Reitmann 1996 B, Verhulst 1990 B] takes place in one-dimensional systems (6.1 - 8a&b) under the following conditions

$$u(0,0) = u_{\mathrm{x}}(0,0) = u_{\mu}(0,0) = 0 \quad . \qquad (6.1 - 20a)$$

In addition, it is assumed that these systems can be approximated at the singular point $x_{\mathrm{S}} = 0$ by

$$\dot{x} = u(x,\mu) \approx u_{\mathrm{x}\mu}(0,0)\mu\, x + \frac{1}{2} u_{\mathrm{xx}}(0,0) x^2 \quad . \qquad (6.1 - 20b)$$

The transcritical bifurcation is usually discussed on the basis of the *standard system*

$$\dot{x} = u(x,\mu) = -U_{\mathrm{x}}(x,\mu) = \mu\, x - x^2 \qquad (6.1 - 21a)$$

with the *potential*

$$U(x,\mu) = -\frac{\mu}{2} x^2 + \frac{1}{3} x^3 \quad . \qquad (6.1 - 21b)$$

The combination of (6.1 -9) and (6.1 - 21a) proves that this system exhibits two singular points as stationary solutions

$$x_{\mathrm{S}1}(\mu) = 0 \quad \text{and} \quad x_{\mathrm{S}2}(\mu) = \mu \quad . \qquad (6.1 - 22a)$$

The corresponding Jacobians (6.1 - 10) are

$$\begin{aligned} J(x_{\mathrm{S}1},\mu) &= \lambda_1(x_{\mathrm{S}1},\mu) = +\mu \quad , \\ J(x_{\mathrm{S}2},\mu) &= \lambda_1(x_{\mathrm{S}2},\mu) = -\mu \quad . \end{aligned} \qquad (6.1 - 22b)$$

The application of (6.1 - 11) to (6.1 - 22b) yields the *bifurcation point* $\mu_{\mathrm{B}} = 0$.

According to the rules α) and β) of Section 6.1.1 the Jacobians (6.1 - 22b) reveal that the singular points (6.1 - 22a) exchange stability and instability at the bifurcation point $\mu_{\mathrm{B}} = 0$. This is illustrated in the *bifurcation diagram* $x_{\mathrm{S}}(\mu)$ shown in Fig. 6.1 - 3. The directions of $\dot{x}$ for $x \neq x_{\mathrm{S}}(\mu)$ are indicated by arrows.

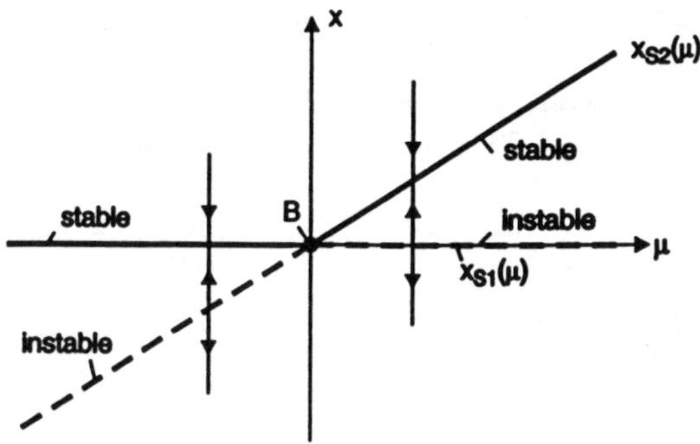

Fig. 6.1 - 3. Bifurcation diagram of the transcritical bifurcation at B

The transcritical bifurcation can be elucidated with the aid of the *two-dimensional system*

$$\begin{aligned} \dot{x} &= u(x,y,\mu) = -U_x(x,y,\mu) = \mu\, x - x^2 \\ \dot{y} &= \upsilon(x,y,\mu) = -U_y(x,y,\mu) = -y \quad , \end{aligned} \tag{6.1 - 23a}$$

which is characterized by the *potential*

$$U(x,y,\mu) = -\frac{\mu}{2}x^2 + \frac{1}{3}x^3 + \frac{1}{2}y^2 \quad . \tag{6.1 - 23b}$$

Its trajectories are determined by

$$dy\,/\,y = \left(x^2 - \mu x\right)^{-1} dx \quad . \tag{6.1 - 23c}$$

The two singular points

$$\vec{r}_{S1}(\mu) = [0,0] \quad \text{and} \quad \vec{r}_{S2}(\mu) = [\mu,0] \tag{6.1 - 24}$$

are defined by (6.1 - 17) and characterized by the two following Jacobian matrices and their determinants

$$\boldsymbol{J}\left(\vec{r}_{S1,2},\mu\right) = \begin{bmatrix} \pm\mu & 0 \\ 0 & -1 \end{bmatrix} \quad , \tag{6.1 - 25a}$$

$$det\, \boldsymbol{J}\left(\vec{r}_{S1,2},\mu\right) = \mp\mu \quad . \tag{6.1 - 25b}$$

Equations (6.1 - 6) and (6.1 - 25b) yield the bifurcation point $\mu_B = 0$. According to rules α) and β) of Section 6.1.1 and (6.1 - 25a) the singular point [0, 0] is stable for

$\mu < 0$ and instable for $\mu > 0$, whereas the singular point $[\mu, 0]$ is instable for $\mu < 0$ and stable for $\mu > 0$.

Fig. 6.1 - 4 shows the trajectories of the system (6.1 - 23a) for $\mu = -1, 0, +1/2$. At the bifurcation point $\mu = \mu_B = 0$ equation (6.1 - 23c) yields the trajectories

$$y = y_\infty\, exp(-1/x) \quad , \tag{6.1 - 26a}$$

while for $\mu \neq 0$ the solutions of (6.1 - 23c) have the form

$$y^{\mu} x(x-\mu)^{-1} = const \quad . \tag{6.1 - 26b}$$

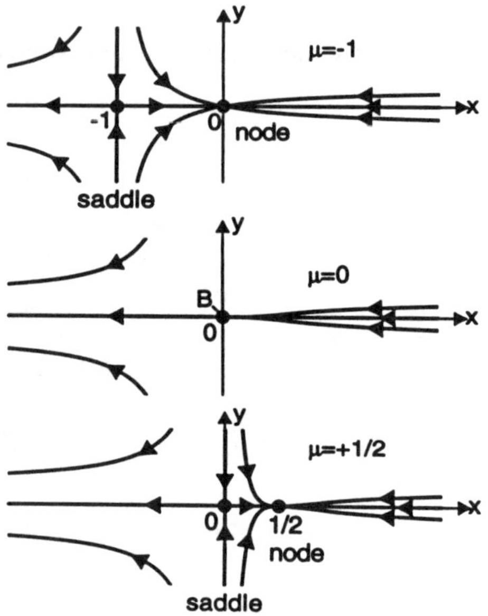

Fig. 6.1 - 4. Trajectories of the two-dimensional system (6.1 - 23a) for $\mu = -1, 0, +1/2$ with a transcritical bifurcation at B

It should be noticed that the trajectories (6.1 - 26a) of the system (6.1 - 23a) at the point $\mu_B = 0$ of its transcritical bifurcation correspond to those (6.1 - 19b) of the system (6.1 - 15a) at the point $\mu_B = 0$ of its saddle-node bifurcation.

6.1.5 Pitchfork Bifurcation

The pitchfork bifurcation [Guckenheimer & Holmes 1983 B, Platschko & Brod 1995 B, Reitmann 1996 B, Verhulst 1990 B] occurs in a one-dimensional system on the following conditions

$$u(0,0) = u_x(0,0) = u_{xx}(0,0) = u_\mu(0,0) = 0 \quad . \tag{6.1 - 27a}$$

Furthermore, it is assumed that these systems can be approximated at the singular point $x_S = 0$ by

$$\dot{x} = u(x,\mu) \approx u_{x\mu}(0,0)\mu x + \frac{1}{3!}u_{xxx}(0,0)x^3 \quad . \tag{6.1 - 27b}$$

The pitchfork bifurcation is usually described with the aid of the *standard system*

$$\dot{x} = u(x,\mu) = -U_x(x,\mu) = \mu x - x^3 \tag{6.1 - 28a}$$

with the *potential*

$$U(x,\mu) = -\frac{\mu}{2}x^2 + \frac{1}{4}x^4 \quad . \tag{6.1 - 28b}$$

The application of (6.1 - 9) to (6.1 - 28a) yields one singular point for $\mu < 0$ and three singular points for $\mu > 0$

$$\begin{aligned} \mu < 0 \quad &: \quad x_{S1}(\mu) = 0 \quad , \\ \mu > 0 \quad &: \quad x_{S1-3}(\mu) = 0, +\mu^{1/2}, -\mu^{1/2} \quad . \end{aligned} \tag{6.1 - 29a}$$

The corresponding Jacobians (6.1 - 10) are

$$\begin{aligned} \mu < 0 \quad &: \quad J(x_{S1},\mu) = \mu \quad , \\ \mu > 0 \quad &: \quad J(x_{S1-3},\mu) = \mu, -2\mu, -2\mu \quad . \end{aligned} \tag{6.1 - 29b}$$

Consequently, the singular point $x_{S1} = 0$ represents a stable solution for $\mu < 0$ and an instable solution for $\mu > 0$. Accordingly, the singular points $x_{S2,3} = \pm\mu^{1/2}$, which exist for $\mu > 0$, are stable solutions. The application of (6.1 - 11) to (6.1 - 29b) shows that $\mu_B = 0$ is the *bifurcation point* as illustrated in the bifurcation diagram Fig. 6.1 - 5. This diagram has the form of a pitchfork.

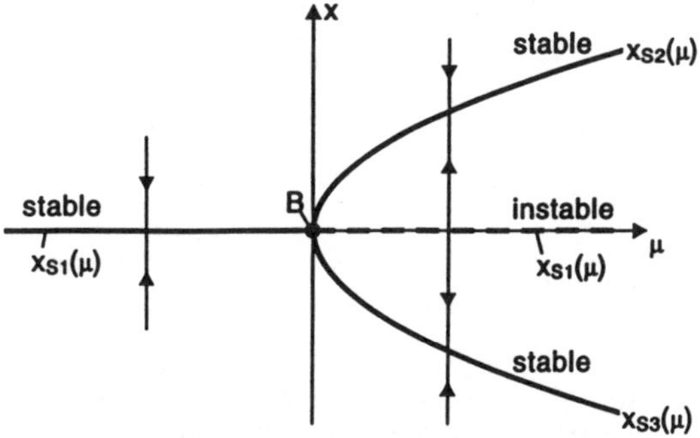

Fig. 6.1 - 5. Bifurcation diagram of the pitchfork bifurcation at *B*

The pitchfork bifurcation can be further elucidated by considering the *two-dimensional system*

$$\dot{x} = u(x,y,\mu) = -U_x(x,y,\mu) = \mu x - x^3$$
$$\dot{y} = \upsilon(x,y,\mu) = -U_y(x,y,\mu) = -y \qquad (6.1 - 30a)$$

with the potential

$$U(x,y,\mu) = -\frac{\mu}{2}x^2 + \frac{1}{4}x^4 + \frac{1}{2}y^2 \quad . \qquad (6.2 - 30b)$$

Its trajectories are determined by

$$dy / y = x^{-1}\left(x^2 - \mu\right)^{-1} dx \quad . \qquad (9.1 - 30c)$$

Application of (6.1 - 4a&b) to (6.1 - 30a) yields the singular points for $\mu > 0$ and $\mu > 0$

$$\mu < 0 \;:\; \vec{r}_{S1} = [0,0] \;,$$
$$\mu > 0 \;:\; \vec{r}_{S1-3} = [0,0], \left[0,+\mu^{1/2}\right], \left[0,-\mu^{1/2}\right] \;. \qquad (9.1 - 31)$$

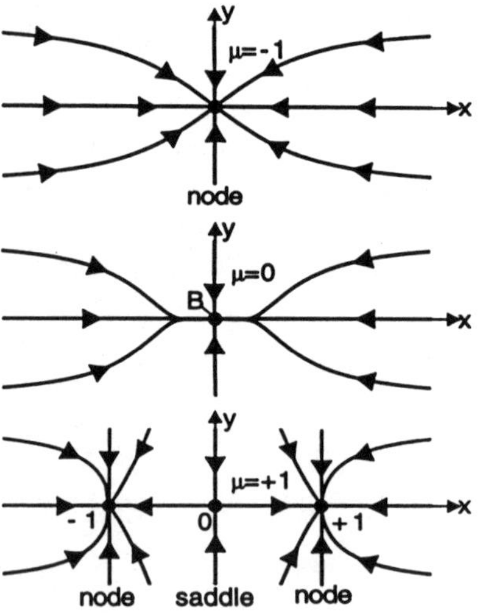

Fig. 6.1 - 6. Trajectories of the two-dimensional system (6.1 - 30a) for $\mu = -1, 0, +1$ with a pitchfork bifurcation at B

By taking into account rules α), β) and γ) of Section 6.1.1 one comes to the conclusion that the critical point $\vec{r}_{S1} = [0,0]$ is stable for $\mu < 0$ and instable for

$\mu > 0$, whereas both critical points $\vec{r}_{S2,3} = \left[\pm\mu^{1/2}, 0\right]$ are stable. The application of (6.1 - 6) to (6.1 - 30a) yields the bifurcation point $\mu_B = 0$.

Fig. 6.1 - 6 shows the trajectories of the system (6.1 - 30a) for $\mu = -1, 0, +1$. For the bifurcation point $\mu = \mu_B = 0$ the solution of (9.1 - 30c) gives as result the trajectories

$$y = y_\infty \, exp\left(-1/2x^2\right) \quad , \tag{9.1 - 32a}$$

while for $\mu \neq 0$ the trajectories have the form

$$y^{2\mu} x^2 \left(x^2 - \mu\right)^{-1} = const \quad . \tag{6.1 - 32b}$$

6.1.6 Rotating Pendulum

The rotating pendulum illustrated in Fig. 6.1 - 7 reflects the basic principle of *James Watt's centrifugal governor* used in stationary steams engines. Its dynamics show a drastic change at a critical circular frequency which is similar to the *pitchfork bifurcation.*

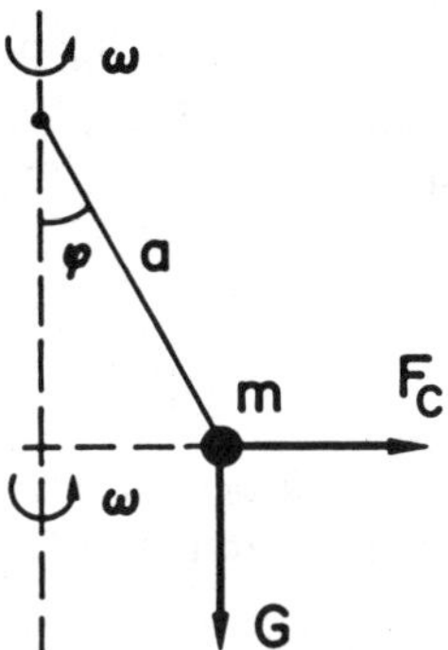

Fig. 6.1 - 7. Rotating pendulum

The rotating pendulum consists of a point of mass m that is suspended with a stiff massless bar of length a and rotates with a constant circular frequency ω around the vertical axis. In the following consideration ω represents the *system parameter* μ. The mass m is subjected to the gravitational force G as well as to the centrifugal force F_c. The motion of the rotating pendulum is governed by the relation between angular momentum L and torsion T. If φ designates the angle between the suspension bar and the vertical, then this relation takes the form

$$\begin{aligned} \dot{L} = ma^2\ddot{\varphi} = T &= -Ga \, sin \, \varphi + F_Z a \, cos \, \varphi \\ &= -mga \, sin \, \varphi + m\omega^2 a \, sin \, \varphi \, cos \, \varphi \quad , \quad \text{or} \end{aligned} \tag{6.1 - 33a}$$

$$\ddot{\varphi} = -\omega_0^2 \, sin \, \varphi \left[1 - (\omega/\omega_0)^2 cos \, \varphi\right] \quad \text{with} \quad \omega_0^2 = g/a \quad , \tag{6.1 - 33b}$$

where $g \approx 9.81$ m/s^2 indicates the acceleration by gravity and ω_0 the circular eigenfrequency of the non-rotating pendulum.

The singular points or stationary solutions of (6.1 - 33b) are

$$\varphi_0 = 0, \quad \text{and} \quad \varphi_{1,2} = \pm arccos(\omega_0 / \omega)^2 \quad \text{for} \quad \omega^2 > \omega_0^2 \quad . \tag{6.1 - 34a}$$

They fulfill the condition

$$\ddot{\varphi}(\varphi_k) = \dot{\varphi}(\varphi_k) = 0 \quad \text{with} \quad k = 0,1,2 \quad . \tag{6.1 - 34b}$$

The decision whether a stationary solution φ_k is stable or instable can be made by considering the approximative equation of motion for small deviations $\delta\varphi = \varphi - \varphi_k$. The approximation of (6.1 - 33b) yields

$$\delta\ddot{\varphi} = -\Omega_0^2\, \delta\varphi = -\left(\omega_0^2 - \omega^2\right)\delta\varphi \qquad \text{for} \quad \varphi_0 \quad , \tag{6.1 - 35a}$$

$$\delta\ddot{\varphi} = -\Omega_1^2 \delta\varphi = -\omega_0^2\left[(\omega/\omega_0)^2 - (\omega_0/\omega)^2\right]\delta\varphi \quad \text{for} \quad \varphi_{1,2} \quad . \tag{6.1 - 35b}$$

Stability occurs for $\Omega_{0,1}^2 > 0$, and instability for $\Omega_{0,1}^2 < 0$. Real $\Omega_{0,1}$ represent the circular eigenfrequencies of the oscillations around the stable equilibra φ_k. The φ_k are plotted as functions of the system parameter $\omega = \mu$ in the bifurcation diagram Fig. 6.1 - 8. There exist two bifurcation points

$$\mu_B = \omega_B = \pm\omega_0 \quad . \tag{6.1 - 36}$$

Fig. 6.1 - 8 demonstrates the similarity between the bifurcation that occurs in the dynamics of the rotating pendulum and the pitchfork bifurcation illustrated in Fig. 6.1 - 6 [Mahnke et al. 1992 B].

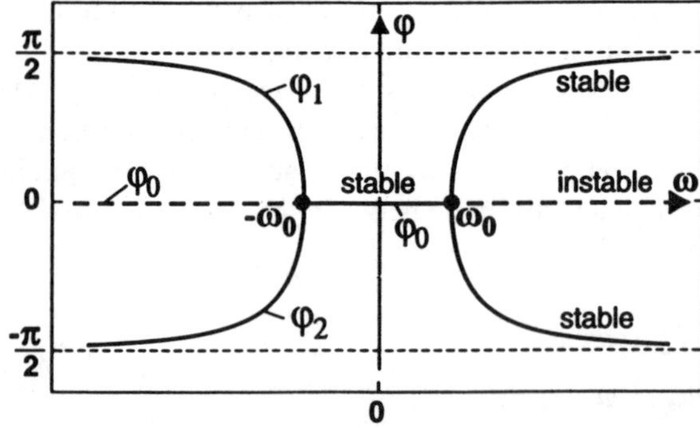

Fig. 6.1 - 8. Bifurcation diagram or catastrophe manifold of the rotating pendulum with the equilibrium positions φ_k, $k = 0, 1, 2$ as functions of the circular frequency ω and the bifurcation points or catastrophe positions $\omega = \pm\omega_0$

The behavior of the rotating pendulum can also be discussed with regard to the *catastrophe theory* [Arnold 1984 B, Poston & Steward 1978 B, Thom 1975 B, Tu 1992 B]. This discussion is based on the *potential* $V(\varphi, \omega)$. It corresponds to the potential energy E_{pot} of the rotating pendulum that can be derived from (6.1 - 33a)

$$V(\varphi,\omega) = E_{pot} = -\int_{\pi/2}^{\varphi} T d\varphi = -ma^2\,\omega_0^2 \cos\varphi\left[1 - \frac{1}{2}(\omega/\omega_0)^2 \cos\varphi\right] \quad . \quad (6.1 - 37)$$

Again, ω represents the system parameter μ. The equilibrium positions or singular points φ_k, $k = 0, 1, 2$ are determined by

$$\frac{\partial}{\partial\varphi} V(\varphi,\omega) = ma^2 \sin\varphi\left[\omega_0^2 - \omega^2 \cos\varphi\right] = 0 \quad . \quad (6.1 - 38)$$

In catastrophe theory, the equilibrium positions φ_k as functions of the system parameter $\omega = \mu$ described by (6.1 - 34a) form the *catastrophe manifold.* Consequently, Fig. 6.1 - 8 represents the catastrophe manifold as well as the bifurcation diagram. The *structurally instable equilibrium positions* φ_k are those equilibrium positions defined by (6.1 - 38), which also fulfill the *catastrophe condition*

$$\frac{\partial^2}{\partial\varphi^2} V(\varphi,\omega) = ma^2\left[\omega_0^2 \cos\varphi + \omega^2\left(1 - 2\cos^2\varphi\right)\right] = 0 \quad . \quad (6.1 - 39)$$

The simultaneous application of (6.1 - 38) and (6.1 - 39) yields the structurally instable equilibrium positions φ_k as well as the *catastrophe positions* $\omega_c = \mu_c$, which correspond to the bifurcation points $\omega_B = \mu_B$

$$\mu_c = \omega_c = \mu_B = \omega_B = \pm\omega_0 \quad \text{with} \quad \omega_0 = +\sqrt{g/a} \quad . \quad (6.1 - 40)$$

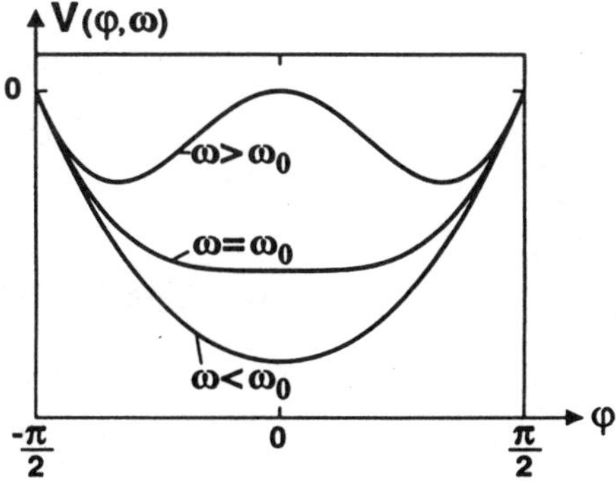

Fig. 6.1 - 9. Potential $V(\varphi, \omega)$ of the rotating pendulum for $0 < \omega < \omega_0$, $\omega = \omega_0 = \omega_B = \omega_c$ and $\omega_0 < \omega$

For illustration Fig. 6.1 - 9 shows the potential $V(\varphi, \omega)$ of (6.1 - 37) for $0 < \omega < \omega_0$, $\omega = \omega_0$ and $\omega > \omega_0$. It demonstrates the splitting of the minimum when $\omega = \omega_0 = \omega_c = \omega_B$.

6.1.7 Hopf Bifurcation

The Hopf bifurcation [Guckenheimer & Holmes 1983 B, Hopf 1942 J, Plaschko & Brod 1995 B, Reitmann 1996 B, Schuster 1984 B, Verhulst 1990 B] that occurs in multidimensional systems is usually exemplified by the following *two-dimensional standard system* in *Cartesian coordinates* (4.2 - 1)

$$\dot{x} = u(x, y, \mu) = -U_x + H_y = -\omega y + x\left(\mu - x^2 - y^2\right) \quad (6.1 - 41a)$$

$$\dot{y} = \upsilon(x, y, \mu) = -U_y - H_x = +\omega x + y\left(\mu - x^2 - y^2\right) \quad (6.1 - 41b)$$

with the following potential U and Hamiltonian H

$$U(x, y, \mu) = -\frac{\mu}{2}\left(x^2 + y^2\right) + \frac{1}{4}\left(x^2 + y^2\right)^2 \quad , \quad (6.1 - 41c)$$

$$H(x, y, \mu) = -\frac{\omega}{2}\left(x^2 + y^2\right) \quad . \quad (6.1 - 41d)$$

In these formulas μ is the system parameter and ω the constant circular frequency. Because of the axial symmetry of the potential U and of the Hamiltonian H it is of advantage to represent the system in *polar coordinates* (4.2 - 6)

$$\dot{r} = \upsilon_r = -U_r = r\left(\mu - r^2\right) \quad (6.1 - 42a)$$

$$\dot{\varphi} = r^{-1}\upsilon_\varphi = -r^{-1}H_r = \omega = const \quad (6.1 - 42b)$$

$$\text{with} \quad r^2 = x^2 + y^2 \quad , \quad \varphi = \arctan(y/x) \quad .$$

Corresponding potential and Hamiltonian are

$$U(r, \varphi) = U(r) = -\frac{\mu}{2}r^2 + \frac{1}{4}r^4 \quad , \quad (6.1 - 42c)$$

$$H(r, \varphi) = H(r) = -\frac{\omega}{2}r^2 \quad . \quad (6.1 - 42d)$$

The *solutions* of the system (6.1 - 42a&b) in polar coordinates are for

$$\mu = 0 \quad : \quad \varphi(t) = \omega t + \varphi_0 \quad , \qquad\qquad (6.1 - 43a)$$
$$r^2(t) = r_0^2\left[1 + 2tr_0^2\right]^{-1} \quad ,$$

$$\mu \neq 0 \quad : \quad \varphi(t) = \omega t + \varphi_0 \quad , \qquad\qquad (6.1 - 43b)$$
$$r^2(t) = r_0^2 \mu \, exp(+2\mu\, t)\left[\mu - r_0^2\{1 - exp(+2\mu\, t)\}\right]^{-1} \quad ,$$

where $\varphi(0) = \varphi_0$ and $r(0) = r(\varphi = \varphi_0) = r_0$.

These solutions demonstrate that

$$\text{for} \quad \mu \leq 0 \quad : \quad r(t \to +\infty) = 0 \quad , \quad \text{and} \qquad\qquad (6.1 - 43c)$$
$$\text{for} \quad \mu > 0 \quad : \quad r(t \to +\infty) = +\sqrt{\mu} \quad .$$

As a consequence the *origin* $r = 0$ represents a stable singular point or attractor for $\mu \leq 0$ that becomes instable for $\mu > 0$. The attractor in the origin $r = 0$ is replaced for $\mu > 0$ by a stable *limit cycle* of radius $r = +\sqrt{\mu}$. An example is presented in Section 4.5.1a in the form of the system (4.5 - 4a-c), which corresponds to system (6.1 - 42a-d) for $\omega = 1$ and $\mu = 1$.

The behavior of the solutions (6.1 - 43c) shows that the system (6.1 - 42a-d) undergoes a Hopf bifurcation at $\mu = \mu_B = 0$. In general, a Hopf bifurcation transforms a stable singular point into a limit cycle [Schuster 1984 B]. The limit cycles are discussed in Section 4.5.

Furthermore, it should be noticed that the *distance* r from the origin of the system considered is governed by (6.1 - 42a), which corresponds to the one-dimensional system (6.1 - 28a&b) with a *pitchfork bifurcation*. Therefore, the singular distances are $r_S(\mu) = 0$ for all μ and $r_S(\mu \geq 0) = +\sqrt{\mu}$ for $\mu \geq 0$. The singular point $r_S(\mu) = 0$ is stable for $\mu < 0$ and instable for $\mu > 0$, whilst the singular distance $r_S(\mu > 0)$ is stable for $\mu > 0$. If $\mu > 0$ and $r = r_S(\mu) = +\sqrt{\mu}$, then the system rotates permanently with the constant circular frequency ω as illustrated in Fig. 6.1 - 10.

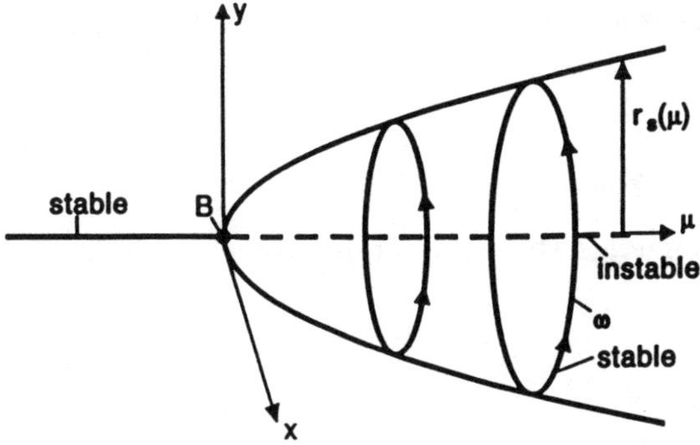

Fig. 6.1 - 10. Hopf bifurcation of the two-dimensional system (6.1 - 41a-d) at *B*

Information on the system described by (6.1 - 41a-d) or (6.1 - 42 a-d) and its Hopf bifurcation is also provided by its *Jacobian matrix* (6.1 - 5a&b)

$$\boldsymbol{J} = \boldsymbol{J}(x,y,\mu) = \begin{Bmatrix} \mu - 3x^2 - y^2 & -\omega - 2xy \\ +\omega - 2xy & \mu - x^2 - 3y^2 \end{Bmatrix} \tag{6.1 - 44}$$

of a point $\vec{r} = [x,y]$. Significant are the Jacobian matrices $\boldsymbol{J}$ and their eigenvalues $\lambda_{1,2}$ for the critical point $r_S(\mu) = 0$ in the origin and for the critical distance $r_S(\mu) = +\sqrt{\mu}$

$$\boldsymbol{J} = \begin{Bmatrix} \mu & -\omega \\ \omega & \mu \end{Bmatrix} \tag{6.1 - 45a}$$

with $\lambda_{1,2} = \mu \pm i\,\omega$ and $\vec{r}_S(\mu) = [0,0]$,

$$\boldsymbol{J} = \begin{Bmatrix} -\mu(1+cos2\varphi) & -\omega - \mu\, sin2\varphi \\ +\omega - \mu sin2\varphi & -\mu(1-cos2\varphi) \end{Bmatrix} \tag{6.1 - 45b}$$

with $\lambda_{1,2} = -\mu \pm i\sqrt{\omega^2 - \mu^2}$ and $\vec{r}_S(\mu) = \left[\sqrt{\mu}\, cos\varphi, \sqrt{\mu}\, sin\varphi\right]$.

At the bifurcation point $\mu = \mu_B = 0$ both matrices are identical

$$\boldsymbol{J} = \begin{Bmatrix} 0 & -\omega \\ \omega & 0 \end{Bmatrix} \tag{6.1 - 45c}$$

with $\lambda_{1,2} = \pm i\,\omega$ and $\vec{r}_S(0) = [0,0]$.

The conjugate imaginary eigenvalues $\lambda_{1,2} = \pm\, i\omega$ of the Jacobian matrix at the bifurcation point $\mu = \mu_B = 0$ are a *characteristic of the Hopf bifurcation.*

A classical example of a one-parametric system with a Hopf bifurcation is the *van der Pol oscillator* described in Section 2.5.5. It is governed by the normalized differential equation

$$\ddot{x} + \mu\left(x^2 - 1\right)\dot{x} + x = 0 \quad , \tag{6.1 - 46}$$

where μ represents the system parameter. This equation is equivalent to the two-dimensional system

$$\begin{aligned} \dot{x} &= u(x,y,\mu) = -y \\ \dot{y} &= \upsilon(x,y,\mu) = +x - \mu\left(x^2 - 1\right)y \end{aligned} \tag{6.1 - 47}$$

Its singular point is in the origin $\vec{r}_S(\mu) = [0,0]$. The corresponding Jacobian matrix (6.1 - 5a&b) and its eigenvalues are

$$J = J(0,0,\mu) = \begin{pmatrix} 0 & -1 \\ +1 & \mu \end{pmatrix} \qquad (6.1 - 48a)$$

$$\text{and} \quad \lambda_{1,2} = \lambda_{1,2}(0,0,\mu) = (\mu/2) \pm i\left[1 - (\mu/2)^2\right]^{1/2} \quad . \qquad (6.1 - 48b)$$

Accordingly, the *Hopf bifurcation* occurs at the point $\mu = \mu_B = 0$, where the two eigenvalues (6.1 - 48b) are conjugate imaginary. This *indicates the presence of a limit cycle*. For $\mu > 0$ the rule α) of Section 6.1.1, when applied to the eigenvalues (6.1 - 48b), postulates that the singular point in the origin $\vec{r}_S(\mu) = [0,0]$ is stable. For $\mu > 0$, however, the rule β) of Section 6.1.1 applied to (6.1 - 48b) shows that the singular point $\vec{r}_S(\mu) = [0,0]$ is instable. On this condition, there occurs a stable limit cycle and the related stationary oscillation of the van der Pol oscillator described in Section 2.5.5.

6.2 Deterministic Chaos

The time-dependence of a system is called *deterministic* if there exists a prescription in terms of differential or difference equations for calculating its future behavior from given initial conditions [Schuster 1984 B]. It could be assumed naively that deterministic motion is rather regular and far from being chaotic because successive states evolve continuously from each other. Yet it was already discovered at the end of last century [Poincaré 1892a B] that certain mechanical systems whose time evolution is governed by Hamilton's equations could display chaotic motion. Most physicists considered this as a mere curiosity. Thus, it took another seventy years until the meteorologist E.N. Lorenz [Lorenz 1963 J] discovered that even a simple system of three coupled first-order nonlinear differential equations can generate completely chaotic trajectories [Sparrow 1982 B]. Thereby he found one of the first examples of deterministic chaos in dissipative systems. *Deterministic chaos* [Bai-Lin 1984 B, Baker & Gollub 1990 B, Bergé et al. 1984 B, Collet & Eckmann 1983 B, Critanovic 1984 B, Devaney 1986 B, Froyland 1992 B, Gallavotti & Zweifel 1988 B, Gleick 1990 B, Guckenheimer & Holmes 1983 B, Halden 1986 B, Infeld & Rowlands 1990 B, Kunick & Steeb 1986 B, Mahnke et al. 1992 B, Mira 1987 B, Moon 1987 B, Moser 1973 B, Ott 1993 B, Percival & Richards 1982 B, Ruelle 1989b B, Schroeder 1991 B, Schuster 1984 B, Tu 1992 B, Verhulst 1990 B] denotes the irregular or chaotic motion generated by a nonlinear system whose dynamical laws determine the time evolution of a state of the system uniquely from the knowledge of its previous history. The chaotic behavior with time is neither due to external sources of noise nor due to an infinite number of degrees of freedom or the uncertainty associated with quantum mechanics. The actual source of the chaotic irregularity is the tendency of the nonlinear system to separate initially adjacent trajectories exponentially with time in a bounded region of the phase space [Schuster 1984 B].

In recent years it has become evident that the phenomenon deterministic chaos is abundant in nature and has an impact on many branches of science. *Examples* of systems producing deterministic chaos are classical many-body systems, fluids near the onset of turbulence, lasers, nonlinear optical devices, Josephson junctions, particle accelerators, plasmas with interacting nonlinear waves, chemical reactions, biological populations, and stimulated heart cells [Schuster 1984].

Here it should be noticed that in *classical mechanics* only a few systems can be integrated. Already H. Poincaré [1892a B] was aware that the nonintegrable three -body problem of classical mechanics implied completely chaotic trajectories. Sixty years later it was demonstrated by the KAM theorem [Arnold 1963 J, Guckenheimer & Holmes 1983 B, Kolmogorov 1954 J, Korsch & Jodl 1994 B, Moser 1967 J, Verhulst 1990 B] that the motion in the phase space of classical mechanics is neither completely regular nor completely irregular, and that the type of trajectory depends sensitively on the selected initial conditions [Schuster 1984 B]. In conclusion, a stable regular classical motion is an exception.

An example of an astonishingly simple system that exhibits deterministic chaos is the *forced Toda oscillator* [Kurz & Lauterborn 1988 J, Lauterborn & Meyer 1986 J] defined by

$$\ddot{x} + r\,\dot{x} + \left[exp(x) - 1\right] = a\;cos\;\omega t \quad . \qquad (6.2 - 1)$$

A simple transformation of the variables demonstrates that this equation is equivalent to the following three-dimensional nonlinear system of coupled differential equations of first order

$$\begin{aligned} \dot{X} &= Y \\ \dot{Y} &= -rY + \left[1 - exp\,X\right] + a\cos Z \\ \dot{Z} &= \omega \quad . \end{aligned} \qquad (6.2 - 2)$$

A characteristic of a nonlinear system of coupled differential equations of first order that generates deterministic chaos is a dimension of three or higher.

Instabilities and chaos were already observed in the stimulated light emission of the *first laser*, i.e. a ruby laser, in 1960 [Maiman 1960 J]. This laser produced an irregular emission with noise and statistical pulses even on quasi-stationary working conditions. For a long time laser scientists showed no interest in this phenomenon mainly because the development and application of these new light sources were more rewarding. Fifteen years later it was demonstrated [Haken 1975 J] that the *Maxwell-Bloch model* of the simple two-level laser with a homogeneously broadened gain profile [Kneubühl & Sigrist 1995 B, Svelto & Hanna 1989 B] in single-mode operation is governed by the same system of differential equations as the Lorenz model [Lorenz 1963 J], which implies deterministic chaos. This discovery initiated today's nonlinear dynamics of lasers [Haken 1983 B, 1984 B, 1985 B, Harrison & Biswas 1985 J, Lugiato & Narducci 1985 J].

6.2.1 Criteria for Chaos

There are various criteria for chaotic motion [Schuster 1984 B], e.g.

α) The *time-dependence* of the signal $x(t)$ looks chaotic.

β) The *power spectrum* $\sigma(\omega)$ defined by (5.1 - 24b) shows broad-band noise at low frequencies.

γ) The *autocorrelation function* $\chi(t)$ defined by (5.1 - 22a) decays rapidly.

δ) The *Poincaré map* [Froyland 1992 B, Guckenheimer & Holmes 1983 B, Korsch & Jodl 1994 B, Plaschko & Brod 1995 B, Schroeder 1991 B, Schuster 1984 B, Verhulst 1990 B] exhibits space filling sets of points.

ε) A quantitative criterium and measure of chaos is the *Lyapunov exponent* [Froyland 1992 B, Guckenheimer & Holmes 1983 B, Korsch & Jodl 1994 B, Plaschko & Brod 1995 B, Reitmann 1996 B, Schroeder 1991 B, Schuster 1984 B]. In order to define this exponent it is assumed that the variables X, Y, Z, ... of the nonlinear system of coupled differential equations, e.g. (6.2 -2), form a vector $\vec{R} = \{X, Y, Z, \ldots\}$. An initial state of the system at time $t = 0$ is represented by $\vec{R}(0)$, whilst a solution of the system forms a trajectory $\vec{R}(t)$ with time t as parameter. Two trajectories $\vec{R}_1(t)$ and $\vec{R}_2(t)$ are adjacent at time t, if their distance

$$D(t) = \left|\Delta\vec{R}(t)\right| = \left|\vec{R}_2(t) - \vec{R}_1(t)\right| > 0 \qquad (6.2 - 3a)$$

is small. Chaos is present if a non-countable set of trajectories adjacent at time $t = 0$ separates exponentially for large times t

$$\lim_{t\to\infty} \left|\Delta\,\vec{R}(t)\right| \approx \left|\Delta\,\vec{R}(0)\right| \, exp(\Lambda + \lambda\, t) \quad \text{with} \quad \lambda > 0 \quad , \qquad (6.2 - 3b)$$

or more precisely

$$\lim_{t\to\infty} \left[t^{-1} \, \ell n \left|\Delta\,\vec{R}(t)\right| \right] = \lambda > 0 \quad , \qquad (6.2 - 3c)$$

where λ designates the Lyapunov exponent. A negative Lyapunov exponent indicates stability, whereas a positive exponent is a sign of chaos. At the *bifurcation points* of a system the Lyapunov exponent is zero.

6.2.2 Routes to Chaos

There exist various models of the route to chaos from regular motion [Schuster 1984 B]. On this route the regular motion is followed by a sequence of instabilities in the form of bifurcations before the motion becomes chaotic. Examples to be mentioned are the following

a) Landau-Hopf model

The historical Landau-Hopf model [Landau 1944 J, Landau & Lifschitz 1959 1959 B, Reitmann 1996 B, Schuster 1984 B] is an attempt to describe the transition from a laminar flow to turbulence, i.e. chaos. This model postulates for an increasing system parameter, e.g. the flow velocity, an infinite cascade of bifurcations, each creating a new oscillation with a specific circular frequency ω_j with $j = 1, 2, \ldots$ and $\omega_j \neq \omega_k$ for $j \neq k$. This model implies that turbulence and chaos consist of an infinite number of oscillations with different circular frequencies ω_j whose ratios are irrational. This model was not successful since experiments have demonstrated that turbulence occurs after the appearance of only a few oscillations with specific frequencies.

b) Model of Ruelle et al.

The model of Ruelle, Takens and Newhouse [Newhouse et al. 1978 J, Ruelle & Takens 1971 J] shows that after only two instabilities a trajectory becomes attracted in a third step to a bounded region of the phase space in which initially close trajectories separate exponentially with time t, such that the motion becomes chaotic [Schuster 1984 B]. The particular regions of the phase space are called *strange attractors*. This route to chaos has been verified experimentally.

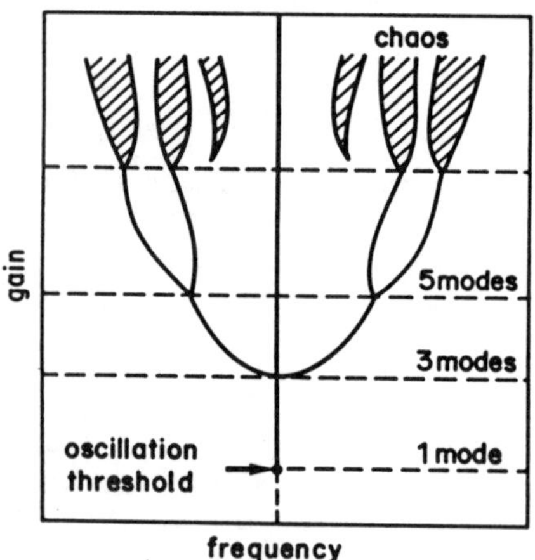

Fig. 6.2 - 1. Route to chaos of a laser via induced mode splitting [Minden & Casperson 1985 J]

The phenomenon of *induced mode-splitting* in lasers with inhomogeneously broadened gain profile [Kneubühl & Sigrist 1995 B, Svelto & Hanna 1989 B] shows an affinity to the models of Landau-Hopf and Ruelle-Takens-Newhouse [Minden & Casperson 1985 J]. Fig. 6.2 - 1 illustrates the route of such a laser from the oscillation threshold to chaos via two bifurcations in form of mode splittings.

c) Infinite sequence of bifurcations

A well-known route to chaos that consists of an infinite sequence of bifurcations accompanied by period doubling was discovered [Coullet & Tresser 1978 J, Feigenbaum 1978 J, Grossmann & Thomae 1977 J] in connection with the logistic map [Froyland 1992 B, Korsch & Jodl 1994 B, Mahnke et al. 1992 B, Plaschko & Brod 1995 B, Reitmann 1996 B, Schroeder 1991 B, Schuster 1984 B] discussed in Section 6.4. The period doublings generating subharmonics occur at an infinite sequence of system parameters r_k with $r_{k+1} > r_k$ and a finite maximum r_∞. For system parameters $r > r_\infty$ the motion of the system becomes chaotic. It was demonstrated [Feigenbaum 1978 J, 1980 B] that these results are not restricted to the logistic map. They are in fact *universal* and valid for a large variety of physical, chemical and biological systems. In these systems the characteristic system parameters r_m fulfill the relation

$$\delta = \lim_{k\to\infty} \left(r_{k+1} - r_k\right) / \left(r_{k+2} - r_{k+1}\right) = 4.669'\,201'\,609'\,1\ldots \quad . \qquad (6.2 - 4a)$$

They reach their upper limit at

$$r_k < r_\infty = 3.569'\,945'\,6\ldots \quad . \qquad (6.2 - 4b)$$

This limit δ is called *Feigenbaum constant.*

Fig. 6.2 - 2 shows the route to chaos via period doubling of the 81.5 μm NH_3 ring-laser emission pumped by a 10.78 μm NO_2 laser whose resonator tuning varies the system parameter r [Weiss et al. 1985 J].

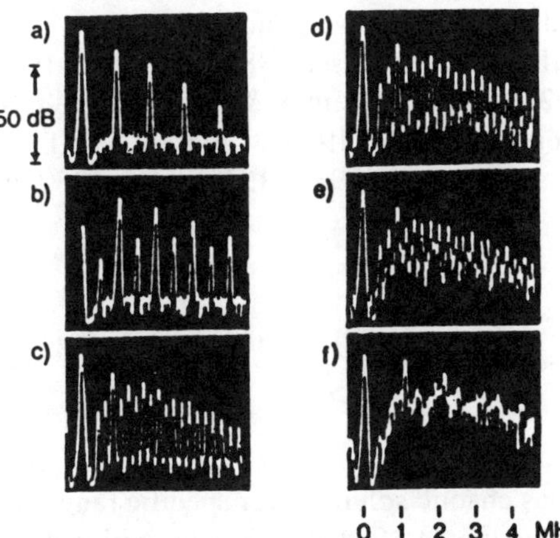

Fig. 6.2 - 2. Route to chaos of a 81.5 μm NH_3 ring laser from the stationary oscillations via period doublings [Weiss et al. 1985 J]

d) Intermittent route to chaos

The intermittent route to chaos [Manneville & Pomeau 1979 J] exhibits a characteristic behavior of a motion or of a signal: An initially regular motion is interrupted statistically by periods of irregular motion, i.e. intermittent bursts. The average number of these bursts increases with the system parameter until the motion becomes completely chaotic. This route to chaos has universal features and represents the fundamental mechanism of *1/f noise* in nonlinear systems [Schuster 1984 B].

Fig. 6.2 - 3 illustrates the intermittent route to chaos of the 3.39 μm HeNe laser emission with the tilting angle ϕ of a laser resonator mirror as system parameter r [Weiss et al. 1983 J].

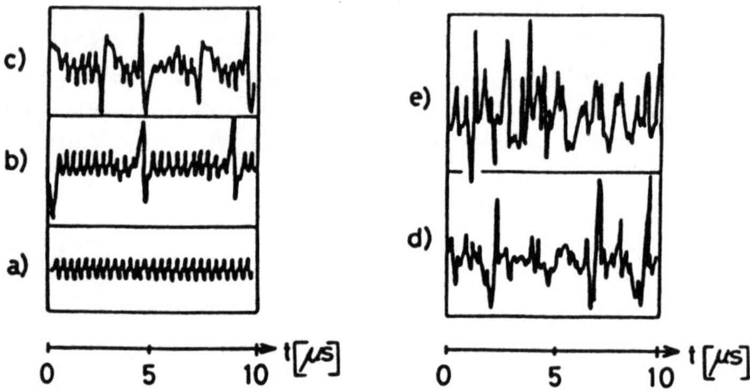

Fig. 6.2 - 3. Intermittent route of the 3.39 μm HeNe laser emission from the stationary oscillation a) to chaos e) [Weiss et al. 1983 J]

6.3 Lorenz Model

The Lorenz model [Lorenz 1963 J, Cvitanovic 1984 B] that describes the dynamics of the terrestrial atmosphere is characterized by the following nonlinear system of coupled differential equations of first order [Froyland & Alfsen 1984 J, Froyland 1992 B, Gaponov-Greknow & Rabinovich 1992 B, Guckenheimer & Holmes 1983 B, Haken 1978 B, Korsch & Jodl 1994 B, Kunick & Steeb 1986 B, Plaschko & Brod 1995 B, Reitmann 1996 B, Ruelle 1989b B, Schuster 1984 B, Sparrow 1982 B, Verhulst 1990 B]

$$\begin{aligned} \dot{x} &= u(x,y,z) = -\sigma x + \sigma y \\ \dot{y} &= \upsilon(x,y,z) = +\rho x - y - z x \quad \text{with} \quad \rho, \beta > 0, \quad \sigma > 1 + \beta \quad . \\ \dot{z} &= w(x,y,z) = -\beta z + x y \end{aligned} \tag{6.3 - 1a}$$

E.N. Lorenz demonstrated that this system has chaotic solutions for specific ranges of its parameters β, ρ and σ [Lorenz 1963 J, Cvitanovic 1984 B]. Thus he initiated the extensive research on deterministic chaos.

The *Lorenz equations* (6.3 - 1a) are equivalent to the *rate equations of a two-level laser* derived from the Maxwell-Bloch model [Haken 1978 B]

$$\begin{aligned} \dot{E} &= -\kappa E + \kappa P \\ \dot{P} &= -\gamma P + \gamma D E \\ \dot{D} &= \gamma_{\mathrm{p}}(\Lambda + 1) - \gamma_{\mathrm{p}} D - \gamma_{\mathrm{p}} \Lambda E P \end{aligned} \quad . \tag{6.3 - 1b}$$

The Lorenz equations (6.3 - 1a) describe also the velocity field of a three-dimensional *stationary flow* according to (4.10 - 12a-d) and (4.10 - 13). The characteristics of this flow are

$$div\ \vec{\upsilon}(\vec{r}) = -(1 + \beta + \sigma) < 0 \quad , \tag{6.3 - 2a}$$

$$2\vec{\omega}(\vec{r}) = curl\ \vec{\upsilon}(\vec{r}) = [2x, -y, \rho - y - z] \quad , \tag{6.3 - 2b}$$

$$\text{where} \quad \vec{\omega}(\vec{r}) = \vec{0} \quad \text{for} \quad \vec{r} = [0, 0, \rho] \quad . \tag{6.3 - 2c}$$

The stationary singular points $\vec{r}_{\mathrm{S}}$ of (6.3 - 1a) characterized by $\vec{\upsilon}(\vec{r}_{\mathrm{S}}) = \vec{0}$ are

$$\vec{r}_{\mathrm{S}1} = \vec{0} = [0, 0, 0] \quad , \tag{6.3 - 3a}$$

$$\vec{r}_{\mathrm{S}2,3} = \left[\pm\sqrt{\beta(\rho - 1)}, \pm\sqrt{\beta(\rho - 1)}, \rho - 1\right] \quad \text{for} \quad \rho > 1 \quad . \tag{6.3 - 3b}$$

6.3.1 Stability Considerations

The stability of the singular points $\vec{r}_{\mathrm{S}}$ is determined by the Jacobian matrix (6.1 - 5a&b) of (6.3 - 1a)

$$J(\vec{r}, \beta, \rho, \sigma) = \begin{pmatrix} -\sigma & +\sigma & 0 \\ (\rho - z) & -1 & -x \\ y & x & -\beta \end{pmatrix} \tag{6.3 - 4}$$

as follows:

α) The three eigenvalues of this matrix at the singular point $\vec{r}_{\mathrm{S}1} = \vec{0}$ are

$$\lambda_{\mathrm{j}}(\vec{r}_{\mathrm{S}1}, \beta, \rho, \sigma) = -\beta, -\frac{1}{2}(1 + \sigma) \pm \frac{1}{2}\left[(1 + \sigma)^2 + 4(\rho - 1)\right]^{1/2} \tag{6.3 - 5}$$

with $\quad j = 1, 2, 3 \quad .$

For $0 < \rho < 1$ all of the eigenvalues (6.3 - 5) are negative. Hence $\vec{r}_{S1} = \vec{0}$ is stable according to rule α) of Section 6.1.1. For $\rho > 1$ however, one of the eigenvalues (6.3 - 5) is positive and $\vec{r}_{S1} = \vec{0}$ instable according to rule β) of Section 6.1.1.

β) The eigenvalues of the Jacobian matrix (6.3 - 4) of the singular points $\vec{r}_{S2,3}$ existing for $\rho > 1$ are determined by the secular equation [Verhulst 1990 B]

$$\begin{aligned} &\lambda_j^3 + \lambda_j^2(1+\beta+\sigma) + \lambda_j\beta(\rho+\sigma) + 2\beta\sigma(\rho-1) = 0 \\ &\text{with} \quad \lambda_j = \lambda_j\left(\vec{r}_{S2,3}, \beta, \rho, \sigma\right) \quad \text{and} \quad j = 1,2,3 \quad . \end{aligned} \tag{6.3 - 6a}$$

For $\rho \approx 1$ these eigenvalues can be approximated by

$$\lambda_j(\rho \approx 1) \approx -\beta, -(1+\sigma), -\frac{2\sigma}{1+\sigma}(\rho-1) \quad \text{with} \quad j = 1,2,3 \quad . \tag{6.3 - 6b}$$

As a consequence the singular points $\vec{r}_{S2,3}$ existing for $\rho > 1$ are stable for $\rho \approx 1$ according to rule α) of Section 6.1.1.

γ) Taking into account the above results on the stability of the singular points $\vec{r}_{S,j}$ with j = 1, 2, 3 one comes to the conclusion that $\rho = 1$ is point of a *pitchfork bifurcation* [Verhulst 1990 B] described in Section 6.1.5.

δ) All three eigenvalues λ_j of (6.3 - 6a) are negative for [Verhulst 1980 B]

$$1 < \rho < \rho_H = \sigma(\sigma+\beta+3)(\sigma-\beta-1)^{-1} \quad . \tag{6.3 - 7}$$

According to rule α) of Section 6.1.1 this statement implies that $\vec{r}_{S2,3}$ are stable on this condition, while $\vec{r}_{S1}$ is instable.

ε) A Hopf bifurcation occurs at $\rho = \rho_H$ because two of the eigenvalues (6.3 - 6a) of the Jacobian matrix (6.3 - 4) of $\vec{r}_{S2,3}$ become imaginary conjugate at this bifurcation point

$$\lambda_j(\rho_H) = -(1+\beta+\sigma) \quad , \quad \pm i\left[2\sigma(\sigma+1)/(\sigma-\beta-1)\right]^{1/2} \quad . \tag{6.3 - 8}$$

The Hopf bifurcation concerning the singular points $\vec{r}_{S2,3}$ does not affect the instable singular point $\vec{r}_{S1} = \vec{0}$.

ζ) For $\rho = \rho_H$ all three singular points $\vec{r}_{S,m}$ with m = 1, 2, 3 are instable in principle. It might be expected that the Hopf bifurcation occurring as ρ passes through ρ_H would give rise to stable periodic orbits around the singular points $\vec{r}_{S2,3}$. Yet, it was demonstrated [Guckenheimer & Holmes 1983 B] that no closed orbits exist near these two points, but only trajectories which describe loops in an unexpected, complicated manner around both of them. These form the so-called *strange Lorenz attractor.*

6.3.2 Strange Lorenz Attractor

The concept of *strange attractors* was introduced by D. Ruelle and F. Takens [1971 J]. Fig. 6.3 - 1 shows trajectories around the strange Lorenz attractor [Hairer et al. 1987 B] characterized by the parameters $\beta = 8/3$, $\rho = 28$, $\sigma = 10$ corresponding to $\vec{r}_{S2,3} = [\pm 8.49,\ \pm 8.49,\ 27]$ and $\rho_H = 24.74$.

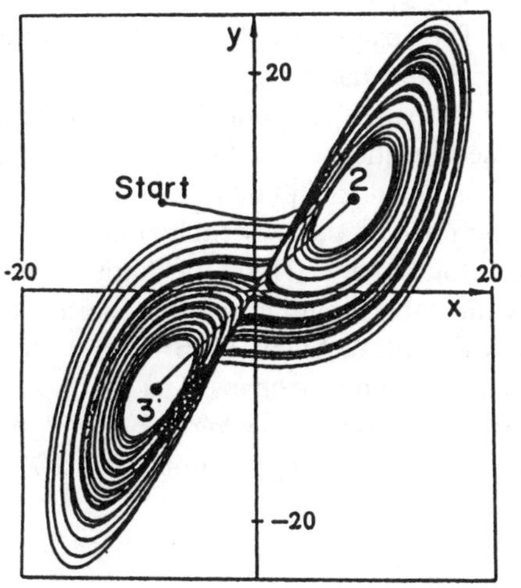

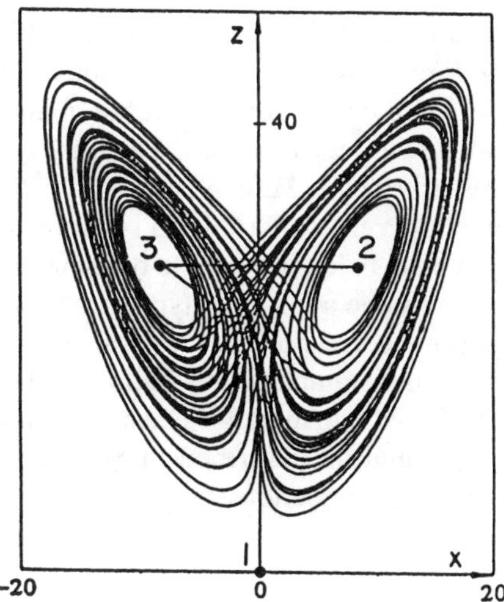

Fig. 6.3 - 1. Strange Lorenz attractor

Orbits of the strange Lorenz attractor show the following features [Verhulst 1980 B]:

α) The orbits are not closed.

β) The orbits describe loops around the right and the left singular point without apparent regularity in the number of loops.

γ) The subsequent number of loops around the right and the left singular point depend extremely on the initial conditions. A change of the initial conditions can produce another alternating series of loops.

δ) The features α) to γ) of these trajectories do not change when the initial conditions are altered. This statement is valid for small as well as for large alternations of the initial conditions.

6.4 Logistic Map

The logistic map [Baker & Grollub 1990 B, Collet & Eckmann 1983 B, Froyland 1992 B, Korsch & Jodl 1994 B, Kunik & Steeb 1986 B, Mahnke et al. 1992 B, Percival & Richards 1982 B, Reitmann 1996 B, Schuster 1984 B, Tu 1992 B, Verhulst 1990 B] represents a standard example of a one-dimensional nonlinear discrete map in the form of the *iteration*

$$x_{n+1} = r\,x_n(1-x_n) = f(x_n) \quad \text{with} \quad n = 0,1,2,3,\dots \quad . \tag{6.4 - 1}$$

It was already introduced by P.F. Verhulst [1838 J] to simulate the growth of a population x_n in closed area [Schuster 1984 B]. The designation "logistic" originates in the French word "logis" for house, lodging. From the feedback mechanism in (6.4 - 1) one could expect that the quantity of interest, i.e. the population x_n, would develop toward mean values. Yet at the end of the seventies it was discovered [Coullet & Tresser 1978 J, Feigenbaum 1978/79 J, Grossmann & Thomae 1977 J] that the iterates x_1, x_2, ... of (6.4 - 1) display as functions of the system parameter r a complicated behavior which becomes chaotic for large r. Thus, the logistic map turned out to be a discrete dynamical system with stable singular points as attractors, oscillations corresponding to strange attractors [Kunik & Steeb 1986 B], pitchfork bifurcations accompanied by period doubling, and deterministic chaos.

The logistic map (6.4 - 1) can be deduced from the *continuous logistic model* of a single population described in Section 4.7.1b. In this model the population $x(t)$ is governed by the relation

$$\dot{x} = \frac{1}{\tau}x\left\{1-\frac{x}{x_m}\right\} = f(x) \quad . \tag{4.7 - 2b}$$

For a given initial population $x(0)$ at time $t = 0$ this differential equation has unique solution $x(t)$ presented in (4.7 - 2c) and illustrated in Fig. 4.7 - 1.

In order to solve (4.7 - 2b) with a digital computer it might be convenient to devide the continuous time t into identical intervals Δt and to make the ansatz

$$t = n\Delta t \quad \text{with} \quad n = 0,1,2,\dots \quad , \tag{6.4 - 2a}$$

$$x(t) = x(n\Delta t) = X_n \quad , \tag{6.4 - 2b}$$

$$\dot{x}(t) \approx \frac{1}{\Delta t}\left[x((n+1)\Delta t) - x(n\Delta t)\right] = \frac{1}{\Delta t}\left[X_{n+1} - X_n\right] \quad . \tag{6.4 - 2c}$$

This ansatz transforms the continuous logistic equation (4.7 - 2b) into the iteration

$$X_{n+1} = \left(1+\frac{\Delta t}{\tau}\right)X_n - \frac{\Delta t}{\tau x_m}X_n^2 \quad . \tag{6.4 - 3a}$$

The transformation

$$X_n = \left(1 + \frac{\tau}{\Delta t}\right) x_m x_n \quad \text{and} \quad r = \left(1 + \frac{\Delta t}{\tau}\right) \qquad (6.4 - 3b)$$

of the variable X_n and the parameters τ and Δt of this iteration yields the logistic map

$$x_{n+1} = r\, x_n \left(1 - x_n\right) = f\left(x_n\right) \quad \text{with} \quad n = 0, 1, 2, \ldots \quad . \qquad (6.4 - 1)$$

The logistic map is *deterministic* because it assigns each x_n exactly one x_{n+1}. The iteration (6.4 - 1) from x_n to x_{n+1} can be performed graphically with the aid of the *logistic parabola* $f(x_n)$ as illustrated in Fig. 6.4 - 1. The logistic map determines the iterative sequences x_1, x_2, x_3, ... for all initial values x_0. These sequences depend on a large extent on the *system parameter* r. This parameter decides on the occurrence of converging, periodic or chaotic sequences and thereby characterizes the logistic map. With regard to the sequences the following has to be mentioned:

α) A *periodic iterative sequence* x_0, x_1, x_2, ... exhibits a period T = 1, 2, 3, ... defined by

$$x_0(r, T) = x_T(r, t) = x_{2T}(r, t) = = = x_{kT}(r, t) = \quad . \qquad (6.4 - 4)$$

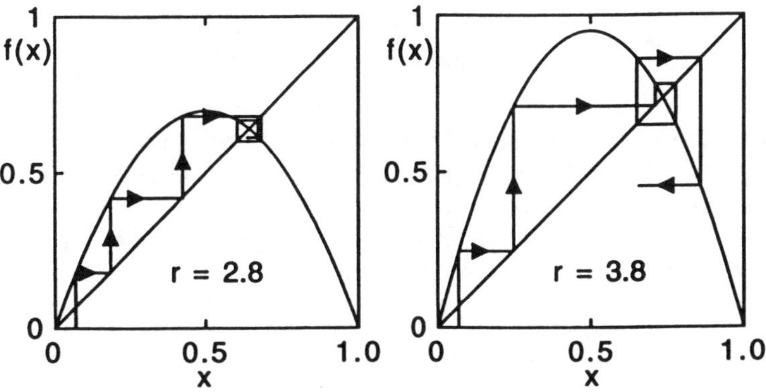

Fig. 6.4 - 1. Iterations of the logistic map for r = 2.8 on the left and r = 3.8 on the right

β) The *unit period* T = 1 marks the fixed singular points $x_0(r, 1) = x_n(r, 1)$ of the logistic map with the system parameter r. They are determined by the equation

$$x_{n+1} = r\, x_n \left(1 - x_n\right) = x_n \quad \text{with} \quad n = 0, 1, \ldots \quad . \qquad (6.4 - 5a)$$

Its solutions are the *two singular points*

$$x_0^{(1)}(r,1)=0 \quad \text{and} \quad x_0^{(2)}(r,1)=1-\frac{1}{r} \quad . \tag{6.4 - 5b}$$

The *stability* of these two points can be determined by iteration of the initially adjacent points defined by

$$y_0 = x_0(r,1)+\varepsilon \quad \text{with} \quad \varepsilon^2 << 1 \quad . \tag{6.4 - 6a}$$

The iteration of such a point y_0 yields

$$\begin{aligned} y_1^{(1)} &\approx \varepsilon r && \text{for} \quad y_0^{(1)}=\varepsilon \quad , \\ y_1^{(2)} &\approx 1-\frac{1}{r}+\varepsilon(2-r) && \text{for} \quad y_0^{(2)}=1-\frac{1}{r}+\varepsilon \quad . \end{aligned} \tag{6.4 - 6b}$$

Accordingly the singular point

$$\begin{aligned} & x_0^{(1)}(r,1)=0 \quad \text{is} \quad \begin{cases} \text{instable} & \text{for} \quad r>1 \\ \text{stable} & \text{for} \quad r=1 \\ \text{strictly stable} & \text{for} \quad 0<r<1 \end{cases} \\ & \text{while} \quad x_0^{(2)}(r,1)=1-\frac{1}{r} \quad \text{is} \quad \begin{cases} \text{instable} & \text{for} \quad 0<r<1; r>3 \\ \text{stable} & \text{for} \quad r=1;3 \\ \text{strictly stable} & \text{for} \quad 1<r<3 \end{cases} \end{aligned} \quad . \tag{6.4 - 6c}$$

This result demonstrates that a *transcritical bifurcation* as described in Section 6.1.4 occurs for $r_0 = 1$, where the initially stable point $x_0^{(1)}(r,1)=0$ becomes instable and the initially instable point $x_1^{(2)}(r,1)$ becomes stable.

γ) The *period two* $T = 2$ concerns the points $x_n(r, 2)$ characterized by

$$x_{n+2} = r^2 x_n(1-x_n)\left[1-r x_n(1-x_n)\right] = x_n \quad . \tag{6.4 - 7a}$$

This equation has four solutions

$$\begin{aligned} x_0^{(1)}(r,2) &= x_0^{(1)}(r,1)=0 \quad , \\ x_0^{(2)}(r,2) &= x_0^{(2)}(r,1)=1-\frac{1}{r} \quad , \\ x_0^{(3,4)}(r,2) &= \frac{1}{2}\left(1+\frac{1}{r}\right)\left(1\pm\left[\frac{r-3}{r+1}\right]^{1/2}\right) \quad \text{for} \quad r\geq 3 \quad . \end{aligned} \tag{6.4 - 7b}$$

The solutions $x_0^{(3,4)}(r,2)$ exist only for $r \geq 3$, where the stable point $x_0^{(2)}(r,1)$ becomes instable according to (6.4 - 6c). Hence, a *pitchfork bifurcation* as described in Section 6.1.5 occurs for $r_1 = 3$ at the point $x_0^{(2)}(3,1) = x_0^{(3)}(3,1) = x_0^{(4)}(3,1) = 2/3$, while the point $x_0^{(1)}(3,1) = 0$ remains instable. The pitchfork bifurcation creates a *subharmonic* by *period doubling*.

δ) The two solutions

$$x_0^{(3,4)}(r,2) \quad \text{are} \quad \begin{cases} \text{instable} & \text{for} \quad r > 1+\sqrt{6} \quad , \\ \text{stable} & \text{for} \quad r = 1+\sqrt{6} \quad , \\ \text{strictly stable} & \text{for} \quad 3 < r < 1+\sqrt{6} \quad . \end{cases} \tag{6.4 - 7c}$$

This gives an indication of the next pitchfork bifurcation at $r_2 = 1+\sqrt{6}$ that generates another subharmonic by period doubling.

The evaluation of the points $x_0^{(m)}(r,T)$ with larger system parameters $r > 3$ and periods $T > 2$ is tedious. For increasing r there occur an *infinite number of pitchfork bifurcations* generating subharmonics by period doubling. The corresponding bifurcation points r_k, where the subharmonics with the periods $T = 2^k$ start, are [Mahnke et al. 1992 B]

$$r_1 = 3 \quad , \quad r_2 \approx 3.4495 \quad , \quad r_3 \approx 3.5441 \quad , \quad r_4 \approx 3.5644 \quad ,$$
$$r_5 \approx 3.5688 \quad , \quad r_6 \approx 3.5697 \quad , \quad r_\infty \approx 3.5699456 \quad .$$

These bifurcation points r_k define the *Feigenbaum constant* δ by the relation mentioned before

$$\delta = \lim_{k\to\infty} (r_{k+1} - r_k)/(r_{k+2} - r_{k+1}) = 4.669'\,201'\,609'\,1\ldots \quad . \tag{6.2 - 4a}$$

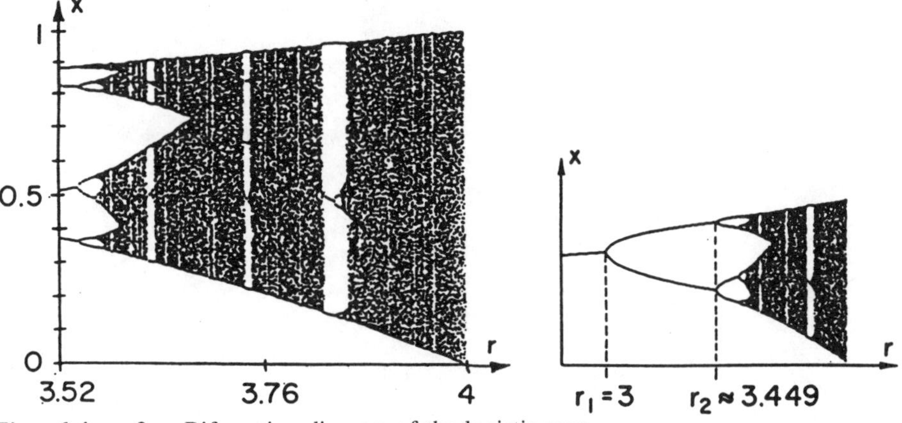

Fig. 6.4 - 2. Bifurcation diagram of the logistic map

ε) *Chaos* occurs for $r > r_\infty$. In the range $r_\infty < r \leq 4$, however, there exist windows with periodic sequences characterized by periods $T \neq 2^k$, namely $T = 3, 5, 6, 7, \ldots$. This phenomenon is called *order in chaos*. These sequences do not occur for $r \geq 4$, where the chaos is fully developed.

Fig. 6.4 - 2 shows the *bifurcation diagram* of the logistic map. It illustrates the sequences x_n for $n \to \infty$ as functions of the system parameter r and reveals the pitchfork bifurcations generating subharmonics by period doubling at r_k as well as the windows in the chaos with periodic sequences in the range $r_\infty < r < 4$. Instable sequences do not appear.

7. Linear Waves

Waves designate the propagation and extension of excitations or distortions in space. The physical or natural processes involved in waves can differ drastically, yet the waves have a common feature. They can be produced at one location, propagated through space and detected at another location. The large variety of different waves demands the systematic classification presented in the following Section 7.1. Fundamental differences exist between the *linear waves*, which permit linear superposition and arbitrary amplitudes, and the *nonlinear waves* with well-determined amplitudes. The linear waves are discussed in Sections 7.2 to 7.6 of this chapter, while the nonlinear waves are topic of Chapter 8. The *standing waves* described in Chapter 9 represent in principle oscillations of extended media with amplitudes and phases depending on the location.

7.1 Classification of Waves

The excitations or distortions propagating in waves can be distinguished by their different geometrical properties. These provide the basis of a classification required for the adequate description of the waves. This classification is performed in the following and elucidated with typical examples.

First discussed are the differences between scalar and vectorial, longitudinal and transversal waves such as sound in liquids and gases, elastic waves in isotropic solids, waves in springs, strings and bars as well as electromagnetic waves in vacuum. The second part of this section is dedicated to the linear waves and their wave equations. The common harmonic waves are described in detail in the subsequent Section 7.2.

7.1.1 Scalar and Vectorial Waves

With respect to the geometry of excitations one distinguishes between scalar and vectorial waves. In *scalar waves* the excitations are scalars

$$u = u(\vec{r},t) \quad , \qquad (7.1 - 1a)$$

while in *vectorial waves* they form vectors

$$\vec{u} = \vec{u}(\vec{r},t) \quad . \qquad (7.1 - 1b)$$

The excitations are also called *fields* because they are functions of position $\vec{r}$ and time t. An *example* of a scalar wave is sound in liquids and gases, whereas elastic waves in a spring and transverse waves in a string represent examples of vectorial waves.

a) Sound in liquids and gases as scalar wave
Sound in liquids and gases represents a scalar pressure and density wave [Alonso & Finn 1967 B, 1970 B, Crawford 1968 B, Kneubühl 1994 B, Resnick & Halliday 1966 B]. This wave implies the propagation of the local departures $\delta p(\vec{r},t)$ of the pressure from the equilibrium pressure p_0 [Pa = $\mathrm{Nm^{-2}}$] as well as of the local departures $\delta \rho(\vec{r},t)$ of the density from the equilibrium density ρ_0 [kg $\mathrm{m^{-3}}$].

Therefore, sound can be described either as scalar *pressure wave*

$$\delta p = \delta p(\vec{r},t) \tag{7.1 - 2a}$$

or as scalar *density wave*

$$\delta \rho = \delta \rho(\vec{r},t) = \upsilon^{-2} \delta p(\vec{r},t) \quad . \tag{7.1 - 2b}$$

In this equation υ [m $\mathrm{s^{-1}}$] indicates the constant sound or *wave velocity*

$$\upsilon = \upsilon\,(\text{sound}) = (\kappa / \rho_0)^{1/2} \quad , \tag{7.1 - 2c}$$

where κ [$\mathrm{Nm^{-2}}$] indicates the *bulk modulus of elasticity*

$$\kappa = \rho_0 \left(\frac{dp}{d\rho} \right)_0 \quad . \tag{7.1 - 2d}$$

For an *ideal gas* this modulus equals

$$\kappa = \gamma p \quad \text{with} \quad \gamma = C_p / C_V \quad . \tag{7.1 - 3a}$$

C_p and C_V [$\mathrm{JK^{-1}\,mol^{-1}}$] represent the molar heat capacities at constant pressure p and constant volume V, respectively. Taking into account the equation of state of ideal gases the velocity of sound in these gases is found to be

$$\upsilon = \upsilon\,(\text{sound, ideal gas}) = (\gamma p_0 / \rho_0)^{1/2} = (\gamma RT / M)^{1/2} \tag{7.1 - 3b}$$

where T [K] is the absolute temperature, $R \approx 8.3145$ $\mathrm{JK^{-1}\,mol^{-1}}$ the gas constant and M [kg] the mass of one mole. This relation demonstrates that the sound velocity of an ideal gas does not depend on the pressure.

b) Elastic waves in a spring

A straight homogeneous spring can support elastic waves [Alonso & Finn 1967 B, 1970 B, Crawford 1968 B, Resnick & Halliday 1966 B] where the axial displacements $\delta \vec{w}$ of the spring segments propagate along the spring axis z. The displacement $\delta \vec{w}$ of the spring segment at the equilibrium position z and time t can be represented by

$$\delta \vec{w} = \delta \vec{w}(z,t) = [0, 0, \delta w(z,t)] \quad . \tag{7.1 - 4a}$$

Since the displacement $\delta \vec{w}$ as excitation $\vec{u}$ is a vector, these waves are termed *vectorial*. In addition they are called *longitudinal* (L) because the displacement $\delta \vec{w}$ is parallel to direction of propagation that is defined by the unit vector $\vec{e}$

$$\vec{u} = \delta \vec{w} = \delta \vec{w}_{\text{L}} \quad \text{parallel to} \quad \vec{e} = [0, 0, 1] \tag{7.1 - 4b}$$

as illustrated in Fig. 7.1 -1. Longitudinal waves are discussed in more detail in Sections 7.1.2&3.

Elastic waves in a straight homogeneous spring with mass m [kg], equilibrium length L [m] and linear force constant f [Nm^{-1}] travel at the constant *wave velocity* [Alonso & Finn 1967 B, 1970 B]

$$\upsilon = (f\, L / m)^{1/2} \quad . \tag{7.1 - 4c}$$

c) Transverse waves in a string

A continuous homogeneous string is subjected to a tension or tangential force. Under equilibrium conditions this string is straight. It can be perturbed by displacements of the string perpendicular to its length. These displacements form waves by traveling along the string [Alonso & Finn 1967 B, 1979 B, Crawford 1968 B, Hazen & Pidd 1965 B, Kneubühl 1994 B, Nettel 1995 B, Resnick & Halliday 1966 B]. If the string in straight equilibrium position defines the z axis of a Cartesian coordinate system, the displacement $\delta \vec{w}$ of the string segment at position z and time t can be represented by

$$\delta \vec{w} = \delta \vec{w}(z,t) = [w_{\text{x}}(z,t), w_{\text{y}}(z,t), 0] \quad . \tag{7.1 - 5a}$$

These displacements $\delta \vec{w}$ are vectors. Consequently, the waves they form are termed *vectorial*. Furthermore, the displacements $\delta \vec{w}$ as excitations $\vec{u}$ are perpendicular to the string in equilibrium and, therefore, perpendicular to the direction $\vec{e}$ of wave propagation.

$$\vec{u} = \delta \vec{w} = \delta \vec{w}_{\text{T}} \quad \text{perpendicular to} \quad \vec{e} = [0, 0, 1] \tag{7.1 - 5b}$$

as illustrated in Figs. 7.1 - 2a-c. For this reason, the waves in a string are called *transverse* (T). Transverse waves are also topic of the following Section 7.1.2.

The transverse waves in a homogeneous continuous string of length L, cross-section A [m^2], mass m [kg] and density ρ [kg m^{-3}] that is subjected to a tension T [Nm^{-2}] or a tangential force F [N] travel at the constant *wave velocity*

$$\upsilon = (T / \rho)^{1/2} = (f / A\rho)^{1/2} = (FL / m)^{1/2} \quad . \tag{7.1 - 5c}$$

7.1.2 Longitudinal and Transversal Waves

Among the vectorial waves one can distinguish between longitudinal and transversal waves. *Examples* are the waves in a spring described in Section 7.1.1b on one hand and the waves in a string discussed in Section 7.1.1c on the other hand. By *definition* the excitations $\vec{u}(\vec{r},t)$ of the *longitudinal* (*L*) waves fulfill the condition

$$curl\, \vec{u}_{\mathrm{L}}(\vec{r},t) = \vec{0} \quad , \tag{7.1 - 6a}$$

while those of the *transversal* (*T*) waves obey

$$div\, \vec{u}_{\mathrm{T}}(\vec{r},t) = 0 \quad . \tag{7.1 - 6b}$$

The designations of these wave types become obvious if the conditions (7.1 - 6a&b) are applied to the plane vectorial waves described in the following.

a) Plane waves

Among the waves in three-dimensional isotropic media the *plane waves* are the simplest. They are characterized by a single direction of propagation that is defined by the unit vector $\vec{e}$ with $|\vec{e}| = 1$. Plane scalar and vectorial waves are represented by

$$u = u(\vec{r},t) = u(\vec{r} \cdot \vec{e},t) \quad \text{and} \tag{7.1 - 7a}$$

$$\vec{u} = \vec{u}(\vec{r},t) = \vec{u}(\vec{r} \cdot \vec{e},t) \quad . \tag{7.1 - 7b}$$

At any time t the plane waves are characterized by *phase planes* that are defined by

$$\vec{r} \cdot \vec{e} = C = const \quad \text{with} \quad |\vec{e}| = 1 \quad . \tag{7.1 - 7c}$$

If the *plane waves* are *traveling in the z direction* of a three-dimensional Cartesian coordinate system *xyz* the plane scalar waves have the form

$$u = u(\vec{r},t) = u(z,t) \quad , \tag{7.1 - 8a}$$

whilst the plane vectorial waves are described by

$$\vec{u} = \vec{u}(\vec{r},t) = \vec{u}(z,t) = \left[u_x(z,t), u_y(z,t), u_z(z,t)\right] \tag{7.1 - 8b}$$

because in this coordinate system

$$\vec{e} = [0,0,1] \quad \text{and} \quad \vec{r}\cdot\vec{e} = z \quad . \tag{7.1 - 8c}$$

The phase planes of these waves are parallel to the xy plane.

The plane vectorial waves (7.1 - 8b) can be separated into longitudinal [L] and transversal [T] waves according to (7.1 - 6a&b). The application of these conditions to (7.1 - 8b) yields the *longitudinal waves*

$$\vec{u}_L = \vec{u}_L(\vec{r},t) = \vec{u}_L(z,t) = \left[0,0,u_z(z,t)\right] \tag{7.1 - 9a}$$

and the proper *transverse waves*

$$\vec{u}_T = \vec{u}_T(\vec{r},t) = \vec{u}_T(z,t) = \left[u_x(z,t), u_y(z,t), 0\right] \quad . \tag{7.1 - 9b}$$

The formulation (7.1 - 9a) corresponds to the representation (7.1 - 4a) of the *longitudinal elastic waves in a spring*, whereas (7.1 - 9b) is equivalent to the description (7.1 - 5a) of the *transverse waves in a string*. Longitudinal and transverse waves propagating in the z direction are illustrated in Fig. 7.1 - 1 and Figs. 7.1 - 2a-c.

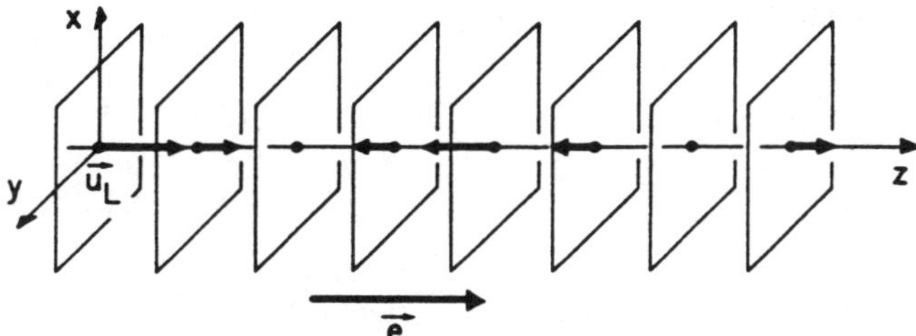

Fig. 7.1 - 1. Longitudinal wave

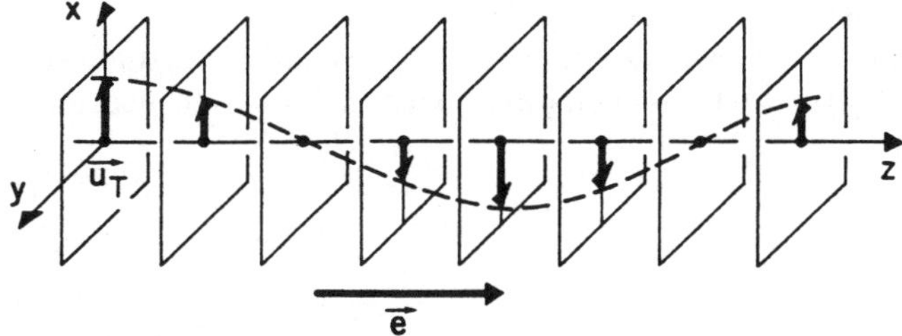

Fig. 7.1 - 2a. Transverse wave with linear polarization

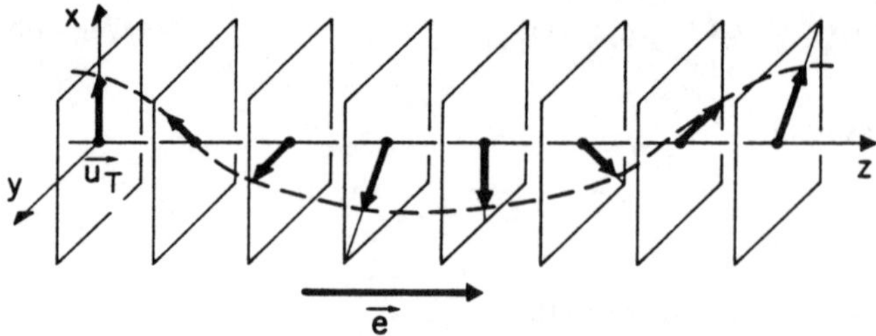

Fig. 7.1 - 2b. Transverse wave with left circular polarization

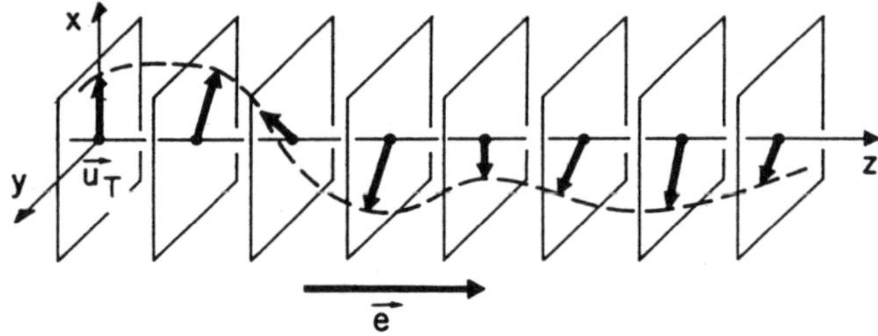

Fig. 7.1 - 2c. Transverse wave without polarization

Linear polarization is a characteristic phenomenon of transverse plane waves. In a linearly polarized wave the transverse vectorial excitation $\vec{u}_T$ of (7.1 - 9b) does not vary its orientation with position z and time t. This orientation is defined by the unit *polarization vector* $\vec{p}$ that is parallel to the phase planes. Consequently, a linearly polarized wave can be written in the form

$$\vec{u}_T = \vec{u}_T(\vec{r},t) = \vec{u}_T(z,t) = u_T(z,t)\,\vec{p} \qquad (7.1 - 10a)$$

$$\text{with} \quad \vec{p} = [\cos\gamma, \sin\gamma, 0] = const \quad . \qquad (7.1 - 10b)$$

This representation demonstrates that this type of vectorial wave $\vec{u}_T(z,t)$ can be interpreted as a scalar wave $u_T(z,t)$. An example of a linearly polarized plane wave (7.1 - 10a) is shown in Fig. 7.1 - 2a.

b) Elastic waves in isotropic solids

In isotropic solids there exist *longitudinal* (*L*) *as well as transversal* (*T*) *elastic waves* [Alonso & Finn 1967 B, 1970 B] whose excitations $\vec{u}$ are the local displacements $\delta\vec{w}(\vec{r},t)$ that either fulfill

$$curl\,\delta\vec{w}_L(\vec{r},t) = \vec{0} \quad \text{or} \qquad (7.1 - 11a)$$

$$div\,\delta\vec{w}_T(\vec{r},t) = 0 \quad . \qquad (7.1 - 11b)$$

The corresponding constant *wave velocities* are

$$\upsilon_{\rm L} = \left(Y / \rho_0\right)^{1/2} \quad \text{and} \tag{7.1 - 12a}$$

$$\upsilon_{\rm T} = \left(G / \rho_0\right)^{1/2} < \upsilon_{\rm L} \tag{7.1 - 12b}$$

where ρ_0 [kg m^{-3}] indicates the equilibrium density, Y [Nm^{-2}] Young's modulus of elasticity and G [Nm^{-2}] the shear modulus of the solid. In solids the shear modulus G is smaller than Young's modulus Y. As explained in Section 7.1.3 the longitudinal (L) elastic waves in isotropic solids correspond to the sound in liquids and gases described in Section 7.1.1a. There exist no transversal (T) elastic shear waves in liquids and gases because these cannot sustain shearing stress [Alonso & Finn 1970 B]. As a consequence their shear modulus G is zero according to Hooke's law.

c) Elastic waves in a thin solid rod

The *longitudinal* [L] and *transverse* [T] *elastic waves* traveling along a thin cylindrical solid rod with a circular cross section [Alonso & Finn 1967 B, 1970 B], whose axis coincides with the z axis of a Cartesian coordinate system, can be written as

$$\delta \vec{w}_{\rm L}(\vec{r},t) = \delta \vec{w}_{\rm L}(z,t) = \left[0, 0, \delta w_z(z,t)\right] \quad , \tag{7.1 - 13a}$$

$$\delta \vec{w}_{\rm T}(\vec{r},t) = \delta \vec{w}_{\rm T}(z,t) = \left[\delta w_{\rm x}(z,t), \delta w_{\rm y}(z,t), 0\right] \tag{7.1 - 13b}$$

in agreement with (7.1 - 9a&b). Since these elastic waves correspond to those in isotropic bulk solids their *wave velocities* are also determined by (7.1 - 12a&b).

If the rod described is clamped at one end and a torque is applied at the other end the rod is twisted. The angle of twist φ, which is called *torsion*, varies along the rod and depends therefore on the position z on the axis. If the applied torque varies with time t, the torsion φ at the position z of the rod axis also depends on time t. This results in a *torsional wave* [Alonso & Finn 1967 B, 1970 B]. For small local torsions $\delta\varphi(z,t)$ this wave can be represented by the following local displacements within the rod

$$\delta \vec{w}_{\rm torsion}(\vec{r},t) = \delta \varphi(z,t)\left[-y, x, 0\right] \quad . \tag{7.1 - 13c}$$

The torsional waves are *transversal* because these displacements fulfill (7.1 - 6b). Yet, they do not represent transverse plane waves as described by (7.1 - 9b) and (7.1 - 13b). They are governed by shearing strain and stress in the same way as the transversal waves (7.1 - 11b) in isotropic bulk solids and the transverse plane waves (7.1 - 13b) in bars. Therefore, the travel at the same constant *wave velocity* $\upsilon_{\rm T}$ given by (7.1 - 12b).

d) Electromagnetic waves in vacuum

Radiowaves, microwaves, infrared, light, ultraviolet, X-rays and γ-rays represent electromagnetic waves. They comprehend two kinds of vectorial excitations $\vec{u}(\vec{r},t)$, namely the electric field $\vec{E}(\vec{r},t)$ and the magnetic field $\vec{H}(\vec{r},t)$. If an electromagnetic wave propagates *in vacuum*, these fields fulfill the following four *Maxwell equations* [Jackson 1957 B]

$$curl\ \vec{H} = +\varepsilon_0\, \partial\vec{E} / \partial t \quad , \qquad (7.1 - 14a)$$

$$curl\ \vec{E} = -\mu_0\, \partial\vec{H} / \partial t \quad , \qquad (7.1 - 14b)$$

$$div\ \vec{E} = 0 \quad , \qquad (7.1 - 14c)$$

$$div\ \vec{H} = 0 \qquad (7.1 - 14d)$$

with $\varepsilon_0 \approx 8.854 \cdot 10^{-12}$ As/ Vm and $\mu_0 = 4\pi \cdot 10^{-7}$ Vs/ Am. The constants ε_0 and μ_0 indicate the *electric permittivity* and the *magnetic permeability* of vacuum.

Since (7.1 - 14c&d) correspond to (7.1 - 6b) the vector superposition of electric field $\vec{E}(\vec{r},t)$ and magnetic field $\vec{H}(\vec{r},t)$ represents a transversal *electromagnetic* or *EM* wave.

Electric and magnetic field of an electromagnetic wave in vacuum fulfill a vectorial wave equation called *Hertz equation*

$$\frac{\partial^2}{\partial t^2}\vec{E} = c^2 \Delta\vec{E} \quad \text{and} \qquad (7.1 - 15a)$$

$$\frac{\partial^2}{\partial t^2}\vec{H} = c^2 \Delta\vec{H} \qquad (7.1 - 15b)$$

with the *velocity c of light in vacuum* determined by

$$c = +(\varepsilon_0\, \mu_0)^{-1/2} \approx 2.998 \cdot 10^8 \text{ m/ s} \qquad (7.1 - 16)$$

The Hertz equation includes the Laplace operator Δ defined by (4.10 - 24) and represents a linear *hyperbolic partial differential equation* of second order [Courant & Hilbert 1968 B, Morse & Feshbach 1953 B, Webster & Szegö 1930 B, Whitham 1974 B]. The Hertz equations (7.1 - 15) can be deduced from the four Maxwell equations (7.1 - 14a-d) by evaluating

$$curl\left(curl\ \vec{E}\right) = grad\left(div\ \vec{E}\right) - \Delta\vec{E} = -\Delta\vec{E} = curl\left(-\mu_0 \partial\vec{H} / \partial t\right) = \ldots \quad \text{and}$$

$$curl\left(curl\ \vec{H}\right) = grad\left(div\ \vec{H}\right) - \Delta\vec{H} = -\Delta\vec{H} = curl\left(+\varepsilon_0 \partial\vec{E} / \partial t\right) = \ldots \quad .$$

In vacuum there exist *linearly polarized plane electromagnetic waves* as solutions of the Hertz equations (7.1 - 15a&b). An example of this type of wave traveling in the z direction is the transverse electromagnetic wave or *TEM wave*

$$\vec{E}(\vec{r},t) = \vec{E}(z,t) = [E(z-ct),0,0] \quad , \tag{7.1 - 17a}$$

$$\vec{H}(\vec{r},t) = \vec{H}(z,t) = [0,H(z-ct),0] \tag{7.1 - 17b}$$

with the velocity

$$\vec{\upsilon} = c\,\vec{e} = c[0,0,1] = [0,0,c] \quad . \tag{7.1 - 17c}$$

The unit *polarization vector* $\vec{p}$ of this wave

$$\vec{p} = [1,0,0] \tag{7.1 - 17d}$$

shows in the x direction because in polarized electromagnetic waves it is *parallel to the electric field* $\vec{E}(\vec{r},t)$ by definition.

In (7.1 - 17a) the electric field $E(z-ct)$ is an arbitrary finite scalar function of $z-ct$. The first two Maxwell equations (7.1 - 14a&b) demand that the magnetic field $\vec{H}(z,t)$ is *perpendicular* to the electric field $\vec{E}(z,t)$ and, in addition, that they are related by

$$H(z-ct) = Z_0^{-1}\,E(z-ct) \quad , \tag{7.1 - 18}$$

where Z_0 is the *wave impedance of vacuum*

$$Z_0 = (\mu_0 / \varepsilon_0)^{1/2} \approx 376.7\ \mathrm{V/A} = 376.7\ \Omega \quad . \tag{7.1 - 19}$$

The relation (7.1 - 18) demonstrates that the electric and the magnetic field $\vec{E}(z,t)$ and $\vec{H}(z,t)$ of traveling plane electromagnetic waves are in phase.

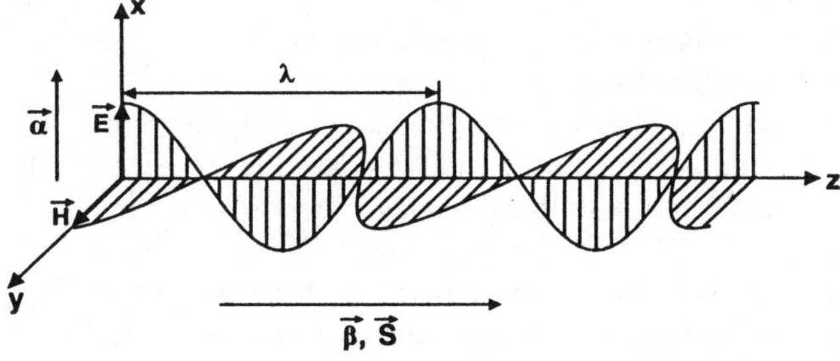

Fig. 7.1 - 3. Harmonic transverse electromagnetic (TEM) wave with linear polarization (7.2 - 17a&b)

Fig. 7.1 - 3 shows an *example* of a TEM wave described by (7.1 - 17a-d). This harmonic TEM wave is discussed in more detail in Section 7.2.4.

7.1.3 Equivalence of Scalar and Longitudinal Waves

Scalar and longitudinal vectorial waves are equivalent. These two types of waves describe the same phenomenon. A scalar wave $u(\vec{r},t)$ can be transformed into a longitudinal (*L*) wave $\vec{u}_{\mathrm{L}}(\vec{r},t)$ e.g. by

$$\vec{u}_{\mathrm{L}}(\vec{r},t) = -grad\, u(\vec{r},t) \quad . \tag{7.1 - 20a}$$

Thus, the excitation $u(\vec{r},t)$ of the scalar wave represents the potential of the excitation $\vec{u}_{\mathrm{L}}(\vec{r},t)$ of the corresponding longitudinal (*L*) wave. This statement is based on the fact that the excitation $\vec{u}_{\mathrm{L}}(\vec{r},t)$ defined by (7.1 - 20a) fulfills (7.1 - 6a) according to

$$curl\, \vec{u}_{\mathrm{L}}(\vec{r},t) = -curl\{grad\, u(\vec{r},t)\} = \vec{0} \quad . \tag{7.1 - 20b}$$

Accordingly, a longitudinal vectorial wave $\vec{u}_{\mathrm{L}}(\vec{r},t)$ can be transformed into a scalar wave $u(\vec{r},t)$ by the integral

$$u(\vec{r},t) = -\int_{\vec{r}_0}^{\vec{r}} \vec{u}_{\mathrm{L}}(\vec{r},t) \cdot d\vec{r} \quad . \tag{7.1 - 20c}$$

Since differential operators on space and time can be exchanged, it is possible to devise additional transformations of scalar waves $u(\vec{r},t)$ into longitudinal (*L*) waves $\vec{u}_{\mathrm{L}}(\vec{r},t)$

$$\frac{\partial^{(\mathrm{n})}}{\partial t^{\mathrm{n}}} \vec{u}_{\mathrm{L}}(\vec{r},t) = -C\, grad\left\{\frac{\partial^{(\mathrm{m})}}{\partial t^{\mathrm{m}}} u(\vec{r},t)\right\} \tag{7.1 - 21}$$

with n, m = 0, 1, 2, ... and C = const.

The corresponding reverse transformations from longitudinal (*L*) waves $u_{\mathrm{L}}(\vec{r},t)$ into scalar waves $u(\vec{r},t)$ can be deduced from (7.1 - 20c).

An example of the equivalence of scalar and longitudinal waves is sound in liquids and gases. In Section 7.1.1a sound in liquids and gases was interpreted as scalar pressure and density wave. In agreement with the above considerations it can also be represented by a longitudinal (*L*) wave with the local displacement $\delta \vec{w}_{\mathrm{L}}(\vec{r},t)$ as excitation. This local displacement is related to the departures $\delta p(\vec{r},t)$ and $\delta \rho(\vec{r},t)$ of pressure and density from their equilibrium values p_0 and ρ_0 by *Euler's equation*

$$\frac{\partial^2}{\partial t^2}\delta\vec{w}_{\mathrm{L}}(\vec{r},t)=\vec{a}_{\mathrm{L}}(\vec{r},t)=-\rho_0^{-1}grad\ \delta p(\vec{r},t)$$
$$=-\rho_0^{-1}\upsilon^2 grad\ \delta\rho(\vec{r},t) \quad , \qquad (7.1 - 22)$$

where $\vec{a}_{\mathrm{L}}$ [m s^{-2}] is the local longitudinal acceleration, ρ_0 equilibrium density and υ the velocity of sound (7.1 - 2c). These relations correspond to a transformation defined by (7.1 - 21). The longitudinal (*L*) waves $\delta\vec{w}_{\mathrm{L}}(\vec{r},t)$ correspond to those in isotropic solids described in Section 7.1.2b.

7.1.4 Linear Waves

Among the scalar and vectorial waves one can distinguish between the relatively simple linear waves and the complicated nonlinear waves [Whitham 1974 B]. The linear waves fulfill *three equivalent requirements*: the superposition principle of Huygens, the principle of unperturbed propagation and the linearity of the characteristic wave equation.

a) Superposition principle of Huygens

The superposition principle of Huygens demands that equivalent linear scalar or vectorial excitations or wave fields superimpose by addition

$$u_{(1+2)}(\vec{r},t)=u_1(\vec{r},t)+u_2(\vec{r},t) \quad , \qquad (7.1 - 23a)$$

$$\vec{u}_{(1+2)}(\vec{r},t)=\vec{u}_1(\vec{r},t)+\vec{u}_2(\vec{r},t) \quad . \qquad (7.1 - 23b)$$

In linear optics this superposition principle is used to calculate diffraction and interference patterns [Möller 1988 B].

b) Principle of unperturbed propagation

The principle of unperturbed superposition requires that a linear scalar or vectorial excitation or wave field $u(\vec{r},t)$ or $\vec{u}(\vec{r},t)$ propagates without perturbation by other excitations or wave fields $u_{\mathrm{k}}(\vec{r},t)$ or $\vec{u}_{\mathrm{k}}(\vec{r},t)$ with k = 1, 2, ...

c) Linearity of wave equation

Linear waves represent solutions of characteristic wave equations in the form of linear partial differential equations. Typical are hyperbolic partial differential equations of second order [Courant & Hilbert 1968 B, Morse & Feshbach 1953 B, Webster & Szegö 1930 B, Whitham 1974 B]. An example is the Hertz equation (7.1 - 15a&b) that is obeyed by the electric and the magnetic field $\vec{E}(\vec{r},t)$ and $\vec{H}(\vec{r},t)$ of electromagnetic waves in vacuum.

The following discussion of wave equations is restricted to *scalar waves* $u(\vec{r},t)$. Wave equations of vectorial waves are similar, yet more complicated in general.

The wave equation of a linear scalar wave $u(\vec{r},t)$ in a one-dimensional time-invariant homogeneous medium is of the form

$$\sum_{k=0}^{\infty}\sum_{p=0}^{\infty} a_{kp} \frac{\partial^{k+p}}{\partial t^{k} \partial z^{p}} u(z,t) = 0 \quad , \tag{7.1 - 24a}$$

whereas that of a linear scalar wave in a one-dimensional time-invariant inhomogeneous medium is described in general by

$$\sum_{k=0}^{\infty}\sum_{p=0}^{\infty} a_{kp}(z) \frac{\partial^{k+p}}{\partial t^{k} \partial z^{p}} u(z,t) = 0 \quad . \tag{7.1 - 24b}$$

A linear scalar wave $u(\vec{r},t)$ in a three-dimensional time-invariant medium obeys a wave equation that in a Cartesian coordinate system xyz can be represented in the form

$$\sum_{k=0}^{\infty}\sum_{m=0}^{\infty}\sum_{n=0}^{\infty}\sum_{p=0}^{\infty} a_{kmnp}(x,y,z) \frac{\partial^{k+m+n+p}}{\partial t^{k} \partial x^{m} \partial y^{n} \partial z^{p}} u(x,y,z,t) = 0 \quad . \tag{7.1 - 25}$$

This equation includes inhomogenuity as well as anisotropy of the three-dimensional medium.

Linear scalar waves as solutions of these types of equations are presented and discussed in the following section of this chapter.

7.2 Harmonic Waves

Harmonic waves designate plane waves whose excitations vary harmonically in space and time. They form the basis of the theory of linear waves because they give the motive for the introduction of the concepts of frequency, wavelength, dispersion relation, phase and group velocity, circular and elliptical polarization, etc.

7.2.1 Description

Harmonic waves can be described either by real or by complex harmonic functions. Real and complex harmonic representations are equivalent.

a) Real representation

In the real representation, a scalar harmonic wave $w(z,t)$ propagating in the z direction of a Cartesian coordinate system xyz is written in the form

$$w = w(z,t) = w^*(z,t) = w(z,t+T) = w(z+\lambda,t) = W\cos\Phi$$
$$= W\cos(\beta z - \omega t + \varphi) = W\cos(\omega t - \beta z - \varphi) \tag{7.2 - 1}$$
$$\text{with} \quad W = W^* \quad \text{and} \quad \Phi = \beta z - \omega t + \varphi \quad ,$$
$$\omega = 2\pi\,\nu = 2\pi / T \quad \text{and} \quad \beta = k = 2\pi\,\tilde{\nu} = 2\pi / \lambda \quad .$$

In these equations W indicates the real amplitude, ν the frequency, ω the circular frequency, T the period, $\tilde{\nu}$ the wavenumber, $\beta = k$ the circular wavenumber or propagation constant, λ the wavelength, φ the phase shift, and Φ the variable phase that determines the phase planes. The designation circular wavenumber k is used by solid-state physicists, while the expression propagation constant β is preferred by electrical engineers.

If ω is assumed positive as usual, the harmonic wave is traveling to the right in the $+z$ direction for positive $\beta > 0$ and to the left in the $-z$ direction for negative $\beta < 0$.

At a *fixed position* z_0 a harmonic wave manifests in a *harmonic oscillation* with the period T as illustrated in Fig. 7.2 - 1a. It shows the oscillation $w(z = z_0 = 0, t)$ at the position $z_0 = 0$ that originates in a harmonic wave with zero phase shift $\varphi = 0$ traveling in the $+z$ direction

$$w = w(z = z_0 = 0, t) = W\cos\omega t = W\cos 2\pi\frac{t}{T} \quad . \tag{7.2 - 2a}$$

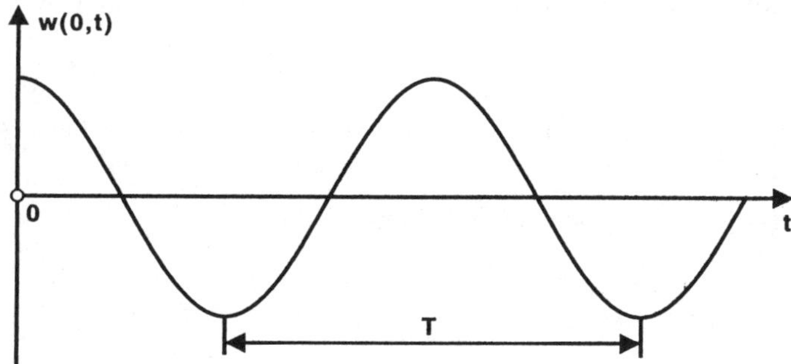

Fig. 7.2 - 1a. Harmonic wave as harmonic oscillation at the position $z = 0$

The momentary picture of a harmonic wave at a *fixed time* $t = t_0$ is a harmonic function with the wavelength λ as period. Fig. 7.2 - 1b presents the picture of a harmonic wave with zero phase shift $\varphi = 0$ traveling in the $+z$ direction taken at time $t_0 = 0$

$$w = w(z, t = t_0 = 0) = W\cos\beta z = W\cos 2\pi\frac{z}{\lambda} \quad . \tag{7.2 - 2b}$$

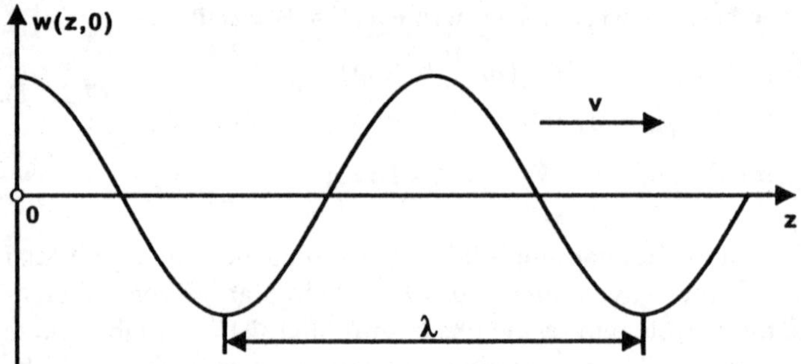

Fig. 7.2 - 1b. Harmonic wave as harmonic wave train at time $t = 0$

b) Complex representation

A scalar harmonic wave traveling in the z direction of a Cartesian coordinate system xyz has the complex representation

$$\begin{aligned} u &= u(z,t) = u(z,t+T) = u(z+\lambda,t) \\ &= U\,exp[i(\beta z - \omega t)] = U\,exp[-i(\omega t - \beta z)] = W\,exp(i\Phi) \qquad (7.2 - 3) \\ &\text{with} \quad U = W\,exp(i\varphi) \quad \text{and} \quad \Phi = \beta z - \omega t + \varphi \quad . \end{aligned}$$

The parameters in these equations are identical to those in (7.2 - 1) except for the complex amplitude U.

The real and complex representations (7.1 - 1) and (7.1 - 3) are related by the equation

$$w(z,t) = Re\{u(z,t)\} = \frac{1}{2}[u(z,t) + u^*(z,t)] = \frac{1}{2}u(z,t) + c.c. \qquad (7.2 - 4)$$

where c.c. and the asterisk * indicate the complex conjugate.

7.2.2 Dispersion Relations

The most relevant characteristic of a linear wave is its *dispersion relation* that relates its frequency ν with its wavelength λ or its circular frequency ω with its propagation constant $\beta = k$

$$D(\omega,\beta) = 0 \quad . \qquad (7.2 - 5a)$$

Circumstances permitting this implicit equation can be either solved for ω or β

$$\omega = \omega(\beta) = \omega(k) \quad , \qquad (7.2 - 5b)$$

$$\beta = \beta(\omega) = k(\omega) \quad . \qquad (7.2 - 5c)$$

In engineering it can be of advantage to replace in these dispersion relations (7.2 - 5b&c) circular frequency ω and propagation constant $\beta = k$ by frequency ν and wavelength λ. This results in the dispersion relations

$$\nu = \nu(\lambda) = \frac{1}{2\pi}\omega(\beta = 2\pi / \lambda) \quad \text{and} \tag{7.2 - 5d}$$

$$\lambda = \lambda(\nu) = \frac{2\pi}{\beta(\omega = 2\pi\,\nu)} \quad . \tag{7.2 - 5e}$$

The dispersion relations (7.2 - 5a-c) can be derived from the linear wave equations of homogeneous media such as (7.1 - 24a) by applying them to the complex representation (7.2 - 3) of harmonic waves. Thus, (7.1 - 24a) is transformed into the dispersion relation

$$D(\omega, \beta) = \sum_{k=0}^{\infty} \sum_{p=0}^{\infty} i^{(p-k)} a_{kp}\, \omega^k \beta^p = 0 \tag{7.2 - 6}$$

that can be *complex*.

a) Waves without dispersion

By definition a wave shows *dispersion* if its wave velocity υ depends either on the frequency $\nu = \omega/2\pi$ or on the wavelength $\lambda = 2\pi/\beta$. Consequently, waves without dispersion propagate with *a constant wave velocity* υ. They obey a Hertz equation of the type (7.1 - 15a&b). In Cartesian coordinates the *Hertz equation* of a *scalar wave* $u(x, y, z, t)$ *without dispersion* has the form

$$u_{tt} = \upsilon^2 \Delta u = \upsilon^2 \left(u_{xx} + u_{yy} + u_{zz}\right) \tag{7.2 - 7}$$

where the indices indicate the partial differentiations and Δ the Laplace operator. The application of this equation to the complex representation (7.2 - 3) of harmonic waves results in the simple dispersion relation of waves without dispersion

$$\omega / \beta = \nu\lambda = \upsilon = const \quad . \tag{7.2 - 8}$$

All linear waves discussed in Section 7.1 travel at a constant velocity υ and, therefore, show no dispersion. Hence, the dispersion relation (7.2 - 8) is valid for elastic waves in a spring, transverse waves in a string, sound in liquids and gases, elastic waves in bulk solids and bars as well as for electromagnetic waves in vacuum with $\upsilon = c$. Waves with dispersion are discussed in Section 7.3.

b) Real dispersion relations

Real dispersion relations characterize a wave *medium without loss or gain*. In a real dispersion relation a real circular frequency ω corresponds to a real propagation constant β. Therefore, frequency ν and wavelength λ are real and well defined for a wave with a real dispersion relation. An *example* are the waves without dispersion because they obey the real dispersion relation (7.2 - 8).

Since circular frequency ω and propagation constant β need to be real simultaneously, the dispersion relation (7.2 - 6) is real if the differences $(p-k)$ are even

$$D(\omega,\beta) \quad \text{real if} \quad p-k=2n,\, n=0,\pm1,\pm2,\ldots \quad . \tag{7.2 - 9}$$

c) Complex dispersion relations

Damping and amplification of a wave by the medium manifests in a complex dispersion relation (7.2 - 5a-e) or (7.2 - 6). Under these circumstances at least one of the circular frequency ω and the propagation constant β is complex. If the dispersion relation $\omega(\beta)$ or $\beta(\omega)$ is an analytic function it represents a conformal mapping of the complex β plane onto the complex ω plane or vice versa [Churchill 1948 B, Kober 1957 B, Levinson & Redheffer 1970 B]. Therefore the question arises whether to choose both, ω and β, as complex variables. In general, however, it is of advantage to assume either ω or β real.

α) For *traveling waves* one usually presumes a *real positive circular frequency* ω. This implies

$$\begin{aligned} &\omega = 2\pi/T = Re\,\omega > 0, \quad Im\,\omega = 0 \quad , \\ &\beta = Re\,\beta + i\,Im\,\beta = \pm 2\pi/\lambda_{\text{eff}} + i\gamma \quad , \end{aligned} \tag{7.2 - 10a}$$

and the traveling-wave representation

$$\begin{aligned} u(z,t) &= U\,exp\left[i\left(\pm\frac{2\pi z}{\lambda_{\text{eff}}} - \omega t\right)\right] exp(-\gamma z) \\ &= U\,exp\left[i\Phi_{\text{eff}}\right] exp(-\gamma z) \quad , \end{aligned} \tag{7.2 - 10b}$$

where λ_{eff} designates the *effective wavelength*, Φ_{eff} the *effective phase*, and γ the *wave damping* or *gain*. These are determined by the complex propagation constant

$$\lambda_{\text{eff}} = \pm 2\pi / Re\beta \geq 0 \quad \text{and} \quad \gamma = Im\beta \quad . \tag{7.2 - 10c}$$

For $Re\,\beta > 0$ the wave (7.2 - 10b) travels in the $+z$ direction. It is damped for $\gamma > 0$ and amplified for $\gamma < 0$.

β) For *standing waves* one assumes a *real positive propagation constant* β and postulates

$$\beta = 2\pi / \lambda = Re\,\beta > 0, \quad Im\,\beta = 0 \quad ,$$
$$\omega = Re\;\omega + i\,Im\,\omega = \pm(2\pi / T_{\text{eff}}) - i\vartheta \quad . \tag{7.2 - 11a}$$

This results in the following representation of a damped or amplified standing wave

$$\begin{aligned} u(z,t) &= \left[U\,exp(i\beta z)\right] exp\left(\mp i\frac{2\pi t}{T_{\text{eff}}}\right) exp(-\vartheta\,t) \\ &= \left[U\,exp(i\beta z)\right] exp(\mp i\omega_{\text{eff}} t)\,exp(-\vartheta\,t) \end{aligned} \tag{7.2 - 11b}$$

with the effective temporal period T_{eff}, the effective circular frequency ω_{eff} and the oscillation damping ϑ. These are determined by the complex circular frequency ω

$$T_{\text{eff}} = 2\pi / \omega_{\text{eff}} = \pm 2\pi / Re\,\omega \geq 0 \quad \text{and} \quad \vartheta = -Im\omega \quad . \tag{7.2 - 11c}$$

For $\vartheta > 0$ the standing wave (7.2 - 11b) is damped whilst for $\vartheta < 0$ it is amplified.

Equation (7.2 - 11b) demonstrates that standing waves represent *oscillations of continuous media* with position-dependent amplitudes, e.g. *standing waves*. They are discussed extensively in Chapter 9.

A standard *example* concerning the application of the formalism described is the interpretation of the *diffusion equation* in Section 7.4.6.

7.2.3 Differential Operators

In general, wave theory and wave mechanics differential operators are used for the *representation of physical quantities* such as wave vector, momentum and energy. The relation between a differential operator and its corresponding physical quantity can usually be elucidated by its application to the complex representation (7.2 - 3) of harmonic waves.

a) General wave theory

The complex representations $u(z, t)$ of harmonic waves propagating in the z direction (7.2 - 3) are *eigenfunctions* of temporal and spatial partial differentiations

$$u_{\text{t}} = \frac{\partial}{\partial t} u = -i\,\omega\,u \quad \text{and} \tag{7.2 - 12a}$$

$$u_{\text{z}} = \frac{\partial}{\partial z} u = +\,i\beta\,u = +ik\,u \quad . \tag{7.2 - 12b}$$

Similarly, the complex representation of a harmonic wave $u(x, y, z, t)$ traveling in an arbitrary direction defined by the unit vector $\vec{e}$ is an *eigenfunction of the gradient*

$$grad\, u = i\vec{\beta}\, u = i\vec{k}\, u$$

$$\text{with} \quad \vec{\beta} = \beta\vec{e} = \vec{k} = k\vec{e} \qquad (7.2 - 12c)$$

$$\text{and} \quad \vec{e} = [e_1, e_2, e_3] \quad , \quad e_1^2 + e_2^2 + e_3^2 = 1$$

where $\vec{\beta} = \vec{k}$ represents the wave vector.

These relations permit the introduction of differential operators for the circular frequency ω and the propagation constant β or circular wavenumber k as well as for the wave vector $\vec{\beta} = \vec{k}$

$$\omega\, u = +i\, u_t = +i \frac{\partial}{\partial t} u \quad , \qquad (7.2 - 13a)$$

$$\beta\, u = \boldsymbol{k}\, u = -i\, u_z = -i \frac{\partial}{\partial z} u \quad , \quad \text{and} \qquad (7.2 - 13b)$$

$$\vec{\beta}\, u = \vec{\boldsymbol{k}}\, u = -i\, grad\, u \quad . \qquad (7.2 - 13c)$$

These differential operators are *complex.* On the contrary, their quadratic and bilinear combinations are *real*

$$\omega^2 u = -u_{tt} = -\frac{\partial^2}{\partial t^2} u \quad , \quad \beta^2 u = \boldsymbol{k}^2 u = -u_{zz} = -\frac{\partial^2}{\partial t^2} u \quad , \qquad (7.2 - 14a)$$

$$\omega\beta\, u = +u_{zt} = +\frac{\partial^2}{\partial z \partial t} u \quad , \qquad (7.2 - 14b)$$

$$\vec{\beta}^2 u = \vec{\boldsymbol{k}}^2 u = -grad \cdot grad\, u = -\Delta u = -\{u_{xx} + u_{yy} + u_{zz}\} \quad . \qquad (7.2 - 14c)$$

These real differential operators can be applied to the real representations (7.2 - 1) as well as to the complex representations (7.2 - 3) of harmonic waves.

b) Wave mechanics

The *de Broglie wave* of wave mechanics [Alonso & Finn 1969 B, 1988 B, Blochinzew 1966 B, Cohen-Tannoudji et al. 1977 B, Kneubühl 1994 B, Messiah 1960 B, 1969 B, 1990 B, Schubert & Weber 1993 B]

$$\psi = \psi(z,t) = \psi_0\, exp\left[\frac{i}{\hbar}(pz - Et)\right] \qquad (7.2 - 15)$$

constitutes a complex representation (7.2 - 3) of a harmonic wave propagating in the z direction. It describes a particle with the energy E and the momentum p that is moving also in the z direction. These physical quantities are linked with the circular frequency ω and the circular wavenumber k or propagation constant β by *Planck's relations*

$$E = \hbar\omega = h\nu \quad , \quad \text{and} \tag{7.2 - 16a}$$

$$p = \hbar k = \hbar\beta = h / \lambda \tag{7.2 - 16b}$$

In these equations ν and λ represent frequency and wavelength of the de Broglie wave. For a de Broglie wave traveling in an arbitrary direction described by the wave vector $\vec{k} = \vec{\beta}$ the relation (7.2 - 16b) has to be replaced by

$$\vec{p} = \hbar\vec{k} = \hbar\vec{\beta} \tag{7.2 - 16c}$$

where $\vec{p}$ indicates the momentum vector. The relevant factor in these relations is *Planck's constant*

$$h = 2\pi\hbar \cong 6.626 \cdot 10^{-34}\ \mathrm{Ws}^2$$

The application of Planck's relations (7.2 - 16a&b) to the de Broglie wave (7.2 - 15) yields a standard complex representation of a harmonic wave

$$\psi = \psi(z,t) = \psi_0\, exp\left[i(\beta z - \omega t)\right] \quad . \tag{7.2 - 17}$$

Furthermore, Planck's relations (7.2 - 16a-c) in combination with the operators (7.2 - 13a-c) give the motive for the introduction of *differential operators for energy and momentum*

$$\boldsymbol{E}\psi = +i\hbar\frac{\partial}{\partial t}\psi \quad , \tag{7.2 - 18a}$$

$$\boldsymbol{p}\psi = -i\hbar\frac{\partial}{\partial z}\psi \quad , \quad \text{and} \tag{7.2 - 18b}$$

$$\vec{\boldsymbol{p}}\psi = -i\hbar\, grad\, \psi \quad . \tag{7.2 - 18c}$$

These operators are essential in wave mechanics, e.g. with regard to the *Schrödinger equation* (7.4 - 46).

7.2.4 Harmonic Electromagnetic Waves in Vacuum

The harmonic transverse electromagnetic (TEM) wave illustrated in Fig. 7.1 - 3 comprises the fields

$$\vec{E}(\vec{r},t)=\vec{E}(z,t)=\left[E_0 cos(\beta z-\omega t-\delta),0,0\right] \quad , \tag{7.2 - 19a}$$

$$\vec{H}(\vec{r},t)=\vec{H}(z,t)=\left[0,H_0 cos(\beta z-\omega t-\delta),0\right] \tag{7.2 - 19b}$$

$$\text{with} \quad \omega/\beta=\nu\lambda=c=\left(\varepsilon_0\mu_0\right)^{-1/2} \tag{7.2 - 19c}$$

$$\text{and} \quad E_0/H_0=Z_0=\left(\mu_0/\varepsilon_0\right)^{1/2} \quad . \tag{7.2 - 19d}$$

In these equations c indicates the velocity of light in vacuum (7.1 - 16) and Z_0 the impedance of vacuum (7.1 - 19). Electric and magnetic field (7.2 - 19a&b) represent solutions (7.1 - 17a&b) of the Maxwell equations (7.1 - 14a-d) and the related Hertz equations (7.1 - 15a&b). The harmonic TEM wave (7.2 - 19a-d) is polarized in the x direction according to the definition (7.1 - 17d) of the polarization vector $\vec{p}$

$$\begin{aligned}&\vec{E}(\vec{r},t)=\vec{E}(z,t)=E_0 cos(\beta z-\omega t-\delta)\vec{p}\\ &\text{with} \qquad \vec{p}=[1,0,0] \quad .\end{aligned} \tag{7.2 - 19e}$$

a) Energy density and intensity

The momentary local energy conservation of electromagnetic radiation is governed by the *Poynting relation* [Jackson 1975 B, Kneubühl 1994 B, Pöschl 1956 B, Sommerfeld 1949 B, 1959 B]

$$\frac{\partial}{\partial t}\rho_{EM}(\vec{r},t)+div\,\vec{S}(\vec{r},t)=0 \quad . \tag{7.2 - 20}$$

It relates the momentary energy density ρ_{EM} [J m^{-3}]

$$\rho_{EM}(\vec{r},t)=\frac{1}{2}\varepsilon_0\,\vec{E}^2(\vec{r},t)+\frac{1}{2}\mu_0\,\vec{H}^2(\vec{r},t) \tag{7.2 - 21}$$

with the Poynting vector $\vec{S}$ [W m^{-2}]

$$\vec{S}(\vec{r},t)=\vec{E}(\vec{r},t)\times\vec{H}(\vec{r},t) \quad . \tag{7.2 - 22}$$

This vector determines momentary amount and direction of the flow of radiation power. Its modulus represents the *momentary intensity* I_m [W m^{-2}], i.e. the momentary amount of energy transported per area and time

$$I_m(\vec{r},t)=\left|\vec{S}(\vec{r},t)\right| \quad . \tag{7.2 - 23}$$

Energy storage and transport by *harmonic TEM waves* are usually measured by the averages of energy density and poynting vector over one period $T = 2\pi/\omega$. The local *average energy density* $\rho_{EM,av}$ [W m^{-3}] is

$$\rho_{Em,av} = \frac{1}{2}\varepsilon_0 E_0^2 = \frac{1}{2}\mu_0 H_0^2 \quad , \tag{7.2 - 24}$$

while the *average intensity* equals

$$I = I_{av} = \frac{1}{2Z_0}E_0^2 = \frac{1}{2}Z_0 H_0^2 \quad . \tag{7.2 - 25}$$

b) Forms of polarization

Monochromatic light or other harmonic electromagnetic radiation in vacuum exhibit linear, circular or elliptical polarization [Gerard & Burch 1975 B, Holister 1979 B, Huard 1994 B, Shewandow 1973 B, Shurcliff 1962 B, Swindel 1995 B, Theocaris & Gdoutos 1979 B]. The various forms of polarization are usually described by the electric field vector $\vec{E}$ that is called light vector $\vec{\alpha}$ [Theocaris & Gdoutos 1979 B] because it defines the momentary polarization in agreement with (7.1 - 17d) and (7.2 - 19e). In optics polarization is historically defined from the point of view of the observer and not with regard to the radiation source. In this respect optics differs from the antenna theory of radio waves and microwaves [Balanis 1982 B, Johnson 1990 B, Johnson & Jasik 1984 B].

In *optics* the *light vector* $\vec{\alpha}$ is defined as the electric field $\vec{E}$ of an electromagnetic wave traveling from $z = -\infty$ in the $+z$ direction to the observer at $z = 0$

$$\vec{\alpha}(z,t) = \left[\alpha_x(z,t),\, \alpha_y(z,t),\, 0\right] = \vec{E}(z,t) = \left[E_x(z,t),\, E_y(z,t),\, 0\right]$$

with

$$\alpha_x(z,t) = A_x \cos(\beta z - \omega t - \delta_x) = A_x\, Re\left[\exp i(\omega t - \beta z + \delta_x)\right] \quad , \tag{7.2 - 26}$$

$$\alpha_y(z,t) = A_y \cos(\beta z - \omega t - \delta_y) = A_y\, Re\left[\exp i(\omega t - \beta z + \delta_y)\right] \quad ,$$

and $\quad A_x, A_y > 0, \quad 0 \le \delta_x, \delta_y < 2\pi \quad .$

This wave is determined by the real parameters A_x, A_y, δ_x and δ_y. The three different types of polarization can be distinguished with the aid of the parameters

$$A_y / A_x = \tan\gamma \quad \text{and} \quad \delta = \delta_y - \delta_x$$
$$\text{with} \quad 0 \le \gamma < \pi/2 \quad \text{and} \quad -\pi < \delta \le \pi \quad . \tag{7.2 - 27}$$

α) *Linearly polarized* waves are characterized by

$$\delta = 0, \pi \quad . \tag{7.2 - 28a}$$

The corresponding light vector has the form

$$\vec{\alpha}(z,t) = A\,cos\left(\beta z - \omega t - \delta_x\right)\vec{p}$$
$$\text{with} \quad \vec{p} = \left[cos\,\gamma,\, sin\,\gamma,\, 0\right] \quad . \tag{7.2 - 28b}$$

β) *Circularly polarized* waves fulfill the conditions

$$\gamma = \pi/4 \quad \text{and} \quad \delta = \pm\pi/2 \quad . \tag{7.2 - 29a}$$

Consequently, their light vector $\vec{\alpha}$ can be represented by

$$\alpha_x(z,t) = +A\,cos\left(\beta z - \omega t - \delta_x\right) \quad ,$$
$$\alpha_y(z,t) = \mp A\,sin\left(\beta z - \omega t - \delta_x\right) \quad . \tag{7.2 - 29b}$$

The two signs of δ and $\alpha_y(z, t)$ in (7.2 - 29a&b) characterize *right- and left-circular polarization*

$$\delta = +\pi/2 \quad : \quad \text{right circular} \quad , \tag{7.2 - 30a}$$

$$\delta = -\pi/2 \quad : \quad \text{left circular} \quad . \tag{7.2 - 30b}$$

When an observer looks towards the source of a right (left)circularly polarized wave he sees the light vector $\vec{\alpha}$ rotating clockwise (counter-clockwise). In a right (left)-circularly polarized wave the light vector $\vec{\alpha}$ moves on a left (right)-hand helix.

γ) *Elliptically polarized* harmonic electromagnetic radiation occurs frequently since its only restrictions are

$$\delta \neq -\pi/2, 0, +\pi/2, +\pi \quad . \tag{7.2 - 31a}$$

The light vector $\vec{\alpha}(z,t)$ of an elliptically polarized wave describes an *ellipse* in the plane $[x, y, 0]$. Its equation can be deduced from (7.2 - 26) by elimination of $(\omega t - \beta z)$

$$\frac{\alpha_x^2}{A_x^2} + \frac{\alpha_y^2}{A_y^2} - 2\frac{\alpha_x \alpha_y}{A_x A_y} cos\,\delta = sin^2\,\delta \quad . \tag{7.2 - 31b}$$

This ellipse is illustrated in Fig. 7.2 - 2. The lengths $a > b$ of its principal semi-axes and the azimuth ψ of this ellipse are determined by

$$A_x^2 + A_y^2 = a^2 + b^2 \quad , \quad A_x A_y sin\,\delta = ab\,sign\,\delta \quad , \tag{7.2 - 32a}$$

$$tan 2\psi = tan 2\gamma \cos\delta \quad \text{with} \quad 0 \le \psi < \pi \quad . \tag{7.2 - 32b}$$

Another parameter of interest is the ellipticity ε defined by

$$sin 2\varepsilon = sin 2\gamma \sin\delta \quad \text{with} \quad -\pi/4 \le \varepsilon \le +\pi/4 \quad . \tag{7.2 - 32c}$$

It is related to the lengths $a > b$ of the principal semi-axes

$$tan\,\varepsilon = (b/a)\, sign\,\delta \quad . \tag{7.2 - 32d}$$

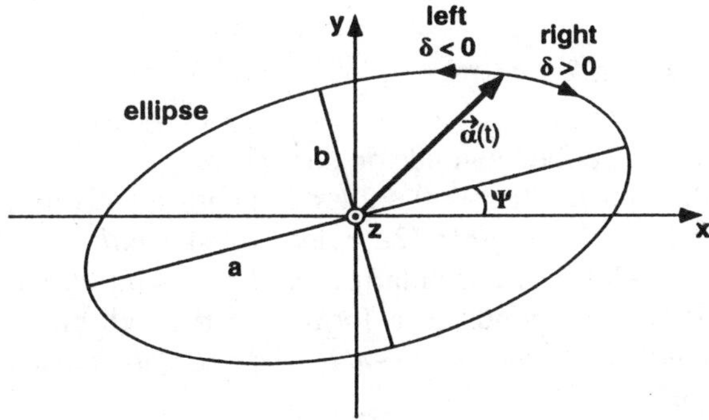

Fig. 7.2 - 2. Elliptically polarized light seen by the observer: $\delta > 0$ right elliptical polarization, $\delta < 0$ left elliptical polarization

In analogy to right- and left-circular polarizations there exists *right- and left-elliptical polarizations*. They can be distinguished by the sign of the phase difference δ (7.2 - 27)

$$0 > \delta < \pi : \text{right-elliptical} \quad , \tag{7.2 - 33a}$$

$$-\pi < \delta < 0 : \text{left-elliptical} \quad . \tag{7.2 - 33b}$$

c) Poincaré sphere

The absolute value of the intensity I (7.2 - 25) of a polarized light ray described by the light vector $\vec{\alpha}$ (7.2 - 26) is

$$I = I_{\text{av}} = \frac{1}{2\,Z_0}\left(A_x^2 + A_y^2\right) = \frac{1}{2\,Z_0}\left(a^2 + b^2\right) \quad . \tag{7.2 - 34}$$

However, in many problems that involve the passage of a polarized light ray through a series of optical elements only the relative value of the intensity I is of interest. In this case the transformation of the shape and not of the size of the ellipse (7.2 - 31b) by each optical element is essential. The shape of this ellipse is determined by the azimuth ψ (7.2 - 32b) and the ellipticity ε (7.2 - 32c&d). Consequently, these two parameters represent the *state of polarization.*

H. Poincaré [1892b B] invented a one-to-one mapping of the polarization states of a light ray onto the surface of a sphere unit radius. A point on this *Poincaré sphere* has the Cartesian coordinates

$$X = \cos 2\varepsilon \cos 2\psi \quad , \tag{7.2 - 35a}$$

$$Y = \cos 2\varepsilon \sin 2\psi \quad , \tag{7.2 - 35b}$$

$$Z = \sin 2\varepsilon \quad , \tag{7.2 - 35c}$$

$$\text{with} \quad -\pi/2 \le 2\varepsilon \le +\pi/2 \quad \text{and} \quad 0 \le 2\psi < 2\pi \quad .$$

Consequently, 2ε equals the latitude and 2ψ the longitude on this sphere.

Each point on the *equator* on this sphere ($2\varepsilon = 0$, $0 \le 2\psi < 2\pi$) corresponds to a different form of linear polarization. *North pole* ($2\varepsilon = +\pi/2$) and *south pole* ($2\varepsilon = -\pi/2$) represent right-circular and left-circular polarization. Each point on the *northern hemisphere* ($0 < 2\varepsilon < \pi/2$) corresponds to a form of a right-elliptical polarization, while a point on the *southern hemisphere* ($-\pi/2 < 2\varepsilon < 0$) corresponds to a form of left elliptical polarization.

d) Stokes vector

The *Stokes vector* $\vec{s}$ [Stokes 1852 J] is a four-dimensional vector that completely characterizes any form of polarization, i.e. shape and size of the light ellipse (7.2 - 31b). It is defined by its four components

$$s_0 = A_x^2 + A_y^2 = 2Z_0 I \quad , \tag{7.2 - 36a}$$

$$s_1 = A_x^2 - A_y^2 \quad , \tag{7.2 - 36b}$$

$$s_2 = 2A_x A_y \sin\delta \quad , \tag{7.2 - 36c}$$

$$s_3 = 2A_x A_y \sin\delta \quad . \tag{7.2 - 36d}$$

These four components are not independent. They satisfy the identity

$$s_0^2 = s_1^2 + s_2^2 + s_3^2 \quad . \tag{7.2 - 36e}$$

In addition, they are related to the Cartesian coordinates (7.2 - 35a-c) of the Poincaré sphere as follows

$$s_1 / s_0 = X \quad , \quad s_2 / s_0 = Y \quad , \quad s_3 / s_0 = Z \quad . \tag{7.2 - 37}$$

The Stokes vector has a fundamental property with respect to the *superposition of incoherent polarized light beams*: The Stokes vector $\vec{s}$ of the superposition of two incoherent polarized light beams is the sum of the corresponding Stokes vectors $\vec{s}_1$ and $\vec{s}_2$ of the two separate beams [Theocaris & Gdoutos 1979 B]

$$\vec{s} = \vec{s}_1 + \vec{s}_2 \quad . \tag{7.2 - 38}$$

e) Jones vector and matrix

The *normalized Jones vector* $\vec{a}$ [Jones 1941 J, Theocaris & Gdoutos 1979 B] is a two-dimensional complex column vector determined by the components α_x and α_y of the light vector $\vec{\alpha}$ (7.2 - 26)

$$\vec{a} = \begin{bmatrix} a_x \\ a_y \end{bmatrix} = \begin{bmatrix} A_x \, exp(i\,\delta_x) \\ A_y \, exp(i\,\delta_y) \end{bmatrix} \quad . \tag{7.2 - 39}$$

It is related to the *components of the Stokes vector* $\vec{s}$ (7.2 - 36a-d) by

$$s_0 = \vec{a}^{\dagger} \vec{a} = 2Z_0 I = A_x^2 + A_y^2 \tag{7.2 - 40a}$$

$$s_1 = \vec{a}^{\dagger} \sigma_z \, \vec{a} \tag{7.2 - 40b}$$

$$s_2 = \vec{a}^{\dagger} \sigma_x \, \vec{a} \tag{7.2 - 40c}$$

$$s_3 = \vec{a}^{\dagger} \sigma_y \, \vec{a} \tag{7.2 - 40d}$$

with $\vec{a}^{\dagger} = \left[a_x^*, \, a_y^* \right]$ and the Pauli spin matrices [Whitney 1971 J]

$$\sigma_x = \begin{bmatrix} 0 & 1 \\ 1 & 0 \end{bmatrix} \; , \quad \sigma_y = \begin{bmatrix} 0 & -i \\ i & 0 \end{bmatrix} \; , \quad \sigma_z = \begin{bmatrix} 1 & 0 \\ 0 & -1 \end{bmatrix} \quad . \tag{7.2 - 41}$$

In these equations the asterisk * indicates the complex conjugate and the cross † the Hermitian conjugate.

Unlike the Stokes vector $\vec{s}$ (7.2 - 36a-d), which is relevant for the superposition of two incoherent polarized light beams, the Jones vector $\vec{a}$ (7.2 - 39) can be used to determine the *superposition of two coherent polarized light beams*. If the two Jones vectors $\vec{a}_1$ and $\vec{a}_2$ characterize the two coherent polarized beams, their superposition is described by the sum

$$\vec{a} = \vec{a}_1 + \vec{a}_2 \quad . \tag{7.2 - 42}$$

The *effect of linear optical elements on polarized light beams* can be described with the aid of *Jones matrices*. The Jones vectors $\vec{a}_{k-1}$ and $\vec{a}_k$ of the light beams entering and emerging the linear optical element *k* are related by the Jones matrix $\boldsymbol{J}_k$

$$\vec{a}_k = \boldsymbol{J}_k \, \vec{a}_{k-1} \tag{7.2 - 43}$$

Since the Jones matrices are associative, the effect of a series of linear optical elements $k = 1, 2, \ldots n$ with the Jones matrices $\boldsymbol{J}_k$ on a light beam characterized by the Jones vector $\vec{a}_0$ is described by the matrix product

$$\vec{a}_n = \boldsymbol{J}_n \, \boldsymbol{J}_{n-1} \cdots \boldsymbol{J}_2 \boldsymbol{J}_1 \, \vec{a}_0 \tag{7.2 - 44}$$

Jones matrices of polarizers and retarders are listed in Tables 4.1 & 4.2 of reference [Theocaris & Gdoutos 1971 B]. A simple example is the *horizontal polarizer* with the Jones matrix

$$\boldsymbol{J}_P(0) = \begin{bmatrix} 1 & 0 \\ 0 & 0 \end{bmatrix} \; . \tag{7.2 - 45}$$

A polarizer with its axis at an angle Θ with the horizontal x axis is described by the Jones matrix

$$\begin{aligned} \boldsymbol{J}_P(\Theta) &= \boldsymbol{R}(-\Theta)\,\boldsymbol{J}_P(0)\,\boldsymbol{R}(\Theta) \\ &= \begin{bmatrix} cos^2\Theta & sin\Theta\, cos\Theta \\ sin\Theta\, cos\Theta & sin^2\Theta \end{bmatrix} \end{aligned} \tag{7.2 - 46}$$

where $R(\Theta)$ indicates the rotation matrix

$$\boldsymbol{R}(\Theta) = \begin{bmatrix} cos\Theta & sin\Theta \\ -sin\Theta & cos\Theta \end{bmatrix} \; . \tag{7.2 - 47}$$

Finally, it should be mentioned that the Jones matrices $\boldsymbol{J}_P(\Theta)$ (7.2 - 46) of the polarizers represent *projection matrices*, which are characterized by

$$\boldsymbol{J}_P(\Theta) = \boldsymbol{J}_P^2(\Theta) = \boldsymbol{J}_P^3(\Theta) = \boldsymbol{J}_P^4(\Theta) = \ldots = \ldots \; . \tag{7.2 - 48}$$

7.3 Wave Velocities and Dispersion

A main characteristic of a traveling wave is its velocity. A thorough study of waves, demonstrates that in general the wave velocity depends on the wave form. In principle, there exist two types of wave velocities, the *phase velocity* υ and the *group or envelope velocity* υ_g. The first is the velocity of harmonic waves (7.2 - 1) and (7.2 - 3), while the second represents the velocity of wave groups and packets.

As mentioned in Section 7.2.2a a wave shows *no dispersion* if the wave velocity depends neither on the circular frequency $\omega = 2\pi\nu$ nor on the propagation constant $\beta = 2\pi / \lambda$. In this case the group velocity υ_g equals the phase velocity υ. This implies that there exists only *one constant wave velocity*

$$\upsilon = \upsilon_g = const \quad . \qquad (7.3 - 1)$$

On the contrary, the group velocity υ_g and the phase velocity υ of waves with dispersion differ and depend either on the circular frequency $\omega = 2\pi\nu$ or on the propagation constant $\beta = k = 2\pi / \lambda$

$$\upsilon = \upsilon(\omega) \neq \upsilon_g = \upsilon_g(\omega) \quad . \qquad (7.3 - 2a)$$

Normal dispersion means that the group velocity υ_g is smaller than the phase velocity υ

$$\upsilon = \upsilon(\omega) > \upsilon_g = \upsilon_g(\omega) \quad , \qquad (7.3 - 2b)$$

whereas *anomalous dispersion* implies that the group velocity υ_g is larger than the phase velocity υ

$$\upsilon = \upsilon(\omega) < \upsilon_g = \upsilon_g(\omega) \quad . \qquad (7.3 - 2c)$$

For the description of *electromagnetic waves* frequency ν and circular frequency ω, phase and group velocity υ and υ_g are often replaced by *vacuum wavelength* λ_v, *index of refraction n* and *group index N*

$$\lambda_v = c / \nu = 2\pi c / \omega = cT \quad , \qquad (7.3 - 3a)$$

$$n(\omega) = c / \upsilon(\omega) \quad , \qquad (7.3 - 3b)$$

$$N(\omega) = c / \upsilon_g(\omega) \quad , \qquad (7.3 - 3c)$$

where c is the velocity of light in vacuum. In this context *phase dispersion* means the frequency dependence of phase velocity υ and index of refraction n while *group dispersion* designates the frequency dependence of group velocity υ_g and group index N.

A wave group or packet can often be separated into a harmonic *carrier wave* with a high circular frequency $\omega_0 = 2\pi\nu_0$ and an *envelope* that represents a relatively slow amplitude modulation. In this situation the envelope can be described by specific *envelope dispersion relations* and *envelope differential equations* as shown in Section

7.3.6. Envelope differential equations are common in the theory of nonlinear waves presented in Chapter 8.

7.3.1 Phase Velocity

The phase velocity υ is the velocity of the harmonic waves represented by (7.2 - 1) or (7.2 - 3) and their phases Φ.

If the harmonic wave travels in a *medium without loss or gain* its dispersion relation (7.2 - 5a-c) is real. This implies that the circular frequency ω, the propagation constant β and the phase Φ can be assumed real. Under these circumstances the phase velocity υ can be deduced from the condition that the phase Φ at time $t+\Delta t$ and position $z+\upsilon\Delta t$ equals that at time t and position z

$$\begin{aligned} \Phi &= \Phi(z,t) = \beta z - \omega t + \varphi \\ &= \Phi(z+\upsilon\Delta t, t+\Delta t) = \beta(z+\upsilon\Delta t) - \omega(t+\Delta t) + \varphi \quad . \end{aligned} \tag{7.3 - 4}$$

The result is the *phase velocity*

$$\upsilon = \omega / \beta = \nu\lambda = \lambda / T = \nu / \tilde{\nu} \quad . \tag{7.3 - 5}$$

The application of (7.3 - 4) to the representations (7.2 - 1) and (7.2 - 3) demonstrates that this velocity υ is the velocity of harmonic waves in the real as well as in the complex representation

$$w = w(z,t) = w(z+\upsilon\Delta t, t+\Delta t) \quad , \tag{7.3 - 6a}$$

$$u = u(z,t) = u(z+\upsilon\Delta t, t+\Delta t) \quad . \tag{7.3 - 6b}$$

Motion and velocity of a real harmonic wave (7.3 - 6a) are illustrated in Fig. 7.3 - 1.

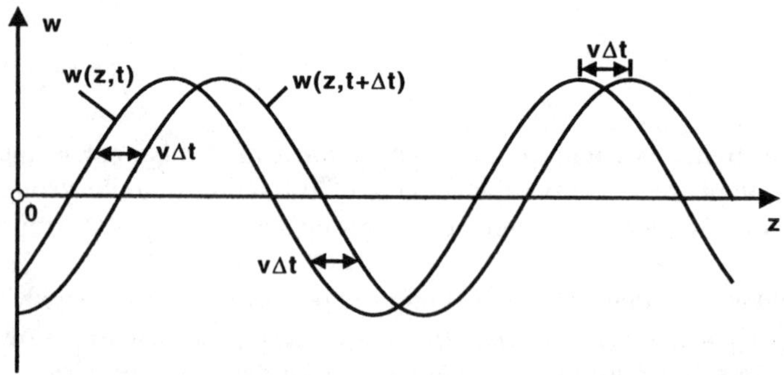

Fig. 7.3 - 1. Motion and velocity of a real harmonic wave (7.3 - 6a)

For a damped or amplified wave traveling in a *medium with loss or gain* the dispersion relation (7.2 - 5a-c) is complex. In this situation one chooses the traveling-wave representation (7.2 - 10a-c) with a real circular frequency ω and a complex propagation constant β. The velocity υ of the effective phase Φ_{eff} defined by (7.2 - 10b) is

$$\upsilon = \omega / \operatorname{Re}\beta(\omega) = \nu \lambda_{eff} = \lambda_{eff} / T \tag{7.3 - 7}$$

where λ_{eff} indicates the effective wavelength (7.2 - 10c).

7.3.2 Group or Envelope Velocity

The group or envelope velocity υ_g is the velocity of the envelope of a wave group or packet [Brillouin 1960 B]. A wave group or packet corresponds to a superposition of harmonic waves with similar circular frequencies ω and propagation constants β. For real as well as complex dispersion relations $\beta(\omega)$ a wave group can therefore be represented by

$$u(z,t) = \int_{\omega_0-\Delta\omega}^{\omega_0+\Delta\omega} U(\omega')\, exp\left[i\left(\beta(\omega')z - \omega' t + \varphi\right)\right] d\omega' \qquad \text{with} \quad \omega_0 >> \Delta\omega > 0 \quad . \tag{7.3 - 8}$$

In order to determine the velocity υ_g of this wave group one approximates the dispersion relation $\beta(\omega)$ by a Taylor series

$$\beta(\omega) = \beta_0 + \Delta\omega \frac{d\beta}{d\omega}(\omega_0) + \ldots \quad \text{with} \quad \beta(\omega_0) = \beta_0 \quad . \tag{7.3 - 9}$$

This approximation permits the decomposition of the wave group (7.3 - 8) into a harmonic *carrier wave* with circular frequency ω_0 and propagation constant β_0 and an *envelope* $E(z, t)$

$$u(z,t) = E(z,t)\, exp\left[i(\beta_0 z - \omega_0 t)\right]$$
$$\text{with} \quad E(z,t) = \int_{-\Delta\omega}^{+\Delta\omega} U(\omega_0 + \Delta\omega')\; exp\left[i\Delta\omega'\left(z\frac{d\beta}{d\omega}(\omega_0) - t\right)\right] d\Delta\omega' \quad . \tag{7.3 - 10}$$

Fig. 7.3 - 2 illustrates this decomposition when applied to a real wave group.

The harmonic *carrier wave* propagates with the phase velocity υ defined by (7.3 - 5) and (7.3 - 7)

$$\upsilon = \omega_0 / \beta_0 \quad \text{for} \quad \beta_0 \; real \quad , \tag{7.3 - 11a}$$

$$\upsilon = \omega_0 / Re\,\beta_0 \quad \text{for} \quad \beta_0 \; complex \quad . \tag{7.3 - 11b}$$

The group velocity υ_g as velocity of the envelope $E(z, t)$ is defined by the condition

$$E = E(z,t) = E\left(z + \upsilon_g \Delta t, t + \Delta t\right) \quad . \tag{7.3 - 12}$$

This condition applied to the representation (7.3 - 10) of $E(z, t)$ yields the group velocity

$$\upsilon_g = \frac{d\omega}{d\beta} = \left[\frac{d\beta}{d\omega}\right]^{-1} \tag{7.3 - 13}$$

for a real dispersion relation $\beta(\omega)$ with ω and β real, and

$$\upsilon_g = \left[\frac{d\,Re\,\beta}{d\omega}\right]^{-1} \tag{7.3 - 14}$$

for a complex dispersion relation $\beta(\omega)$ with ω real and β complex.

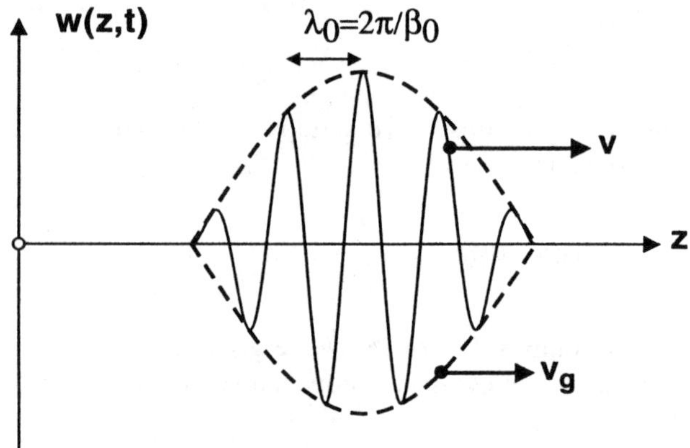

Fig. 7.3 - 2. Decomposition of a wave group into carrier wave and envelope (7.3 - 10)

7.3.3 Phase Dispersion

Phase dispersion or simply *dispersion* designates the frequency dependence of phase velocity υ and refractive index n

$$\frac{d\upsilon(\omega)}{d\omega} \neq 0 \quad \text{and} \quad \frac{dn(\omega)}{d\omega} \neq 0 \tag{7.3 - 15}$$

where the refractive index n equals

$$n = c / \upsilon(\omega) = c\beta(\omega) / \omega = \lambda_v / \lambda \tag{7.3 - 16a}$$

for electromagnetic waves in media without loss or gain, and

$$n = c / \upsilon(\omega) = c\frac{Re\,\beta(\omega)}{\omega} = \lambda_v / \lambda_{eff} \tag{7.3 - 16b}$$

for those in media with loss or gain characterized by ω real and β complex. The frequency dependence (7.3 - 15) of υ and n implies that they differ from group velocity υ_g and group index N

$$\upsilon(\omega) \neq \upsilon_g(\omega) \quad \text{and} \quad n(\omega) \neq N(\omega) \tag{7.3 - 17}$$

in agreement with (7.3 - 2a). This statement can be proved with the aid of the equations

$$\upsilon_g = \upsilon\left[1 - \frac{\omega}{\upsilon}\frac{d\upsilon}{d\omega}\right]^{-1} = \upsilon\left[1 + \frac{\lambda_v}{\upsilon}\frac{d\upsilon}{d\lambda_v}\right]^{-1} \quad , \tag{7.3 - 18a}$$

$$N = n + \omega\frac{dn}{d\omega} = n - \lambda_v\frac{dn}{d\lambda_v} \quad . \tag{7.3 - 18b}$$

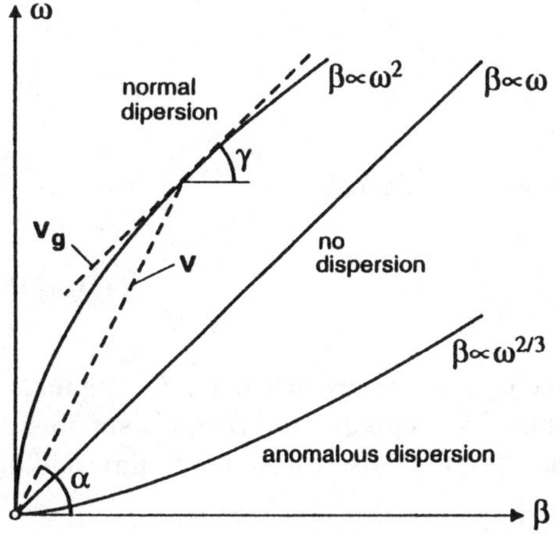

Fig. 7.3 - 3. Dispersion relations $\omega(\beta)$ of surface waves: Gravity waves on shallow liquids without dispersion (7.3 - 26b), gravity waves on deep liquids with normal dispersion (7.3 - 27b) and capillary waves with anomalous dispersion (7.3 - 28b)

According to (7.3 - 1) and (7.3 - 2b&c) *three types of dispersion* can be distinguished. The dispersion is

$$\textit{normal} \quad \text{if} \quad \upsilon_g < \upsilon, \quad N > n, \quad \frac{dn}{d\omega} > 0, \quad \frac{dn}{d\lambda_v} < 0 \quad , \tag{7.3 - 19a}$$

$$\textit{missing} \quad \text{if} \quad \upsilon_g = \upsilon, \quad N = n, \quad \frac{dn}{d\omega} = \frac{dn}{d\lambda_v} = 0 \quad , \tag{7.3 - 19b}$$

$$\textit{anomalous} \quad \text{if} \quad \upsilon_g > \upsilon, \quad N < n, \quad \frac{dn}{d\omega} < 0, \quad \frac{dn}{d\lambda_v} > 0 \quad . \tag{7.3 - 19c}$$

Examples of the dispersion relations $\omega(\beta)$ of these three types of dispersion are illustrated in Fig. 7.3 - 3. They represent the dispersions of the different surface waves of liquids discussed in Section 7.3.5.

7.3.4 Group Dispersion

Group dispersion means the frequency dependence of the group velocity υ_g and the group index N

$$\frac{d\upsilon_g(\omega)}{d\omega} \neq 0 \quad \text{and} \quad \frac{dN(\omega)}{d\omega} \neq 0 \quad , \tag{7.3 - 20}$$

where the group index N is determined by

$$N = c \,/\, \upsilon_g(\omega) = c\frac{d\beta(\omega)}{d\omega} \tag{7.3 - 21a}$$

for electromagnetic waves in media without loss or gain and

$$N = c \,/\, \upsilon_g(\omega) = c\frac{d\, Re\,\beta(\omega)}{d\omega} \tag{7.3 - 21b}$$

for those in media with loss or gain that are characterized by ω real and β complex.

In *photonics* the group dispersion in optical waveguides and fibers is suppressed in order to eliminate the frequency dependence of the pulse-transmission time and the related deterioration of the pulse shape.

In general, the group dispersion can be described by

$$\frac{dN}{d\omega} = -c\;\upsilon_g^{-2}\frac{d\upsilon_g}{d\omega} = 2\frac{dn}{d\omega} + \omega\frac{d^2n}{d\omega^2} \quad , \tag{7.3 - 22a}$$

$$\frac{dN}{d\lambda_v} = -\lambda_v\frac{d^2n}{d\lambda_v^{\;2}} \quad . \tag{7.3 - 22b}$$

For electromagnetic waves in media without loss or gain the group dispersion can be represented by

$$\frac{d\upsilon_g}{d\omega} = -\upsilon_g^2 \frac{d^2\beta}{d\omega^2} \quad , \qquad (7.3 - 23a)$$

$$\frac{dN}{d\omega} = c\frac{d^2\beta}{d\omega^2} \quad , \qquad (7.3 - 23b)$$

and for those in media with loss or gain characterized by ω real and β complex

$$\frac{d\upsilon_g}{d\omega} = -\upsilon_g^2 \frac{d^2 Re\beta}{d\omega^2} \quad , \qquad (7.3 - 24a)$$

$$\frac{dN}{d\omega} = c\frac{d^2 Re\beta}{d\omega^2} \quad . \qquad (7.3 - 24b)$$

7.3.5 Surface Waves of Liquids

The three types (7.3 - 19a-c) of dispersion occur in surface waves of liquids, e.g. *water waves* [Alonso & Finn 1967 B, 1970 B, Crawford 1968 B, Kneubühl 1994 B, Lüst 1978 B]. These waves obey the *general dispersion relation* [Lüst 1978 B]

$$\omega^2 = \beta\, g\left(1+\frac{\sigma}{\rho g}\beta^2\right)tanh(h\beta) \qquad (7.3 - 25a)$$

with $\omega = 2\pi\nu$ and $\beta = 2\pi/\lambda$.

In this equation $g \approx 9.81\ ms^{-2}$ indicates the acceleration due to gravity, σ [Nm^{-1}] the surface tension, ρ [$kg\ m^{-3}$] the density and h [m] the depth of the liquid. According to (7.3 - 5) the phase velocity υ [$m\ s^{-1}$] of these waves is determined by

$$\upsilon^2 = \left(\frac{g}{\beta}+\frac{\sigma\beta}{\rho}\right)tanh(h\beta) \quad . \qquad (7.3 - 25b)$$

If the parameters σ, ρ and h meet specific conditions, these general relations (7.3 - 25a&b) can be replaced by simple *approximations*, which imply *different types of dispersion.*

a) Gravity waves on shallow liquids

Gravity waves on shallow liquids, e.g. water, occur on the two following conditions

$$\lambda << h \quad \text{and} \quad \frac{\sigma}{\lambda^2} << \rho g \quad . \qquad (7.3 - 26a)$$

The first inequality indicates a small depth h, while the second requires that the acceleration of gravity g dominates over the surface tension σ. These two conditions reduce (7.3 - 25a) to the dispersion relation

$$\beta = \omega[gh]^{-1/2} \quad . \tag{7.3 - 26b}$$

The application of (7.3 - 5) and (7.3 - 13) to this relation demonstrates that phase and group velocity are identical and constant

$$\upsilon = \upsilon_g = [gh]^{1/2} = const \quad . \tag{7.3 - 26c}$$

Since these velocities fulfill condition (7.3 - 1) the gravity waves on shallow liquids show *no dispersion.* In addition (7.3 - 26c) reveals that these waves *do not depend on the nature of the liquid* that is specified by the density ρ and the surface tension σ.

b) Gravity waves on deep liquids

Gravity waves on deep liquids, e.g. water are characterized by the two following conditions

$$\lambda << h \quad \text{and} \quad \frac{\sigma}{\lambda^2} << \rho g \quad . \tag{7.3 - 27a}$$

The first condition requires a large depth h, whilst the second an acceleration of gravity g that dominates over the surface tension. The two conditions permit the approximation of (7.3 - 25a) by the dispersion relation

$$\beta = g^{-1}\omega^2 \quad . \tag{7.3 - 27b}$$

The application of (7.3 - 5) and (7.3 - 13) yields the wave velocities

$$\upsilon = 2\upsilon_g = g / \omega \quad . \tag{7.3 - 27c}$$

Since the phase velocity υ exceeds the group the velocity υ_g, the gravity waves on deep liquids exhibit *normal dispersion.* According to (7.3 - 27b&c) these velocities *depend neither on the depth h nor on the nature of the liquid* that is characterized by the density ρ and the surface tension σ.

c) Capillary waves

Capillary waves or *ripples* obey the conditions

$$\lambda << h \quad \text{and} \quad \frac{\sigma}{\lambda^2} >> \rho g \quad . \tag{7.3 - 28a}$$

Consequently, their wavelengths λ are small. Capillary waves are observed when a gentle wind blows over water. The second condition of (7.3 - 28a) requires that the surface tension σ dominates over the gravity. If both conditions (7.3 - 28a) are applied to (7.3 - 25a) it yields the approximation

$$\sigma \beta^3 = \rho \omega^2 \quad . \tag{7.3 - 28b}$$

The corresponding wave velocities derived with the aid of (7.3 - 5) and (7.3 - 13) are determined by

$$\upsilon = \frac{2}{3} \upsilon_g = (\sigma / \rho)^{5/6} \omega^{1/3} \quad . \tag{7.3 - 28c}$$

Since their group velocity υ_g surpasses their phase velocity υ the capillary waves show *anomalous dispersion*. They are *not influenced by gravity*.

Fig. 7.3 - 3 on page 361 presents the dispersion relations of capillary waves and gravity waves on shallow and deep liquids. With regard to its interpretation it should be mentioned that in the $\omega\beta$ plane the phase velocity υ corresponds to the direction α defined by the ratio ω/β whilst the group velocity υ_g is represented by the direction γ of the tangent determined by $d\omega/d\beta$

$$tg\,\alpha = \omega / \beta = \upsilon \quad \text{and} \quad tg\,\gamma = d\omega / d\beta = \upsilon_g \quad . \tag{7.3 - 29}$$

7.3.6 Envelope Equations

As demonstrated in Section 7.3.2 a wave group or packet can be decomposed into a product of a *harmonic carrier wave* and an *envelope* $E(z, t)$

$$u(z,t) = E(z,t) \exp\left[i(\beta_0 z - \omega_0 t)\right] \quad . \tag{7.3 - 10}$$

The carrier wave is characterized by the real circular frequency ω_0 and the real propagation constant β_0.

If the variation of the envelope $E(z, t)$ with position z and time t is considerably less than that of the carrier wave with wavelength $\lambda_0 = 2\pi / \beta_0$ and period $T_0 = 2\pi / \omega_0$, then the entire wave group $u(z, t)$ can be described essentially by the envelope. This requires, however, a differential equation that determines the envelope $E(z, t)$, i.e. the so-called envelope equation.

The linear envelope equations considered in this section form the basis of nonlinear envelope equations used for the description of specific phenomena of nonlinear waves such as the solitons discussed in Chapter 8.

a) Envelope dispersion relations

The envelope equation can be derived from the wave equation of the field $u(z, t)$ and the corresponding dispersion relation (7.2 - 5a-c) with the aid of the envelope dispersion relation. This relation is deduced by assuming a complex harmonic envelope of the form

$$E(z,t) = E_0 \, exp\left[i(\Delta\beta \, z - \Delta\omega \, t)\right] \quad . \tag{7.3 - 30a}$$

Its combination with the carrier wave introduced in (7.3 - 10) yields the complex harmonic wave

$$u(z,t) = E_0 \, exp\left[i\{(\beta_0 + \Delta\beta)z - (\omega_0 + \Delta\omega)t\}\right] \tag{7.3 - 30b}$$

with the circular frequency $\omega = \omega_0 + \Delta\omega$ and the propagation constant $\beta = \beta_0 + \Delta\beta$. Thus, the decomposition of ω and β corresponds to the splitting of the wave group into carrier wave and envelope. The decomposed ω and β permit to approximate the dispersion relations (7.2 - 5a-c) near ω_0 and β_0 by Taylor series. These relate $\Delta\omega$ and $\Delta\beta$ and thus form the *envelope dispersion relations*

$$0 = \Delta\omega \frac{\partial D}{\partial \omega}(\omega_0, \beta_0) + \Delta\beta \frac{\partial D}{\partial \beta}(\omega_0, \beta_0) + \Delta\omega \, \Delta\beta \frac{\partial^2 D}{\partial \omega \, \partial \beta}(\omega_0, \beta_0) + \dots \tag{7.3 - 31a}$$

$$\text{with} \quad D(\omega_0 \; \beta_0) = 0 \quad , \quad \text{or}$$

$$\Delta\omega = \Delta\beta \frac{d\omega}{d\beta}(\beta_0) + \frac{\Delta\beta^2}{2} \frac{d^2\omega}{d\beta^2}(\beta_0) + \dots \tag{7.3 - 31b}$$

$$\text{with} \quad \omega(\beta_0) = \omega_0 \quad , \quad \text{or}$$

$$\Delta\beta = \Delta\omega \frac{d\beta}{d\omega}(\omega_0) + \frac{\Delta\omega^2}{2} \frac{d^2\beta}{d\omega^2}(\omega_0) + \dots \tag{7.3 - 31c}$$

$$\text{with} \quad \beta(\omega_0) = \beta_0 \quad .$$

These envelope dispersion relations are transformed into the *corresponding envelope equations* by replacing the variables $\Delta\omega$ and $\Delta\beta$ by the corresponding differential operators. These can be determined by partial differentiations of the complex harmonic wave (7.3 - 30b) or its envelope (7.3 - 30a) with respect to position z and time t

$$E_z = i \, \Delta\beta \, E \quad \text{and} \quad E_t = -i \, \Delta\omega \, E \quad . \tag{7.3 - 32a}$$

This result justifies the introduction of the *envelope differential operators* for $\Delta\beta$ and $\Delta\omega$

$$\Delta\beta \, E = -i \frac{\partial}{\partial z} E \quad \text{and} \quad \Delta\omega \, E = +i \frac{\partial}{\partial t} E \quad . \tag{7.3 - 32b}$$

In the following these operators are used to derive the envelope equations corresponding to the explicit dispersion relations (7.3 - 31b&c).

b) Schrödinger envelope equation

The introduction of the operators (7.3 - 32b) and the group velocity (7.3 - 13) in the explicit envelope dispersion relation (7.3 - 31b) yields the *Schrödinger envelope equation*

$$iE_{\mathrm{t}} = -i\,\upsilon_{\mathrm{g}}\,E_{\mathrm{z}} - \frac{1}{4}\left\{\frac{d}{d\omega}\,\upsilon_{\mathrm{g}}^{2}\right\}E_{\mathrm{zz}} + \ldots \qquad (7.3 - 33)$$

where all coefficients refer to $\omega = \omega_0$ and $\beta = \beta_0$.

The terms with E_{zz}, E_{zzz}, ... vanish if there is no group dispersion (7.3 - 20). On this condition (7.3 - 33) corresponds to the *reduced Hertz equation* discussed in Section 7.4.2. Then, the envelope propagates undisturbed with the group velocity υ_g. Therefore, it is of advantage to replace in the Schrödinger envelope equation (7.3 - 33) the stationary coordinate system *zt* by a coordinate system *ZT* that travels with the constant group velocity υ_g

$$Z = z - \upsilon_{\mathrm{g}} t \quad \text{and} \quad T = t \quad . \qquad (7.3 - 34a)$$

For the envelope *E* this procedure yields the relations

$$E_{\mathrm{z}} = E_{\mathrm{Z}} \quad \text{and} \quad E_{\mathrm{t}} = E_{\mathrm{T}} - \upsilon_{\mathrm{g}} E_{\mathrm{Z}} \quad . \qquad (7.3 - 34b)$$

Thus, (7.3 - 33) is transformed into *reduced* Schrödinger envelope equation

$$i\,E_{\mathrm{T}} = -\frac{1}{4}\left\{\frac{d}{d\omega}\,\upsilon_{\mathrm{g}}^{2}\right\}E_{\mathrm{ZZ}} + \ldots \quad . \qquad (7.3 - 35)$$

In a first approximation this equation represents the linear *Schrödinger equation* to be discussed in Section 7.4.7.

c) Modified Schrödinger envelope equation

The introduction of the differential operators (7.3 - 32b), of the group velocity (7.3 - 13) and the group index (7.3 - 21a) in the explicit envelope dispersion relation (7.3 - 31c) results in the *modified Schrödinger envelope equation* [Hasegawa 1989 B]

$$\begin{aligned} -i\,E_{\mathrm{z}} &= +i\,\upsilon_{\mathrm{g}}^{-1} E_{\mathrm{t}} - \frac{1}{2}\,\upsilon_{\mathrm{g}}^{-2}\,\frac{d\upsilon_{\mathrm{g}}}{d\omega}\,E_{\mathrm{tt}} + \ldots \quad \text{or} \\ -icE_{\mathrm{z}} &= +iN\,E_{\mathrm{t}} + \frac{1}{2}\frac{dN}{d\omega}\,E_{\mathrm{tt}} + \ldots \end{aligned} \qquad (7.3 - 36)$$

where all coefficients refer to $\omega = \omega_0$ and $\beta = \beta_0$. The terms with E_{tt}, E_{ttt}, ... vanish if there is no group dispersion (7.3 - 20). Then (7.3 - 36) represents the *reduced Hertz equation* discussed in Section 7.4.2. This implies that the envelope travels propagates without distortion with the constant group velocity υ_g. Consequently, one substitutes the stationary coordinate system *zt* by a coordinate system *ZT* with a retarded time *T*

$$Z = z \quad \text{and} \quad T = t - \frac{1}{\upsilon_g} z \quad . \tag{7.3 - 37a}$$

This results in the following relations between the partial derivates of *E*

$$E_z = E_Z - \frac{1}{\upsilon_g} E_T \quad \text{and} \quad E_t = E_T \quad . \tag{7.3 - 37b}$$

The transformation of (7.3 - 36) according to (7.3 - 37a&b) yields the *reduced* modified Schrödinger envelope equation

$$i\,E_Z = -\frac{1}{2} \upsilon_g^{-2} \frac{d\upsilon_g}{d\omega} E_{TT} + \ldots \quad = +\frac{1}{2c} \frac{dN}{d\omega} E_{TT} + \ldots \quad . \tag{7.3 - 38}$$

In a first approximation this equation corresponds to the linear Schrödinger equation of Section 7.4.7 with time and position exchanged.

The reduced envelope equations (7.3 - 35) and (7.3 - 38) determine the deformation of the envelope of a wave group observed from a point of view that travels with the group velocity υ_g.

7.4 Linear Plane Waves in Homogeneous Isotropic Media

This chapter is dedicated to the best known plane linear waves in homogeneous isotropic media. They are characterized by specific wave equations. Taken into consideration are the Hertz, the reduced Hertz, the linear Klein-Gordon, the telegraph, the linear diffusion, the linearized Korteweg-de Vries and the linear Schrödinger equation. Each type of wave is discussed with regard to the wave equation, the dispersion relation, phase and group velocity as well as to general, harmonic and nonharmonic solutions of the wave equation.

Harmonic waves are determined by the dispersion relations (7.2 - 5a-c). The subsequent discussion of plane linear harmonic waves is based on the assumption that they move in the *z* direction and on the complex and real representations

$$u(\vec{r},t) = u(z,t) = U\, exp\left[i(\beta(\omega)z - \omega t)\right] \quad , \quad \text{and} \tag{7.4 - 1a}$$

$$w(\vec{r},t) = w(z,t) = W\, cos\left[\beta(\omega)z - \omega t + \varphi\right] \tag{7.4 - 1b}$$

with the dispersion relation $\beta(\omega)$ according to (7.2 - 5c).

In many situations harmonic or even general solutions of a wave equation do not satisfy the demands. Required may be solutions that fulfill given initial or boundary conditions. *Local or boundary conditions* are relevant for the description of the propagation of local time-dependent excitations in a medium as well as for the study of the standing waves discussed in Chapter 9. Thus, a time-dependent local excitation at the position $z = 0$ can be described by the local or boundary conditions

$$u(0,t) = L(t) \quad , \quad \text{and} \tag{7.4 - 2a}$$

$$u_z(0,t) = M(t) = dN(t)/dt \quad . \tag{7.4 - 2b}$$

Initial conditions are of importance in the dynamics of extended wave trains. For instance the initial conditions for a wave train at time $t = 0$ may be

$$u(z,0) = F(z) \quad , \quad \text{and} \tag{7.4 - 3a}$$

$$u_t(z,0) = G(z) = d\,H(z)/dz \quad . \tag{7.4 - 3b}$$

With respect to these equations it should be remembered that the indices z and t in the above and following equations indicate the partial differentiations with respect to position z and time t.

7.4.1 Hertz Equation

The Hertz equation is characteristic of *waves without dispersion* (7.2 - 7), e.g. *electromagnetic waves in vacuum* (7.1 - 15a&b). For a *plane wave* propagating in the z direction it takes the form

$$u_{tt} - \upsilon^2 u_{zz} = 0 \quad \text{with} \quad \upsilon = const \quad . \tag{7.4 - 4}$$

In agreement with (7.2 - 8) its dispersion relation is

$$\omega/\beta = \upsilon\lambda = \upsilon = \upsilon_g = const \quad \text{with} \quad \omega \geq 0 \quad . \tag{7.4 - 5}$$

The absence of dispersion implies equal phase and group velocity υ and υ_g.

According to *d'Alembert's law* [Webster 1927 B, Webster & Szegö 1930 B] the *general solution* of the Hertz equation (7.4 - 4) consists of a first arbitrary wave propagating without distortion with the constant velocity $+\upsilon$ in the z direction and a second arbitrary wave traveling without distortion with the constant velocity υ in the $-z$ direction

$$u(z,t) = f(z - \upsilon t) + g(z + \upsilon t) \quad . \tag{7.4 - 6}$$

In this equation $f(z)$ and $g(z)$ are arbitrary functions that can be differentiated twice. The two waves of (7.4 - 6) show no dispersion because they do not change shape.

According to (7.4 - 6) also a linear combination of two counter-running *harmonic waves* with the same circular frequency $\omega = \upsilon/\beta$ forms a solution of the Hertz equation (7.4 - 4)

$$w(z,t) = W_1 cos(\beta z - \omega t + \varphi_1) + W_2 cos(\beta z + \omega t - \varphi_2) \quad . \qquad (7.4 - 7)$$

The time-dependent *local conditions* (7.4 - 2a&b) at the position $z = 0$ yield the following solution of the Hertz equation (7.4 - 4) [Webster 1927 B, Webster & Szegö 1930 B]

$$u(z,t) = \frac{1}{2}L(t - \frac{z}{\upsilon}) - \frac{\upsilon}{2}N(t - \frac{z}{\upsilon}) + \frac{1}{2}L(t + \frac{z}{\upsilon}) + \frac{\upsilon}{2}N(t + \frac{z}{\upsilon}) \qquad (7.4 - 8)$$

$$\text{with} \quad u(0,t) = L(t) \quad \text{and} \quad u_z(0,t) = dN(t)/dt \quad ,$$

while the *initial conditions* (7.4 - 3a&b) at time $t = 0$ are fulfilled by the solution

$$u(z,t) = \frac{1}{2}F(z - \upsilon t) - \frac{1}{2\upsilon}H(z - \upsilon t) + \frac{1}{2}F(z + \upsilon t) + \frac{1}{2\upsilon}H(z + \upsilon t) \qquad (7.4 - 9)$$

$$\text{with} \quad u(z,0) = F(z) \quad \text{and} \quad u_t(z,0) = dH(z)/dz \quad .$$

7.4.2 Reduced Hertz Equation

This equation forms the basis for the description of nonlinear waves without dispersion in Section 8.2. The solution of this equation is determined by the time-dependence of the wave velocity.

a) Constant wave velocity

The reduced Hertz equation with a constant wave velocity υ

$$u_t + \upsilon u_z = 0 \quad \text{with} \quad \upsilon = const \qquad (7.4 - 10)$$

represents arbitrary propagating *waves without dispersion*. It serves as basis of many nonlinear wave equations.

The complex and real representations (7.4 - 1a&b) of harmonic waves are solutions of the reduced Hertz equation (7.4 - 10) if they fulfill the dispersion relation of waves without dispersion

$$\omega / \beta = \upsilon\lambda = \upsilon = const \quad . \qquad (7.2 - 8)$$

Consequently, the phase velocity equals the group velocity

$$\upsilon = \upsilon_g = const \quad . \qquad (7.3 - 1)$$

The general solution of the reduced Hertz equation (7.4 - 10) has the form

$$u(z,t) = f(z - \upsilon t) \tag{7.4 - 11}$$

where $f(z)$ represents an arbitrary function that can be differentiated. This solution demonstrates that a corresponding wave, wave group or wave train travels without distortion.

For a *local excitation* (7.4 - 2) at position $z = 0$ the solution of the reduced Hertz equation (7.4 - 10) yields the wave

$$u(z,t) = L(t - \frac{z}{\upsilon}) \quad \text{with} \quad u(0,t) = L(t) \quad , \tag{7.4 - 12}$$

while for a *given waveform* (7.4 - 3a) at time $t = 0$ the solution has the form

$$u(z,t) = F(z - \upsilon t) \quad \text{with} \quad u(z,0) = F(z) \quad . \tag{7.4 - 13}$$

b) Time-dependent wave velocity

The reduced Hertz equation with the time-dependent wave velocity $\upsilon(t)$

$$u_t + \upsilon(t) u_z = 0 \tag{7.4 - 14}$$

has the *general solution*

$$u(z,t) = f(z - \int_0^t \upsilon(t')dt') \tag{7.4 - 15}$$

with an arbitrary differentiable function $f(z)$. For a *given waveform* (7.4 - 3a) at time $t = 0$ the solution of (7.4 - 14) is

$$u(z,t) = F(z - \int_0^t \upsilon(t')dt') \quad \text{with} \quad u(z,0) = F(z) \quad . \tag{7.4 - 16}$$

7.4.3 Linear Klein-Gordon Equation

The linear Klein-Gordon equation

$$u_{tt} - c^2 u_{zz} + \omega_C^2 u = 0 \tag{7.4 - 17}$$

with the velocity c of light in vacuum and the *cut-off circular frequency* ω_C is well known in microwave technology as well as in elementary particle physics.

Hollow metallic waveguides [Borgnis & Papas 1958 J, Klages 1956B, Kneubühl & Sigrist 1995 B, Marcuvitz 1948 B, Montgomery et al. 1948 B] constitute basic passive elements of *microwave techniques*. The electromagnetic fields in these waveguides are usually represented by linear combinations of orthogonal field distributions called *modes*. Each of these modes fulfills (7.3 - 17) with usually different characteristic cut-off circular frequencies ω_C. Electromagnetic waves with circular frequencies ω below a specific ω_C cannot pass the waveguide in the form of the corresponding mode.

In *particle physics* the linear Klein-Gordon equation (7.3 - 17) represents the *wave equation of a free relativistic particle*. It can be derived from the *energy-momentum relation* of such a particle with mass m and energy E

$$E^2 = m^2c^4 + c^2p^2 \quad . \tag{7.4 - 18}$$

The introduction of the quantum mechanical operators (7.2 - 18a&b) for energy and momentum in (7.4 - 18) yield (7.4 - 17). In quantum mechanics

$$\omega_C = c\beta_C = 2\pi c / \lambda_C = mc^2 / \hbar \tag{7.4 - 19}$$

is called Compton circular frequency, while λ_C is named as *Compton wavelength*.

a) Dispersion

The *dispersion relation* (7.2 - 5a) of (7.4 - 17) can be derived with the aid of the real (7.2 - 1) or complex (7.2 - 3) representation of harmonic waves. This results in

$$\omega^2 - \omega_C^2 + c^2\beta^2 = 0 \quad \text{with} \quad \omega \geq \omega_C \geq 0 \quad . \tag{7.4 - 20}$$

The waves governed by (7.4 - 17) show *normal dispersion* according to

$$\upsilon = c^2 / \upsilon_g = c\left[1 - (\omega_C / \omega)^2\right]^{-1/2} = c\left[1 - (\lambda_v / \lambda_{vC})^2\right]^{-1/2} \geq \upsilon_g \quad , \tag{7.4 - 21a}$$

$$n = 1 / N = \left[1 - (\omega_C / \omega)^2\right]^{1/2} = \left[1 - (\lambda_v / \lambda_{vC})^2\right]^{1/2} \leq N \quad . \tag{7.4 - 21b}$$

These equations imply

$$\upsilon(\omega = \omega_C) = \infty \quad , \quad \upsilon_g(\omega = \omega_C) = 0 \quad \text{and} \tag{7.4 - 21c}$$

$$\upsilon(\omega = \infty) = c \quad , \quad \upsilon_g(\omega = \infty) = c \quad . \tag{7.4 - 21d}$$

At the low-frequency limit the phase velocity υ goes to infinity, whereas the group velocity υ_g tends to zero. At high frequencies phase and group velocity υ and υ_g approach the velocity c of light in vacuum.

Finally, it should be noticed that the waves determined by (7.4 - 17) exhibit *group dispersion* because

$$\frac{dN}{d\omega} = -N^3 \omega_C{}^2 / \omega^3 \neq 0 \quad . \tag{7.4 - 22}$$

b) Solutions on given conditions

Under the time-dependent local conditions (7.4 - 2a&b) the linear Klein-Gordon equation (7.4 - 17) has the solution [Webster 1927 B, Webster & Szegö 1930 B]

$$\begin{aligned} u(z,t) &= \frac{1}{2} L(t - \frac{z}{c}) + \frac{1}{2} L(t + \frac{z}{c}) + \\ &+ \frac{1}{2} \omega_C z \int\limits_{t-\frac{z}{c}}^{t+\frac{z}{2}} d\tau \, L(\tau) \frac{J_1(\beta_C \left[c^2 (t-\tau)^2 - z^2\right]^{1/2})}{\left[c^2 (t-\tau)^2 - z^2\right]^{1/2}} \\ &+ \frac{1}{2} c \int\limits_{t-\frac{z}{c}}^{t+\frac{z}{c}} d\tau \, M(\tau) J_0(\beta_C \left[c^2 \left(t - \tau^2\right) - z^2\right]^{1/2}) \end{aligned} \tag{7.4 - 23}$$

with $u(0,t) = L(t)$ and $u_z(0,t) = M(t)$.

In this equation $J_k(x)$, $k = 0, 1, 2, \ldots$ represent Bessel functions of the first kind [Abramowitz & Stegun 1965 B]. The first two terms correspond to the propagation of the local excitation (7.4 - 2a&b) in the $\pm z$ directions without distortion whereas the third and the fourth term describe the effect of dispersion.

The solution of (7.4 - 17) under the *initial conditions* (7.4 - 3a&b) reveals the dynamics of the wave field [Webster 1927 B, Webster & Szegö 1930 B, Wyld 1976 B]

$$\begin{aligned} u(z,t) &= \frac{1}{2} F(z - ct) + \frac{1}{2} F(z + ct) \\ &- \frac{1}{2} \omega_c t \int\limits_{z-ct}^{z+ct} dy \, F(y) \frac{I_1(\beta_C \left[(z-y)^2 - c^2 t^2\right]^{1/2})}{\left[(z-y)^2 - c^2 t^2\right]^{1/2}} \\ &+ \frac{1}{2c} \int\limits_{z-ct}^{z+ct} dy \, G(y) \, I_0(\beta_C \left[(z-y)^2 - c^2 t^2\right]^{1/2}) \end{aligned} \tag{7.4 - 24}$$

with $u(z,0) = F(z)$ and $u_t(z,0) = G(z)$.

In this equation $I_k(x) = (i)^{-k} J_k(ix)$, $k = 0, 1, 2, \ldots$ are the modified Bessel functions of the first kind [Abramowitz & Stegun 1965 B]. As in (7.4 - 23) the first two terms represent the wave propagation without distortion, while the third and fourth term reflect the distortion by dispersion.

7.4.4 Telegraph Equation

The telegraph equation [Bronstein et al. 1993 B, Courant & Hilbert 1968 B, Sommerfeld 1949 B, Webster 1927 B, Webster & Szegö 1930 B]

$$u_{tt} + (2/\tau)u_t + \Omega^2 u = \upsilon_0^2 u_{zz} \qquad (7.4 - 25)$$

describes the variation of the potential difference U[V = volt] and the current I [A = amp] on a *telegraph or transmission line* formed by a twin conductor or a coaxial cable. In this equation υ_0 indicates a wave velocity, Ω a characteristic circular frequency and τ a characteristic time of damping. For a telegraph or transmission line with the capacitance per length C_L [farad/m], the self-inductance per length L_L [henry/m], the resistance per length R_L [ohm/m] and the current leakage per length G_L [ohm^{-1} m^{-1}], the current I on the line is determined by the corresponding equation

$$L_L C_L I_{tt} + (R_L C_L + L_L G_L) I_t + R_L G_L I = I_{zz} \quad . \qquad (7.4 - 26a)$$

The parameters of this equation are related to those of (7.4 - 25) according to

$$\upsilon_0 = (L_L C_L)^{-1/2} \quad , \quad \Omega = (R_L G_L / L_L C_L)^{1/2} \quad , \quad \text{and}$$
$$\tau = 2(R_L / L_L + G_L / C_L)^{-1} \quad . \qquad (7.4 - 26b)$$

If $\tau = \infty$, $\Omega = \omega_C$ and $\upsilon_0 = c$ the telegraph equation (7.4 - 25) is identical with the *linear Klein-Gordon equation* (7.4 - 17), and when $\tau = \infty$ and $\Omega = 0$ it corresponds to the *Hertz equation* (7.4 - 4).

a) Dispersion relation and limit solutions

The complex representations of harmonic waves (7.4 - 1a) that obey the telegraph equation (7.4 - 25) are characterized by the dispersion relation

$$\upsilon_0^2 \beta^2 = \omega^2 + 2i(\omega/\tau) - \Omega^2 \quad . \qquad (7.4 - 27)$$

Three limits of this equation are of interest.

α) For *extremely high frequencies* the dispersion relation (7.4 - 27) takes the form

$$\upsilon_0 = \omega / \beta \quad \text{for} \quad \omega \to \infty \quad . \qquad (7.4 - 28a)$$

Consequently, phase and group velocity become equal and the dispersion vanishes

$$\upsilon = \upsilon_g = \upsilon_0 = const \quad . \tag{7.4 - 28b}$$

β) The *stationary solutions* of (7.4 - 25) are characterized by the dispersion relation

$$\beta = \pm i\Omega / \upsilon_0 = \pm i\alpha \quad \text{for} \quad \omega = 0 \quad . \tag{7.4 - 29a}$$

The corresponding general stationary solution has the form

$$u(z,t) = s(z) = C_1 cosh\,\alpha z + C_2 sinh\,\alpha z \tag{7.4 - 29b}$$

with arbitrary constants $C_{1,2}$.

γ) Spatially *homogeneous solutions* of (7.4 - 25) imply $\beta = 2\pi/\lambda = 0$. This condition reduces (7.4 - 27) to

$$\omega^2 + 2i(\omega / \tau) - \Omega^2 = 0 \quad . \tag{7.4 - 30a}$$

Accordingly, the telegraph equation (7.4 - 25) is transformed into the oscillation equation (2.2 - 28) of the damped harmonic oscillator

$$r_{tt} + (2 / \tau) r_t + \Omega^2 r = 0 \quad \text{with} \quad u(z,t) = r(t) \quad . \tag{7.4 - 30b}$$

According to Table 2.2 - 1 these oscillations are

$$\left.\begin{array}{l} \text{subcritically} \\ \text{critically} \\ \text{supercritically} \end{array}\right\} \text{damped for} \begin{cases} \Omega\tau > 1 \\ \Omega\tau = 1 \\ \Omega\tau < 1 \end{cases} . \tag{7.4 - 30c}$$

b) Ideal telegraph lines

Ideal telegraph and transmission lines transmit signals without distortion of their shapes except for damping. This goal is achieved by telegraph lines with the *critical damping* defined by (7.4 - 30c)

$$\Omega\tau = 1 \quad . \tag{7.4 - 31a}$$

According to (7.4 - 26b) this requirement is fulfilled by a telegraph line if

$$G_L L_L = C_L R_L \quad . \tag{7.4 - 31b}$$

In real telegraph cables C_L, R_L and G_L cannot be varied substantially. Therefore L_L has to be matched by introducing self-induction coils at periodic intervals.

The solution of telegraph equation (7.4 - 25) that fulfills (7.4 - 31a) can be written either as

$$u(z,t) = exp(-\alpha z)\, f(z - \upsilon_0 t) \quad \text{with} \quad \alpha = \Omega / \upsilon_0 \quad \text{or as} \tag{7.4 - 32a}$$

$$u(z,t) = exp(-\Omega t)\, g(z - \upsilon_0 t) \tag{7.4 - 32b}$$

where $f(z)$ and $g(z)$ are arbitrary functions that can be differentiated twice. The solutions (7.4 - 32a&b) are equivalent because

$$f(z - \upsilon_0 t) = exp[\alpha(z - \upsilon_0 t)]\, g(z - \upsilon_0 t) \quad . \tag{7.4 - 32c}$$

The general solution (7.4 - 32a) demonstrates that the original signal is damped yet *not distorted* during transmission through an ideal telegraph line. Thus *no proper dispersion* occurs. This phenomenon can be explained by taking into account the *dispersion relation* of the ideal telegraph lines as defined by (7.4 - 31a)

$$\beta = Re\,\beta + i\, Im\,\beta = (\omega / \upsilon_0) + i\,\alpha \quad . \tag{7.4 - 33a}$$

This relation implies equal phase and group velocity of the signals propagating along the ideal telegraph lines

$$\upsilon = \omega / Re\,\beta = \upsilon_g = d\omega / dRe\,\beta = \upsilon_0 \quad . \tag{7.4 - 33b}$$

Therefore, these lines show no dispersion in accordance with the general solution (7.4 - 32a).

c) Solutions on given initial conditions

The solution of the telegraph equation (7.4 - 25) with the initial conditions (7.4 - 3a&b) can be represented as follows [Bronstein et al. 1993 B]

$$u(z,t) = \frac{1}{2} F(z - \upsilon_0 t) + \frac{1}{2} F(z - \upsilon_0 t)$$
$$+ \frac{1}{2} \upsilon_0^{-1} \int_{z-\upsilon_0 t}^{z+\upsilon_0 t} ds [G(s) I_0(r) + F(s) \omega_0^2 t\, r^{-1} I_1(r)] \tag{7.4 - 34a}$$

with $\omega_0^2 = \Omega^2 - (1/\tau)^2 \quad , \quad r^2 = (\omega_0 / \upsilon_0)^2 [(s - z)^2 - \upsilon_0^2 t^2]$

and $u(z,0) = F(z) \quad , \quad u_t(z,0) = G(z) \quad .$

The integral includes modified Bessel functions of the first kind [Abramowitz & Stegun 1965 B]

$$I_k(x) = (i)^{-k} J_k(ix), \quad k = 0,1,2 \ldots \quad . \tag{7.4 - 34b}$$

In addition, it should be noticed that $\omega_0^2 > 0$ for subcritical damping and $\omega_0^2 < 0$ for supercritical damping.

7.4.5 Linear Diffusion Equation

The linear diffusion equation or second Fick equation [Alonso & Finn 1967 B, Kneubühl 1994 B]

$$u_t = D u_{zz} \quad \text{with} \quad D > 0 \tag{7.4 - 35}$$

describes diffusion and heat conduction. It is determined by the diffusion constant D [$m^2\ s^{-1}$].

The application of the complex representation (7.4 - 1a) of a harmonic wave to (7.4 - 35) yields the complex dispersion relation

$$\omega = -i D\beta^2 \quad \text{or} \quad \beta = (1+i)\sqrt{\omega / 2D} \quad . \tag{7.4 - 36}$$

a) Damped harmonic waves

Since the dispersion relation (7.4 - 36) is complex the corresponding harmonic waves are damped. The *traveling wave* has the form

$$u(z,t) = U\, exp\left[i\left(2\pi\frac{z}{\lambda_{\text{eff}}} - \omega t\right)\right] exp(-\gamma z)$$
$$\text{with} \quad \lambda_{\text{eff}} = 2\pi\sqrt{2D/\omega} \quad \text{and} \quad \gamma = \sqrt{\omega / 2D} \quad , \tag{7.4 - 37a}$$

while the *standing wave* can be represented by

$$u(z,t) = U\, exp(i\beta z) exp(-\vartheta t)$$
$$\text{with} \quad \beta = 2\pi / \lambda \quad \text{and} \quad \vartheta = D\beta^2 \quad . \tag{7.4 - 37b}$$

Heat or temperature waves $T(z, t)$ are also determined by the linear diffusion equation (7.4 - 35). They can be derived from the complex waves (7.4 - 37a) and (7.4 - 37b) by taking

$$T(z,t) = T_0 + Re\, u(z,t) \quad \text{and} \quad \Delta T = U \quad . \tag{7.4 - 38a}$$

Thus, (7.4 - 37a) is transformed into the *damped heat wave*

$$T(z,t) = T_0 + \Delta T\, cos\left(z\sqrt{\omega / 2D} - \omega t\right) exp\left(-z\sqrt{\omega / 2D}\right) \tag{7.4 - 38b}$$

where D is the thermal diffusion constant. The exponential decline of a harmonic *spatial temperature variation* is described by the modification (7.4 - 38a) of (7.4 - 37b)

$$T(z,t) = T_0 + \Delta T \cos \beta z \, exp\left(-D\beta^2 t\right) \quad . \tag{7.4 - 38c}$$

b) Solutions on given conditions

According to Duhamel [Duhamel 1833 Z, Webster & Szegö 1930 B] the solution of (7.4 - 35) for a given time-dependent *local excitation* (7.4 - 2a) can be written as

$$\begin{aligned} u(z,t) &= \frac{2}{\sqrt{\pi}} \int_0^\infty L(t - \frac{z^2}{4Ds^2}) exp\left(-s^2\right) ds \\ &= \frac{z}{\sqrt{4\pi D}} \int_0^\infty L(t-s) exp\left[-\frac{z^2}{4Ds}\right] s^{-3/2} ds \\ &= \frac{z}{\sqrt{4\pi D}} \int_{-\infty}^{t} L(s) exp\left[-\frac{z^2}{4D(t-s)}\right] (t-s)^{-3/2} ds \end{aligned} \tag{7.4 - 39a}$$

with $u(0,t) = L(t)$.

The solution of (7.4 - 35) under the initial condition (7.4 - 3a), which may represent an initial concentration or a temperature inhomogenuity, has the form [Margenau & Murphy 1956 B, Webster & Szegö 1938 B]

$$\begin{aligned} u(z,t) &= \int_{-\infty}^{+\infty} \Gamma(z,t,y) F(y) dy = \int_{-\infty}^{+\infty} \frac{1}{\sqrt{4\pi Dt}} exp\left[-\frac{(z-y)^2}{4Dt}\right] F(y) dy \\ &= \frac{1}{\sqrt{\pi}} \int_{-\infty}^{+\infty} exp\left(-y^2\right) F\left(z + \sqrt{4Dt}\, y\right) dy \end{aligned} \tag{7.4 - 39b}$$

with $u(z,0) = F(z)$.

In the first integral $\Gamma(z, t, y)$ represents the *Green function*. This equation describes the diffusion of momentary inhomogenuities of concentrations or heat.

c) Concentration or heat pole

The Green function $\Gamma(z, t, y)$ of (7.4 - 39b) is related to a well-known solution $u(z, t)$ of (7.4. - 35) that is called concentration or heat pole. It describes the diffusion of a Dirac δ-function inhomogenuity of concentration or heat at time $t = 0$

$$u(z,t) = \frac{F_0}{\sqrt{4\pi Dt}} exp\left[-\frac{z^2}{4Dt}\right] \tag{7.4 - 40a}$$

with $u(z,0) = F(z) = F_0 \, \delta(z)$.

The diffusion of the δ-function pole can be characterized by the increase of the halfwidth Δz of the Gauss function in (7.4 - 40a) with time t

$$\Delta z(t) = 4(\ell n 2)^{1/2} (Dt)^{1/2} \quad . \tag{7.4 - 40b}$$

7.4.6 Linearized Korteweg-de Vries Equation

The linearized Korteweg-de Vries (KdV) equation [Drazin 1983 B, Karpman 1975 B]

$$u_t + \upsilon_0 u_z + K u_{zzz} = 0 \quad \text{with} \quad \upsilon_0 \geq 0 ; K > 0 \tag{7.4 - 41}$$

comprehends the linear dispersive part of the nonlinear KdV equation [Dodd et al. 1982 B, Drazin 1983 B] discussed in Section 8.4.

The *dispersion relation* of (7.4 - 41) is real and cubic

$$\omega = \beta(\upsilon_0 - K\beta^2) \quad . \tag{7.4 - 42a}$$

The corresponding *phase and group velocity* are

$$\upsilon = \upsilon_0 [1 - K\beta^2] \tag{7.4 - 42b}$$

$$\upsilon_g = \upsilon_0 [1 - 3K\beta^2] < \upsilon \quad . \tag{7.4 - 42c}$$

Since υ_g is smaller than υ the linear KdV waves show *normal dispersion.*

a) Solution on given initial condition

The *Chauchy method* [Webster & Szegö 1930 B] yields a solution of (7.4 - 41) for the initial condition (7.4 - 3a) [Karpman 1975 B] with the integral

$$\begin{aligned} u(z,t) &= \int_{-\infty}^{+\infty} \Gamma(z,t,y) F(y) dy \\ &= \int_{-\infty}^{+\infty} (3Kt)^{-1/3} Ai([3Kt]^{-1/3}[(z-y) - \upsilon_0 t]) F(y) dy \end{aligned} \tag{7.4 - 43}$$

with $\quad u(z,0) = F(z) \quad .$

In this equation $\Gamma(z, t, y)$ is the Green function and $Ai(x)$ an Airy function [Abramowitz & Stegun 1965 B]. The KdV wave (7.4 - 43) propagates in the $+z$ direction.

b) Airy pulse

A special solution of (7.4 - 41) is the Airy pulse that represents the propagation and dispersion of a Dirac δ-pulse excitation at time $t = 0$

$$u(z,t) = F_0[3Kt]^{-1/3} Ai([3Kt]^{-1/3}[z - v_0 t]) \quad \text{with} \quad u(z,0) = F(z) = F_0\, \delta(z) \quad . \tag{7.4 - 44}$$

The Airy pulse (7.4 - 44) as solution of (7.4 - 41) is equivalent to the Green function of (7.4 - 43). This pulse is illustrated in Fig. 7.4 - 1. On its way it becomes weaker and longer because of dispersion.

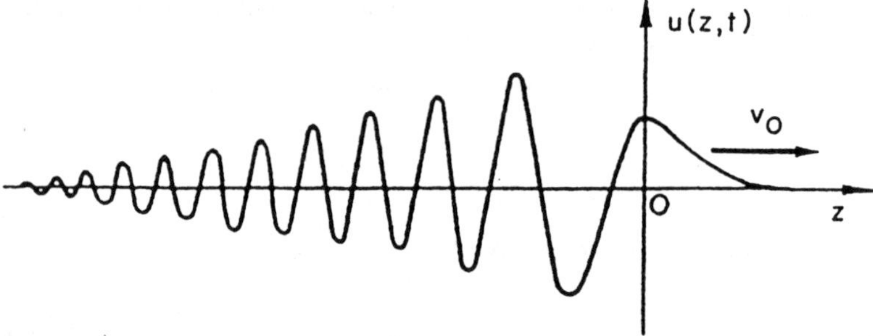

Fig. 7.4 - 1. Airy pulse (7.4 - 44)

7.4.7 Linear Schrödinger Equation

In classical nonrelativistic mechanics the *momentum-energy relation* of a free particle with mass in and momentum p in the z direction is

$$E = p^2 / 2m \quad . \tag{7.4 - 45}$$

The substitution of energy E and momentum p by the corresponding quantum mechanical operators (7.2 -18a&b) yields the *linear Schrödinger equation* of the free particle

$$i\hbar\psi_t = -\frac{\hbar^2}{2m}\,\psi_{zz} \tag{7.4 - 46}$$

where $\psi(z, t)$ represents the wave function. This equation can be reduced to its *normal form*

$$i\,u_t = -S\,u_{zz} \quad \text{with} \quad S = \hbar / 2m \tag{7.4 - 47}$$

Of the same form are the reduced *envelope equations* (7.3 - 35) and (7.3 - 38). The normal form (7.4 - 47) is converted into the *linear diffusion equation* (7.4 - 35) when $S = -i\,D$.

The *dispersion relation* (7.2 - 5b) of the normal form (7.4 - 47) is

$$\omega = S\beta^2 \quad . \tag{7.4 - 48a}$$

This relation implies the following *phase velocity* and *group velocity*

$$\upsilon = \omega / \beta = S\beta = \sqrt{S\,\omega} \quad , \tag{7.4 - 48b}$$

$$\upsilon_g = d\omega / d\beta = 2S\beta = 2\sqrt{S\omega} = 2\,\upsilon = \hbar\beta / m = p / m = \upsilon_m > \upsilon \quad . \tag{7.4 - 48c}$$

Since $\upsilon_g > \upsilon$ the Schrödinger waves show *anomalous dispersion*. Furthermore, (7.4 - 48c) demonstrates that the group velocity υ_g corresponds to the *particle velocity* υ_m.

a) Solution on given initial condition

The solution of (7.4 - 47) under an initial condition (7.4 - 3a) can be deduced from the corresponding solution (7.4 - 39b) of the linear diffusion equation (7.4 - 35) by assuming an imaginary diffusion constant $D = i\,S$. This procedure results in

$$u(z,t) = \int_{-\infty}^{+\infty} \Gamma(z,t,y) F(y)\, dy = \int_{-\infty}^{+\infty} \frac{1}{\sqrt{4\pi i\, St}}\, exp\left[i \frac{(z-y)^2}{4St} \right] F(y) dy \tag{7.4 - 49}$$

$$\text{with} \quad u(z,0) = F(z) \quad .$$

In the first integral $\Gamma(z, t, y)$ designates the Green function.

b) Wave packet

A particular solution (7.4 - 49) of (7.4 - 47) is the wave packet [Flügge 1990 B, Margenau & Murphy 1956 B]

$$u(z,t) = U\left[1 + 2i\frac{St}{a^2}\right]^{-1/2} exp - \left[\frac{z^2 - 2i\,a^2\beta_0 z + 2i\,a^2\, St\,\beta_0{}^2}{2\left(a^2 + 2i\,St\right)} \right] \tag{7.4 - 50}$$

$$\text{with} \quad u(z,0) = U\, exp\left(-\frac{z^2}{2a^2} \right) exp\left(i\,\beta_0\, z\right) \quad .$$

The propagation and broadening of this wave packet subjected to anomalous dispersion has to be considered from the point of view of *quantum mechanics* [Alonso & Finn 1970 B, Baym 1969 B, Fick 1968 B, Landau & Lifschitz 1969 B, Messiah 1960 B, 1969 B, 1990 B, Pauli 1950 B]. The wave function $\psi(z, t)$ of the

Schrödinger equation e.g. (7.4 - 46), determines the probability density $\rho(z, t)$ of finding a particle at position z and time t

$$\rho(z,t) = \psi^*(z,t)\,\psi(z,t) = |\psi(z,t)|^2 \quad . \tag{7.4 - 51}$$

In this equation the asterisk * indicates the complex conjugate.

Application of (7.4 - 51) to the wave function (7.4 - 50) yields the following probability density of the wave packet

$$\rho(z,t) = \psi^*(z,t)\,\psi(z,t) = u^*(z,t)u(z,t)$$

$$= \frac{U^*U}{\left[1+\left(2\,St/a^2\right)^2\right]^{1/2}}\,exp-\left[\frac{\left(z-\upsilon_g t\right)^2}{a^2\left[1+\left(2\,St/a^2\right)^2\right]}\right] \tag{7.4 - 52}$$

with $\upsilon_g = 2\,S\beta_0 = \hbar\beta_0/m = p/m = \upsilon_m$.

Thus, the wave packet moves with the group velocity υ_g (7.4 - 48c) that equals the particle velocity υ_m and broadens due to dispersion.

7.5 Electromagnetic Waves in Linear Media

As demonstrated in Section 7.1.2d electromagnetic waves in vacuum show no dispersion. This section is dedicated to electromagnetic waves in simple media that involve classical dispersion.

7.5.1 Characterization of the Media

The media considered are electromagnetically linear, neutral, homogeneous and isotropic. They are characterized by specific relations and functions.

By definition a medium is electrical neutral if its charge density ρ_{el} [As m^{-3}] is zero

$$\rho_{el}(\vec{r},t) = 0 \quad . \tag{7.5 - 1}$$

The electromagnetic properties of a linear homogenous and isotropic medium can be described by three equations and scalar functions of time t that relate electric field $\vec{E}$ [V m^{-1}], electric displacement $\vec{D}$ [As m^{-2}], current density $\vec{j}$ [A m^{-2}], magnetic field $\vec{H}$ [A m^{-1}] and magnetic induction $\vec{B}$ [T = Vs m^{-2}]. These equations represent the *convolutions* (3.2 - 24a) characteristic of the linear transfer systems discussed in Chapter 5

$$\vec{j}(\vec{r},t) = \sigma(t) * \vec{E}(\vec{r},t) \quad , \tag{7.5 - 2a}$$

$$\vec{D}(\vec{r},t) = \varepsilon(t) * \varepsilon_0 \vec{E}(\vec{r},t) \quad , \tag{7.5 - 2b}$$

$$\vec{B}(\vec{r},t) = \mu(t) * \mu_0 \vec{H}(\vec{r},t) \quad \text{with} \quad \varepsilon_0 \, \mu_0 = c^{-2} \quad . \tag{7.5 - 2c}$$

In these equations the asterisk * indicates the convolution, c the velocity of light in vacuum, ε_0 and μ_0 the electric permittivity and the magnetic permeability of vacuum. $\sigma(t)$, $\varepsilon(t)$ and $\mu(t)$ are the *impulse response functions* (5.1 - 5a&b) of the *electrical conductivity* σ [Ω^{-1} m^{-1} = AV^{-1} m^{-1}], the *relative permittivity* ε [1] and the *relative permeability* μ [1].

The principle of *causality* formulated in Section 5.1.3a requires according to (5.1 - 8a&b) that

$$\sigma(t<0) = \varepsilon(t<0) = \mu(t<0) = 0 \quad . \tag{7.5 - 3}$$

The following discussion makes use of *convolution calculus*, e.g.

$$f_1(t) * f_2(t) = f_2(t) * f_1(t) \quad , \tag{3.2 - 24c}$$

$$\begin{aligned} f_1(t) * f_2(t) * f_3(t) &= \{f_1(t) * f_2(t)\} * f_3(t) \\ &= f_1(t) * \{f_2(t) * f_3(t)\} \quad , \end{aligned} \tag{5.1 - 7c}$$

$$\frac{d}{dt}\{f_1(t) * f_2(t)\} = \frac{d}{dt} f_1(t) * f_2(t) = f_2(t) * \frac{d}{dt} f_2(t) \quad , \tag{5.1 - 6a}$$

$$curl\{f(t) * \vec{F}(\vec{r},t)\} = f(t) * curl\ \vec{F}(\vec{r},t) \quad , \tag{7.5 - 4a}$$

$$div\{f(t) * \vec{F}(\vec{r},t)\} = f(t) * div\ \vec{F}(\vec{r},t) \quad , \tag{7.5 - 4b}$$

$$\Delta\{f(t) * \vec{F}(\vec{r},t)\} = f(t) * \Delta\ \vec{F}(\vec{r},t) \quad . \tag{7.5 - 4c}$$

a) Maxwell equations

Electromagnetic phenomena in arbitrary media are determined by the *general Maxwell equations*

$$curl\ \vec{H} = \vec{j} + \frac{\partial}{\partial t}\vec{D} \quad , \tag{7.5 - 5a}$$

$$curl\,\vec{E} = -\frac{\partial}{\partial t}\vec{B} \quad , \tag{7.5 - 5b}$$

$$div\,\vec{D} = \rho_{\text{el}} \quad , \tag{7.5 - 5c}$$

$$div\,\vec{B} = 0 \quad . \tag{7.5 - 5d}$$

The introduction of (7.5 -1) and (7.5 - 2a-c) in these relations yield the Maxwell equations of the linear neutral, homogeneous and isotropic media

$$curl\,\vec{H} = \left\{\sigma + \varepsilon_0 \frac{\partial}{\partial t}\varepsilon\right\} * \vec{E} = \sigma * \vec{E} + \varepsilon_0\,\varepsilon * \frac{\partial}{\partial t}\vec{E} \quad , \tag{7.5 - 6a}$$

$$curl\,\vec{E} = -\mu_0 \frac{\partial}{\partial t}\mu * \vec{H} = -\mu_0\,\mu * \frac{\partial}{\partial t}\vec{H} \quad , \tag{7.5 - 6b}$$

$$div\,\vec{E} = 0 \quad , \tag{7.5 - 6c}$$

$$div\,\vec{H} = 0 \quad . \tag{7.5 - 6d}$$

The combination of (7.5 - 2a) and (7.5 - 6c) results in

$$div\,\vec{j} = 0 \tag{7.5 -7a}$$

in agreement with the general principle of *conservation of the electric charge*

$$\frac{\partial}{\partial t}\rho_{\text{el}} + div\,\vec{j} = 0 \tag{7.5 - 7b}$$

because according to (7.5 - 1) the charge density ρ_{el} in the media considered is zero.

b) Wave equation

The wave equation that determines the propagation of electromagnetic waves in linear neutral, homogeneous and isotropic media can be derived from the Maxwell equations (7.5 - 6a-d) by evaluating

$$curl\left(curl\,\vec{E}\right) = -\Delta\vec{E} = \ldots \quad , \quad \text{or} \quad curl\left(curl\,\vec{H}\right) = -\Delta\vec{E} = \ldots \quad .$$

The result is

$$c^2 \Delta \vec{E} \left[\text{or } \vec{H}\right] = \eta_{\text{op}}^2 \vec{E} \left[\text{or } \vec{H}\right]$$

$$= \mu * \left\{ \varepsilon_0^{-1} \frac{\partial}{\partial t} \sigma + \frac{\partial^2}{\partial t^2} \varepsilon \right\} * \vec{E} \left[\text{or } \vec{H}\right] \tag{7.5 - 8}$$

$$= \mu * \left\{ \varepsilon_0^{-1} \sigma \frac{\partial}{\partial t} \vec{E} \left[\text{or } \vec{H}\right] + \varepsilon * \frac{\partial^2}{\partial t^2} \vec{E} \left[\text{or } \vec{H}\right] \right\}$$

where η_{op}^2 represents the operator of the square of the complex index of refraction η.

7.5.2 Dispersion of Harmonic Waves

Electromagnetic waves oscillating harmonically with time t are governed by the *Fourier transform* (2.2 - 46a) of the basic wave equation (7.5 - 8). When performing the Fourier transformation of (7.5 - 8) one has to take the following rules of Appendix A.2.1 into account

$$\boldsymbol{F}\{x_1(t) * x_2(t)\} = x_1(\omega)\, x_2(\omega) \quad , \quad \text{and} \tag{7.5 - 9a}$$

$$\boldsymbol{F}\left\{ \frac{d^{\text{n}}}{dt} x(t) \right\} = -(i\omega)^{\text{n}}\, x(\omega) \quad . \tag{7.5 - 9b}$$

The result is the *modified wave equation*

$$\Delta \vec{E}(\vec{r},\omega) \left[\text{or } \vec{H}(\vec{r},\omega)\right] + \beta^2(\omega) \vec{E}(\vec{r},\omega) \left[\text{or } \vec{H}(\vec{r},\omega)\right] = 0 \tag{7.5 - 10a}$$

with the propagation constant

$$\beta(\omega) = \omega\, \eta(\omega)\, c^{-1} \quad , \tag{7.5 - 10b}$$

and the complex index of refraction

$$\eta(\omega) = \{n(\omega) + i\kappa(\omega)\} = \left\{ \mu(\omega) \left(\varepsilon(\omega) + i \frac{\sigma(\omega)}{\varepsilon_0 \omega} \right) \right\}^{1/2} \quad . \tag{7.5 - 10c}$$

By application of (7.5 - 10a) to the complex representation (7.4 - 1a) of a plane harmonic wave propagating in the z direction it can be demonstrated that (7.5 - 10b) represents the *dispersion relation*. The corresponding *phase and group velocities* are

$$\upsilon(\omega) = c\, n^{-1}(\omega) \quad , \tag{7.5 - 11a}$$

$$\upsilon_g(\omega) = c\left[n(\omega) + \omega\frac{d}{d\omega}n(\omega)\right]^{-1} \qquad (7.5 - 11b)$$

where $n(\omega)$ indicates the real part of the complex refractive index (7.5 - 10c).

a) Dispersion in dielectric media

By definition dielectric media are linear nonmagnetic insulators. Homogeneous and isotropic dielectric media fulfill the relations

$$\vec{j}(\vec{r},t) = \vec{0} \quad \text{and} \quad \sigma(t) = 0 \quad , \qquad (7.5 - 12a)$$

$$\vec{D}(\vec{r},t) = \varepsilon(t) * \varepsilon_0\, \vec{E}(\vec{r},t) \quad , \qquad (7.5 - 12b)$$

$$\vec{B}(\vec{r},t) = \mu_0\, \vec{H}(\vec{r},t) \quad . \qquad (7.5 - 12c)$$

For any medium, linear as well as nonlinear, the dielectric displacement $\vec{D}$ can be split into the contribution from the vacuum and that from the medium

$$\vec{D}(\vec{r},t) = \varepsilon_0\, \vec{E}(\vec{r},t) + \vec{P}(\vec{r},t) \qquad (7.5 - 13)$$

where $\vec{P}(\vec{r},t)$ indicates the *dielectric polarization*. For homogeneous and isotropic dielectric media $\vec{P}(\vec{r},t)$ can be represented by

$$\vec{P}(\vec{r},t) = \chi_{\mathrm{el}}(t) * \varepsilon_0\, \vec{E}(\vec{r},t) \quad . \qquad (7.5 - 14a)$$

$\chi_{\mathrm{el}}(t)$ is the impulse response function of the *electric susceptibility*. Taking into account (7.5 - 12b), (7.5 - 13) and (7.5 - 14a) the corresponding impulse response function of the permittivity can be written as

$$\varepsilon(t) = \delta(t) + \chi_{\mathrm{el}}(t) \quad . \qquad (7.5 - 14b)$$

The Fourier transformation (2.2 - 46a) of this relation yields

$$\varepsilon(\omega) = 1 + \chi_{\mathrm{el}}(\omega) \quad . \qquad (7.5 - 15)$$

For homogeneous and isotropic dielectric media the permittivity $\varepsilon(\omega)$ determines the index of refraction $\eta(\omega)$ by *Debye's law*

$$\eta(\omega) = n(\omega) + i\kappa(\omega) = (\varepsilon(\omega))^{1/2} \quad . \qquad (7.5 - 16)$$

This law can be derived from (7.5 - 10c) and (7.5 - 12a-c).

b) Debye dispersion

The interaction of electromagnetic radiation with dielectric media at audio, radio and microwave frequencies υ that cover the range from 0 Hz to 10^{12} Hz can often be described by the Debye dispersion discussed in Section 2.2.5b. This dispersion can be represented by the impulse response function of the electric susceptibility

$$\chi_{el}(t) = \chi_0 \frac{t}{\tau^2} exp(-t/\tau)\, H(t) \quad . \tag{7.5 - 17}$$

The Fourier transformation (2.2 - 46a) of this function results in the electric susceptibility

$$\chi_{el}(\omega) = \chi_0 (1 - i\,\omega\,\tau)^{-2} \quad . \tag{7.5 - 18a}$$

The static permittivities $\varepsilon_{st} = \varepsilon(\omega = 0)$ of dielectric media are usually considerably larger than unity, e.g. $\varepsilon_{st}(H_2O) = 81.6$ and $\varepsilon_{st}(NaCl) = 5.9$. Therefore, one can assume

$$\chi_0 >> 1 \quad \text{and} \quad \chi_{el}(\omega) \approx \varepsilon(\omega) \quad . \tag{7.5 - 18b}$$

The application of Debye's law (7.5 - 16) to (7.5 - 18a&b) yields the complex refractive index

$$\begin{aligned} \eta(\omega) &= n(\omega) + i\kappa(\omega) \approx n_0 (1 - i\,\omega\,\tau)^{-1} \\ \text{with} \quad n_0 &= \chi_0^{1/2} \quad , \\ n(\omega) &= n_0 \left[1 + (\omega\tau)^2\right]^{-1} = n_0\, DR(\omega;\tau) \quad , \\ \kappa(\omega) &= n_0\, \omega\tau \left[1 + (\omega\tau)^2\right]^{-1} = n_0\, DA(\omega;\tau) \quad . \end{aligned} \tag{7.5 - 19}$$

The functions $DR(\omega;\ \tau)$ and $DA(\omega;\ \tau)$ are defined by (2.2 - 48) and illustrated in Fig. 2.2 - 9.

c) Lorentz line shape and dispersion

In the infrared, visible and ultraviolet at frequencies ν above 10^{12} Hz many narrow spectral absorptions and emissions can be approximated by the Lorentz line shape and dispersion discussed in Section 2.2.5c. This line shape and the corresponding dispersion can be modeled with the aid of the following impulse response function of the electric susceptibility

$$\begin{aligned} &\chi_{el}(t) = \chi_0\, \tau^{-1} sin\, \omega_0 t\, exp(-t/\tau)\, H(t) \\ &\text{with} \quad \omega_0 \tau >> 1 \quad \text{and} \quad |\chi_0| << 1 \quad . \end{aligned} \tag{7.5 - 20}$$

For $\omega \approx \omega_0 > 0$ the Fourier transform (2.2 - 46a) of this function can be approximated by

$$\chi_{el}(\omega) \approx \frac{i}{2}\chi_0[1 - i(\omega - \omega_0)\tau]^{-1} \quad . \tag{7.5 - 21}$$

The terms with $(\omega + \omega_0)$ are neglected in this approximation. Because $|\chi_0|$ is small according to (7.5 - 20), the complex index of refraction $\eta(\omega)$ is approximately

$$\eta(\omega) = (1 + \chi_{el}(\omega))^{1/2} \approx 1 + \frac{1}{2}\chi_{el}(\omega) \quad \text{or} \tag{7.5 - 22}$$

$$\eta(\omega) = n(\omega) + i\kappa(\omega) \approx 1 + \frac{i}{4}\chi_0[1 - i(\omega - \omega_0)\tau]^{-1}$$

with

$$n(\omega) = \frac{1}{4}\chi_0(\omega_0 - \omega)\left[1 + (\omega_0 - \omega)^2\tau^2\right]^{-1} = +\frac{1}{4}\chi_0\, LD(\omega;\tau,\omega_0) \quad , \tag{7.5 - 23}$$

$$\kappa(\omega) = \frac{1}{4}\chi_0\left[1 + (\omega_0 - \omega)^2\tau^2\right]^{-1} = +\frac{1}{4}\chi_0\, LL(\omega;\tau,\omega_0) \quad .$$

The functions *LL* and *LD* are the normalized Lorentz line shape and dispersion defined by (2.2 - 50a-c) and illustrated in Fig. 2.2 - 10.

d) Kramers-Kronig relations

The Kramers-Kronig relations [Kneubühl 1989 J, Kramers & Kronig 1919 J, Kronig 1926, Mills 1991 B, Römer 1994 B] connect the imaginary part $\chi_{el}''(\omega)$ with the real part $\chi_{el}'(\omega)$ of the electric susceptibility $\chi_{el}(\omega)$

$$\chi_{el}(\omega) = Re\,\chi_{el}(\omega) + i\,Im\,\chi_{el}(\omega) = \chi_{el}'(\omega) + i\chi_{el}''(\omega) \tag{7.5 - 24}$$

Thus, they also relate imaginary and real part of the relative electric permittivity $\varepsilon(\omega)$.

These relations can be derived by writing the impulse response function of the electric susceptibility in the same form as (7.5 - 17) and (7.5 - 20)

$$\chi_{el}(t) = \alpha(t)\, H(t) \tag{7.5 - 25a}$$

with the assumption that $\alpha(t)$ is antisymmetric

$$\alpha(-t) = -\alpha(t) \quad . \tag{7.5 - 25b}$$

This assumption is permitted because $\alpha(t)$ of (7.5 - 25a) is defined only for $t > 0$.

The Fourier transformation (2.2 - 46a) of (7.5 - 25a) yields the electric susceptibility $\chi_{el}(\omega)$. Because $\alpha(t)$ is asymmetric according to (7.5 - 25b) its Fourier transform $\alpha(\omega)$ is imaginary and therefore fulfills

$$Re\,\alpha(\omega) = 0 \quad . \tag{7.5 - 26a}$$

According to Appendix A.2 the Fourier transform of (7.5 - 25a) is

$$\begin{aligned}\chi_{el}(\omega) &= \frac{1}{2\pi} H(\omega) * \alpha(\omega) = \left(\frac{i}{2\pi\omega} + \frac{1}{2}\delta(\omega)\right) * \alpha(\omega)\\ &= \left[-\frac{1}{\pi\omega} * \left(-\frac{i}{2}\alpha(\omega)\right)\right] + i\left(-\frac{i}{2}\alpha(\omega)\right)\\ &= \chi_{el}'(\omega) + i\chi_{el}''(\omega)\end{aligned} \tag{7.5 - 26b}$$

$$\text{where} \quad \chi_{el}''(\omega) = -\frac{i}{2}\alpha(\omega) \tag{7.5 - 26c}$$

because $\alpha(\omega)$ is imaginary according to (7.5 - 26a).

Equations (7.5 - 26b&c) imply the Kramers-Kronig relations

$$\begin{aligned}\chi_{el}'(\omega) &= \left[-\frac{1}{\pi\omega} * \chi_{el}''(\omega)\right] = Hi\,\chi_{el}''(\omega) \quad ,\\ \chi_{el}''(\omega) &= \left[+\frac{1}{\pi\omega} * \chi_{el}'(\omega)\right] = (Hi)^{-1}\chi_{el}'(\omega)\end{aligned} \tag{7.5 - 27a}$$

where Hi indicates the *Hilbert transformation* listed in Appendix A.4.4. The integral form of the Kramers-Kronig relations is

$$\begin{aligned}\chi_{el}'(\omega) &= \frac{1}{\pi} P \int_{-\infty}^{+\infty} \frac{\chi_{el}''(\omega')d\omega'}{\omega' - \omega} \quad ,\\ \chi_{el}''(\omega) &= -\frac{1}{\pi} P \int_{-\infty}^{+\infty} \frac{\chi_{el}'(\omega')d\omega'}{\omega' - \omega}\end{aligned} \tag{7.5 - 27b}$$

where P represents Cauchy's principal part of the integral.

Pairs of functions related by the Hilbert transformation and Kramers-Kronig relations are listed in Appendix A.4.4.

7.5.3 Transient Waves

Transient electromagnetic waves in the media which are defined in Section 7.5.1 can be determined with the aid of the *Laplace transformation* (3.2 - 23a) of the basic wave equation (7.5 - 8)

$$c^2\Delta\vec{E}(\vec{r},p)=\varepsilon(p)\mu(p)\left[p^2\vec{E}(\vec{r},p)-p\,\vec{E}(\vec{r},t=+0)-\frac{\partial}{\partial t}\vec{E}(\vec{r},t=+0)\right]$$
$$+\varepsilon_0^{-1}\sigma(p)\mu(p)\left[p\vec{E}(\vec{r},p)-\vec{E}(\vec{r},t=+0)\right] \qquad (7.5 - 28)$$

The same equation is valid for the magnetic field $\vec{H}(\vec{r},t)$, respectively $\vec{H}(\vec{r},p)$. Furthermore, it is assumed that the electric and the magnetic field obey the restrictions

$$\vec{E}(\vec{r},t\leq 0)=\frac{\partial}{\partial t}\vec{E}(\vec{r},t\leq 0)=\vec{0} \quad ,$$
$$\vec{H}(\vec{r},t\leq 0)=\frac{\partial}{\partial t}\vec{H}(\vec{r},t\leq 0)=\vec{0} \quad . \qquad (7.5 - 29)$$

Thus, the electromagnetic waves considered represent causal systems in the sense of (3.1 - 6). In addition, this assumption avoids specific problems of the Laplace transformation at time $t = 0$.

The simplest transient waves are those in *media without dispersion*. These can be characterized by the following impulse response functions (7.5 - 2a-c)

$$\sigma(t)=0 \quad , \quad \varepsilon(t)=\varepsilon_{\mathrm{r}}\delta(t-0) \quad , \quad \mu(t)=\mu_{\mathrm{r}}\delta(t-0) \quad . \qquad (7.5 - 30a)$$

In these equations $\delta(t - 0)$ describes a Dirac delta pulse immediately after time $t = 0$. According to Appendix A.3.2 the Laplace transforms of these functions are

$$\sigma(p)=0 \quad , \quad \varepsilon(p)=\varepsilon_{\mathrm{r}} \quad , \quad \mu(p)=\mu_{\mathrm{r}} \quad . \qquad (7.5 - 30b)$$

The application of these functions to the Laplace transform (7.5 - 28) of the basic wave equation yields

$$c^2\Delta\vec{E}(\vec{r},p)=n^2\left[p^2\vec{E}(\vec{r},p)-p\,\vec{E}(\vec{r},t=+0)-\frac{\partial}{\partial t}\vec{E}(\vec{r},t=+0)\right]$$
$$\text{with} \quad n^2=\varepsilon_{\mathrm{r}}\mu_{\mathrm{r}} \quad . \qquad (7.5 - 31)$$

For a plane wave $\vec{E}(z,t)$ traveling in the z direction and with the initial conditions

$$\vec{E}(z,t=+0)=\frac{\partial}{\partial t}\vec{E}(z,t=+0)=\vec{0} \qquad (7.5 - 32a)$$

equation (7.5 - 31) is reduced to

$$c^2\frac{\partial^2}{\partial z^2}\vec{E}(z,p)=n^2p^2\vec{E}(z,p) \qquad (7.5 - 32b)$$

with the solution

$$\vec{E}(z,p) = \vec{E}_1(p)\,exp(-pzn/c) + \vec{E}_2(p)\,exp(+pzn/c) \quad , \qquad (7.5 - 33a)$$

where $\vec{E}_1(p)$ and $\vec{E}_2(p)$ are arbitrary functions of p. The inverse Laplace transformation (3.2 - 23b) yields according to Appendix A.3.1

$$\vec{E}(z,t) = \vec{E}_1(t - zn/c)\,H(t - zn/c) \quad \text{for} \quad t > 0 \quad , \qquad (7.5 - 33b)$$

where $\vec{E}_1(t)$ indicates an arbitrary function. This equation represents a wave without dispersion propagating in the z direction with the velocity c/n.

7.6 Linear Waves in Periodic Media and Structures

The concept of waves comprises the propagation of excitations or distortions in discrete structures as well as in continuous media. Examples of discrete structures are periodic linear chains and crystal lattices. The theory of waves in these structures [Brillouin 1946 B] is equivalent to that of electromagnetic waves in continuous periodic media and to the wave mechanics of a particle in a periodic potential, e.g. an electron in a semiconductor. Characteristic of waves in periodic media and structures are periodic dispersion relations, Brillouin zones and frequency or energy gaps.

7.6.1 Infinite Chains with Identical Springs and Masses

Infinite chains with identical springs and masses are used as models for lattice vibrations in homoeopolar crystals, e.g. germanium, silicon and diamond [Ashcroft & Mermin 1976 B, Blakemore 1974 B, Brillouin 1946 B, Kittel 1963 B, 1971 B, Wang 1966 B, Ziman 1964 B]. This type of chain is illustrated in Fig. 7.6 - 1.

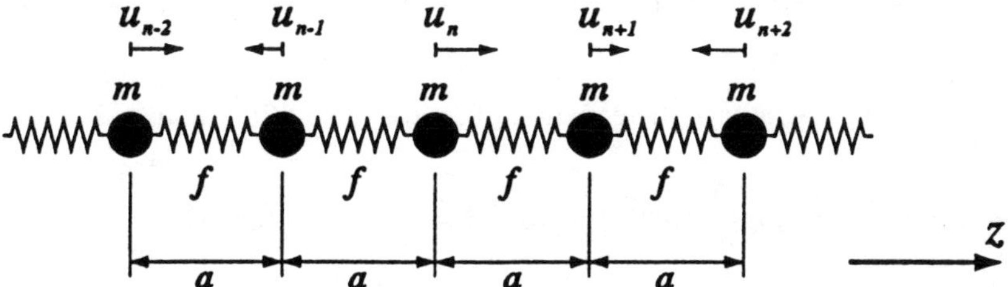

Fig. 7.6 - 1. Infinite chain with identical springs and masses

In equilibrium all masses m of this chain have the same distance a. They are connected with identical springs characterized by the linear force constant f. The main interest is in the *longitudinal* [L] *waves* of this chain. They correspond to the propagation of the longitudinal displacements of the masses m from their equilibrium positions. Each mass m is assigned the number n of its equilibrium position $z_e(n) = na$. As a consequence its longitudinal displacement is described by the function

$$u_n(t) = u(z_e = na, t) \quad \text{with} \quad n = 0, \pm 1, \pm 2, \ldots \quad . \tag{7.6 - 1a}$$

Accordingly, its momentary position can be represented by

$$z_n(t) = z(z_e = na, t) = na + u(z_e = na, t) = na + u_n(t) \quad . \tag{7.6 - 1b}$$

The *dynamics* of the longitudinal displacements $u_n(t)$ is determined by Newton's law of mechanics. In this context it should be noticed that the springs exert force only between the nearest neighbors among the masses. This can be assumed approximately for most forces acting between atoms or ions in crystals. On these conditions the longitudinal displacements $u_n(t)$ of the masses m obey the following *equations of motion*

$$m\ddot{u}_n = -f(u_n - u_{n-1}) + f(u_{n+1} - u_n) \quad \text{or} \tag{7.6 - 2a}$$

$$4\omega_0^{-2} u_{tt}(na, t) = u(na - a, t) - 2u(na, t) + u(na + a, t)$$
$$\text{with} \quad \omega_0 = 2\sqrt{f/m} \quad . \tag{7.6 - 2b}$$

The dispersion relation of the harmonic waves on the chain can be derived with the aid of the *translation operator* defined by

$$\boldsymbol{T}(\boldsymbol{a})\{u(z,t)\} = u(z + a, t) \quad . \tag{7.6 - 3a}$$

Another representation of this operator is based on the Taylor series

$$\boldsymbol{T}(\boldsymbol{a})\{u(z,t)\} = \sum_{r=0}^{\infty} \frac{1}{r!}\left(a\frac{\partial}{\partial z}\right)^r u(z,t) = exp\left(a\frac{\partial}{\partial z}\right)u(z,t) \quad . \tag{7.6 - 3b}$$

The application of this operator to the equation of motion (7.6 - 2b) yields

$$4\omega_0^{-2}\frac{\partial^2}{\partial t^2} u(z,t) = exp\left(a\frac{\partial}{\partial z}\right)u(z,t) + exp\left(-a\frac{\partial}{\partial z}\right)u(z,t) - 2u(z,t)$$
$$\text{with} \quad z = na\,,\, n = 0, \pm 1, \pm 2, \ldots \tag{7.6 - 4}$$

A *longitudinal harmonic wave* of the chain has the complex form

$$u(z,t) = u(na,t) = U\,exp[i(\beta z - \omega t)] = U\,exp[i(\beta\, na - \omega t)] \quad . \tag{7.6 - 5}$$

Its application to (7.6 - 4) results in the *dispersion relation* illustrated in Fig. 7.6 - 2.

$$4(\omega / \omega_0)^2 = 2 - exp(+i\beta a) - exp(-i\beta a) \quad \text{or} \tag{7.6 - 6a}$$

$$\omega = \omega_0 \, sin(\beta a / 2) \, sign \beta > 0 \quad . \tag{7.6 - 6b}$$

This dispersion relation is *periodic* in the propagation constant β because the chain is periodic in z. The spatial period a of the chain entails a period $2\pi/a$ of the dispersion relation. A single period of the dispersion relation is called *Brillouin zone* [Brillouin 1946 B]. Fig. 7.6 - 2 shows the central or zeroth Brillouin zone of the dispersion relation (7.6 - 6a&b) that covers the range $-\pi/a < \beta \leq \pi/a$.

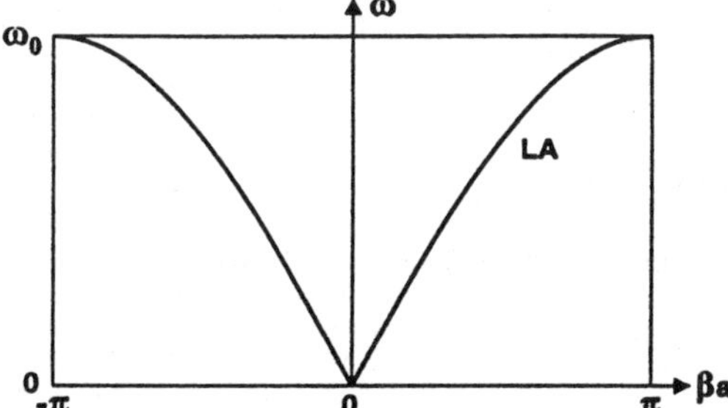

Fig. 7.6 - 2. Dispersion relation $\omega(\beta)$ of a chain with identical springs and masses (7.6 - 6b)

Phase and group velocity of the waves of the chain can be evaluated from (7.6 - 6b) by making use of (7.3 - 5) and (7.3 - 13)

$$\upsilon = \omega / \beta = \upsilon_0 \frac{sin(\beta a / 2)}{\beta a / 2} sign \beta \quad \text{with} \quad \upsilon_0 = \frac{1}{2} \omega_0 a = \sqrt{f / m} \quad , \tag{7.6 - 7a}$$

$$\upsilon_g = d\omega / d\beta = \upsilon_0 \, cos(\beta a / 2) \, sign \beta \quad . \tag{7.6 - 7b}$$

These velocities are plotted in Fig. 7.6 - 3 for $0 \leq \beta \leq \pi/a$. The group velocity υ_g is zero for $\beta = \pi/a$. Since $|\upsilon_g| \leq |\upsilon|$ the longitudinal waves of the chain exhibit *normal dispersion*.

The circular frequency ω of the waves varies from $\omega = 0$ for $\beta = 0$ to $\omega = \omega_0$ for $\beta = \pi/a$. This phenomenon is related to the phase shift $\Delta\Phi$ between the oscillations of two adjacent masses m

$$\Delta\Phi = \beta a \quad . \tag{7.6 - 8}$$

According to this equation two adjacent masses m of the chain oscillate in phase for $\beta = 0$ and in opposition for $\beta = \pi/a$. For $\beta = 0$ the spring between the two adjacent masses m is not stressed, while for $\beta = \pi/a$ it is subjected to maximum stress.

Therefore, one finds $\omega = 0$ for $\beta = 0$ and $\omega = \omega_0$ for $\beta = \pi/a$. Since for $\omega \approx 0$ the oscillations of two adjacent masses m are almost in phase, the waves of the chain with equal springs and masses are called *longitudinal acoustic* or LA waves.

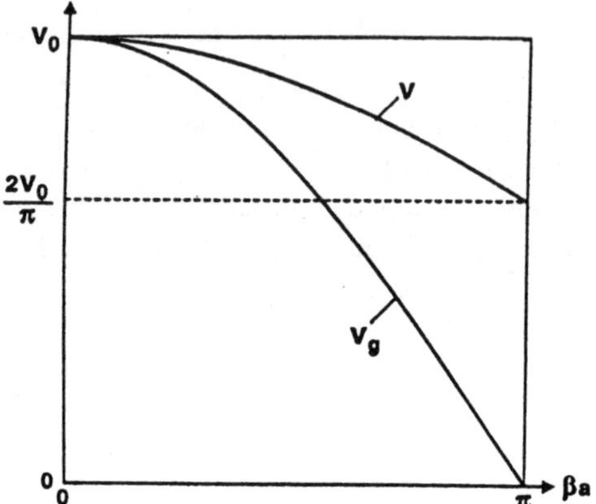

Fig. 7.6 - 3. Phase velocity $v(\beta)$ and group velocity $v_g(\beta)$ of a chain with identical springs and masses (7.6 - 7a&b)

7.6.2 Infinite Chains with Identical Springs and Alternating Masses

Infinite chains of equal springs and alternating masses serve as simple models for studying the dynamics of the lattices of heteropolar crystals, e.g. sodium chloride, potassium chloride and lithium fluoride. This type of chain is illustrated in Fig. 7.6 - 4.

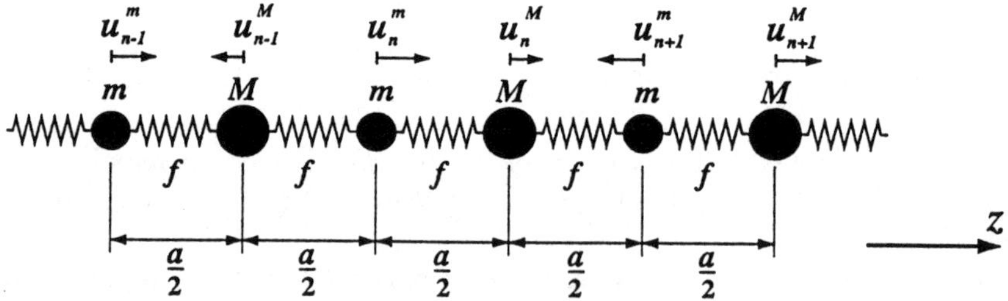

Fig. 7.6 - 4. Infinite chain with identical springs and alternating masses

The chain consists of identical springs with the force constant f and two different masses m and $M > m$ arranged alternatively. If the chain is in equilibrium, then the masses have the distance $a/2$. Consequently, the *period of the chain is a.*
The masses m and M are labeled by their equilibrium positions $z_e = na$ with $n = 0$, ±1, ±2, ... and $z_e = (2n + 1)a/2$ with $n = 0$, ±1, ±2, ... Their momentary

longitudinal displacements from the equilibrium positions are described by the functions

$$u_n^m(t) = u(z_e = na, t) \quad \text{and} \quad u_n^M(t) = u(z_e = (n+\frac{1}{2})a, t) \quad . \tag{7.6 - 9a}$$

The corresponding momentary positions are

$$\begin{aligned} z_n^m(t) &= na + u(z_e = na, t) \quad \text{and} \\ z_n^M(t) &= (n+\frac{1}{2})a + u(z_e = (n+\frac{1}{2})a, t) \quad . \end{aligned} \tag{7.6 - 9b}$$

The dynamics of the longitudinal displacements (7.6 - 9a) is governed by Newton's law of mechanics on one hand, and by the assumption that the springs exert forces only between adjacent masses m and M on the other hand. This yields the following *equations of motion* for these displacements

$$\begin{aligned} m\,\ddot{u}_n^m &= -f\left(u_n^m - u_{n-1}^M\right) - f\left(u_n^m - u_n^M\right) \quad \text{and} \\ M\,\ddot{u}_n^M &= -f\left(u_n^M - u_n^m\right) - f\left(u_n^M - u_{n+1}^m\right) \quad \text{with} \quad M > m \quad . \end{aligned} \tag{7.6 - 10a}$$

These equations can be transformed into

$$\begin{aligned} 2\omega_1^{-2} u_{tt}(na,t) &= u((n-\frac{1}{2})a,t) + u((n+\frac{1}{2})a,t) - 2u(na,t) \quad , \\ 2\omega_2^{-2} u_{tt}((n+\frac{1}{2})a,t) &= u(na,t) + u((n+1)a,t) - 2u((n+\frac{1}{2})a,t) \\ \text{with} \quad \omega_1 &= +\sqrt{2f/m} \geq \omega_2 = +\sqrt{2f/M} \quad . \end{aligned} \tag{7.6 - 10b}$$

The introduction of the translation operator (7.6 - 3a&b) yields the final modification of these coupled equations

$$\begin{aligned} 2\omega_1^{-2} \frac{\partial^2}{\partial t^2} u(z,t) &= \left\{1 + exp\left(-a\frac{\partial}{\partial z}\right)\right\} u(z+\frac{a}{2},t) - 2u(z,t) \quad , \\ 2\omega_2^{-2} \frac{\partial^2}{\partial t^2} u(z+\frac{a}{2},t) &= \left\{1 + exp\left(+a\frac{\partial}{\partial z}\right)\right\} u(z,t) - 2u(z+\frac{a}{2},t) \\ \text{with} \quad z &= na, n = 0, \pm 1, \pm 2, \ldots \quad . \end{aligned} \tag{7.6 - 10c}$$

In a *longitudinal harmonic wave* of the chain considered the masses m and M have different amplitudes U^m and U^M

$$u(z,t) = U^{\mathrm{m}}\, exp\left[i(\beta z - \omega t)\right] \quad \text{and}$$

$$u(z+\frac{a}{2},t) = U^{\mathrm{M}}\, exp\left[i(\beta z - \omega t)\right] \tag{7.6 - 11}$$

$$\text{with} \quad z = na,\, n = 0, \pm 1, \pm 2, \ldots \quad .$$

The introduction of this wave ansatz into (7.6 - 10c) yields a system of linear equations for the amplitudes U^{m} and U^{M}

$$\begin{aligned} 2\left\{(\omega/\omega_1)^2 - 1\right\}U^{\mathrm{m}} + \left\{1 + exp(-i\beta a)\right\}U^{\mathrm{M}} &= 0 \\ \left\{1 + exp(+i\beta a)\right\}U^{\mathrm{m}} + 2\left\{(\omega/\omega_2)^2 - 1\right\}U^{\mathrm{M}} &= 0 \end{aligned} \tag{7.6 - 12}$$

This system of equations has nontrivial solutions only if the determinant of the coefficients is zero. This condition yields the *secular equation*

$$\begin{aligned} \left[(\omega/\omega_1)^2 - 1\right]\left[(\omega/\omega_2)^2 - 1\right] &= \frac{1}{4}\left[1 + exp(-i\beta a)\right]\left[1 + exp(+i\beta a)\right] \\ &= cos^2(\beta a/2) \end{aligned} \tag{7.6 - 13a}$$

which constitutes the *dispersion relation* of the waves of the chain under consideration. It can be transformed into a quadratic equation for ω^2

$$D(\beta,\omega) = \omega^4 - \left(\omega_1^2 + \omega_2^2\right)\omega^2 + \omega_1^2\omega_2^2 sin^2(\beta a/2) = 0 \quad . \tag{7.6 - 13b}$$

Its solution is

$$\omega^2 = \omega^2(\beta) = \frac{\omega_1^2 + \omega_2^2}{2} \pm \frac{1}{2}\left[\omega_1^4 + \omega_2^4 + 2\omega_1^2\omega_2^2 cos\beta a\right]^{1/2} \quad . \tag{7.6 - 13c}$$

The dispersion relation (7.6 - 13a-c) is illustrated in Fig. 7.6 - 5. According to (7.6 - 13c) it has two branches, a *longitudinal acoustic (LA) branch* characterized by the minus sign and a *longitudinal optical (LO) branch* characterized by the plus sign. The labels "acoustic" and "optical" have been chosen by taking account of the real ratio r of the amplitudes and the phase difference $\Delta\Phi$ of the displacements of adjacent mass m and M. These are determined by (7.6 - 12)

$$U^{\mathrm{M}}/U^{\mathrm{m}} = r\, exp(i\Delta\Phi) = \frac{1}{2}\frac{1 + exp(i\beta a)}{1 - (\omega/\omega_2)^2} \quad . \tag{7.6 - 14a}$$

For long wavelengths with $\beta = 2\pi/\lambda \approx 0$ the ratios r and phase differences $\Delta\Phi$ of the branches *LA* and *LO* differ considerably

$$LA: \quad r(\beta = 0, \omega = 0) = 1 \quad , \qquad \Delta\Phi(\beta = 0, \omega = 0) = 0 \quad ,$$

$$LO: \quad r(\beta = 0, \omega = \omega_0) = \frac{m}{M} \quad , \quad \Delta\Phi(\beta = 0, \omega = \omega_0) = \pi \quad . \tag{7.6 - 14b}$$

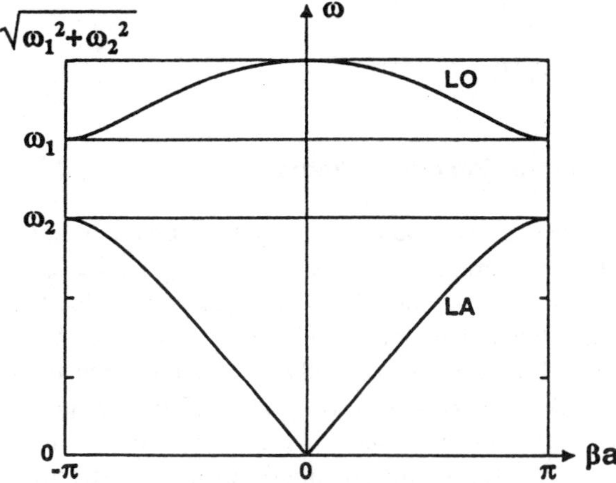

Fig. 7.6 - 5. Dispersion relation $\omega(\beta)$ of a chain with identical springs and alternating masses

In the high-frequency *LO* branch with $\beta \approx 0$ adjacent masses m and M oscillate in opposition with $\Delta\phi \approx \pi$. In ionic heteropolar crystals, e.g. sodium chloride, the two different masses m and M correspond to different ions with opposite electric charges. If they oscillate in opposition they represent an oscillating electric dipole that absorbs or emits optical, i.e. electromagnetic radiation. On the contrary adjacent different masses m and M and the related ions of opposite electric charges oscillate in phase in the low-frequency *LA* branch with $\beta \approx 0$. Thus, there results no oscillating electric dipole interacting with electromagnetic radiation. The *LA* branch represents sound in crystals, while the *LO* branch is responsible for the interaction of ionic crystals with "light", i.e. electromagnetic radiation in the infrared frequency range.

The dispersion relation (7.6 - 13c) and Fig. 7.6 - 5 reveal a *frequency gap* between the *LA* and the *LO* branch. It is determined by the equation

$$\omega_{\rm LO}^2(\beta = 0) - \omega_{\rm LA}^2(\beta = 0) = \left[\omega_1^4 + \omega_2^4 + 2\omega_1^2\omega_2^2 cos\,\beta\, a\right]^{1/2} \tag{7.6 - 15}$$

This gap closes at $\beta = \pm\pi / a$ for $\omega_1 = \omega_2$, respectively for $m = M$.

Since the chain considered exhibits the geometrical period a, the corresponding *dispersion relation* (7.6 - 13a-c) *is periodic* in β with the period $2\pi/a$. As central or zeroth *Brillouin zone* one usually chooses the range $-\pi/a < \beta \leq + \pi/a$.

Phase and group velocity of the waves in consideration can be evaluated by application of (7.3 - 5) and (7.3 - 13) to (7.6 - 13b&c). Essential are the velocities at the limits $\beta = 0$ and $\beta = \pi/a$

$$LA: \quad \upsilon(\beta=0)=\upsilon_g(\beta=0)=\frac{a}{2}\omega_1\,\omega_2\left(\omega_1^2+\omega_2^2\right)^{-1/2}$$
$$LO: \quad \upsilon(\beta=0)=\infty \quad , \quad \upsilon_g(\beta=0)=0 \quad , \tag{7.6 - 16a}$$

$$LA: \quad \upsilon(\beta=\pi/a)=\omega_2 a/\pi \quad , \quad \upsilon_g(\beta=\pi/a)=0 \quad ,$$
$$LO: \quad \upsilon(\beta=\pi/a)=\omega_1 a/\pi \quad , \quad \upsilon_g(\beta=\pi/a)=0 \quad . \tag{7.6 - 16b}$$

7.6.3 Electromagnetic Waves in Periodic Media

The behavior of electromagnetic waves in periodic media and structures, e.g. hollow metallic waveguides with periodic corrugations and cross-sections, periodic optical fibers and waveguides, distributed Bragg reflectors (DBR) and distributed feedback (DFB) lasers [Gnepf & Kneubühl 1986 J, Kneubühl 1993 B, Kneubühl & Sigrist 1995 B, Kogelnik & Shank 1971 J, Preiswerk et al. 1984 J] is determined by the Bragg effect [Alonso & Finn 1967 B, 1970 B, Hazen & Pidd 1965 B, Ziman 1964 B]. All these devices make use of the narrow-band characteristics of the Bragg reflection and related phenomena. The following considerations are restricted to periodic linear media without dispersion.

a) Wave equation

In a first approximation the effect of periodic linear media on electromagnetic radiation can be described by a modified scalar Hertz equation with periodic real or complex coefficients. This equation can be deduced [Kneubühl 1993 B, Kneubühl & Sigrist 1995 B] from the Maxwell equations of linear media without dispersion characterized by a periodic relative electric permittivity $\varepsilon(z)$ and a periodic electric conductivity $\sigma(z)$. For simplicity, the relative magnetic permeability μ is assumed to be constant. Thus, the Maxwell equations take the form

$$curl\,\vec{H} = \sigma(z)\vec{E} + \varepsilon_0\,\varepsilon(z)\dot{\vec{E}} \quad , \tag{7.6 - 17a}$$

$$curl\,\vec{E} = -\mu\,\mu_0\,\dot{\vec{H}}_0 \quad , \tag{7.6 - 17b}$$

$$div\,\vec{H} = 0 \quad , \tag{7.6 - 17c}$$

$$div\,\varepsilon(z)\vec{E} = 0 \quad \text{or} \quad div\,\vec{E} = -\vec{E}\cdot grad(\ell n\,\varepsilon(z)) \tag{7.6 - 17d}$$

$$\text{with} \quad \varepsilon(z)=\varepsilon(z+L) \quad \text{and} \quad \sigma(z)=\sigma(z+L) \quad . \tag{7.6 - 17e}$$

In these and the following equations L indicates the spatial period.

For a transverse electromagnetic wave propagating in the z direction with the electric field in the x direction (7.6 - 17d) takes the form

$$div\,\vec{E} = 0 \quad \text{for} \quad \vec{E} = \vec{E}(z,t) = [E(z,t),0,0] \quad . \tag{7.6 - 17f}$$

With these Maxwell equations, the wave equation of the field $\vec{E}(z,t)$ can be derived as follows

$$\begin{aligned} curl\left(curl\,\vec{E}\right) &= -\Delta\vec{E} + grad\,div\,\vec{E} = -\Delta\vec{E} = curl\left(-\mu\,\mu_0\,\vec{H}_t\right) \\ &= -\mu\,\mu_0 \frac{\partial}{\partial t} curl\,\vec{H} = -\mu\,\mu_0\,\sigma(z)\vec{E}_t - \mu\,\varepsilon(z)\mu_0\,\varepsilon_0\,\vec{E}_{tt} \quad . \end{aligned}$$

The result is the *periodic modified Hertz equation* for the scalar field $E(z, t)$

$$\begin{aligned} & n^2(z)c^{-2}E_{tt} + s(z)E_t = E_{zz} \\ & \text{with} \quad n^2(z) = \mu\varepsilon(z) = n^2(z+L) \quad ; \quad s(z) = \mu\,\mu_0\,\sigma(z) = s(z+L) \\ & \text{and} \quad c^{-2} = \mu_0\,\varepsilon_0 \quad . \end{aligned} \tag{7.6 - 18}$$

In this equation $s(z) > 0$ means loss and $s(z) < 0$ gain.

The relevant solutions of (7.6 - 18) are *harmonically oscillating waves* of the form

$$E(z,t) = u(z)\,exp(-i\,\omega\,t) \tag{7.6 - 19a}$$

where $u(z)$ fulfills the real or complex *Hill differential equation* [Jakubovic & Starzinski 1975 B, Magnus & Winkler 1966 B, Strutt 1932 B]

$$u_{zz} + \left[(n(z)\omega / c)^2 + i\,\omega\,s(z)\right]u = 0 \quad . \tag{7.6 - 19b}$$

This equation is equivalent to the *wave equation of periodic linear optical media*

$$u_{zz} + K^2(\omega,z)u = 0 \quad \text{with} \quad K(\omega,z) = K(\omega,z+L) \quad . \tag{7.6 - 20a}$$

For small $|s(z)|$ the coefficient $K(\omega, z)$ can be approximated by

$$\begin{aligned} & K(\omega,z) \cong (n(z)\omega / c) - i\,\alpha(z) \\ & \text{with} \quad \alpha(z) = -\gamma(z) = -i\,c\,s(z) / 2n(z) \quad , \end{aligned} \tag{7.6 - 20b}$$

where $\alpha(z)$ indicates the periodic gain and $\gamma(z)$ the periodic loss.

The wave equation (7.6 - 20a&b) serves as basis of standard theories of continuous-wave (cw) *distributed-feedback* (DFB) *lasers* [Kneubühl 1993 B, Kogelnik & Shank 1971 J]. These lasers are usually classified according to the spatial periodic modulations of refractive index $n(z)$ and gain $\alpha(z)$. *Index modulation* means $\alpha(z) = \alpha$

constant and $n(z) = n(z + L)$ periodic, while *gain modulation* designates the case where $\alpha(z) = \alpha(z + L)$ is periodic and $n(z) = n$ remains constant. *Mixed or hybrid modulation* requires that both, $\alpha(z) = \alpha(z + L)$ and $n(z) = n(z + L)$, are periodic.

Harmonic solutions of (7.6 - 20a&b) without boundary conditions yield the *dispersion relations* of electromagnetic waves in periodic passive media and cw lasers, whereas harmonic solutions with boundary conditions yield the laser *modes* [Gnepf & Kneubühl 1986 J, Kneubühl 1993 B, Kneubühl & Sigrist 1995 B, Kogelnik & Shank 1971 J]. Modes represent standing waves. Since this chapter is dedicated to traveling waves, the modes will not be discussed in the following.

Periodic optical media and structures without loss or gain are characterized by a real $K(\omega, z)$. In this case the wave equation (7.6 - 20a&b) corresponds to the normalized oscillation equation (2.3 - 57a) of *periodically modulated* or *parametric oscillators* if position z and time t are exchanged. As a consequence, the equations and solutions of the Section 2.3.7 dedicated to parametric linear oscillators can also be applied to electromagnetic waves in periodic media and structures.

The best-known methods to solve wave equation (7.6 - 20a&b) for complex as well as for real $K(\omega, z)$ are based either on the *Floquet matrix theory* [Gnepf & Kneubühl 1986 J, Kneubühl 1993 B, Kneubühl & Sigrist 1995 B] or on the *theory of coupled waves* [Kneubühl 1993 B, Kneubühl & Sigrist 1995 B, Kogelnik & Shank 1972 J].

b) Floquet matrix theory

The *Floquet matrix theory* or *Bloch- wave theory* [Bateman 1959 B, Birkhoff & Rota 1989 B, Coddington & Levinson 1955 B, Kaplan 1962 B, Magnus & Winkler 1966 B, Zwillinger 1989 B] yields exact solutions in contrary to the theory of coupled waves. It is well suited for square-wave and other periodic step modulations. Consequently, it was early applied to real Hill equations with square-wave and other periodic step modulations in mechanics [Meissner 1918 J], electrical engineering [Brillouin 1946 B] and solid state physics [Kronig & Penney 1931 J, Brillouin 1946 B].

The concepts of the Floquet matrix theory can be elucidated by solving (7.6 - 20a&b) for a *homogenous medium with extremely small periodic distortions* separated by the period L. Within this period $K(\omega, z)$ is independent of z according to

$$K(\omega,z) = K(\omega) = (n\omega / c) - i\alpha \qquad (7.6 - 21a)$$

where n and α are constant. Since $K(\omega, z) = K(\omega)$ does not depend of z the solution of (7.6 - 20a) is simply

$$u(z) = E_{\pm}\, exp(\pm i\beta z) \quad , \qquad (7.6 - 21b)$$

where the propagation constant β is determined by the dispersion relation

$$\beta = K(\omega) = (n\omega / c) - i\alpha \quad . \qquad (7.6 - 21c)$$

The solution (7.6 - 21b) corresponds the two counter-running waves

$$E(z,t) = E_+ exp\left[i(\beta z - \omega t)\right] + E_- exp\left[-i(\beta z + \omega t)\right] \qquad (7.6 - 21d)$$

with E_+ as amplitude of the wave traveling in the $+z$ direction and E_- as amplitude of that propagating in the $-z$ direction.

The solution (7.6 - 21b&c) can also represented with the translation operator $\boldsymbol{T}(\boldsymbol{L})$ introduced in Section 7.6.1. This operator forms a matrix and relates the field $E(z, t)$ and the gradient $E_z(z, t)$ at position z with those at the position $z + L$ according to

$$\begin{pmatrix} E(z+L,t) \\ E_z(z+L,t) \end{pmatrix} = \boldsymbol{T}(\boldsymbol{L}) \begin{pmatrix} E(z,t) \\ E_z(z,t) \end{pmatrix} \quad . \qquad (7.6 - 22a)$$

In general, this matrix is *unimodular*

$$det\, \boldsymbol{T}(\boldsymbol{L}) = 1 \qquad (7.6 - 22b)$$

and its *eigenvalues* Λ are related to the *propagation constant* β [Brillouin 1946 B]

$$\Lambda(L) = exp(\pm i\beta L) \quad . \qquad (7.6 - 22c)$$

Consequently, the propagation constant β can be evaluated with the aid of the *trace* of the translation matrix (7.6 - 22a)

$$cos\, \beta L = \frac{1}{2} tr\, \boldsymbol{T}(\boldsymbol{L}) \quad . \qquad (7.6 - 22d)$$

This equation represents the *dispersion relation* with the characteristic *band structure* and its *frequency gaps* of a periodic linear optical medium. It separates the range of real positive circular frequencies ω into

α) *allowed bands,* where the electromagnetic waves propagate without loss or gain. They are characterized by real propagation constants β that imply the conditions

$$Im\, \beta = 0 \quad \text{and} \quad Im\, cos\, \beta L = 0, \quad cos^2 \beta L \le 1 \qquad (7.6 - 23a)$$

β) *forbidden bands* or *frequency gaps,* where the electromagnetic waves are damped or amplified. They correspond to imaginary or complex propagation constants β that fulfill the conditions

$$Im\, \beta \ne 0 \quad \text{and} \quad Im\, cos\, \beta L \ne 0 \quad \text{or} \quad cos^2 \beta L > 0 \quad . \qquad (7.6 - 23b)$$

For $K(\omega, z) = K(\omega)$ of (7.6 - 21a) the translation matrix (7.6 - 22a) takes the form

$$T(L) = \begin{pmatrix} cosK(\omega)L & +K^{-1}(\omega)\, sinK(\omega)L \\ -K(\omega)\, sinK(\omega L) & cosK(\omega)L \end{pmatrix} \quad . \tag{7.6 - 24a}$$

When (7.6 - 22d) is applied to this matrix it becomes

$$cos\, \beta\, L = cos\, K(\omega)L \quad . \tag{7.6 - 24b}$$

The solution β of this equation is

$$Re\, \beta = \pm(n\omega / c) + \frac{2\pi}{L} m \quad \text{with} \quad m = 0, \pm 1, \pm 2, \ldots \quad , \tag{7.6 - 24c}$$

$$Im\, \beta = -\alpha \quad . \tag{7.6 - 24d}$$

For $m = 0$ this solution is identical with (7.6 - 21b&c). It is also meaningful for $m \neq 0$ because of the extremely small periodic distortions of the medium which determine the period L. This period causes the *periodicity of the dispersion relation* with the period $\Delta\beta = 2\pi/L$. The periodic intervals of the dispersion relation are named *Brillouin zones* [Bethe & Sommerfeld 1967 B, Brillouin 1946 B, Kittel 1971 B]. Fig. 7.6 - 6 shows the dispersion relation (7.6 - 24c) and its Brillouin zones. The different branches of this dispersion relation are labeled by the integer m. These branches have different *phase velocities* v, yet identical *group velocities* v_g.

$$\upsilon = \pm(c/n)\left[1 \pm (\lambda / L)m\right]^{-1} \text{ with } \lambda = 2\pi c / n\omega,\, m = 0, \pm 1, \pm 2, \ldots \quad , \tag{7.6 - 25a}$$

$$\upsilon_g = c / n \quad . \tag{7.6 - 25b}$$

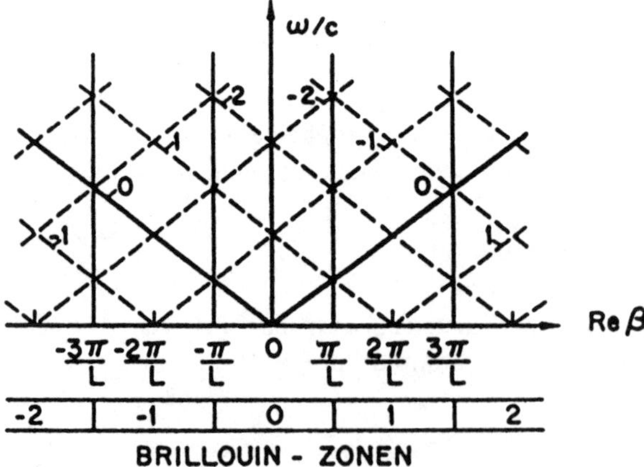

Fig. 7.6 - 6. Dispersion relation $\omega(Re\, \beta)$ of a medium with extremely weak periodic distortions (7.6 - 24c)

The intersections of the branches of the dispersion relation (7.6 - 24c) fulfill the *Bragg condition*

$$Re\,\beta = 2\pi / \lambda_{\mathrm{eff}} = r(\pi / L) \quad . \tag{7.6 - 26}$$

The parameter r is the order of the Bragg effect and related phenomena, e.g. *Bragg reflection* (DBR) and *distributed feedback* (DFB) in lasers.

The *relevant aspects* of electromagnetic waves propagating in media with periodic weak and strong modulations of refractive index, gain or loss can be demonstrated with the aid of the *square-wave modulation* with the period L illustrated in Fig. 7.6 - 7. It is defined by the following $K(\omega, z)$

$$\begin{aligned} &K(\omega,z) = K_1 = K - \Delta K \quad \text{for} \quad 0 < z < L/2 \;, \\ &K(\omega,z) = K_2 = K + \Delta K \quad \text{for} \quad L/2 < z < L \;, \\ &\text{where} \quad K = n\omega / c + i\alpha, \; \Delta K = \frac{1}{2}\{\Delta n\omega / c + i\Delta\alpha\} \;. \end{aligned} \tag{7.6 - 27a}$$

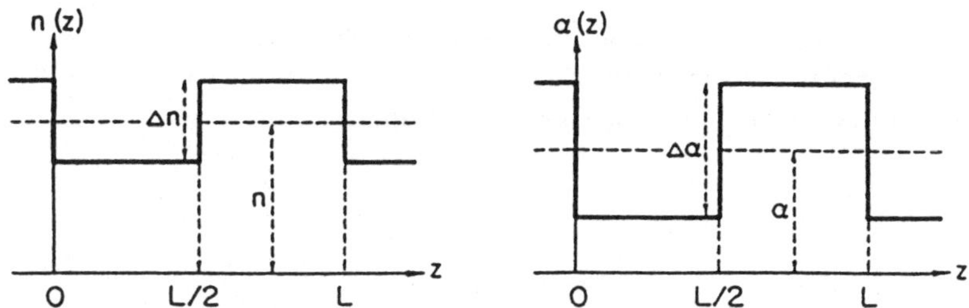

Fig. 7.6 - 7. Square-wave modulation of the refractive index n and the gain α

The corresponding translation matrix is

$$\boldsymbol{T}(\boldsymbol{L}) = \prod_{\mathrm{j}=1}^{2} \begin{pmatrix} cos\left(K_{\mathrm{j}}L/2\right) & +K_{\mathrm{j}}^{-1}\, sin\left(K_{\mathrm{j}}L/2\right) \\ -K_{\mathrm{j}} sin\left(K_{\mathrm{j}}L/2\right) & cos\left(K_{\mathrm{j}}L/2\right) \end{pmatrix} \quad . \tag{7.6 - 27b}$$

The application of (7.6 - 24d) to this matrix yields the *dispersion relation*

$$cos\,\beta L = \frac{K^2\, cosKL - \Delta K^2 cos\Delta\, KL}{K^2 - \Delta K^2} \quad , \tag{7.6 - 27c}$$

where K and ΔK are the functions of ω described by (7.6 - 27a). This dispersion relation is valid for large and small, real and complex K and ΔK, i.e. for all kinds of modulations.

For weak modulations with $|\Delta K| << |K|$ the dispersion relation (7.6 - 27c) can be approximated by

$$(\beta-\beta_r)^2=\{n\omega/c+i\alpha-n\omega_r/c\}^2-\left\{\frac{n}{2}(\Delta\omega_r/c)\right\}^2$$

$$\text{with}\quad \left\{\frac{n}{2}(\Delta\omega_r/c)\right\}^2=\frac{1}{2}\kappa_r^2\left\{1-(-1)^r\cos\left(\frac{\Delta n}{2n}\pi r\right)\right\}\quad , \tag{7.6 - 28a}$$

$$\kappa_r=2\left\{\frac{\Delta n}{2nL}+\frac{i\Delta\alpha}{2\pi r}\right\}\quad ,$$

$$\beta_r=\frac{n}{c}\omega_r=r\frac{\pi}{L};\, r=0,\pm1,\pm2,\ldots\quad .$$

In these equations $\Delta\omega_r$ indicate the *frequency gaps* and κ_r the *coupling constants*. The frequency gaps are observed at the Bragg propagation constants $\beta=\beta_r=r\pi/L$, when the periodic medium shows *neither loss or gain*. This condition means $\alpha=\Delta\alpha=0$. The frequency gaps $\Delta\omega_r$ at the Bragg circular frequencies ω_r are

$$\omega(\beta_r)=\omega_r\pm\frac{1}{2}\Delta\omega_r\quad\text{where}\quad \omega_r=r(c/n)(\pi/L)\quad . \tag{7.6 - 28b}$$

The second equation of (7.6 - 28a) reveals that for weak index modulations with small $|\Delta n|$ the frequency gaps $\Delta\omega_r$ of even order with r = 2, 4, 6, ... vanish in a first approximation. Furthermore the frequency gap $\Delta\omega_r$ disappears for

$$\cos\frac{\Delta n}{2n}\pi r=(-1)^r\quad . \tag{7.6 - 28c}$$

Fig. 7.6 - 8 shows positions and widths of the *frequency gaps* $\Delta\omega_r$ of Bragg order r = 1, 2, 3 as functions of the relative rectangular modulation $\Delta n/2n$ of the refractive index for $\alpha=\Delta\alpha=0$, which have been calculated with (7.6 - 27c).

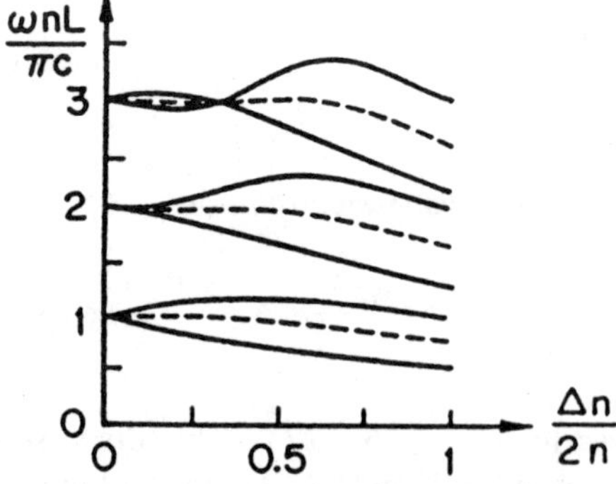

Fig. 7.6 - 8. Band structure of the square-wave modulation $\Delta n/2n$ of the refractive index n. Shown are positions and widths of the frequency gaps $\Delta\omega_r$ as well as the gap centers (dashed lines) as functions of the modulation $\Delta n/2n$

Equations (7.6 - 27c) and (7.6 - 28a) permit to determine the *dispersion relations of all types of square-wave modulations*: pure index, pure gain, mixed index and gain modulation. Fig. 7.6 - 9 illustrates the dispersion relations, i.e. *Re* $L\beta$ and *Im* $L\beta$ as functions of the normalized frequency $nL\omega/c$, of pure index modulation ($\Delta\alpha = 0$) without ($\alpha = 0$) and with ($\alpha > 0$) gain, as well as of weak ($\Delta\alpha \geq 0$) and strong ($\Delta\alpha >> 0$) pure gain modulation ($\Delta n = 0$). This figure shows the following relevant phenomena:

α) The frequency gaps $\Delta\omega_r$ of passive periodic media ($\alpha = \Delta\alpha = 0$) are closed when gain ($\alpha > 0$) is introduced.

β) For *pure index modulation* ($\Delta n \neq 0$; $\Delta\alpha = 0$) *with gain* ($\alpha > 0$) the frequency gaps $\Delta\omega_r$ of passive periodic media are replaced by a resonant increase of the wave amplification (*Im* $\beta < 0$).

γ) For *pure gain modulation* ($\Delta n = 0$; $\Delta\alpha \neq 0$; $\alpha < 0$) the frequency gaps $\Delta\omega_r$ of passive periodic media are substituted by a resonant decrease of the wave amplification (*Im* $\beta < 0$).

δ) For *strong pure gain modulation* ($\Delta n = 0$; $\Delta\alpha >> 0$; $\alpha >> 0$) there exist gaps in the real part *Re* β of the propagation constant β. This means that an entire range of effective wavelengths $\lambda_{\text{eff}} = 2\pi/Re\,\beta$ is assigned to each Bragg frequency ω_r.

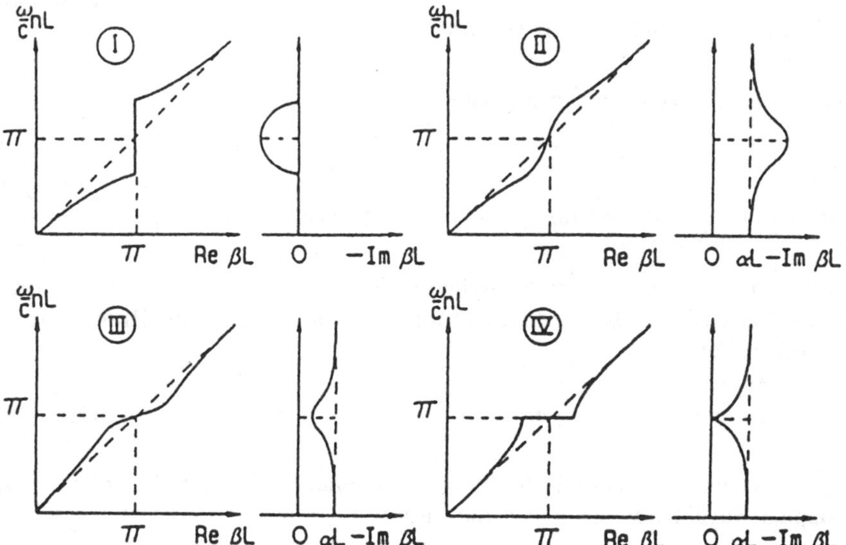

Fig. 7.6 - 9. Dispersion relations of the first-order Bragg or DFB effect for (I) pure index modulation without gain, (II) pure index modulation with gain, (III) weak pure gain modulation, (IV) strong pure gain modulation

ε) For *mixed index and gain modulation* ($\Delta n \neq 0$, $\Delta\alpha \neq 0$) *with gain* ($\alpha > 0$) the frequency gaps $\Delta\omega_r$ of passive periodic media are closed according to α) and replaced by maxima and/or minima of the wave amplification ($Im\ \beta < 0$).

Dispersion relations alone do not suffice to explain all characteristics of *distributed feedback (DFB) lasers*. They need to be supplemented by the resonance frequencies ω_q and threshold gains α_q of their modes q as well as by a formalism of nonlinear saturation effects in cw and pulsed operation [Kneubühl 1993 B]. Resonance frequencies and threshold gains are essentially determined by the finite lengths R of the lasers [Gnepf & Kneubühl 1984 J, Kneubühl 1993 B, Kneubühl & Sigrist 1995 B].

c) Coupled-wave theory

The coupled-wave theory [Brillouin 1946 B, Kneubühl 1993 B, Kneubühl & Sigrist 1995 B, Kogelnik & Shank 1972 J] yields an *approximate solution* of the wave equation (7.6 - 20a&b) of periodic linear optical media. Essentially it is restricted to weak harmonic modulations of refractive index n, gain α or loss $\gamma = -\alpha$. On these conditions the wave equation (7.6 - 20a&b) corresponds to a real or a complex *Matthieu differential equation* [Abramowith & Stegun 1965 B, Magnus & Winkler 1966 B, Morse 1930 J, Strutt 1932 B].

The coupled-wave theory of (7.6 - 20a&b) starts with the ansatz

$$K(\omega,z) = K(\omega,z+L) = \{n\omega / c + i\alpha\} + \{\Delta n\omega / c + i\Delta\alpha\} cos 2\pi z / L$$
$$\text{with } |\Delta n| << n \text{ and } |\alpha|,|\Delta\alpha| << n\omega / c \quad . \qquad (7.6 - 29a)$$

Consequently $K^2(\omega, z)$ can be approximated by

$$K^2(\omega,z) \approx$$
$$n^2(\omega / c)^2 + i\alpha 2n\omega / c + (n\omega / c)\{\Delta n\omega / c + i\Delta\alpha\} 2 cos\, 2\pi z / L \quad . \qquad (7.6 - 29b)$$

Usually, the coupled-wave theory is applied to *Bragg and DFB effects of first order* ($r = 1$). In this case the Bragg propagation constant β_B and the Bragg circular frequency ω_B are defined as

$$\beta_B = \pi / L \quad \text{and} \quad \omega_B = \pi c / nL \quad . \qquad (7.6 - 29c)$$

In order to solve wave equation (7.6 - 20a) with the harmonic modulation (7.6 - 29b) by the coupled-wave theory one makes the ansatz

$$u(z) = E_+(z)\, exp(+i\beta_B z) + E_-(z) exp(-i\beta_B z) \qquad (7.6 - 30a)$$

that represents the superposition of two counter-propagating waves with the propagation constant β_B. The harmonic modulation of $K(\omega, z)$ of (7.6 - 29a) is proportional to

$$2\cos(2\pi z / L) = 2\cos 2\beta_B z = exp(-2i\beta_B z) + exp(+2i\beta_B z) \quad . \qquad (7.6 - 30b)$$

Therefore, it *couples the two waves* (7.6 - 30a) traveling in opposite directions. This coupling causes a variation of the two amplitudes E_+ and E_- along the periodic medium. Thus, E_+ and E_- become functions of z. This implies an *energy exchange between the two waves* along the z axis.

The functions $E_+(z)$ and $E_-(z)$ are evaluated on the assumption that the second derivatives $d^2E_\pm(z)/d^2z$ can be neglected. Then, the application of (7.6 - 20a), (7.6 - 29b) and (7.6 - 30a) yields the *coupled-wave system of equations*

$$\frac{d}{dz}\begin{pmatrix} E_-(z) \\ E_+(z) \end{pmatrix} = \begin{pmatrix} -i[\Delta\omega(n/c) + i\alpha] & -i\kappa \\ +i\kappa & +i[\Delta\omega(n/c) + i\alpha] \end{pmatrix} \begin{pmatrix} E_-(z) \\ E_+(z) \end{pmatrix} \qquad (7.6 - 30c)$$

$$\text{with} \quad \Delta\omega = (\omega - \omega_B) \quad \text{and} \quad 2\kappa = [(\Delta n/n)(\pi/L) + i\Delta\alpha] \quad .$$

In this system κ indicates the coupling constant. The *dispersion relation* of the periodic medium described by (7.6 - 20a) and (7.6 - 29b) can be derived from (7.6 - 30c) by evaluating the eigenvalues $i\Delta\beta$ of its characteristic matrix. These are determined by the equation

$$det\begin{pmatrix} -i[\Delta\omega(n/c) + i\alpha + \Delta\beta] & -i\kappa \\ +i\kappa & +i[\Delta\omega(n/c) + i\alpha - \Delta\beta] \end{pmatrix} = 0 \quad , \qquad (7.6 - 31a)$$

which can be transformed into

$$\{\Delta\omega(n/c) + i\alpha\}^2 = \kappa^2 + \Delta\beta^2$$
$$\text{with} \quad \Delta\beta = \beta - \beta_B \quad \text{and} \quad \Delta\omega = \omega - \omega_B \quad . \qquad (7.6 - 31b)$$

This equation represents the *dispersion relation in the vicinity of the Bragg or DFB effect of first order* characterized by (7.6 - 29c). The variable $\Delta\beta$ defines on one hand the derivation of the propagation constant β from $\beta_B = \pi/L$, and on the other hand the period $\Lambda_M = 2\pi/Re\Delta\beta$ of the modulation of the amplitudes $E_\pm(z)$ by the wave coupling. The coupled-wave dispersion relation (7.6 - 31b) corresponds to a approximation of the dispersion relation (7.6 - 28a) determined with Floquet matrices. It also shows a *frequency gap* for pure index modulation ($\Delta\alpha = 0$) without gain or loss ($\alpha = 0$) that is related to the frequency gap $\Delta\omega_1$ of (7.6 - 28a)

$$\omega(\beta_B) = \omega_B \pm (\Delta\omega_B / 2) \quad \text{with} \quad \Delta\omega_B = 2(c/n)\kappa \quad . \qquad (7.6 - 31c)$$

With respect to *distributed feedback (DFB) lasers* it should be noticed that the dispersion relations (7.6 - 28a) and (7.6 - 31b) describe periodic laser structures of infinite length. As mentioned at the end of Section 7.6.3b the resonance frequencies ω_q and the threshold gains α_q of the laser modes q are essentially determined by the laser length R and the boundary conditions at both ends of the laser [Gnepf & Kneubühl 1986 J, Kneubühl 1993 B]. In the coupled-wave theory of lasers it is often assumed that the laser length R is an integer multiple M of the period L, i.e. $R = ML$, and that the laser extends from $z = -R/2 = -ML/2$ to $z = +R/2 = +ML/2$. On these assumptions the most common conditions are

$$E_-(-ML/2) = E_+(+ML/2) = 0 \quad . \qquad (7.6 - 32)$$

They permit an easy evaluation of resonance frequencies ω_q and threshold gains α_q of the DFB modes [Gnepf & Kneubühl 1986 J, Kneubühl 1993 B, Kneubühl & Sigrist 1995 B, Kogelnick & Shank 1972 J].

7.6.4 Wave Mechanics of a Particle in a Periodic Potential

The wave mechanics of a particle in a periodic potential show the same features as the theory of electromagnetic waves in periodic media discussed in the previous section. Such features are the Brillouin zones and the frequency or energy gaps.

a) Schrödinger equations

The wave mechanics [Alonso & Finn 1988 B, Baym 1969 B, Blochinzew 1966 B, Cohen-Tannoudji 1977 B, Fick 1968 B, Landau & Lifschitz 1979 B, Messiah 1960 B, 1969 B, 1990 B, Pauli 1950 B, Schubert & Weber 1980 B, 1993 B] of a particle with mass m in a one-dimensional periodic potential $V(z)$ is characterized by the time-dependent Schrödinger equation [Brillouin 1946 B, Flügge 1990 B, Kittel 1963 B]

$$\begin{aligned} i\hbar\, \psi_t(z,t) &= \frac{-\hbar^2}{2m}\, \psi_{zz}(z,t) + V(z)\, \psi(z,t) \\ \text{with} \quad V(z) &= V(z+L) \quad \text{and} \quad F(z) = F(z+L) = -V_z(z) \quad , \end{aligned} \qquad (7.6 - 33)$$

where $F(z)$ indicates the periodic force acting on the particle. Since the potential $V(z)$ is time-invariant, there exists a stationary solution of (7.6 --33) in the form

$$\psi(z,t) = \Psi(z)\, exp(-i\omega t) = \Psi(z)\, exp\left(-\frac{iE}{\hbar}t\right) \quad . \qquad (7.6 - 34a)$$

The energy E is related to the circular frequency ω by Planck's relation (7.2 - 16a). The corresponding probability density (7.4 - 51) does not vary with time t

$$\rho(z,t) = \psi^*(z,t)\, \psi(z,t) = \Psi^*(z)\, \Psi(z) = \rho(z) \quad . \qquad (7.6 - 34b)$$

This fact confirms that (7.6 - 34a) *represents a stationary solution*. The application of (7.6 - 33) to (7.6 - 34a) yields the *time-independent Schrödinger equation*

$$\Psi_{zz}(z) + \frac{2m}{\hbar^2}(E - V(z))\,\Psi(z) = 0$$
$$\text{with} \quad V(z) = V(z+L) \quad . \qquad (7.6 - 37)$$

It is analogous to the wave equation (7.6 - 20a) of periodic linear optical media according to

$$u_{zz}(z) + K^2(\omega, z)\,u(z) = 0$$
$$\text{with} \quad K^2(\omega, z) = \frac{2m}{\hbar^2}(\hbar\omega - V(z)) \quad \text{and} \quad \omega = E/\hbar \quad . \qquad (7.6 - 38)$$

As a consequence the solutions of (7.6 - 20a) also represent those of (7.6 - 37) when Planck's relations (7.2 - 16a&b) are taken into account. In this context it should be noticed that the dispersion relations $\omega(\beta)$ related to (7.6 - 20a) correspond to the energy-momentum relations $E(p)$ associated with (7.6 - 37).

b) Bloch waves

The wave functions $\Psi(z)$ as solutions of the time-independent Schrödinger equation (7.6 - 37) with periodic potential show a number of specific features.

According to the *theorem of Floquet* there exist two solutions $\Psi_r(z)$, $r = 1, 2$ of (7.6 - 37) for each value of the energy E. They fulfill the relations [Flügge 1990 B]

$$\Psi_r(z+L) = \lambda_r\,\Psi_r(z) \quad \text{with} \quad r = 1, 2 \quad \text{and} \quad \lambda_1\lambda_2 = 1 \quad , \qquad (7.6 - 39)$$

where L is the period of the potential $V(z)$.

The *theorem of Bloch* postulates that these functions have the form

$$\Psi_r(z) = \Psi(z, \pm\beta) = \Phi(z, \pm\beta)\,exp(\pm i\beta z) \qquad (7.6 - 40a)$$

with the periodic function

$$\Phi(z+L, \pm\beta) = \Phi(z, \pm\beta) \quad . \qquad (7.6 - 40b)$$

These *Bloch functions* or *Bloch waves* fulfill the equation

$$\Psi(z+L, \pm\beta) = \Psi(z, \pm\beta)\,exp(i\beta L) \quad . \qquad (7.6 - 40c)$$

Since

$$exp(i2\pi n) = 1 \quad \text{with} \quad n = 0, \pm 1, \pm 2, \ldots \quad , \qquad (7.6 - 41a)$$

the Bloch functions are periodic in β

$$\Psi(z,\beta) = \Psi(z,\beta + n2\pi / L) \quad \text{with} \quad n = 0, \pm 1, \pm 2, \ldots \quad . \qquad (7.6 - 41b)$$

As a consequence the propagation constant β is determined only to a multiple of the period

$$\Delta\beta = \beta_{n+1} - \beta_n = 2\pi / L \quad . \qquad (7.6 - 42a)$$

According to Planck's relations (7.2 - 16a&b) this period entails the p periodicity of the energy-momentum relation $E(p)$ with the period

$$\Delta p = p_{n+1} - p_n = h / L \quad . \qquad (7.6 - 42b)$$

This permits to divide the energy-momentum relation into an infinite number of sections of width Δp, which are called *Brillouin zones* [Brillouin 1946 B].

In the phase space (z, p) each Brillouin zone has the volume

$$L\,\Delta p = h \quad . \qquad (7.6 - 43)$$

This equation is relevant in *statistical mechanics*.

c) Particle in a comb potential

The dynamics of particles in periodic potentials are well illustrated by the wave mechanics of a particle in a comb or shah potential that consists of an infinite series of equidistant Dirac delta functions (2.3 - 46c)

$$V(z) = V(z+L) = \frac{\hbar^2 \kappa}{m} \sum_{n=-\infty}^{n=+\infty} \delta(z - nL) \quad . \qquad (7.6 - 44)$$

The sampling or shah function III(t) [Bracewell 1986 B] is elucidated in Appendix A.4.1. The dispersion relation $\beta(\omega)$ and energy-momentum relation $E(p)$ corresponding to (7.6 - 44) are determined by the relation [Flügge 1990 B]

$$\cos\beta L = \cos k L + (\kappa / k) \sin k L = \frac{\cos(k L - \varphi)}{\cos\varphi} \qquad (7.6 - 45)$$

$$\text{with} \quad \beta = p / \hbar \quad , \quad k^2 = \frac{2m}{\hbar}\omega = \frac{2m}{\hbar^2} E \quad \text{and} \quad \tan\varphi = \kappa / k \quad .$$

In *solid state physics* the ranges of E or $\omega = E / \hbar$ are designated as *allowed bands* if the corresponding p or $\beta = p / \hbar$ are real, and *forbidden bands* or *gaps* if the corresponding p or $\beta = p / \hbar$ are complex. In the forbidden bands the Bloch waves are damped because $Im\ \beta > 0$. According to these definitions the *edges of the allowed bands* are determined by the condition

$$\cos \beta L = \pm 1 \quad . \tag{7.6 - 46a}$$

The application of this equation to the energy-momentum relation (7.6 - 45) yields the conditions for the lower and the upper edges of the allowed bands

$$\begin{aligned} &k L = n\pi + 2\varphi \quad \text{for the lower edges} \quad , \\ &k L = (n+1)\pi \quad \text{for the upper edges} \quad , \\ &\text{with} \quad n = 0,1,2,3,\ldots \quad . \end{aligned} \tag{7.6 - 46b}$$

The characteristics of the energy-momentum relation (7.6 - 45) of a particle in the comb potential (7.6 - 44) are illustrated for $\kappa L = 4$ in the following three figures 7.6 - 10-12. Fig. 7.6 - 10 presents the allowed (white) and the forbidden (hatched) energy bands. The energy-momentum relations $E(p)$ of the three lowest allowed energy bands are plotted in Fig. 7.6 - 11. They are separated by energy gaps. For comparison this figure also shows the energy-momentum relation $E(p) = p^2/2\,m$ of a free particle (dashed line). Finally, Fig. 7.6 - 12 represents the energy-momentum relation $E(p)$ with the momentum p reduced by multiples of h/L. It contains the same information as Fig. 7.6 - 11.

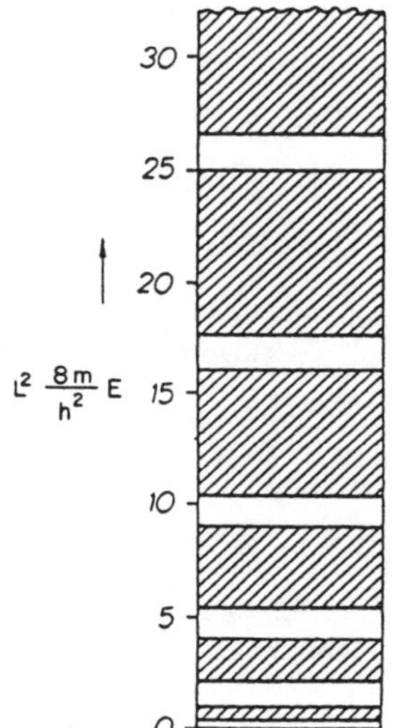

Fig. 7.6 - 10. Allowed (white) and forbidden (hatched) energy bands of a particle in the periodic comb potential (7.6 - 44) with $\kappa L = 4$

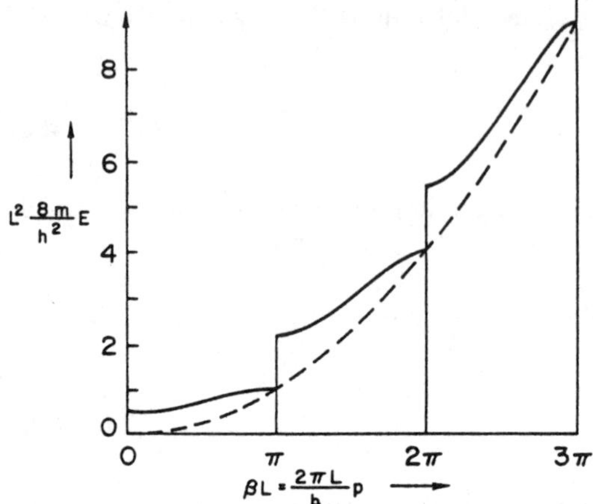

Fig. 7.6 - 11. Energy-momentum relation $E(p)$ of a particle in the periodic comb potential (7.6 - 44) with $\kappa L = 4$. This figure includes the three lowest allowed energy bands and the two lowest energy gaps

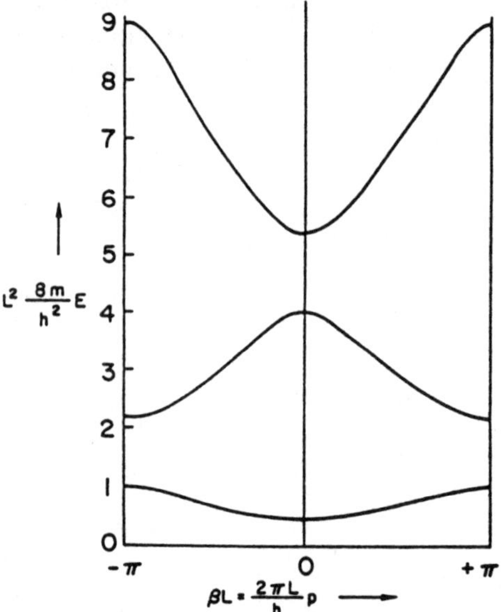

Fig. 7.6 - 12. The energy E of a particle in a periodic comb potential (7.6 - 44) with $\kappa L = 4$ as a function of the momentum p reduced by multiples of $2\pi/L$

8. Nonlinear Waves

Nonlinear waves are governed by nonlinear partial differential equations. In general, neither the superposition principle of Huygens described in Section 7.1.4a nor the principle of unperturbed propagation summarized in Section 7.1.4b can be applied to these waves. The amplitudes of linear waves as solutions of the basic linear partial differential equations can be chosen arbitrarly. On the contrary, the amplitudes of the nonlinear waves are determined by their basic differential equations. As a consequence periods and wavelengths of periodic nonlinear waves depend on their amplitude. Typical forms of nonlinear waves are the *solitary waves* where the deterioration of the wave by dispersion is compensated by nonlinearity. Examples are *solitons* and periodic solitary waves.

This chapter starts with comments on general aspects of nonlinear waves and continues with the discussion of nonlinear waves without dispersion, nonlinear diffusion, waves governed by the nonlinear equations of Korteweg-de Vries, Klein-Gordon, Schrödinger and Maxwell-Bloch. It concludes with the description of waves of the nonlinear periodic Toda chain.

8.1 Wave Equations and Solitary Waves

The most characteristic among the partial differential equations [Aimes 1972 B, Bateman 1959 B, Bhatnagar 1979 B, Courant & Hilbert 1968 B, Davis 1962 B, Dodd et al. 1982 B, Infeld & Rowlands 1990 B, John 1982 B, Karpman 1975 B, Lamb 1995 B, Rauch 1991 B, Schechter 1977 B, Smoller 1994, Sommerfeld 1947 B, Webster 1927 B, Webster & Szegö 1930 B, Whitham 1974 B, Zauderer 1983 B, Zwillinger 1989 B] that govern one-dimensional nonlinear waves are the *hyperbolic equations* with the standard form

$$u_{tt} - u_{zz} = \Psi(t, z, u_t, u_z, u) \tag{8.1 - 1}$$

where Ψ is a nonlinear function of u_t, u_z and/or u. The indices t and z indicate partial differentiations with respect to t and z. The hyperbolic equations (8.1 - 1) require initial as well as boundary conditions. The initial conditions define where a wave starts from, while the boundary conditions describe how wave and boundary interact, e.g. by absorption or scattering. Hyperbolic equations can be solved in principle by the *method of characteristics* [Courant & Hilbert 1968 B, Landau & Lifshitz 1959 B,

Whitham 1974 B, Zauderer 1983 B, Zwillinger 1989 B]. Examples of hyperbolic equations are the nonlinear *Klein-Gordon equation* (Section 8.5) and the *Maxwell-Bloch equation* (Section 8.7).

Evolution equations [Ablowitz & Clarkson 1991 B, Smoller 1994 B, Zwillinger 1989 B] represent another category of partial differential equations describing one-dimensional nonlinear waves. They have the standard form

$$u_t = \Theta(u, u_z, u_{zz}, \ldots) \quad . \tag{8.1 - 2}$$

Examples are the equations of *nonlinear waves without dispersion* (Section 8.2) and *nonlinear diffusion* (Section 8.3), the *Korteweg-de Vries* (Section 8.4) and the *nonlinear Schrödinger* (Section 8.6) equation.

These equations may obey *conservation laws* of the form [Zwillinger 1989 B]

$$\frac{\partial}{\partial t} T(u, u_z, u_{zz}, \ldots) + \frac{\partial}{\partial z} X(u, u_z, u_{zz}, \ldots) \tag{8.1 - 3a}$$

where T designates the *conserved density* and X the *flux*. Another statement of (8.1 - 3a) is that

$$\frac{d}{dt} \int T(u, u_z, u_{zz}, \ldots)\, dz = 0 \tag{8.1 - 3b}$$

for the solutions $u(t, z)$ of (8.1 - 2) whenever the integral converges.

Examples of conservation laws are those of the *Korteweg-de Vries* equation discussed in Section 8.4.1.

Wave groups, wave packets or pulses that are governed by linear partial differential equations deteriorate by dispersion during propagation. On the contrary those determined by nonlinear partial differential equations can propagate without change if the nonlinearity compensates the dispersion. These are called solitary waves [Dodd et al. 1982 B, Drazin 1983 B, Russel 1844 J]. Real *solitary waves* $u(z, t)$ can be characterized by the following conditions [Dodd et al. 1982 B, Drazin 1983 B]

$$u(z,t) = U(z - \upsilon_g t) = U(Z) \quad , \tag{8.1 - 4a}$$

$$\lim_{Z \to -\infty} U(Z) = c_1 = const \quad \text{and} \quad \lim_{Z \to +\infty} U(Z) = c_2 = const \quad , \tag{8.1 - 4b}$$

$$u^2(z,t) = U^2(Z) \le c_3^2 < \infty \quad . \tag{8.1 - 4c}$$

In (8.4 - 4a) υ_g indicates the group velocity. Typical relations between c_1 and c_2 are $c_1 = c_2 = 0$ as well as $c_1 - c_2 = \pm 2\pi$.

A solitary wave was first observed on the Edinburgh to Glasgow canal by J. Scott Russell [Dodd et al. 1982 B, Drazin 1983 B, Russell 1844 J]. He described it as a large solitary elevation of the water in the channel in the form of a round smooth and well-defined heap of water that rolled forward with great velocity apparently without change of form or diminution of speed. This wave is illustrated in Fig. 8.1 - 1.

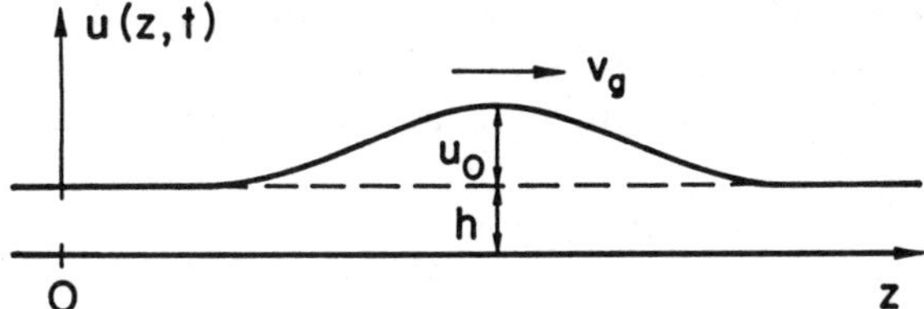

Fig. 8.1 - 1. Solitary wave observed by J. Scott Russel [Russel 1844 J]

By subsequent experiments J. Scott Russell discovered that the constant group velocity υ_g of this wave depends on the maximum elevation or amplitude $u_0 \geq 0$ of the wave and on the depth h of the undisturbed water in the channel

$$\upsilon_g^2 = g(h + u_0) \quad \text{with} \quad g = 9.81\,\mathrm{m\,s^{-2}} \quad , \tag{8.1 - 5a}$$

where g indicates the acceleration by gravity. Later, it was found [Boussinesq 1871 J, Rayleigh 1876 J] that this solitary wave can be described by

$$u(z,t) = u_0 \; sech^2 \beta(z - \upsilon_g t) = u_0 \; cosh^{-2} \beta(z - \upsilon_g t) \tag{8.1 -- 5b}$$

with the propagation constant β determined by the relation

$$\beta^2 = 3\,u_0 \left[4h^2(h + u_0)\right]^{-1} \quad . \tag{8.1 - 5c}$$

The solitary wave (8.1 - 5b) is nonlinear because its amplitude u_0 is related to the group velocity υ_g and to the propagation constant β according to (8.1 - 5a) and (8.1 - 5c).

The general theory of this type of surface waves was published in 1895 [Korteweg & de Vries 1895 J]. It is based on the *Korteweg-de Vries equation* [Dodd et al. 1982 B, Drazin 1983 B]

$$u_t = \frac{3}{2}\sqrt{g/h}\left\{u u_z + \frac{1}{3}\sigma u_{zzz} + \frac{2}{3}\alpha u_z\right\} \tag{8.1 - 6}$$
$$\text{with} \quad 3\sigma = h^3 - 3hT/g\rho \quad .$$

In this equation u[m] is the local elevation, T[N m^{-1}] the surface tension of the liquid of density ρ[kg m^{-3}] and α[m] a small but otherwise arbitrary constant.

Seventy years later, Zabusky & Kruskal [1965 J] demonstrated that solitary waves as solutions of the Korteweg-de Vries equation (8.1 - 6) can interact and carry on thereafter almost as if they never had interacted. Therefore the word "*soliton*" was coined for such a wave after "proton", "photon" to emphasize that it is a localized entity that may keep its identity after interaction [Drazin 1983 B].

A soliton is *not precisely defined.* Today, it is used [Drazin 1983 B] to describe any solution of a nonlinear equation that (i) represents a wave of permanent form; (ii) is localized, decaying or becoming constant at infinity; and (iii) may interact strongly with other solitons yet after interaction it retains its form, almost as if the principle of superposition is valid.

8.2 Nonlinear Waves without Dispersion

The *linear reduced Hertz equation*

$$u_t + \upsilon u_z = 0 \quad \text{with} \quad \upsilon = const \tag{7.4 - 10}$$

discussed in Section 7.4.2 describes waves without dispersion. Its general solution has the form

$$u(z,t) = f(z - \upsilon t) \quad \text{with} \quad u(z,0) = f(z) \quad , \tag{7.4 - 11}$$

where $f(z)$ is an arbitrary differentiable function that defines the wave form at time $t = 0$. This solution represents waves without dispersion because it can be demonstrated that their phase velocity υ equals the group velocity υ_g

$$\upsilon = \upsilon_g = const \quad . \tag{7.3 - 1}$$

The linear reduced Hertz equation (7.4.2) becomes *nonlinear* when the constant velocity υ is replaced by a velocity that depends on the wave function $u(z, t)$

$$u_t + \upsilon(u) u_z = 0 \quad . \tag{8.2 - 1}$$

This equation has an implicit solution of the form [Dodd et al. 1982 B]

$$u = u(z,t) = f(z - \upsilon(u)t) \quad \text{with} \quad u(z,0) = f(z) \quad . \tag{8.2 - 2}$$

This solution also represents waves without dispersion because the nonlinear reduced Hertz equation (8.2 - 1) has *no dispersive term* u_{zzz}, etc. [Dodd et al. 1982 B].

The determination of the wave function $u(z, t)$ with the aid of the implicit solution (8.1 - 2) is often as difficult as by solving the wave equation (8.1 - 1) directly. As a consequence, the discussion of nonlinear waves without dispersion is usually restricted to the *simple case* [Dodd et al. 1982 B, Drazin 1983 B] where

$$\upsilon(u) = \upsilon_0 + u \quad \text{with} \quad \upsilon_0 = const \quad . \tag{8.2 - 3a}$$

The corresponding wave equation

$$u_t + v_0 u_z + u u_z = 0 \tag{8.2 - 3b}$$

has the implicit solution

$$u(z,t) = f\big(z - v_0 t - u(z,t)t\big) \quad \text{with} \quad u(z,0) = f(z) \quad . \tag{8.2 - 4}$$

This equation can be simplified by the *Galilei transformation*

$$Z = z - v_0 t \quad \text{and} \quad T = t \tag{8.2 - 5}$$

that results in the wave equation [Bhatnagar 1979 B]

$$u_T + u u_Z = 0 \tag{8.2 - 6}$$

and its implicit solution

$$u(Z,T) = f\big(Z - u(Z,T)T\big) \quad \text{with} \quad u(Z,0) = f(Z) \quad . \tag{8.2 - 7}$$

All the nonlinear reduced Hertz equations (8.2 - 1), (8.2 - 3b) and (8.2 - 6) imply *nonlinearity without dispersion.* Its effect can be easily demonstrated by considering the propagation and deformation of a *triangle wave packet* defined by the initial condition

$$u(z,0) = f(z) = u_0\, \Lambda(z / z_0) \quad \text{with} \quad z_0 > 0, u_0 > 0 \quad , \tag{8.2 - 8}$$

where

$$\Lambda(x) = \begin{cases} 0 & \text{for} \quad |x| \geq 1 \\ 1 - |x| & \text{for} \quad |x| \leq 1 \end{cases} \tag{8.2 - 9}$$

designates the triangle function [Bracewell 1986 B] mentioned in Appendix A.2.2. The solution of (8.2 - 6) with the initial condition (8.2 - 8) is the wave function

$$\begin{aligned} u(Z,T) &= u_0 \frac{1 + (Z / z_0)}{1 + (T / t_0)} \quad && \text{for} \quad -z_0 \leq Z - uT \leq 0 \\ u(Z,T) &= u_0 \frac{1 - (Z / z_0)}{1 - (T / t_0)} \quad && \text{for} \quad 0 \leq Z - uT \leq z_0 \\ u(Z,T) &= 0 && \text{for} \quad |Z - uT| \geq z_0 \quad \text{with} \quad t_0 = z_0 / u_0 \quad . \end{aligned} \tag{8.2 - 10}$$

This solution is illustrated in Fig. 8.2 - 1 for $T = t = 0$, $t_0/2$, t_0, $2\,t_0$. It demonstrates that the higher parts of the wave propagate faster than the lower parts. At $T = t = t_0$ the solution $u(Z, T)$ forms a *shock wave*, while the *wave breaks* for $T = t > t_0$. Then, the function $u(Z, T)$ becomes multi-valued [Dodd et al. 1982 B].

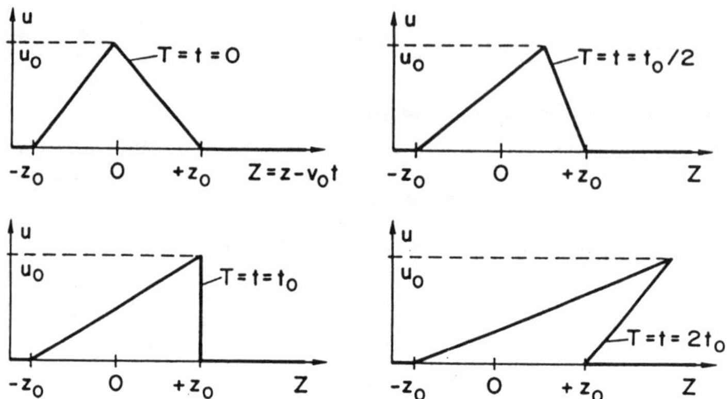

Fig. 8.2 - 1. Nonlinear wave without dispersion (8.2 - 10). The shock wave occurs at $T = t = t_0$, whilst the wave breaks for $T = t > t_0$

Shock waves and wave breaking are characteristic of nonlinear waves without dispersion and dissipation. Dispersion and dissipation terms u_{zzz}, etc. and u_{zz}, etc. in a wave equation can impede wave breaking [Landau & Lifshitz 1959 B, Ehitham 1974 B] as demonstrated in Section 8.3.

8.3 Nonlinear Diffusion

The simplest nonlinear diffusion is described by *Burgers equation* [Burgers 1948 J, Dodd et al. 1982 B, Drazin 1983 B, Smoller 1994 B]. It can be deduced from the *nonlinear reduced Hertz equation* (8.2 - 6) by adding a dissipation or diffusion term u_{zz}

$$u_t + u\,u_z = D\,u_{zz} \quad . \qquad (8.3 - 1)$$

The parameter D represents the diffusion constant. The elimination of the nonlinear term uu_z yields the linear diffusion equation (7.4 - 35).

Burgers equation (8.3 - 1) can be transformed into the linear diffusion equation (7.4 - 35) in two ways:

α) by the *Cole-Hopf-Transformation* [Cole 1951 J, Dodd et al. 1982 B, Drazin 1983 B, Hopf 1950 J]

$$u(z,t) = -2D\frac{\partial}{\partial z}\left(\ell n\, W(z,t)\right) \quad . \qquad (8.3 - 2a)$$

The application of this transformation to (8.3 - 1) and subsequent integration of u over z results in the equation

$$W_t - f(t)\,W = D\,W_{zz} \tag{8.3 - 2b}$$

with the arbitrary function $f(t)$. The ansatz

$$W(z,t) = w(z,t)\,exp\int_{t_0}^{t} f(\tau)\,d\,\tau \tag{8.3 - 2c}$$

yields a function $w(z, t)$ that fulfills the linear diffusion equation

$$w_t = D\,w_{zz} \quad . \tag{7.4 - 35}$$

β) by the *Bäcklund Transformation* [Bäcklund 1875 J, Dodd et al. 1982 B, Drazin 1983 B]

$$w_z = -u\,w\,/\,2\,D \quad , \tag{8.3 - 3a}$$

$$w_t = \left(u^2 - 2D\,u_z\right)w\,/\,4D \quad . \tag{8.3 - 3b}$$

Also this transformation results in a function $w(z, t)$ that obeys the linear diffusion equation (7.4 - 35), whereas the original function $u(z, t)$ is determined by Burgers equation (8.3 - 1).

Two *solutions of Burgers equation* (8.3 - 1) should be noticed. First, there exists a *time-independent solution*

$$u(z,t) = u(z) = -2D\left[z - z_0\right]^{-1} \quad . \tag{8.3 - 4}$$

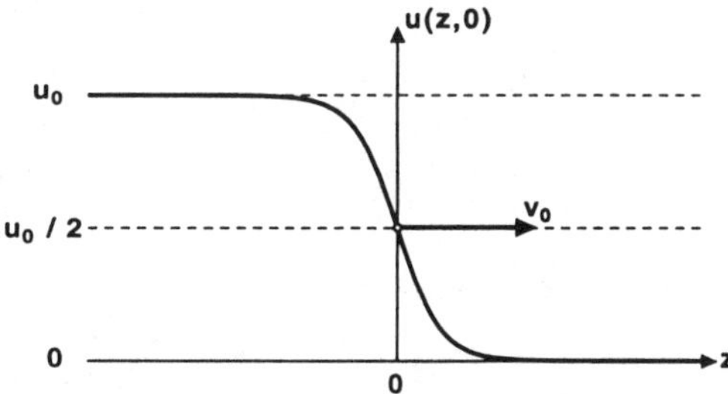

Fig. 8.3 - 1. Weak shock wave with Taylor profile (8.3 - 5)

Secondly, there is a time-dependent solution that represents a weak *shock wave* with *Taylor profile* [Dodd et al. 1982 B, Drazin 1983 B, Taylor 1910 J] that moves with the constant velocity υ_0 (Fig. 8.3 - 1)

$$u(z,t) = \upsilon_0\left[1 - tanh\left\{\upsilon_0\left(z - \upsilon_0\, t\right) / 2D\right\}\right] \quad . \tag{8.3 - 5}$$

Consequently, arbitrarly *small diffusion* ($D \geq 0$) *impedes the wave breaking* that occurs without diffusion ($D = 0$) according to (8.2 - 10) when $T = t > t_0$.

8.4 Korteweg-de Vries Equation

Replacement of u by $6u$ in the nonlinear reduced Hertz equation (8.2 - 6) and subsequent addition of the dispersive term u_{zzz} yields the *standard form* of the Korteweg-de Vries or KdV equation [Korteweg & de Vries 1895 J]

$$u_t + 6\,u u_z + u_{zzz} = 0 \quad . \tag{8.4 - 1}$$

However, various *other forms* of the KdV equation are cited in the literature [Ablowitz & Clarkson 1991 B, Dodd et al. 1982 B, Drazin 1983 B, Drazin & Johnson 1989 B]. These can be deduced from the standard form (8.4 - 1) by the following *linear transformation* of the variables

$$u = \alpha\left(w + \upsilon_0\right), \; z = \beta\, Z, t = \gamma\, T \quad . \tag{8.4 - 2a}$$

The result is the general KdV equation

$$w_T + 6\left(\alpha\,\gamma / \beta\right)\left(w + \upsilon_0\right) w_Z + \left(\gamma / \beta^3\right) w_{ZZZ} = 0 \tag{8.4 - 2b}$$

Another form of the KdV equation is gained by the introduction of a pseudo-potential w(z, t) according to

$$u(z,t) = \frac{\partial}{\partial z} w(z,t) \quad , \tag{8.4 - 3a}$$

which fulfills the equation

$$w_t + 3\left(w_z^2\right) + w_{zzz} = g(t) \quad , \tag{8.4 - 3b}$$

where $g(t)$ is an arbitrary function of time t. If $g(t)$ is zero

$$g(t) = 0 \quad , \tag{8.4 - 4a}$$

then the *Cole-Hopf transformation* [Cole 1951 J, Dodd et al. 1982 B, Hopf 1950 J]

$$w(z,t) = 2\frac{\partial}{\partial z}\ell n\, f(z,t) \quad \text{and} \quad u(z,t) = 2\frac{\partial^2}{\partial z^2}\ell n\, f(z,t) \tag{8.4 - 4b}$$

transforms (8.4 - 3b) as well as (8.4 - 1) into the homogeneous *Hirota equation* [Dodd et al. 1982 B, Hirota 1971 J]

$$f\, f_{zt} - f_z f_t + f f_{zzzz} + 3 f_{zz}^2 - 4 f_z\, f_{zzz} = 0 \quad . \qquad (8.4 - 4c)$$

8.4.1 Conservation Laws

The KdV equation (8.4 - 1) represents an *evolution equation* of the form (8.1 - 2). Therefore, one can formulate *conservation laws* of the type (8.1 - 3a&b). The comparison of (8.1 - 3a) and (8.4 - 1) gives the motive to define the following conserved density T and flux X

$$T_1 = u \quad \text{and} \quad X_1 = 3u^2 + u_{zz} \qquad (8.4 - 5a)$$

$$\text{with} \quad \int_{-\infty}^{+\infty} u(z,t)\, dz = const \quad . \qquad (8.4 - 5b)$$

After multiplication of (8.4 - 1) with $2u$ one can also define

$$T_2 = u^2 \quad \text{and} \quad X_2 = 2u^3 - u_z^2 + 2uu_{zz} \qquad (8.4 - 6a)$$

$$\text{with} \quad \int_{-\infty}^{+\infty} u^2(z,t)\, dz = const \quad . \qquad (8.4 - 6b)$$

Finally, it should be noticed that the KdV equation (8.4 - 1) implies an *infinite number of conservation laws* (8.1 - 3a&b) with corresponding conserved densities T_k and fluxes X_k where k = 1, 2, 3, ... [Zwillinger 1989 B].

8.4.2 Traveling Waves

Relevant solutions of the KdV equation (8.4 - 1) and the equivalent differential equations (8.4 - 2b), (8.4 - 3b) and (8.4 - 4c) are the so-called *traveling waves*. They represent waves with finite amplitudes traveling at a constant speed υ_0 without changing their form. Mathematically they are described by a function $U(\Theta)$ of a single variable or phase Θ that forms a linear combination of time t and position z

$$\begin{aligned} &u(z,t) = U(\Theta) \quad \text{with} \quad U^2(\Theta) < \infty \quad , \\ &\Theta = \beta z - \omega t + \delta = \beta(z - \upsilon_0 t) + \delta \quad \text{and} \quad \upsilon_0 = \omega / \beta \quad . \end{aligned} \qquad (8.4 - 7)$$

If one applies this ansatz to solve (8.4 - 1), then one finds the following differential equation for $U(\Theta)$

$$(6U - \upsilon_0)U' + \beta^2 U''' = 0 \quad . \tag{8.4 - 8a}$$

The integration of this equation over Θ yields

$$\beta^2 U'' - \upsilon_0 U + 3U^2 = A \quad , \quad \text{or} \tag{8.4 - 8b}$$

$$\frac{1}{2}\beta^2 \frac{d}{dU}(U')^2 - \upsilon_0 U + 3U^2 = A \quad . \tag{8.4 - 8c}$$

The integration of (8.4 - 8c) over U results in

$$\frac{1}{2}\beta^2 (U')^2 = B + AU + \frac{1}{2}\upsilon_0 U^2 - U^3 \quad , \tag{8.4 - 8d}$$

where A and B designate arbitrary constants. On the condition that

$$(U')^2 \geq 0 \quad , \tag{8.4 - 8e}$$

equation (8.4 - 8d) can be transformed into

$$\left[B + AU + \frac{1}{2}\upsilon_0 U^2 - U^3\right]^{-1/2} \beta\, dU = \sqrt{2}\, d\Theta \quad , \text{or} \tag{8.4 - 8f}$$

$$[-(U-a)(U-b)(U-c)]^{-1/2} \sqrt{2}\,\beta\, dU = \sqrt{2}\, d\Theta \tag{8.4 - 8g}$$

with $\quad a \leq b \leq c \quad .$

The solutions of (8.4 - 8f&g) are *elliptical integrals of the first kind* [Abramowitz & Stegun 1965 B, Gradshteyn & Ryzhik 1965 B]. Solutions $U(\Theta)$ according (8.4 - 7) exist only for real a, b, c. *Three types of solutions* $U(\Theta)$ exist [Ablowitz & Clarkson 1991 B]:

$$\begin{aligned} &a < b < c\text{: periodic cnoidal waves} \\ &a = b < c\text{: solitons, solitary waves} \\ &a \leq b = c\text{: constants} \end{aligned} \tag{8.4 - 9}$$

Solitons and cnoidal waves will be discussed in Sections 8.4.3&4.

The traveling waves can also be determined with the aid of the Hirota equation (8.4 - 4c). The application of the ansatz

$$\begin{aligned} &f(z,t) = F(\Theta) \quad , \\ &\text{with} \quad \Theta = \beta z - \omega t + \delta = \beta(z - \upsilon_0 t) + \delta \quad \text{and} \quad \upsilon_0 = \omega / \beta \end{aligned} \tag{8.4 - 10a}$$

to (8.4 - 4c) yields the equations

$$v_0 = \omega / \beta = \beta^2 \left[FF^{(4)} + 3\left(F''\right)^2 - 4F'F''' \right] \left[FF'' - \left(F'\right)^2 \right]^{-1} \quad , \qquad (8.4 - 10b)$$

$$U(\Theta) = U(\beta z - \omega t + \delta) = \frac{2\beta^4}{v_0 F^2} \left[FF^{(4)} + 3\left(F''\right)^2 - 4F' F''' \right] \quad . \qquad (8.4 - 10c)$$

The first equation (8.4 - 10b) decides whether $F(\Theta)$ is a solution. If yes, it determines the "dispersion relation" $v_0(\beta)$ or $\omega(\beta)$. The second equation (8.4 - 10c) permits to derive $U(\Theta)$ from the solution $F(\Theta)$ of (8.4 - 10b). These relations are mainly used to evaluate the soliton solutions of the KdV equation (8.4 - 1) [Ablowitz & Clarkson 1991 B, Dodd et al. 1982 B, Hirota 1971 J].

8.4.3 Korteweg-de Vries Solitons

The solutions of the KdV equation (8.4 - 1) form solitons if the parameters a, b, c of (8.4 - 8g) are real and $a \geq b > c$.

The simplest method to derive soliton solutions of (8.4 - 1) makes use of the modified Hirota equations (8.4 - 10a-c). A *single soliton* is found with the ansatz

$$F(\Theta) = \cosh \Theta \quad . \qquad (8.4 - 11a)$$

If this ansatz is applied to (8.4 - 10b), it yields the "dispersion relation"

$$v_0 = \omega / \beta = 4\beta^2 \quad . \qquad (8.4 - 11b)$$

The form of the corresponding soliton is determined by (8.4 - 10c)

$$u(z,t) = U(\Theta) = \frac{1}{2} v_0 \, sech^2 \left[\frac{1}{2} \sqrt{v_0} \left(z - v_0 t - z_0 \right) \right] \quad . \qquad (8.4 - 11c)$$

This soliton is shown in Fig. 8.4 - 1. Its height u_{max} and halfwidth $\Delta z_{1/2}$ depend on the velocity v_0

$$u_{max} = v_0 / 2 \text{ and } \Delta z_{1/2} \approx \frac{3.524}{\sqrt{v_0}} \quad . \qquad (8.4 - 11d)$$

The *same soliton* (8.4 - 11c) results for the ansatz

$$F(\Theta) = 1 + exp\, \Theta \quad . \qquad (8.4 - 12a)$$

The application of (8.4 - 10b) to this ansatz results in the "dispersion relation"

$$v_0 = \omega / \beta = \beta^2 \quad . \qquad (8.4 - 12b)$$

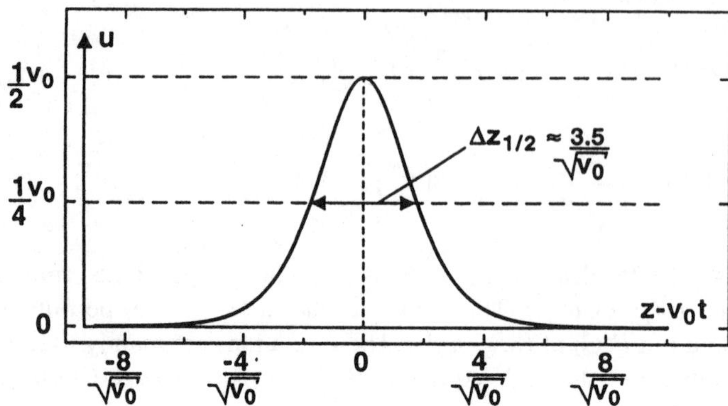

Fig. 8.4 - 1. Korteweg-de Vries soliton (8.4 - 11c)

The corresponding soliton derived with the aid of (8.4 - 10c) equals that of (8.4 - 11c) in spite of the fact that the "dispersion relation" (8.4 - 12b) differs from (8.4 - 11b).

A *general form* of the soliton (8.4 - 11c) is found by integration of (8.4 - 8g) on the condition that a, b, c are real and $a = b < c$ [Ablowitz & Clarkson 1991 B]. The result is a "dispersion relation"

$$\upsilon_0 = \omega / \beta = 4\beta^2 \tag{8.4 - 13a}$$

that corresponds to (8.4 - 11a) and the soliton

$$u(z,t) = u_0 + \frac{1}{2}\upsilon_0\, sech^2\left\{\frac{1}{2}\sqrt{\upsilon_0}\left[z - (\upsilon_0 + 6u_0)t - z_0\right]\right\} \quad . \tag{8.4 - 13b}$$

Its velocity is

$$\upsilon = \upsilon_0 + 6u_0 \quad . \tag{8.4 - 13c}$$

The original purpose of the Hirota equation (8.4 - 4c) was the derivation of multi-soliton solutions of the KdV equation (8.4 - 1) [Hirota 1971 J]. A *two-soliton solution* is found with the ansatz [Dodd et al. 1982 B]

$$\begin{aligned} u &= 2\frac{\partial^2}{\partial x^2} f \quad , \\ f &= 1 + exp\,\Theta_1 + exp\,\Theta_2 + A\,exp(\Theta_1 + \Theta_2) \quad , \\ \Theta_k &= \beta_k\left(z - \beta_k^2 t\right) + \delta_k = \beta_k\left(z - \beta_k^2 t - z_k\right) \quad , \\ A &= \left[(\beta_1 - \beta_2)/(\beta_1 + \beta_2)\right]^2 \quad . \end{aligned} \tag{8.4 - 14a}$$

This ansatz results in a solution that represents a combination of two solitons [Ablowitz & Clarkson 1991 B]

$$u(z,t) = \left(\beta_2^2-\beta_1^2\right)\frac{\left(\beta_2^2-\beta_1^2\right)+\beta_2^2 \cosh\Theta_1+\beta_1^2 \cosh\Theta_2}{\left(\beta_2-\beta_1\right)\cosh\left(\Theta_2+\Theta_1\right)+\left(\beta_2+\beta_1\right)\cosh\left(\Theta_2-\Theta_1\right)} \quad . \qquad (8.4 - 14b)$$

A *collision* of two such KdV solitons is illustrated in Fig. 8.4 - 2. They interact strongly, yet after interaction they retain their form. This is a characteristic of solitons.

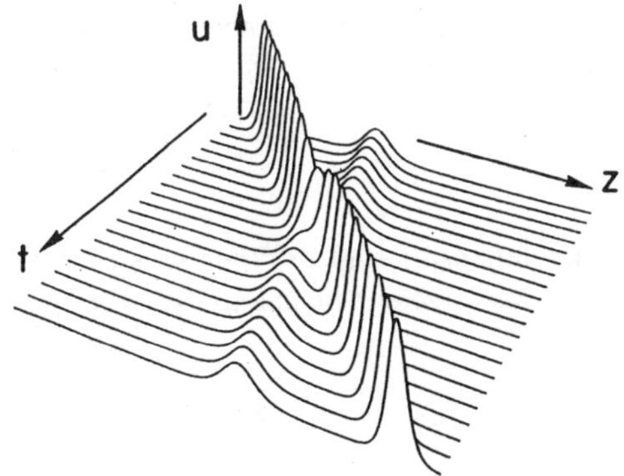

Fig. 8.4 - 2. Collision of two Korteweg-de Vries solitons (8.4 - 14b)

8.4.4 Cnoidal Korteweg-de Vries Waves

The KdV equation (8.4 - 1) has solutions in the form of periodic non-harmonic waves called *cnoidal waves* [Korteweg & de Vries 1985 J]. These solutions exist on the conditions (8.4 - 9) that a, b, c of (8.4 - 8g) are real and obey the inequalities $a < b < c$. They have the form [Ablowitz & Clarkson 199 B, Drazin 1983 B]

$$u(z,t) = u_0 + \frac{m}{2}\,\upsilon_0\, cn^2\!\left(\frac{1}{2}\sqrt{\upsilon_0}\left[z-\upsilon t - z_0\right] \,\middle|\, m\right) \quad , \qquad (8.4 - 15a)$$

the velocity

$$\upsilon = (2m-1)\,\upsilon_0 + 6u_0 \quad , \qquad (8.4 - 15b)$$

the wavelength

$$\lambda = 4K(m) / \sqrt{\upsilon_0} \quad , \qquad (8.4 - 15c)$$

and the period

$$T = \lambda / \upsilon \quad . \tag{8.4 - 15d}$$

In these equations $cn(x|m)$ represents Jacobi's elliptic cosine and $K(m)$ the complete elliptic integral of the first kind [Abramowitz & Stegun 1965 B, Gradshteyn & Ryzhik 1965 B]. They are discussed in Section 2.5 and illustrated in Figs. 2.5 - 2&3.

Another representation of the cnoidal wave is

$$u(z,t) = u_0 + u_1\, cn^2\left(\frac{2\,K(m)}{\lambda}\left[z - \upsilon t - z_0\right] \,|\, m\right) \quad , \tag{8.4 - 16a}$$

where the amplitudes u_0, u_1 and the velocity υ are arbitrary parameters. The relevant parameter m that defines the shape of the wave is determined by

$$0 \le m = 2u_1\left[6u_0 + 4u_1 - \upsilon\right]^{-1} \le 1 \quad . \tag{8.4 - 16b}$$

This relation restricts the choice of the parameters u_0, u_1 and υ. When m is evaluated, the wavelength λ and the period T of the cnoidal wave can be calculated with the aid of the equations

$$\lambda = K(m)\sqrt{8m / u_1} \quad , \tag{8.4 - 16c}$$

$$T = K(m)\sqrt{8m / u_1 \upsilon^2} \quad . \tag{8.4 - 16d}$$

8.4.5 Generalized Korteweg-de Vries Equations

The generalized KdV equations have the *standard form* [Drazin 1983 B]

$$u_t + (n+1)(n+2)u^n\, u_z + u_{zzz} = 0 \quad \text{with} \quad n = 1, 2, 3, \dots \quad . \tag{8.4 - 17}$$

The original standard KdV equation (8.4 - 1) is characterized by $n = 1$. These equations have soliton or *solitary-wave solutions* in the form

$$u^n(z,t) = \frac{1}{2}\,\upsilon_0\, sech^2\left\{\frac{n}{2}\sqrt{\upsilon_0}\left(z - \upsilon_0 t - z_0\right)\right\} \quad . \tag{8.4 - 18}$$

The generalized KdV equation with $n = 2$

$$u_t + 12u^2\, u_z + u_{zzz} = 0 \tag{8.4 - 19}$$

has solutions that represent traveling oscillatory pulses called *breathers* or *bions* [Drazin 1983 B]

$$u(z,t) = -\sqrt{2}\,\frac{\partial}{\partial z}\, arc\,tan\left\{\frac{\beta_2\, sin\left[\beta_1\left(z - \upsilon_1 t - z_1\right)\right]}{\beta_1\, cosh\left[\beta_2\left(z - \upsilon_2 t - z_2\right)\right]}\right\} \quad , \qquad (8.4 - 20)$$

$$\text{with} \quad \upsilon_1 = 3\beta_2^2 - \beta_1^2 \quad \text{and} \quad \upsilon_2 = \beta_2^2 - 3\beta_1^2 \quad .$$

This type of pulse corresponds to a periodic wave with the velocity υ_1 that is confined by an envelope with the velocity υ_2 [Drazin 1989 B].

8.5 Nonlinear Klein-Gordon Equations

The nonlinear Klein-Gordon equations represent generalizations of the linear Klein-Gordon equation (7.4 - 17). In physics they are written either as

$$u_{tt} - c^2\, u_{zz} + \omega_C^2\, \frac{d}{du}\, \Phi(u) = 0 \qquad (8.5 - 1a)$$

or as

$$u_{zz} - c^{-2}\, u_{tt} = \beta_C^2\, \frac{d}{du}\, \Phi(u) \qquad (8.5 - 1b)$$

with the potential Φ and the Compton circular frequency

$$\omega_C = c\beta_C = 2\pi c\, /\, \lambda_C = mc^2\, /\, \hbar \quad . \qquad (7.4 - 19)$$

In these equations λ_C indicates the *Compton wavelength* and β_C the Compton propagation constant.

Examples of nonlinear Klein-Gordon equations and *solitary solutions* are listed in Table 8.5 - 1. Variables and parameters are

$$y = z \pm \beta\, c\, t - z_0 \quad , \quad \gamma = (1 - \beta^2)^{-1/2} \quad , \quad \beta = \upsilon\, /\, c \quad . \qquad (8.5 - 2)$$

The two representations (8.5 - 1a&b) can be normalized by the transformation

$$T = \omega_C t \quad \text{and} \quad Z = \beta_C z \qquad (8.5 - 3a)$$

that results in the *first standard form*

$$u_{TT} - u_{ZZ} + \frac{d}{du}\, \Phi(u) = 0 \quad . \qquad (8.5 - 3b)$$

Table 8.5 - 1. Solitary solutions of nonlinear Klein-Gordon equations [Kneubühl & Feng 1991 J]. The parameters y and γ correspond to (8.5 - 2).

Type	$\Phi(u)$	$\frac{d\Phi}{du}(u)$	$\frac{d^2\Phi}{du^2}(0)$	$\frac{d^3\Phi}{du^3}(0)$	Soliton u
Klein Gordon	$\frac{1}{2}u^2$	u	1	0	--
Sine-Gordon	$2\sin^2\frac{u}{2}$	$\sin u$	1	0	$4\tan^{-1}\left[\exp\left(\pm\beta_c\gamma y\right)\right]+n\pi$ $(n=0,\pm2,\pm4,\ldots)$
	$-2\sin^2\frac{u}{2}$	$-\sin u$	-1	0	$4\tan^{-1}\left[\exp\left(\pm\beta_c\gamma y\right)\right]+n\pi$ $(n=\pm1,\pm3,\pm5,\ldots)$
arctan	$\frac{1}{2}\sin^4 u$	$\frac{1}{2}\sin(2u)-\frac{1}{4}\sin(4u)$	0	0	$\frac{\pi}{2}\pm\tan^{-1}\left(\beta_c\gamma y\right)+n\pi \quad (n=0,\pm1,\pm2,\ldots.)$
Φ^4 - model	$\frac{1}{2}u^2(2-u)^2$	$2u^3-6u^2+4u$	4	-12	$1\pm\tanh\left(\beta_c\gamma y\right)$
	$-\frac{1}{2}u^2+\frac{1}{4}u^4$	$-u+u^3$	-1	0	$\pm\tanh\left(\frac{\beta_c\gamma y}{\sqrt{2}}\right)$
	$\frac{1}{2}u^2-\frac{1}{4}u^4$	$u-u^3$	1	0	$\pm\sqrt{2}\,\mathrm{sech}\left(\beta_c\gamma y\right)$
Φ^6 - model	$\frac{1}{8}u^2\left(2-u^2\right)^2$	$\frac{3}{4}u^5-2u^3+u$	1	0	$\left[1\pm\tanh\left(\beta_c\gamma y\right)\right]^{1/2}$
Casahorran $(n = 3, 4, 5)$	$\frac{1}{2n^2}u^2\left(2-u^n\right)^2$	$\frac{1}{n^2}u\left(2-u^n\right)\left(2-(n+1)u^n\right)$	$\frac{4}{n^2}$	0	$\left[1\pm\tanh\left(\beta_c\gamma y\right)\right]^{1/n}$
sech $(n = 1, 2, 3, \ldots)$	$\frac{1}{2}u^2\left(1-u^n\right)$	$u-\frac{n+2}{2}u^{n+1}$	1	0	$(\pm1)^{n+1}\,\mathrm{sech}^{2/n}\left(\frac{n}{2}\beta_c\gamma y\right)$
Lorentz	$2u^3-2u^4$	$6u^2-8u^3$	0	12	$\left[1+\left(\beta_c\gamma y\right)^2\right]^{-1}$

The additional transformation

$$2X = Z - T \quad \text{and} \quad 2Y = Z + T \tag{8.5 - 4a}$$

yields the *second standard form*

$$u_{XY} = \frac{d}{du}\Phi(u) \quad . \tag{8.5 - 4b}$$

Both standard forms are characteristic of *hyperbolic partial differential equations* (8.1 - 1) [Courant & Hilbert 1968 B, Landau & Lifshitz 1959 B, Witham 1974 B, Zauderer 1983 B, Zwillinger 1989 B].

A nonlinear Klein-Gordon equation of special interest is the normalized *Sine-Gordon equation*

$$u_{tt} - u_{zz} + \sin u = 0 \tag{8.5 - 5a}$$

with the potential

$$\begin{aligned} &\Phi(u) = 1 - \cos u = 2\sin^2(u/2) \quad , \\ &\text{where} \quad \Phi(0) = 0, \quad \frac{d}{du}\Phi(0) = 0 \quad . \end{aligned} \tag{8.5 - 5b}$$

The nonlinear Klein-Gordon equations (8.5 - 1a&b) describe nonlinear waves in homogeneous media. Equation (8.5 - 1b) can be easily modified to characterize nonlinear waves in aperiodic and periodic inhomogeneous media [Kneubühl & Feng 1991 J] by introduction of a variable refractive index $n(z)$ as follows

$$u_{zz} - n^2(z)c^{-2}\,u_{tt} = \beta_C^2 \frac{d}{du}\Phi(u,z) \quad . \tag{8.5 - 6}$$

8.5.1 Traveling Waves

Traveling-wave solutions of the Korteweg-de Vries equation have been characterized in the previous Section 8.4.2. The differential equation that determines traveling waves governed by Klein-Gordon equations can be derived from (8.5 - 3b) with the ansatz

$$u(Z,T) = U(\Theta) \quad , \quad U^2(\Theta) < \infty \quad , \quad \Theta = Z - VT \quad , \quad V = \upsilon / c \quad . \tag{8.5 - 7a}$$

The result is

$$\left(1 - V^2\right)U'' = \frac{1}{2}\left(1 - V^2\right)\frac{d}{dU}\left(U'\right)^2 = \frac{d}{dU}\Phi(U) \quad . \tag{8.5 - 7b}$$

The integration of this equation over U yields

$$(U')^2 = 2[\Phi(U) + \Phi_0](1 - V^2)^{-1} \quad . \tag{8.5 - 7c}$$

In this equation Φ_0 is an arbitrary constant. A real solution of (8.5 - 7c) requires that both factors on the right have the same sign. If yes (8.5 - 7c) can be written as

$$(1 - V^2)^{1/2} [\Phi(U) + \Phi_0]^{-1/2} \, dU = \sqrt{2} \, d\Theta \quad . \tag{8.5 - 7d}$$

The following calculation is restricted to the *Sine-Gordon equation* (8.5 - 5a&b). The corresponding traveling-wave equation (8.5 - 7d) is

$$(1 - V^2)^{1/2} [\Phi_0 + 2\, sin^2(U/2)]^{-1/2} \, d(U/2) = \frac{1}{\sqrt{2}} d\Theta \tag{8.5 - 8a}$$

with $\Theta = Z - VT$.

The introduction of the new variable

$$s = sin(U/2) \tag{8.5 - 8b}$$

transforms (8.5 - 8a) into

$$(1 - V^2)^{1/2} \left[(1 - s^2) \{ \Phi_0 / 2 + s^2 \} \right]^{-1/2} ds = d\Theta \quad . \tag{8.5 - 8c}$$

The solutions of this equation are discussed in the two subsequent Sections 8.5.2 & 3. They are determined by the values of Φ_0 and V^2.

8.5.2 Sine-Gordon Solitons

Soliton solutions of the Sine-Gordon equation (8.5 - 5a&b) exist on the condition

$$V^2 < 1 \quad \text{and} \quad \Phi_0 = 0 \quad . \tag{8.5 - 9a}$$

The integration of (8.5 - 8a) under these circumstances yields

$$u_{\alpha\sigma}(z,t) = U_{\alpha\sigma}(\Theta) = 4\alpha \; arc \; tan \left\{ exp\left[\sigma (1 - V^2)^{-1/2} \Theta \right] \right\} \tag{8.5 - 9b}$$

with $\alpha = \pm 1, \sigma = \pm 1$ and $\Theta = Z - VT$.

These solutions imply

$$\Delta U = U_{\alpha\sigma}(+\infty) - U_{\alpha\sigma}(-\infty) = \alpha \sigma 2\pi \quad . \tag{8.5 - 9c}$$

Accordingly, they are called

$$\begin{array}{llll} \text{"}kink\text{" solitons} & \text{for} \quad \alpha\sigma = +1 & \text{and} & \Delta U = +2\pi \quad , \\ \text{"}antikink\text{" solitons} & \text{for} \quad \alpha\sigma = -1 & \text{and} & \Delta U = -2\pi \quad . \end{array} \tag{8.5 - 9d}$$

Fig. 8.5 - 1 shows a "kink" soliton for $\alpha = \sigma = +1$ and $V^2 = 3/4$.

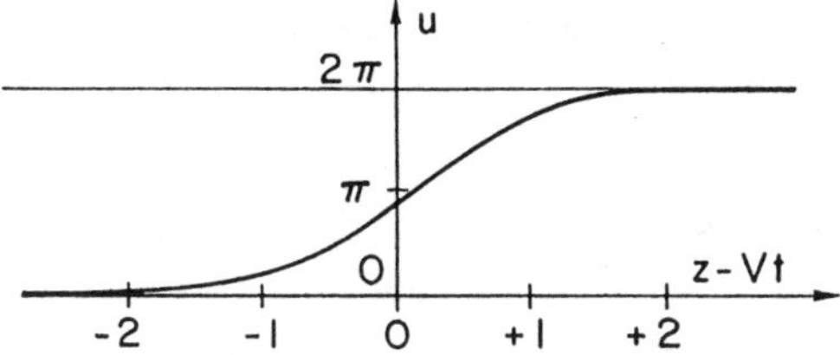

Fig. 8.5 - 1. Sine-Gordon "kink" soliton with $\alpha = \sigma = 1$ and $V^2 = 1$ (8.5 - 9b-d)

These and other Sine-Gordon solitons can be derived with the ansatz [Drazin 1983 B]

$$tan\left(\frac{1}{4}u(z,t)\right) = w(z,t) = f(z)\,/\,g(t) \quad . \tag{8.5 - 10a}$$

Because of the identity

$$sin\ u = 4\,w\left(1 - w^2\right)\left(1 + w^2\right)^{-2} \tag{8.5 - 10b}$$

the Sine-Gordon equation (8.5 - 5a&b) can be transformed into the differential equation for $w(z, t)$

$$\left(1 + w^2\right)\left(w_{tt} - w_{zz} + w\right) = 2w\left(w_t^2 - w_z^2 + w^2\right) \quad . \tag{8.5 - 10c}$$

It yields the following relation for $f(z)$ and $g(t)$ [Drazin 1983 B]

$$\frac{\left(f''/f\right)'}{\left(f^2\right)'} = -\frac{\left(\ddot{g}/g\right)^{\cdot}}{\left(g^2\right)^{\cdot}} = 2\mu = const$$

$$\text{with} \quad f' = \frac{d}{dz}f \quad \text{and} \quad \dot{g} = \frac{d}{dt}g \quad . \tag{8.5 - 11a}$$

Partial integration of the separated differential equations (8.5 - 11a) for $f(z)$ and $g(t)$ and the comparison with (8.5 - 10c) results in the ordinary differential equations

$$f'^2 = \mu f^4 + (1 + \lambda) f^2 + \nu \quad \text{and} \quad \dot{g}^2 = -\mu g^4 + \lambda g^2 - \nu \quad , \tag{8.5 - 11b}$$

whose solutions are Jacobi's elliptical functions [Abramowitz & Stegun 1965 B].

With respect to this method of solution it should be noticed that also the solution (8.5 - 9b) of (8.5 - 5a&b) can be split into the factors $f(z)$ and $g(t)$. This becomes obvious when this solution is represented by

$$tan\left(\alpha u / 4\right) = exp\, \sigma\left(1 - V^2\right)^{1/2}\left(Z - V T\right) \quad . \tag{8.5 - 11c}$$

In addition this method also yields solutions in the form of *two interacting identical solitons*. If $\mu = 0$ and $\lambda > 0$ in (8.5 - 11b), then there exists a solution of the form [Drazin 1983 B]

$$u(z,t) = 4\, arc\ tan\left\{V \frac{sinh\, \gamma Z}{cosh \gamma\, V T}\right\} \text{ with } \gamma = \left(1 - V^2\right)^{-1/2} \quad . \tag{8.5 - 12a}$$

This solution is illustrated in Fig. 8.5 - 2 with T = t, Z = z, c = 1. The two *identical "kink" solitons* approach each other for times $t >> -1$, interact at $t \approx 0$ and separate at $t >> +1$ without change of shape. This is characteristic of proper solitons.

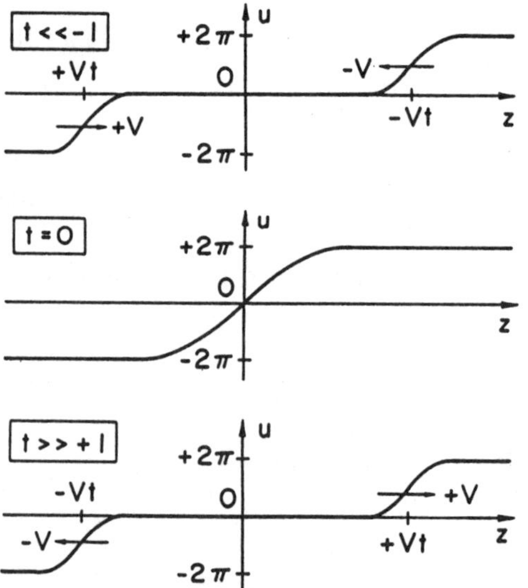

Fig. 8.5 - 2. Two interacting Klein-Gordon "kink"solitons (8.5 - 12a)

The soliton interaction is even better demonstrated by the interaction of *two identical "hump" solitons* whose representation is found by partial differentiation of the solution (8.5 - 12a)

$$u_z(z,t) = 4 \gamma V \frac{cosh\, \gamma Z\, cosh\, \gamma V T}{sinh^2\, \gamma Z + cosh^2\, \gamma V T} \quad . \tag{8.5 - 12b}$$

This interaction is shown in Fig. 8.5 - 3 with T = t, Z = z, c = 1.

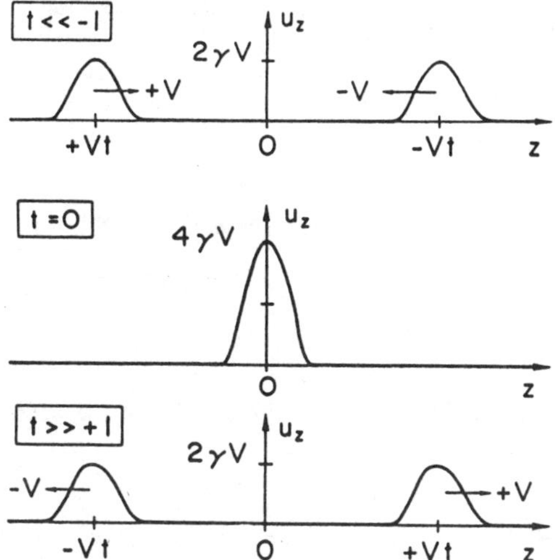

Fig. 8.5 - 3. Two interacting Klein-Gordon "hump" solitons (8.5 - 12b)

The Sine-Gordon equation (8.5 - 5a&b) also implies solutions in the form of oscillating pulses called *breathers* or *bions*. These solutions occur for $\lambda = -\omega^2 < 0$ and $v > 0$ in (8.5 - 11b). The breather [Drazin 1983 B]

$$u(z,t) = 4\, arc\ tan\left\{ \frac{\left(1-\omega^2\right)^{1/2}}{\omega} sech\left[\left(1-\omega^2\right)^{1/2} Z\right] sin\, \omega T \right\} \qquad (8.5 - 13)$$

is shown for different times t in Fig. 8.5 - 4 with T = t, Z = z, c = 1.

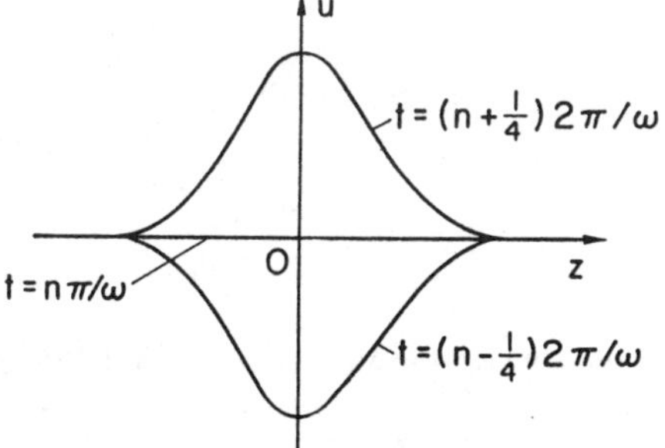

Fig. 8.5 - 4. Sine-Gordon breather or bion (8.5 - 13) at different times t

8.5.3 Cnoidal Sine-Gordon Waves

Besides solitons the Sine-Gordon equation (8.5 - 5a&b) has solutions that represent periodic anharmonic waves with finite amplitudes. They are designated as cnoidal waves since they include Jacobi's elliptical functions. Three different kinds can be distinguished with the aid of the parameters of (8.5 - 7d):

α) $V^2 < 1\,,\ \Phi_0 > 0$
The corresponding cnoidal waves are written as

$$u(z,t) = U(\Theta) = 2\,arc\ sin\left\{\left[m\Phi_0 / 2\right]^{1/2}\ sd\ (\left[m(1-V^2)\right]^{-1/2}\,\Theta|m)\right\} \qquad (8.5 - 14a)$$

$$\text{with} \quad \Theta = Z - VT \quad \text{and} \quad m = \left(1 + \Phi_0 / 2\right)^{-1} \quad .$$

The function $sd(x|m)$ represents a *periodic Jacobian elliptical function* [Abramowitz & Stegun 1965 B] with the limit functions

$$sd\left(x\,|\,0\right) = sin\ x \quad \text{and} \quad sd\left(x\,|\,1\right) = sinh\ x \quad . \qquad (8.5 - 14b)$$

Corresponding wavelength λ and period T are

$$\lambda = 4K(m)\left[m\left(1-V^2\right)\right]^{1/2} \quad \text{and} \qquad (8.5 - 14c)$$

$$T = 4K(m)\left[m\left(V^{-2}-1\right)\right]^{1/2} \quad , \qquad (8.5 - 14d)$$

where $K(m)$ indicates the complete elliptical integral of the first kind [Abramowitz & Stegun 1965 B, Gradshteyn & Ryzhik 1965 B] discussed in Section 2.5 and illustrated in Fig. 2.5 - 2.

β) $V^2 > 1\,;\ -2 \le \Phi_0 \le 0$
The corresponding cnoidal waves are

$$u(z,t) = U(\Theta) = 2arc\ sin\left\{m^{1/2} sn((V^2-1)^{-1/2}\,\Theta|m)\right\} \qquad (8.5 - 15a)$$

$$\text{with} \quad \Theta = Z - VT \quad \text{and} \quad m = -\Phi_0 / 2 \quad .$$

The function $sn(x|m)$ is the *Jacobian elliptic sine* [Abramowitz & Stegun 1965 B] with the limit functions

$$sn(x,0) = sin\ x \quad \text{and} \quad sn\left(x\,|\,1\right) = tanh\ x \quad . \qquad (8.5 - 15b)$$

Wavelength λ and period T of these cnoidal waves are

$$\lambda = 4K(m)\left(V^2-1\right)^{1/2} \quad \text{and} \qquad (8.5 - 15c)$$

$$T = 4K(m)\left(1 - V^{-2}\right)^{1/2} \quad . \tag{8.5 - 15d}$$

γ) $V^2 > 1\,; -\infty < \Phi_0 < -2$
The corresponding cnoidal waves have the form

$$u(z,t) = U(\Theta) = 2arc\ sin\left\{ sn\left(\left[m\left(V^2 - 1\right)\right]^{-1/2} \Theta \middle| m\right)\right\} \tag{8.5 - 16a}$$

with $\Theta = Z - VT$ and $m = -2 / \Phi_0$.

Their wavelength λ and period T are

$$\lambda = 4K(m)\left[m\left(V^2 - 1\right)\right]^{1/2} \quad \text{and} \tag{8.5 - 16b}$$

$$T = 4K(m)\left[m\left(1 - V^{-2}\right)\right]^{1/2} \quad . \tag{8.5 - 16c}$$

8.6 Nonlinear Schrödinger Equation

In wave mechanics the dynamics of a non-relativistic particle of mass m in a one-dimensional potential V is governed by the *time-dependent Schrödinger equation*

$$i\hbar\psi_t + \frac{\hbar^2}{2m}\psi_{zz} - V\psi = 0 \quad \text{with} \quad \psi = \psi(z,t) \quad , \tag{8.6 - 1}$$

where ψ indicates the wave function. In many cases the potential V is a function of position z and time t. Yet, it is possible that the potential V depends on the probability density $|\psi|^2$, e.g. in the form

$$V = V(|\psi|^2) = -\hbar\gamma|\psi|^2 \quad . \tag{8.6 - 2}$$

With this potential (8.6 - 1) takes the form of the cubic *nonlinear Schrödinger (NLS) equation*

$$i\psi_t + \frac{\hbar}{2m}\psi_{zz} + \gamma|\psi|^2\psi = 0 \quad . \tag{8.6 - 3}$$

By application of the transformation

$$T = (\gamma\, sign\, \gamma)t \quad \text{and} \quad Z = (m\,\gamma\, sign\, \gamma / \hbar)^{1/2} z \tag{8.6 - 4}$$

this equation is reduced to its *standard form*

$$i\,\psi_T + \frac{1}{2}\psi_{ZZ} + |\psi|^2\,\psi\, sign\,\gamma = 0 \quad . \tag{8.6 - 5}$$

The potential (8.6 - 2) of (8.6 - 3) decreases with increasing probability density $|\psi|^2$ when $\gamma > 0$. If, by accident, this density is at a certain position higher than in its vicinity, then the potential with $\gamma > 0$ causes its further increase, i.e. *self trapping of the particle*. With respect to solitary waves this results in *bright solitons*.

On the contrary the potential (8.6 - 2) decreases for decreasing probability density $|\psi|^2$ when $\gamma < 0$. If this density is at a certain position lower than in its neighborhood, the potential (8.6 - 2) with $\gamma < 0$ causes its further decrease. This implies *self trapping of holes* in the probability density and the corresponding formation of *dark solitons*.

The NLS equation (8.6 - 3) and its standard form (8.6 - 5) are relevant in wave mechanics. Yet, it is not the aim of this book to describe significance and application of the NLS equation in wave mechanics. Instead, the following sections of this book are dedicated to the *nonlinear Kerr media*, which are important in today's nonlinear optics, fiber optics [Agrawal 1989 B, Hasegawa 1989 B], quantum optics and optoelectronics. They are characterized by the *modified NLS equation* that is derived from the usual NLS equation (8.6 - 3) by the exchange of position z and time t as variables.

8.6.1 Nonlinear Kerr Media

A Kerr medium is defined by a *nonlinear optical refractive index* n of the form

$$n\left(\omega,|E|^2\right) = n_0(\omega) + n_2|E|^2 \quad , \tag{8.6 - 6a}$$

where ω indicates the circular frequency. The corresponding propagation constant β is

$$\beta = \beta\left(\omega,|E|^2\right) = c^{-1}\omega\, n\left(\omega,|E|^2\right) = c^{-1}\omega\, n_0(\omega) + c^{-1}\omega\, n_2|E|^2 \quad . \tag{8.6 - 6b}$$

For an electromagnetic wave whose frequency spectrum comprises only circular frequencies ω close to a *carrier circular frequency* ω_0 the propagation constant (8.6 - 6b) can be approximated by

$$\Delta\beta = \beta\left(\omega,|E|^2\right) - \beta(\omega_0,0) \approx \beta'\,\Delta\omega + \frac{1}{2}\beta''\Delta\omega^2 + \gamma|E|^2 \tag{8.6 - 7a}$$

with the parameters

$$\beta' = (\upsilon_g)^{-1} = c^{-1}N = c^{-1}\left[n_0(\omega_0) + \omega_0\, n_0'(\omega_0)\right] \quad ,$$

$$\beta'' = c^{-1}(dN/d\omega) = c^{-1}\left[2n_0'(\omega_0) + \omega_0\, n_0''(\omega_0)\right] \quad , \qquad (8.6 - 7b)$$

$$\gamma = c^{-1}\omega_0\, n_2 \quad .$$

These parameters determine group velocity υ_g, group index N and group dispersion $dN/d\omega$ by

$$N = c/\upsilon_g = c\left[\partial\beta/\partial\omega\right]_{\omega=\omega_0, E=0} \quad ,$$

$$dN/d\omega = c\left[\partial^2\beta/\partial\omega^2\right]_{\omega=\omega_0, E=0} \quad . \qquad (8.6 - 7c)$$

The replacement of the differences $\Delta\beta$ and $\Delta\omega$ in (8.6 - 7a) by the corresponding envelope operators

$$\boldsymbol{\Delta\beta}\, E = -i\, E_z \quad \text{and} \quad \boldsymbol{\Delta\omega}\, E = +i\, E_t \qquad (7.3 - 32b)$$

yields the *envelope equation* [Hasegawa 1989 B]

$$i\, E_z + i\beta' E_t - \frac{1}{2}\beta''\, E_{tt} + \gamma|E|^2 = 0 \quad . \qquad (8.6 - 8)$$

The elimination of the nonlinear term in this equation results in the modified linear Schrödinger equation (7.3 - 36).

Equation (8.6 - 8) describes an envelope that propagates with the group velocity υ_g. This permits the introduction of the *retarded time*

$$\tau = t - \beta' z = t - (z/\upsilon_g) \quad . \qquad (8.6 - 9a)$$

The transformation of the zt into the $z\tau$ coordinate system eliminates the E_t term of (8.6 - 8) with the result

$$i\, E_z - \frac{1}{2}\beta'' E_{\tau\tau} + \gamma|E|^2 E = 0 \quad . \qquad (8.6 - 9b)$$

This is the *modified NLS equation*. Its *standard form* is obtained by the transformation

$$Z = |\gamma|\, z \quad , \quad T = |\gamma/\beta''|\, \tau \quad , \qquad (8.6 - 10a)$$

which yields

$$i\,E_{\mathrm{Z}} - \frac{1}{2}\left(sign\frac{dN}{d\omega}\right)E_{\mathrm{TT}} + (sign\ \gamma)|E|^2\,E = 0 \quad . \tag{8.6 - 10b}$$

The dispersion relation of an NLS envelope can be derived with the aid of the ansatz

$$E(z,t) = E_0\ exp\left[i(\Delta\beta_{\mathrm{NLS}}\,z - \Delta\omega\,t)\right] \quad . \tag{8.6 - 11a}$$

The application of this ansatz to (8.6 - 8) results in the *nonlinear dispersion relation*

$$\Delta\beta_{\mathrm{NLS}} = \frac{N}{c}\Delta\omega + \frac{1}{2c}\left(\frac{dN}{d\omega}\right)\Delta\omega^2 + \gamma|E_0|^2 \tag{8.6 - 11b}$$

that includes the amplitude E_0. The corresponding *phase and group velocities* are determined by

$$c\,/\,\upsilon_{\mathrm{NLS}} = N + \frac{1}{2}\left(\frac{dN}{d\omega}\right)\Delta\omega + \gamma\,/\,\Delta\omega|E_0|^2 \quad , \tag{8.6 - 12a}$$

$$\text{and} \quad c\,/\,\upsilon_{\mathrm{gNLS}} = N + \left(\frac{dN}{d\omega}\right)\Delta\omega \quad . \tag{8.6 - 12b}$$

Pulses passing nonlinear Kerr media show *self-phase modulation* or *SPM* [Agrawal 1989 B] caused by the nonlinearity. This effect occurs independent of phase or group dispersion. Therefore β'', which according to (8.6 - 7b) represents the group dispersion in the NLS equation (8.6 - 9b), can be neglected in the description of *SPM*. As a consequence (8.6 - 9b) can be reduced to

$$\frac{\partial}{\partial z}E(z,\tau) = i\gamma\,|E(z,\tau)|^2\,E(z,\tau) \tag{8.6 - 13}$$

where τ is the retarded time defined by (8.6 - 9a). This equation has a solution of the form

$$E(z,\tau) = E(0,\tau)\,exp[i\varphi(z,\tau)] = E(0,\tau)\ exp\left[i\gamma|E(0,\tau)|^2\,z\right] \tag{8.6 - 14}$$

$$\text{with} \quad E(0,\tau) = E(0,t) \quad \text{and} \quad \varphi(z,\tau) = \varphi(z,t) \quad ,$$

where $\varphi(z,t)$ is the *self-modulated phase*. The corresponding *momentary frequency* ω_{m} is determined by

$$\omega_{\mathrm{m}}(z,t) = \frac{\partial}{\partial t}\varphi(z,t) = \gamma\,z\frac{d}{dt}|E(0,t)|^2 \quad . \tag{8.6 - 15}$$

The effect of self-phase modulation can be demonstrated with a *Gaussian pulse* propagating through a nonlinear Kerr medium. At the position $z = 0$ it has the form

$$E(0,t) = E_0\, exp\left(-t^2 / 2t_0^2\right) \quad . \tag{8.6 - 16a}$$

When this ansatz is applied to (8.6 - 14) and (8.6 - 15) it yields the following phase and frequency

$$\varphi(z,t) = \gamma\, E_0^2 z\, exp\left(-t^2 / t_0^2\right) \quad , \tag{8.6 - 16b}$$

$$\omega_{\mathrm{m}}(z,t) = -2\gamma\left(E_0 / t_0\right)^2 z\, t\, exp\left(-t^2 / t_0^2\right) \quad . \tag{8.6 - 16c}$$

While at the start position $z = 0$ of the pulse the phase and momentary frequency $\varphi = \omega_{\mathrm{m}} = 0$ do not depend on time t, they vary with time t at any other position $z = z_0 \neq 0$ on the path of the pulse.

8.6.2 Solitons in Kerr Media

There exist bright and dark solitons in nonlinear Kerr media as solutions of the modified NLS equation (8.6 - 9b). Because the Kerr coefficient n_2 of (8.6 - 6a) is usually positive [Hasegawa 1989 B] γ is also assumed to be positive.

If the Kerr medium exhibits *anomalous group dispersion* the modified NLS equation (8.6 - 9b) is characterized by

$$dN / d\omega < 0, \quad \beta'' < 0, \quad \text{and} \quad \gamma > 0 \quad . \tag{8.6 - 17a}$$

The corresponding standard form (8.6 - 10b) is

$$i\, E_{\mathrm{Z}} = -\frac{1}{2} E_{\mathrm{TT}} - |E|^2 E \quad . \tag{8.6 - 17b}$$

The negative sign of the potential term implies *self trapping of particles* according to the notice following (8.6 - 5). Therefore, one expects the existence of *bright solitons.* The simplest has the form

$$\begin{aligned} &E_{\mathrm{b}}(Z,T) = E_0\, exp\left(\frac{i}{2} E_0^2 Z\right) sech\left(E_0 T\right) \\ &\text{with} \quad Z = \gamma z, \quad T = \left[-c\gamma\left(\frac{dN}{d\omega}\right)\right]^{1/2} \left(t - (z / v_{\mathrm{g}})\right) \quad , \end{aligned} \tag{8.6 - 18a}$$

and the real amplitude $E_0 > 0$. The modulus and the phase of its field $E_0(Z, T)$ are determined by

$$|E_b(Z,T)|^2 = E_0^2 \, sech^2 \, E_0 T \quad , \tag{8.6 - 18b}$$

$$\varphi = \varphi(Z) = \frac{1}{2} E_0^2 \, Z \quad . \tag{8.6 - 18c}$$

This bright soliton (8.6 - 18b) is illustrated on the left side of Fig. 8.6 - 1.

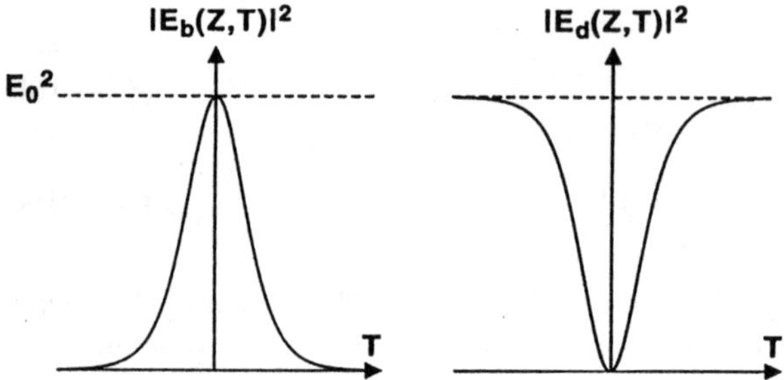

Fig. 8.6 - 1. Bright (8.6 - 18b) and dark (8.6 - 20b) soliton in a nonlinear Kerr medium

The modified NLS equation (8.6 - 9b) of a Kerr medium with *normal group dispersion* is characterized by

$$dN/d\omega > 0, \quad \beta'' > 0, \quad \text{and} \quad \gamma > 0 \quad . \tag{8.6 - 19a}$$

This implies the standard form (8.6 - 10b)

$$i\,E_Z = -\frac{1}{2} E_{TT} + |E|^2 E \quad . \tag{8.6 - 19b}$$

The positive potential term yields self trapping of holes according to the notice after (8.6 - 5). This results in the formation of *dark solitons*, e.g. a soliton in the form

$$E_d(Z,T) = E_0 \, exp\left(\frac{i}{2} E_0^2 \, Z\right) tanh \, E_0 T \tag{8.6 - 20a}$$

$$\text{with} \quad Z = \gamma z \quad , \quad T = \left[c\gamma / \frac{dN}{d\omega}\right]^{1/2} \left(z - (t/\upsilon_g)\right) \quad ,$$

and the real amplitude $E_0 > 0$. Modulus and phase of the field $E_d(Z, T)$ are determined by

$$|E_d(Z,T)|^2 = E_0^2 \left(1 - sech^2 \, E_0 T\right) = E_0^2 \left(tanh \, E_0 T\right)^2 \quad , \tag{8.6 - 20b}$$

$$\varphi = \varphi(Z) = \frac{1}{2} E_0^2 Z \quad . \tag{8.6 - 20c}$$

The dark soliton (8.6 - 20b) illustrated on the right side of Fig. 8.6 - 1 represents a hole in the otherwise constant level E_0^2 with the shape of the bright soliton (8.6 - 18b). This is confirmed by the fact that the sum of the bright and the dark soliton shown in Fig. 8.6 - 1 is constant

$$|E_b(Z,T)|^2 + |E_d(Z,T)|^2 = E_0^2 \quad . \tag{8.6 - 21}$$

More complicated soliton solutions of the NLS equation (8.6 - 9b) are discussed in the literature [Dodd et al. 1982 B, Hasegawa 1989 B, etc.].

8.6.3 Kerr Media with Gain

Nonlinear Kerr media with gain form the basis of models of fiber lasers [Agrawal 1989 B, Hasegawa 1989 B]. The generation of solitons by these lasers is essential for communication with optical fibers because they travel extreme distances without deterioration and with little attenuation. The nonlinear Kerr media can be described with an *extended modified NLS equation* [Hasegawa 1989 B, Zakharov & Shabat 1972 J]. A Kerr medium with anomalous dispersion ($\beta'' < 0$), a positive Kerr coefficient ($\gamma > 0$) and gain or loss is represented by the *standard equation*

$$i E_Z = -E_{TT} - |E|^2 E + i g E \quad , \tag{8.6 - 22}$$

where $g > 0$ signifies gain and $g < 0$ indicates loss.

An approximate solution of (8.6 - 22) is the *pseudo-soliton*

$$E(Z,T) \approx E_0(Z) \, exp\big(i \varphi_g(Z)\big) \, sech\big(E_0(Z) T\big) \quad \text{with} \tag{8.6 - 23a}$$

$$E_0(Z) = E_0 \, exp \, 2(g Z) \quad , \tag{8.6 - 23b}$$

$$\varphi_g(Z) = \frac{1}{8g} E_0^2 \{ exp(4 g Z) - 1 \} \quad , \tag{8.6 - 23c}$$

$$Z = \gamma z \ \text{ and } \ T = \left[-c\gamma / \frac{dN}{d\omega} \right]^{1/2} \big(t - (z / \upsilon_g) \big) \quad . \tag{8.6 - 23d}$$

This approximation is valid if

$$|g| << |E(Z,T)|^2 \quad . \tag{8.6 - 23e}$$

The approximation of the exponential functions of (8.6 - 23b&c) yields

$$E_0(Z) \approx E_0(1+2g\,Z) \quad , \tag{8.6 - 24a}$$

$$\varphi_g(Z) \approx \frac{1}{2} E_0^2 Z(1+2g\,Z) \quad . \tag{8.6 - 24b}$$

For an interpretation of the solution (8.6 - 23a-d) it is of advantage to take into account the carrier wave with the circular frequency ω_0 and the propagation constant $\beta_0 = \beta(\omega_0, E = 0)$. Then the solution takes the form

$$E(z,t) = E_g(z,t)\,exp\left(i\,\Phi_g(z,t)\right) \quad \text{with} \tag{8.6 - 25a}$$

$$E_g(z,t) \approx E_0(1+2\gamma g\,z)$$
$$sech\left\{E_0(1+2\,\gamma g z)\left[-c\gamma / \frac{dN}{d\omega}\right]^{1/2}\left(t-(z/\upsilon_g)\right)\right\} \quad , \tag{8.6 - 25b}$$

$$\Phi_g(z,t) \approx \beta_0 z - \omega_0 t + \varphi_g(z,t) = \left(\beta_0 + \frac{\gamma}{2}E_0^2\right)z + g\,\gamma^2 E_0^2\, z^2 - \omega_0 t \quad . \tag{8.6 - 25c}$$

The corresponding *propagation constant* and *phase velocity* depend on the amplitude E_0 and on the position z

$$\beta(z) = \frac{\partial}{\partial z} = \Phi_g = \beta_0 + \frac{\gamma}{2}E_0^2 + 2g\,\gamma^2\, E_0^2\, z \quad , \tag{8.6 - 26a}$$

$$\upsilon(z) = \upsilon_0\left[1+\frac{1}{2}(\gamma/\beta_0)E_0^2(1+4g\gamma z)\right]^{-1} \quad \text{with} \quad \upsilon_0 = \omega_0/\beta_0 \quad . \tag{8.6 - 26b}$$

The pseudo-soliton (8.6 - 25a-c) exhibits a *chirp* because it travels at a constant group velocity υ_g whilst its phase velocity υ varies with the position z. Equation (8.6 - 26b) demonstrates that the chirp of this pseudo-soliton originates in the gain g.

8.7 Maxwell-Bloch Equations

In nonlinear optics and quantum optics the dynamics of short optical pulses in dielectric media with nonlinear polarization $\vec{P}$ are described by the *Maxwell-Bloch equations* or *optical Bloch equations* [Allen & Eberly 1975/87 B, Dicke 1953 J, Dodd et al. 1982 B, Feynman et al. 1957 J, Lamb 1971 J, McCall & Hahn 1967 J, 1969 J, Meystre & Sargent 1990 B].

In the theory of *self-induced transparency* one considers a short optical pulse passing an absorbing medium that contains a homogeneous concentration n_0 of almost identical atomic *two-level systems*. The energy difference ΔE between these two levels is assumed to be

$$\Delta E = E_a - E_b = \hbar(\omega_{ab} + \Delta\omega) \quad \text{with} \quad |\Delta\omega| << \omega_{ab} \quad , \qquad (8.7 - 1)$$

where $\Delta\omega$ is individual for each system. This assumption means *inhomogeneous broadening* [Kneubühl & Sigrist 1995 B, Meystre & Sargent 1990 B, Svelto & Hanna 1989 B] of the resonance line at the circular frequency ω_{ab}.

The transitions between the two levels E_a and E_b are induced by an electrical dipole moment $\vec{p}$. Consequently, the *polarization* $\vec{P}$ caused by the two-level systems can be approximated by

$$\vec{P} = n_0 \vec{p} \quad . \qquad (8.7 - 2)$$

The electric field $\vec{E}$ in the medium is determined by the *wave equation*

$$\Delta\vec{E} - c^{-2}\vec{E}_{tt} - (\sigma / \varepsilon_0)\vec{E}_t = \varepsilon_0^{-1}\vec{P} \quad . \qquad (8.7 - 3)$$

In this equation the absorption by the medium is described by the electrical conductivity σ.

A *TEM wave* with the electrical field $\vec{E}$ in the x direction that is propagating in the z direction can be represented by the scalar ansatz

$$E_x(z,t) = E(z,t) cos\, \Phi(z,t) = E(z,t) cos(\beta_0 z - \omega_0 t + \varphi(z,t)) \quad , \qquad (8.7 - 4a)$$

where $\varphi(z, t)$ and $E(z, t)$ indicate phase and envelope. The carrier is characterized by the circular frequency ω_0 and the propagation constant $\beta_0 = \omega_0/c$. It is assumed that $\varphi(z, t)$ and $E(z, t)$ vary slowly with time t and position z. This implies

$$E_t << \omega_0 E \quad , \quad E_z << \beta_0 E \quad \text{and} \quad \varphi_t << \omega_0 \varphi \quad , \quad \varphi_z << \beta_0 \varphi \quad . \qquad (8.7 - 4b)$$

In order to formulate the dynamics of the electric field (8.7 - 4a) it is of advantage to presume that the carrier frequency ω_0 of the incoming short optical pulse equals the central resonance frequency ω_{ab} of the two-level systems

$$\omega_0 = \omega_{ab} \quad . \qquad (8.7 - 5)$$

The electric dipole moment $\vec{p} = [p_x, 0, 0]$ of a two-level system with the resonance frequency $\omega_{ab} + \Delta\omega$ varies with position z and time t. It can be decomposed into a component in phase with the electric field $E(z, t)$ and a component with a phase shift of $\pi/2$

$$p_x = p(z,t,\Delta\omega) = p_M \left\{ P(z,t,\Delta\omega) sin\, \Phi(z,t) + Q(P,t,\Delta\omega) cos\, \Phi(z,t) \right\} \ . \quad (8.7 - 6)$$

Here p_M indicates the *matrix element* of the dipole with respect to the transition *ab*. The corresponding *Rabi circular frequency* is defined as

$$\Omega_R(z,t) = p_M E(z,t)/\hbar \quad . \quad (8.7 - 7)$$

On these conditions $P(z, t, \Delta\omega)$ and $Q(z, t, \Delta\omega)$ fulfill the *Maxwell-Bloch equations*

$$\dot{N} = -(N \pm 1)/T_1 - \Omega_R P \quad , \quad (8.7 - 8a)$$

$$\dot{P} = -P/T_2 + (\Delta\omega + \varphi_t)Q + \Omega_R N \quad , \quad (8.7 - 8b)$$

$$\dot{Q} = -Q/T_2 - (\Delta\omega + \varphi_t)P \quad , \quad (8.7 - 8c)$$

where $N(z, t, \Delta\omega)$ represents the normalized *population inversion*, while T_1 and T_2 indicate the longitudinal and transversal *relaxation times* of the classical *Bloch equations* of nuclear magnetic resonance or nmr. The factor $(N \pm 1)$ originates in the initial condition $N(z, -\infty) = \pm 1$.

Self-induced transparency works only with extremely short optical pulses. Their pulse width Δt is considerably shorter that the relaxation times T_1 and T_2. Therefore T_1 and T_2 do not influence the pulse dynamics. This can be taken into account by assuming

$$T_1 = T_2 = \infty \quad . \quad (8.7 - 9)$$

On this assumption the Maxwell-Bloch equations (8.7 - 8a-c) are reduced to

$$\dot{N} = -\Omega_R P \quad , \quad (8.7 - 10a)$$

$$\dot{P} = (\Delta\omega + \varphi_t)Q + \Omega_R N \quad , \quad (8.7 - 10b)$$

$$\dot{Q} = -(\Delta\omega + \varphi_t)P \quad . \quad (8.7 - 10c)$$

If $\varphi_{tt} \approx 0$ equations (8.7 - 10b&c) can be replaced by [Dodd et al. 1982 B]

$$\ddot{P} + (\Delta\omega + \varphi_t)^2 P = \frac{\partial}{\partial t}(\Omega_R N) \quad . \quad (8.7 - 11)$$

Multiplication of (8.7 - 10a-c) with N, P and Q and subsequent addition yields

$$N\dot{N} + P\dot{P} + Q\dot{Q} = 0 \quad . \quad (8.7 - 12)$$

Significant initial conditions for (8.7 - 10a-c) are

$$N(z,-\infty)=\pm 1 \quad \text{and} \quad P(z,-\infty)=Q(z,-\infty)=0 \quad . \tag{8.7 - 13a}$$

They imply that at time $t = -\infty$ the electric dipole (8.7 - 6) is zero, and level a or b is fully occupied. On these conditions the two-level systems neither absorb nor emit radiation at time $t = -\infty$.

The integration of (8.7 - 12) over time t with the initial conditions (8.7 - 13a) yields

$$N^2 + P^2 + Q^2 = 1 \quad . \tag{8.7 - 13b}$$

According to (8.7 - 13a&b) the vector $[N, P, Q]$ starts at $[\pm 1, 0, 0]$ and then moves on the *unit sphere*.

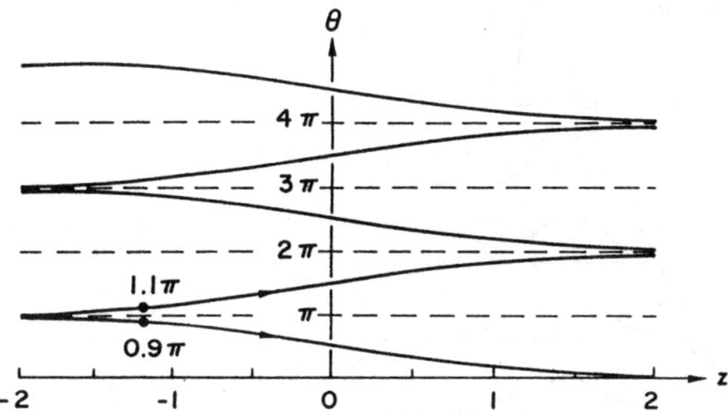

Fig. 8.7 - 1. Pulse area $\Theta(z)$ of an attenuator characterized by a negative sign in (8.7 - 15a&b) [McCall & Hahn 1967 J]

With respect to short optical pulses the solution of the reduced Maxwell-Bloch equations (8.7 - 10a-c) is governed by the *pulse-area theorem*. The pulse area is defined as the angle

$$\Theta = \Theta(z) = \int_{-\infty}^{+\infty} \Omega_{\mathrm{R}}(z,t)dt = (p_{\mathrm{M}} / \hbar) \int_{-\infty}^{+\infty} E(z,t)dt \quad . \tag{8.7 - 14}$$

For a non-absorbing optical medium with zero electrical conductivity $\sigma = 0$ the Maxwell-Bloch equations (8.7 - 10a-c) yield the following differential equation for the pulse area $\Theta(z)$

$$d\Theta / dz = \pm(\alpha / 2) sin\, \Theta$$
$$\text{with} \quad \alpha = const\ \omega_0 n_0 p_{\mathrm{M}}^2 \quad . \tag{8.7 - 15a}$$

Its solution

$$tan\left(\frac{1}{2}\Theta(z)\right)=tan\left(\frac{1}{2}\Theta(z_0)\right)exp\left[\pm\alpha(z-z_0)\right] \tag{8.7 - 15b}$$

is illustrated in Fig. 8.7 - 1 for the minus sign.

The optical medium is called *attenuator* if the sign in (8.7 - 15a&b) is negative according to Fig. 8.7 - 1 [Lamb 1971 J]. If on this condition an optical pulse starts at z_0 with the area $\Theta(z_0) < \pi$ e.g. 0.9π, then its area decreases with increasing z. If, however, the pulse starts at z_0 with the area $\pi > \Theta(z_0) < 2\pi$, e.g. 1.1π, then its area increases towards 2π with increasing z and a stable pulse with $\Theta = 2\pi$ is formed. These phenomena are illustrated in Fig. 8.7 - 2.

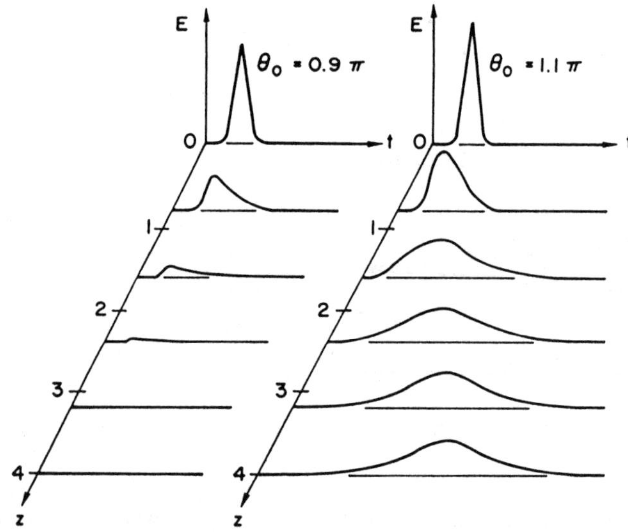

Fig. 8.7 - 2. Development of pulses in an attenuator with $\Theta(0) = 0.9\pi$ and $\Theta(0) = 1.1\pi$ [McCall & Hahn 1967 J]

The stable *2π pulse* with

$$\Theta(z) = 2\pi = const \tag{8.7 - 16a}$$

has the form

$$E(z,t) = 2\beta\upsilon\, sech\left[\beta(z-\upsilon t)-\varphi\right] = 2\omega\, sech\left[\beta z - \omega t - \varphi\right] \quad , \tag{8.7 - 16b}$$

the dispersion relation [Dodd et al. 1982 B]

$$\beta^2 = \omega^2\left[1+\alpha\left(\omega_0^2+\omega^2\right)^{-1}\right] \quad , \tag{8.7 - 16c}$$

and the velocity

$$\upsilon = \pm\left[1+\alpha\left(\omega_0^2+\omega^2\right)^{-1}\right]^{-1/2} \quad . \tag{8.7 - 16d}$$

If the sign in (8.7 - 15a&b) is positive, then the optical medium is designated *amplifier* [Lamb 1971 J]. This situation is represented by Fig. 8.7 - 1 when the sign of z is inverted. If an optical pulse in an amplifier starts at z_0 with an area $\Theta < \pi$, e.g. 0.9π, then this area increases with z and approaches π. Thus a stable pulse with $\Theta = \pi$ is formed. if, on the other hand, this pulse starts at z_0 with an area $\pi < \Theta < 2\pi$, this area decreases with increasing z and also approaches π. This also results in the formation of the pulse with $\Theta = \pi$.

The stable π *pulse* with

$$\Theta(z) = \pi = const \tag{8.7 - 17a}$$

has the form [Lamb 1971 J]

$$E(z,t) = \beta\,\upsilon\,sech\left[\beta(z-\upsilon t)-\psi\right] = \omega\,sech\left[\beta z-\omega t-\psi\right] \quad . \tag{8.7 - 17b}$$

The π pulse as well as the 2π pulse represent *solitary waves*. However, they are *no solitons* because there exist no stable $n\pi$ pulses for $n > 2$ on one hand and neither the π pulse nor the 2π pulse survives a collision [Dodd et al. 1982 B].

8.8 Toda Chain

The Toda chain [Toda 1967 J, 1981 B] constitutes a discrete periodic and nonlinear medium. It consists of identical masses m connected by identical *nonlinear springs* as illustrated in Fig. 7.6 - 1. In equilibrium the springs have the length a. The characteristic of the springs is nonlinear, yet its form permits the analytical solution of the chain's equation of motion.

The *general equation of motion* of identical masses and springs has the form

$$m\ddot{u}_n = F(r_{n-1}) - F(r_n) = \Phi'(r_n) - \Phi'(r_{n-1}) \quad \text{with} \quad n = 0, \pm 1, \pm 2, \ldots \quad . \tag{8.8 - 1a}$$

The variable u_n designates the displacement of the mass with the equilibrium position na. $F(r_n)$ is the force excerted by the spring between the masses n and $n+1$ and $\Phi(r_n)$ represents its potential according to

$$F(r_n) = -\Phi'(r_n) \quad \text{where} \quad r_n = u_{n+1} - u_n \quad . \tag{8.8 - 1b}$$

A chain with *linear* springs, which is discussed in Section 7.6 and elucidated by Figs. 7.6 - 1 to 7.6 - 3, is defined by the potential and the corresponding linear force

$$\Phi(r_n) = (f/2)r_n^2 \quad \text{and} \quad F(r_n) = -f\,r_n \quad , \tag{8.8 - 2}$$

while the *Toda chain* is characterized by the following potential and nonlinear force

$$\begin{aligned} &\Phi(r_n) = (f/b^2)\left[exp(-b\,r_n) + b\,r_n\right] \quad \text{and} \\ &F(r_n) = (f/b)\left[\exp(-b\,r_n) - 1\right] \\ &\text{with} \quad f > 0; -\infty < b < +\infty \quad . \end{aligned} \tag{8.8 - 3}$$

For *small* r_n^2 the forces in the Toda chain correspond to those in the chain with linear springs because

$$\begin{aligned} &\Phi(r_n \to 0) \approx (f/b^2) + (f/2)r_n^2 \quad \text{and} \\ &F(r_n \to 0) \approx -f\,r_n \quad . \end{aligned} \tag{8.8 - 4}$$

For large r_n^2 the forces $F(\pm r_n)$ differ according to

$$\lim_{br_n \to +\infty} F(r_n) = -(f/b) \quad \text{and} \quad \lim_{br_n \to -\infty} F(r_n) = \pm\infty \quad . \tag{8.8 - 5}$$

The solutions of the equation of motion (8.8 - 1a&b) with Toda potential and force (8.8 - 3) comprehend solitons as well as cnoidal waves. These are presented in the next two sections.

8.8.1 Toda Solitons

The Toda solitons as solutions of (8.8 - 1a&b) and (8.8 - 3) have the form [Toda 1981 B]

$$exp(-br_n) - 1 = A^2\,sech^2\left[\beta(an \pm \upsilon t)\right] \tag{8.8 - 6a}$$

with the amplitude

$$A = sinh(\beta a) \quad , \tag{8.8 - 6b}$$

and the velocity

$$\begin{aligned} &\upsilon = \omega_0 A/\beta = (\omega_0/\beta)\,sinh(\beta a) \geq \upsilon_0 = \omega_0 a \quad , \\ &where \quad \omega_0 = +\sqrt{f/m} \quad . \end{aligned} \tag{8.8 - 6c}$$

The solution (8.8 - 6a-c) represents *bright solitons* [Toda 1981 B]. Dark Toda solitons of the form

$$exp(-br_n) = C^2 - B^2 sech^2[\beta(an \pm \upsilon t)] \qquad (8.8 - 7)$$

do not exist.

8.8.2 Cnoidal Toda Waves

Cnoidal Toda waves as solutions of (8.8 - 1a&b) and (8.8 - 3) exist in the form [Toda 1981 B]

$$exp(-br_n) - 1 = A^2 \left\{ dn^2 (2K(m)\left[\frac{na}{\lambda} \pm \frac{t}{T}\right] \mid m) - \frac{E(m)}{K(m)} \right\} \quad . \qquad (8.8 - 8a)$$

In this equation $K(m)$ and $E(m)$ signify the complete elliptical integrals of the first and the second kind, whilst $dn(x|m)$ designates a Jacobian elliptical function [Abramowitz & Stegun 1965 B]. Wavelength λ and period T of this wave are linked by the dispersion relation

$$\omega_0 T = 2K(m)\left[sn^{-2}(2K(m)a / \lambda \mid m) - 1 - \frac{E(m)}{K(m)} \right]^{1/2} \quad , \qquad (8.8 - 8b)$$

where $sn(x|m)$ indicates Jacobi's elliptical sine [Abramowitz & Stegun 1965 B]. Amplitude A^2 and period T determine the parameter m in $K(m)$, $E(m)$, etc. by the relation

$$2K(m) = \omega_0 T A \geq \pi \quad . \qquad (8.8 - 8c)$$

Thus, the complete elliptical integral $K(m)$ that is illustrated in Fig. 2.2 - 3 also limits the ranges of A and T.

The following approximations are valid for small parameters $m \to 0$ [Abramowitz & Stegun 1965 B]

$$sn(x|m) \approx sin\, x \quad , \qquad (8.8 - 9a)$$

$$dn^2(x|m) \approx 1 - \frac{m}{2} + \frac{m}{2} cos\, 2x \quad , \qquad (8.8 - 9b)$$

$$E(m) / K(m) \approx 1 - \frac{m}{2} \quad . \qquad (8.8 - 9c)$$

As a consequence, a cnoidal Toda wave (8.8 - 8a) becomes a *harmonic wave* when m approaches zero

$$-b\,r_{\mathrm{n}} \approx \frac{m}{8}(a/\lambda)^2 \cos 2\pi\left[\frac{na}{\lambda} \pm \frac{t}{T}\right] \quad \text{for} \quad m \to 0 \quad . \qquad (8.8 - 10a)$$

Wavelength λ and period T of this wave are related by [Toda 1981 B]

$$\lambda / T \approx \upsilon_0 = \omega_0\, a \quad . \qquad (8.8 - 10b)$$

This demonstrates that the velocity υ (8.8 - 6c) of the Toda solitons (8.8 - 6a) is higher than the velocity υ_0 of the approximately harmonic waves (8.8 - 10a).

9. Standing Waves

A *standard example* of a *one-dimensional standing wave* [Alonso & Finn 1967 B, 1970 B, 1992 B, Courant & Hilbert 1968 B, Halliday et al. 1993 B, Hazen & Pidd 1965 B, Kneubühl 1994 B] is the *motion of a string* in a guitar, harp, violin or piano that is stretched between two clamps separated by a fixed distance. Such a string is somehow made to oscillate. The driving force does not need to have a particular frequency of its own. The string selects automatically a frequency such that an integer number of half wavelengths fits into the distance between the two fixed supports. This phenomenon is called *resonance*. The lowest resonant frequency usually predominates, while higher resonant frequencies called *overtones* contribute various amounts. In the case of a string it is obvious that the overtones are integral multiples of the lowest frequency and are therefore named *harmonics*.

If the string is *in resonance*, then the standing-wave patterns or *modes* corresponding to the resonant frequencies show a number of *nodes*. These are the points on the string where the amplitude of oscillation is zero. Both ends of the string represent nodes according to the boundary condition that they are fixed. Such *boundary conditions* are essential in the theory of standing waves. It should also be noticed that even a small driving force can give rise to a large amplitude of a resonant standing wave.

For harmonic waves there exists a *relation between one-dimensional standing and traveling harmonic waves*. According to this relation the interference of two harmonic waves with identical amplitudes and phases yet opposite directions produces a standing wave [Halliday et al. 1993 B].

Another example of one-dimensional standing waves are those of air inside a pipe [Alonso & Finn 1967 B, 1970 B]. Standing waves *in two dimensions* [Alonso & Finn 1967 B 1970 B, Kneubühl 1974 B] occur on a drumskin, a membrane in a microphone and on the tympanic membrane, whereas those *in three dimensions* are found in the resonant cavities of acoustics [Helmholtz 1882 J, 1896 B, Hess 1983 J, Morse 1936 B, Mores & Ingard 1986 B, Rayleigh 1870 B, Rosencwaig 1980 B] and of microwave techniques [Atwater 1962 B, Baden-Fuller 1969 B, Borgnis & Papas 1958 J, Pöschl 1955 B, Ramo et al. 1965 B].

Standing waves are also relevant in non-relativistic wave mechanics as *solutions of the time-independent Schrödinger equation* [Baym 1969 B, Blochinzew 1966 B, Cohen-Tannoudji et al. 1977 B, Fick 1968 B, Flügge 1990 B, Landau & Lifschitz 1979 B, Messiah 1960 B, 1969 B, 1990 B, Schubert & Weber 1980 B]. In this equation, the energy eigenvalues replace the resonant frequencies according to Planck's relation (7.2 - 16a).

This chapter is dedicated mainly to linear and nonlinear standing waves in one dimension. Also discussed are two- and three-dimensional standing waves governed by the Hertz equation (7.2 - 7).

9.1 Wave Form and Boundary Conditions

In general, a standing wave is a solution $u(\vec{r},t)$ of a wave equation that can be represented as a product of a function $g(t)$ of time t and a function of position $\vec{r}$ [Courant & Hilbert 1968 B]

$$u(\vec{r},t) = g(t)U(\vec{r}) \quad . \qquad (9.1 - 1a)$$

For standing waves in one-dimensional media this equation is reduced to

$$u(z,t) = g(t)U(z) \quad , \qquad (9.1 - 1b)$$

where z indicates the position.

In many cases the factor $g(t)$ describes a harmonic oscillation. The factor $U(\vec{r})$ or $U(z)$ represents the corresponding *standing-wave pattern*. Thus, a standing wave constitutes an *oscillation of an extended medium* with amplitude and phase depending on the position $\vec{r}$ or z. Positions $\vec{r}_N$ or z_N with zero amplitudes

$$U(\vec{r}_N) = 0 \quad \text{or} \quad U(z_N) = 0 \qquad (9.1 - 2)$$

are called *nodes, nodal lines or surfaces*.

The standing-wave pattern $U(\vec{r})$ or $U(z)$ is determined by the wave equation on one hand and by the *boundary conditions* on the other hand. The definition of a standing wave requires therefore the knowledge of both, the wave equation and the boundary conditions. The boundary conditions also determine the domain of a standing wave.

For a *standing wave in one dimension* the boundary conditions describe the situations at the ends $z = a$ and $z = b$ of its domain $a \le z \le b$. In mathematics they are therefore called *endpoint conditions* [Birkhoff & Rota 1989 B]. Typical endpoint conditions for wave equations in the form of second order differential equations are the *homogeneous conditions*:

α) fixed ends

$$u(a,t) = u(b,t) = 0 \quad , \qquad (9.1 - 3a)$$

β) free ends

$$u_z(a,t) = u_z(b,t) = 0 \quad , \tag{9.1 - 3b}$$

γ) a fixed and a free end

$$u(a,t) = u_z(b,t) = 0 \quad , \tag{9.1 - 3c}$$

δ) the Sturm-Liouville conditions

$$A\,u(a,t) + A'\,u_z(a,t) = 0 \quad \text{and} \quad B\,u(b,t) + B'\,u_z(b,t) = 0 \quad , \tag{9.1 - 3d}$$

and the *inhomogeneous conditions*

$$\varepsilon)\quad U(a) \neq 0 \quad \text{and/ or} \quad U(b) \neq 0 \quad . \tag{9.1 - 4}$$

In the situation where the domain of a standing wave is infinite, e.g. $-\infty, \leq z \leq +\infty$, the endpoint conditions (9.1 - 3a-d) or (9.1 - 4) are usually replaced by a *normalization* or standardization in the form

$$\int_{-\infty}^{+\infty} U(z)^2 \, d\,z = 1 \quad , \tag{9.1 - 5a}$$

or with a *weight function* $\rho(z)$ [Abramowitz & Stegun 1965 B]

$$\int_{-\infty}^{+\infty} U(z)^2 \, \rho(z) \, d\,z = 1 \quad . \tag{9.1 - 5b}$$

An example is the boundary condition for the wave functions of the harmonic oscillator discussed in Section 9.5.2c.

9.2 Free Vibrations of a String

As mentioned before, the free vibrations of a homogeneous string stretched between two clamps separated by the fixed distance L represent a standard example of *standing waves in one dimension* [Alonso & Finn 1967 B, 1970 B, 1992 B, Courant & Hilbert 1968 B, Halliday et al. 1993 B, Hazen & Pidd 1965 B, Kneubühl 1994 B]. Waves of a string are governed by the *Hertz equation*

$$u_{tt} - \upsilon^2 u_{zz} = 0 \tag{7.4 - 4}$$

with the constant velocity

$$\upsilon = (T/\rho)^{1/2} \tag{7.1 - 5c}$$

that is determined by the density ρ [kg m^{-3}] and the tension T [Nm^{-2}] of the string.
With the ansatz

$$u(z,t) = g(t)U(z) \tag{9.1 - 1b}$$

for standing waves the Hertz equation (7.4 - 4) yields

$$\dot{g}(t)/g(t) = \upsilon^2 U''(z)/U(z) = -\omega^2 = const \quad . \tag{9.2 - 1a}$$

These two quotients are constant because the quotient on the left is a function of time t, whereas that on the right represents a function of position z. The relation (9.2 - 1a) can therefore be split into the two ordinary differential equations

$$\ddot{g} + \omega^2 g = 0 \quad \text{and} \quad U'' + \beta^2 U = 0 \quad \text{with} \quad \beta = \omega/\upsilon \quad . \tag{9.2 - 1b}$$

The solutions of these equations are

$$g(t) = sin(\omega t - \alpha) \quad , \tag{9.2 - 2a}$$

$$U(z) = R\,sin(\beta z - \phi) = S\,sin\,\beta z + C\,cos\,\beta z \quad , \tag{9.2 - 2b}$$

$$\text{with} \quad R^2 = S^2 + C^2 \quad , \quad tan\Phi = -C/S \quad .$$

The circular frequency ω of (9.2 - 2a&b) is an arbitrary parameter because these solutions do not take account of the boundary conditions of the vibrating string.

a) Fixed-end conditions

The homogeneous fixed-end conditions

$$u(0,t) = U(0) = 0 \quad \text{and} \quad u(L,t) = U(L) = 0 \tag{9.2 - 3a}$$

of the type (9.1 - 3a) for the solutions (9.2 - 2a&b) agree with the assumption that the string is fixed by two clamps separated by the distance L. The application of these conditions results in

$$C = 0 \quad \text{and} \quad \beta L = n\pi \quad \text{with} \quad n = 0,1,2,3,\ldots \quad . \tag{9.2 - 3b}$$

If $n = 0$ the solution (9.2 - 2a&b) is trivial

$$\beta_0 = \omega_0 \quad \text{and} \quad u_0(z,t) = U_0(z) = 0 \quad . \tag{9.2 - 4}$$

It describes the *string at rest.*

For n = 1, 2, 3, ... the solutions (9.2 - 2a&b) represent the *resonant modes* or *eigenfunctions*

$$u_n(z,t) = sin(\omega_n t - \alpha_n) U_n(z) = U_n \, sin(\omega_n t - \alpha_n) sin \, \beta_n z \qquad (9.2 - 5a)$$

with the resonant propagation constants and the resonant wavelengths

$$\beta_n = n\pi / L \quad \text{and} \quad \lambda_n = 2\pi / \beta_n = 2L / n \qquad (9.2 - 5b)$$

and the *resonant circular frequency* or *eigenfrequency*

$$\omega_n = n\pi \upsilon / L \quad \text{with} \quad n = 1,2,3,\ldots \quad . \qquad (9.2 - 5c)$$

The standing-wave patterns $U(z)$ of the resonant modes (9.2 - 5a) with n = 1, 2, 3, 4 are illustrated in Fig. 9.2 - 1.

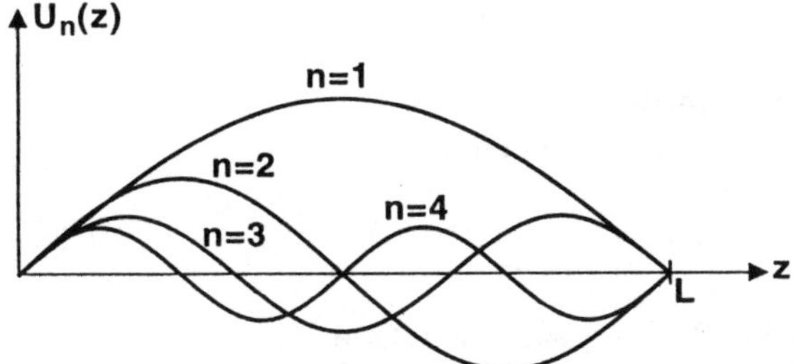

Fig. 9.2 - 1. Standing-wave patterns $U_n(z)$ of the resonant modes n = 1 - 4 of a homogeneous string fixed by two clamps separated by the distance L

The *general solution* of (7.4 - 4) with the boundary conditions (9.2 - 3b) is formed by the linear superposition of the resonant modes (9.2 - 5a-c)

$$u(z,t) = \sum_{n=1}^{\infty} U_n \, sin(\omega_n t - \alpha_n) sin \, \beta_n z \quad . \qquad (9.2 - 6a)$$

with the arbitrary parameters U_n and α_n. They are determined by the initial conditions. According to (9.2 - 5b&c) the general solution (9.2 - 6a) constitutes *Fourier series* in time t as well as with respect to the position z

$$u(z,t) = \sum_{n=1}^{\infty} U_n \, sin(n\,\omega_1 t - \alpha_n) \, sin\,(n\beta_1 z) \quad . \qquad (9.2 - 6b)$$

Consequently, it is *periodic in space and time.* Inherent wavelength λ and period T are

$$\lambda = \lambda_1 = 2\pi / \beta_1 = 2L \quad \text{and} \quad T = 2\pi / \omega_1 = 2L / \upsilon \quad . \qquad (9.2 - 6c)$$

b) Sturm-Liouville conditions
The Sturm-Liouville conditions

$$u(0) = U(0) = 0 \quad \text{and} \quad u_z(L) / u(L) = U'(L) / U(L) = \beta_C \tag{9.2 - 7a}$$

of the type (9.1 - 3d) imply the following parameter conditions for the solutions (9.2 - 2b)

$$C = 0 \quad \text{and} \quad tan\, \beta\, L = \beta / \beta_C \quad . \tag{9.2 - 7b}$$

If $\beta_m > 0$ with m = 1, 2, 3, ... are the solutions of the second equation (9.2 - 7b), then the resonant modes and frequencies of the string can be written in the form

$$\begin{aligned} &u_m(z,t) = U_m sin(\omega_m t - \alpha_m) sin\, \beta_m\, z \\ &\text{with} \quad \omega_m = \beta_m / \upsilon \quad \text{and} \quad m = 1, 2, 3, \dots \quad . \end{aligned} \tag{9.2 - 7c}$$

The parameters U_m and α_m depend on initial conditions.

c) Inhomogeneous boundary conditions
The application of the inhomogeneous boundary conditions

$$U(0) = U_0 \neq 0 \quad \text{and} \quad U(L) = U_L \neq 0 \tag{9.2 - 8a}$$

of the type (9.1 - 4) to the solution (9.2 - 2b) yields the relations

$$U_0 = -R\, sin\Phi \quad \text{and} \quad U_L = R\, sin(\beta L - \Phi) \quad . \tag{9.2 - 8b}$$

As a consequence, the eigenfrequencies ω_m depend on the amplitude R

$$\omega_m(R) = \upsilon \beta_m(R) = \frac{\upsilon}{L}\left[2m\pi + arc\,sin(U_L / R) - arc\,sin(U_0 / R)\right] \quad . \tag{9.2 - 8c}$$

The corresponding resonant modes of amplitude R are

$$\begin{aligned} &u_m(z,t) = R\, sin(\omega_m(R)t - \alpha)\, sin[\beta_m(R) z - \Phi(R)] \\ &\text{with} \quad \Phi(R) = -arc\,sin(U_0 / R) \quad . \end{aligned} \tag{9.2 - 8d}$$

9.3 Free Standing Waves on Membranes

A homogeneous thin membrane with density ρ [kg m^{-3}] and thickness d [m] is stretched in the xy plane by a constant surface tension or force per unit length

τ [N m^{-1} = kg s^{-2}]. It can be perturbed by transverse displacements δw_{T} in the z direction. These displacements constitute waves [Alonso & Finn 1967 B, Kneubühl 1994 B] on the membrane in the form

$$\delta w_{\mathrm{T}} = \delta w_{\mathrm{T}}(x,y,t) = u(x,y,t) \quad , \tag{9.3 - 1}$$

which obey the *two-dimensional Hertz equation*

$$u_{\mathrm{tt}} - \upsilon^2 \Delta u = u_{\mathrm{tt}} - \upsilon^2 \left(u_{\mathrm{xx}} + u_{\mathrm{yy}}\right) = 0 \quad . \tag{9.3 - 2}$$

Since the Hertz equation does not imply dispersion, phase velocity υ and group velocity υ_{g} are identical

$$\upsilon = \upsilon_{\mathrm{g}} = \left(\tau / \rho d\right)^{1/2} \quad . \tag{9.3 - 3}$$

Specific solutions of (9.3 - 2) represent *"plane" traveling waves* on the membrane in the form

$$\begin{aligned} & u = u(\vec{r},t) = f(\vec{e}\cdot\vec{r} - \upsilon t) \\ & \text{with} \quad \vec{e} = \left[e_{\mathrm{x}}, e_{\mathrm{y}}, 0\right] \quad , \quad |\vec{e}| = 1 \quad , \quad \vec{r} = [x,y,0] \quad . \end{aligned} \tag{9.3 - 4}$$

In this equation $\vec{e}$ indicates the direction of propagation and $f(x)$ an arbitrary function. Examples of this type of waves are the *harmonic waves*

$$\begin{aligned} & u = u(\vec{r},t) = U_0 \cos\left(\vec{k}\cdot\vec{r} - \omega t + \alpha\right) \\ & \text{with} \quad \vec{k} = k\vec{e} = \left[k_{\mathrm{x}}, k_{\mathrm{y}}, 0\right] \quad \text{and} \quad k = 2\pi / \lambda = \omega / \upsilon \quad . \end{aligned} \tag{9.3 - 5}$$

Amplitudes U_0 and phases α of these waves are arbitrary. The propagation vector $\vec{k}$ is defined by the direction $\vec{e}$ of propagation and the circular wavenumber k.

Standing waves on a membrane are solutions of (9.3 - 2) of the form

$$u = u(\vec{r},t) = \sin(\omega t - \alpha)\, U(\vec{r}) \quad , \tag{9.3 - 6}$$

where $U(\vec{r})$ fulfills the *Helmholtz equation*

$$\Delta U(\vec{r}) + \vec{k}^2 U(\vec{r}) = 0 \quad \text{with} \quad \vec{k}^2 = k^2 = \omega^2 / \upsilon^2 \quad . \tag{9.3 - 7}$$

9.3.1 Modes of a Rectangular Membrane

The standing waves on a rectangular membrane of width a and length b are determined by the two-dimensional Helmholtz equation (9.3 - 7) in Cartesian coordinates xy

$$U_{xx} + U_{yy} + k^2 U = 0 \quad \text{with} \quad U = U(x,y) \tag{9.3 - 8}$$

on one hand, and by the boundary conditions

$$\begin{aligned} u(0,y,t) = U(0,y) = 0 \quad , \quad u(x,0,t) = U(x,0) = 0 \quad , \\ u(a,y,t) = U(a,y) = 0 \quad , \quad u(x,b,t) = U(x,b) = 0 \quad , \end{aligned} \tag{9.3 - 9}$$

on the other hand. These two conditions are fulfilled by the *resonant oscillation modes*

$$u_{mn}(x,y,t) = U_{mn} sin(\omega_{mn} t - \alpha_{mn}) sin(m\pi x / a) sin(n\pi y / b) \tag{9.3 - 10a}$$

with the *resonant circular frequencies*

$$\omega_{mn} = \upsilon k_{mn} = \pi \upsilon \left[\left(\frac{m}{a} \right)^2 + \left(\frac{n}{b} \right)^2 \right]^{1/2} \quad , \tag{9.3 - 10b}$$

$$\text{where} \quad \upsilon = (\tau / \rho d)^{1/2} \quad \text{and} \quad m, n = 1, 2, 3, \ldots \quad .$$

Amplitudes U_{mn} and phases α_{mn} are arbitrary.

The *general solution* of (9.3 - 7) and (9.3 - 8) with the boundary condition (9.3 - 9) is formed by the linear superposition of the resonant oscillation modes (9.3 - 10a)

$$u(x,y,t) = \sum_{m=1}^{\infty} \sum_{n=1}^{\infty} u_{mn}(x,y,t) \quad . \tag{9.3 - 11}$$

9.3.2 Modes of a Circular Membrane

A standard example of a circular membrane is the *drumskin*. The standing waves on a circular membrane of radius a are governed by the two-dimensional Helmholtz *equation in polar coordinates* $r\varphi$

$$\begin{aligned} & U_{rr} + r^{-1} U_r + r^{-2} U_{\varphi\varphi} + k^2 U = 0 \\ & \text{with} \quad U = U(r,\varphi) \quad , \\ & \text{and} \quad r^2 = x^2 + y^2 \quad , \quad \varphi = arctan(y / x) \quad , \end{aligned} \tag{9.2 - 12}$$

on one hand, and by the boundary conditions

$$U(a,\varphi)=0 \quad , \tag{9.2 - 13a}$$

$$U(r,\varphi+2\pi)=U(r,\varphi) \quad , \tag{9.2 - 13b}$$

$$U(0,\varphi)=U(0,0)\,\text{finite} \tag{9.2 - 13c}$$

on the other hand. The ansatz

$$U(r,\varphi)=R(s)cos\big(m(\varphi-\varphi_{\rm m})\big) \tag{9.3 - 14}$$

$$\text{with} \quad s=kr \quad \text{and} \quad m=0,1,2,3...$$

fulfills the boundary condition (9.2 - 13b). In addition, its application to (9.2 - 12) results in *Bessel's differential equation* [Abramowitz & Stegun 1965 B] for *R*(*s*)

$$s^2R_{\rm ss}+sR_{\rm s}+\left(s^2-m^2\right)R=0 \quad . \tag{9.3 - 15}$$

Taking into account the boundary conditions (9.3 - 13a&b) yields the following solutions of this equation

$$\begin{aligned} &R_{\rm mn}(k_{\rm mn}r)=U_{\rm mn}J_{\rm m}(j_{\rm mn}r/a) \\ &\text{with} \quad J_{\rm m}(j_{\rm mn})=0 \quad , \quad m=0,1,2,3,... \quad , \quad n=1,2,3,... \quad , \end{aligned} \tag{9.3 - 16}$$

where the $J_{\rm m}(x)$ designate the *Bessel functions of the first kind* [Abramowitz & Stegun 1965 B] and $j_{\rm mn}$ their zeros.

Consequently, the resonant *oscillation modes* of the circular membrane are

$$u_{\rm mn}(r,\varphi,t)=U_{\rm nm}sin(\omega_{\rm mn}t-\alpha_{\rm mn})J_{\rm m}(j_{\rm mn}r/a)cos\big(m(\varphi-\varphi_{\rm m})\big) \tag{9.3 - 17a}$$

$$\text{with} \quad \omega_{\rm mn}=j_{\rm mn}\upsilon/a \quad . \tag{9.3 - 17b}$$

Again, amplitudes $U_{\rm mn}$ and phases $\alpha_{\rm mn}$ are arbitrary. They have to be fixed by initial conditions.

The *general solution* of (9.3 - 7) and (9.3 - 8) with the boundary conditions (9.3 - 13a-c) is given by the linear combination of the resonant oscillation modes 9.3 - 17a)

$$u(r,\varphi,t)=\sum_{\rm m=0}^{\infty}\sum_{\rm n=1}^{\infty}u_{\rm mn}(r,\varphi,t) \quad . \tag{9.3 - 18}$$

9.4 Free Acoustic Standing Waves in Cavities

Acoustic waves in gases and liquids can be represented as scalar pressure waves with the momentary local departure $\delta p(\vec{r},t)$ of the pressure from the equilibrium pressure p_0 [Pa = Nm^{-2}] as wave function

$$\delta p = \delta p(\vec{r},t) = u(\vec{r},t) \quad . \tag{9.4 - 1}$$

These waves are governed by the *three-dimensional Hertz equation*

$$u_{tt} - \upsilon^2 \Delta u = u_{tt} - \upsilon^2 \left(u_{xx} + u_{yy} + u_{zz}\right) = 0 \quad , \tag{9.4 - 2}$$

where υ represents the velocity of sound (7.1 - 2c). Since these waves show no dispersion, the group velocity υ_g equals the phase velocity υ. For an ideal gas this yields

$$\upsilon = \upsilon_g = \left(\gamma RT / M\right)^{1/2} \tag{9.4 - 3}$$

in accordance with (7.1 - 3b).

The best known solutions of (9.4 - 2) constitute the *plane traveling waves* with the characteristic form

$$u = u(\vec{r},t) = f(\vec{e} \cdot \vec{r} - \upsilon t)$$
$$\text{with} \quad \vec{e} = \left[e_x, e_y, e_z\right] \quad , \quad |\vec{e}| = 1 \quad \text{and} \quad \vec{r} = [x, y, z] \quad . \tag{9.4 - 4}$$

In this representation $\vec{e}$ defines the direction of propagation, whereas $f(x)$ is an arbitrary function.

Examples of plane traveling waves are the *harmonic waves* of the form

$$u(\vec{r},t) = U_0 cos\left(\vec{k} \cdot \vec{r} - \omega t + \alpha\right)$$
$$\text{with} \quad \vec{k} = k\vec{e} = \left[k_x, k_y, k_z\right] \quad \text{and} \quad k = 2\pi / \lambda = \omega / \upsilon \quad . \tag{9.4 - 5}$$

Wavelength λ and direction $\vec{e}$ of propagation of such a wave are determined by the propagation vector $\vec{k}$.

9.4.1 Traveling and Standing Spherical Waves

For scalar waves of spherical symmetry the three-dimensional Hertz equation (9.4 - 2) takes the form

$$u_{tt} - \upsilon^2 \left\{ r^{-2} \frac{\partial}{\partial r} \left(r^2 u_r \right) \right\} = 0 \quad , \tag{9.4 - 6a}$$

$$\text{or} \quad (ru)_{tt} - \upsilon^2 (ru)_{rr} = 0 \quad , \tag{9.4 - 6b}$$

with $u = u(r,t)$ and $r^2 = x^2 + y^2 + z^2$.

As a consequence the *general traveling wave solution* can be written as

$$u(r,t) = r^{-1} \{ f(r - \upsilon t) + g(r + \upsilon t) \} \tag{9.4 - 7}$$

where $f(r)$ and $g(r)$ are arbitrary functions.

A specific solution is the spherical wave emitted by a *harmonically oscillating point source* in the center $r = 0$

$$u(r,t) = U_0 r^{-1} cos(kr - \omega t + \alpha) \tag{9.4 - 8a}$$

$$\text{or} \quad u(r,t) = U_c r^{-1} exp\{ i(kr - \omega t) \} \tag{9.4 - 8b}$$

with $k = 2\pi / \lambda = \omega / \upsilon$ and $U_c = |U_c| exp(i\alpha)$.

In analogy to (9.3 - 6) and (9.3 - 7) *standing spherical waves* are solutions of (9.4 - 2) in the form

$$u = u(r,t) = sin(\omega t - \alpha) U(r) \quad , \tag{9.4 - 9}$$

where $U(r)$ is determined by the *Helmholtz equation*

$$r^{-2} \frac{\partial}{\partial r} \left(r^2 U_r \right) + k^2 U = 0 \quad , \tag{9.4 - 10a}$$

$$\text{or} \quad (rU)_{rr} + k^2 (rU) = 0 \quad \text{with} \quad k = 2\pi / \lambda = \omega / \upsilon \quad , \tag{9.4 - 10b}$$

and the boundary conditions.

For *spherical acoustic standing waves* or modes in a sphere of radius a the momentary local departure of the pressure

$$\delta p = \delta p(r,t) = u(r,t) \tag{9.4 - 11a}$$

from the equilibrium pressure p_0 varies with the distance r from the center and fulfills the *boundary conditions*

$$\delta\, p(0,t) = u(0,t) \quad \text{finite} \quad , \quad \text{and} \qquad (9.4 - 11b)$$

$$\frac{\partial}{\partial r}\delta\, p(a,t) = u_{\mathrm{r}}(a,t) = 0 \quad . \qquad (9.4 - 11c)$$

The last condition (9.4 - 11c) originates in by the requirement that the component of the local gas velocity normal to a *rigid wall* vanishes.

The resonant spherical *oscillation modes* corresponding to the boundary conditions (9.4 - 11b&c) are

$$\delta\, p_{\mathrm{n}}(r,t) = \Delta p_{\mathrm{n}}\, sin(\omega_{\mathrm{n}} t - \alpha_{\mathrm{n}})\, r^{-1} sin(k_{\mathrm{n}} r) \qquad (9.4 - 12a)$$

$$\text{with} \quad k_{\mathrm{n}} a = tan(k_{\mathrm{n}} a) = \omega_{\mathrm{n}} a / \upsilon \quad , \quad n = 1,2,3,\ldots \quad , \qquad (9.4 - 12b)$$

$$\text{where} \quad k_{\mathrm{n}} a < \pi\left(n + \frac{1}{2}\right) \quad \text{and} \quad k_{\mathrm{n}} a \approx \pi\left(n + \frac{1}{2}\right) \quad \text{for} \quad n >> 1 \quad .$$

The phases α_{n} are arbitrary. In accordance with (9.4 - 11b) the amplitudes of these modes are finite at the center $r = 0$ because

$$\delta\, p(0,t) = k_{\mathrm{n}} \Delta p_{\mathrm{n}} sin(\omega_{\mathrm{n}} t - \alpha_{\mathrm{n}}) \quad . \qquad (9.4 - 13)$$

9.4.2 Acoustic Modes of a Spherical Cavity

Today, spherical cavities as *acoustic Helmholtz resonators* [Helmholtz 1882 J, 1886 B, Morse 1936 B, Morse & Ingard 1986 B, Rayleigh 1870 B, Rosencwaig 1980 B] find application in laser photoacoustics [Karbach & Hess 1985 B]. The acoustic standing waves or modes of spherical symmetry in a spherical cavity have been described in the previous Section 9.4.1. This section comprehends all possible types of modes in a sperical cavity.

The acoustic waves in a spherical cavity are governed by the *Hertz equation in polar coordinates*

$$x = r\, sin\theta\, cos\phi \quad , \quad y = r\, sin\theta\, sin\phi \quad , \quad z = r\, cos\theta \quad , \qquad (9.4 - 14)$$

$$\text{with} \quad 0 \le \theta \le \pi \quad , \quad 0 \le \phi < 2\pi \quad .$$

In these coordinates the Hertz equation takes the form [Jeffrey 1995 B, Kneubühl 1994 B]

$$u_{\mathrm{tt}} = \upsilon^2 \Delta_{\mathrm{r},\theta,\phi} u = \upsilon^2 r^{-2} \left\{ \frac{\partial}{\partial r}\left(r^2 u_{\mathrm{r}}\right) + \Delta_{\theta,\phi} u \right\} \qquad (9.4 - 15a)$$

with $\Delta_{\theta,\phi} u = \frac{1}{sin\theta} \frac{\partial}{\partial\theta} \left(sin\theta \, u_{\theta} \right) + \frac{1}{sin^2\theta} u_{\phi\phi}$, (9.4 - 15b)

where Δ indicates Laplace operators. *Standing waves* are represented in spherical coordinates by

$$u(r, \theta, \phi, t) = sin(\omega t - \alpha) U(r, \theta, \phi) \quad . \tag{9.4 - 16a}$$

The combination of (9.4 - 15a) with (9.4 - 16a) yields the *Helmholtz equation in spherical coordinates*

$$\Delta_{r,\theta,\phi} U(r, \theta, \phi) = k^2 U(r, \theta, \phi) \quad \text{with} \quad k = \omega / \upsilon \quad . \tag{9.4 - 16b}$$

This equation has to be solved with the boundary conditions

$$U(0, \theta, \phi) = U(0,0,0) \quad \text{finite} \quad , \quad \text{and} \tag{9.4 - 17a}$$

$$U_r(a, \theta, \phi) = \frac{\partial}{\partial r} U(a, \theta, \phi) = 0 \quad , \tag{9.4 - 17b}$$

which correspond to (9.4 - 11b&c).

The *general solution* of (9.4 - 16a&b) with the boundary conditions (9.4 - 17a&b) consists of the linear combination of the *resonant acoustic modes*. As *real pressure waves* these have the form

$$\begin{aligned} \delta p_{n\ell|m|}(r, \theta, \phi, t) &= p_{n\ell|m|}(r, \theta, \phi, t) - p_0 \\ &= \Delta p \, sin(\omega_{n\ell} t - \alpha_{n\ell}) \, j_\ell(s_{\ell n} r / a) \, S_{\ell|m|}(\theta, \phi) \end{aligned} \tag{9.4 - 18}$$

with $\omega_{n\ell} = s_{\ell n} \upsilon / a$

and $n = 1,2,3,\ldots$, $\ell = 0,1,2,\ldots$, $|m| = 0,1,\ldots\ell$.

In this formula $j_\ell(s)$ is the ℓ-th order *spherical Bessel function* of the first kind [Abramowith & Stegun 1965 B] and $s_{\ell n}$ is the n-th positive zero of its first derivative

$$\frac{d}{ds} j_\ell(s_{\ell n}) = 0 \tag{9.4 - 19a}$$

that determines the *resonant circular frequency* or eigenfrequency $\omega_{n\ell}$ of the corresponding mode. The $s_{\ell n}$ with the smallest indices ℓ and n are listed in Table

9.4 - 1. It demonstrates that the mode with $\ell = 1$, $n = 1$ oscillates with the *lowest eigenfrequency* ω_{11}.

Table 9.4 - 1. Positive zeros $s_{\ell n}$ of the first derivatives (9.4 - 19a) of the spherical Bessel functions [Karbach & Hess 1985 J, Roy. Soc. Math. Tab. 1960 B]

n =	1	2	3
$\ell = 0$	4.493	7.725	10.904
1	**2.082**	5.940	9.206
2	3.342	7.290	...
3	4.514	8.584	...
4	5.647	9.840	...
5	6.756	...	...

For $\ell >> 1$ the first positive zero $s_{\ell n}$ can be approximated by [Abramowitz & Stegun 1965 B, Roy. Soc. Math. Tab. 1960 B]

$$s_{\ell 1} \approx \left(\ell + \frac{1}{2}\right)\left\{1 + 0.806\left(\ell + \frac{1}{2}\right)^{-2/3} - 0.237\left(\ell + \frac{1}{2}\right)^{-4/3} - +\right\} \quad . \qquad (9.4 - 19b)$$

The spherical Bessel functions of the first kind $j_\ell(s)$, which can be deduced from Rayleigh's formula

$$j_\ell(s) = (-s)^\ell \left(\frac{1}{s}\frac{d}{ds}\right)^\ell \frac{\sin s}{s} \quad \text{with} \quad \ell = 0, 1, 2, \ldots \quad , \qquad (9.4 - 20a)$$

fulfill the differential equations [Abramowitz & Stegun 1965 B]

$$s^2 \frac{d^2}{ds^2} j_\ell + 2s \frac{d}{ds} j_\ell + \left[s^2 - \ell(\ell+1)\right] = 0 \quad . \qquad (9.4 - 20b)$$

The *surface harmonics* $S_{\ell|m|}(\theta, \phi)$ have the form [Abramowitz & Stegun 1965 B]

$$S_{\ell|m|}(\theta, \phi) = P_\ell^{|m|}(\cos\theta) \begin{matrix} \cos|m|\phi \\ \sin|m|\phi \end{matrix} \qquad (9.4 - 21)$$

$$\text{with} \quad \ell = 0, 1, 2, \ldots \quad , \quad |m| = 0, 1, \ldots, \ell \quad .$$

Therefore, there exist two linearly independent surface harmonics for each $|m| > 0$. As a consequence the eigenfrequencies $\omega_{n\ell}$ of the resonant modes are $(2\ell+1)$*-fold degenerate.*

The *Legendre functions* $P_\ell^{|m|}(z)$ are defined by the differential equation [Abramowitz & Stegun 1965 B]

$$\left(1-2^2\right)\frac{d^2}{dz^2}P_\ell^{|m|}+2z\frac{d}{dz}P_\ell^{|m|}+\left[\ell(\ell+1)-\frac{m^2}{1-z^2}\right]P_\ell^{|m|}=0 \quad . \tag{9.4 - 22a}$$

For $|m| = m = 0$ the Legendre functions equal the *Legendre polynomials* determined by Rodrigues' formula

$$P_\ell^0(z)=P_\ell(z)=\frac{1}{2^{\mathrm{n}}n!}\frac{d^{\mathrm{n}}}{dz^{\mathrm{n}}}\left(z^2-1\right)^{\mathrm{n}} \quad , \tag{9.4 - 22b}$$

while for $|m| > 0$ these functions represent the *associated Legendre functions of the first kind* which can be deduced from the Legendre polynomials by

$$P_\ell^{|m|}(z)=\left(z^2-1\right)^{\frac{1}{2}|m|}\frac{d^{|m|}}{dz^{|m|}}P_\ell(z) \quad . \tag{9.4 - 22c}$$

The Legendre polynomials (9.4 - 22b) are relevant for the axial or *azimuthal resonant modes* characterized by $m = 0$ [Karbach & Hess 1985 J].

There exists an *equivalent complex representation* of the real resonant modes (9.4 - 18)

$$u_{\mathrm{n}\ell\mathrm{m}}(r,\theta,\phi,t)=sin(\omega_{\mathrm{n}\ell}t-\alpha_{\mathrm{n}\ell})U_{\mathrm{n}\ell\mathrm{m}}(r,\theta,\phi) \tag{9.4 - 23}$$

with $U_{\mathrm{n}\ell\mathrm{m}}(r,\theta,\phi)=U_0\ j_\ell(s_{\ell\mathrm{n}}r/a)\ Y_{\ell\mathrm{m}}(\theta,\phi)$

and $n=1,2,3,\ldots \quad , \quad \ell=0,1,2,\ldots \quad , \quad m=0,\pm1,\ldots,\pm\ell \quad .$

The $Y_{\ell\mathrm{m}}(\theta,\phi)$ indicate the spherical harmonics [Blochinzew 1966 B, Kneubühl 1994 B, McGervey 1971 B, Messiah 1960 B, 1969 B, 1990 B]

$$Y_{\ell\mathrm{m}}(\theta,\phi)=N_{\ell|\mathrm{m}|}\ P_\ell^{|\mathrm{m}|}(cos\theta)\ exp(im\phi) \tag{9.4 - 24a}$$

with $\ell=0,1,2,\ldots \quad , \quad m=0,\pm1,\ldots\pm\ell \quad .$

The normalization factors $N_{\ell|\mathrm{m}|}$ are determined by the condition

$$\int_0^{2\pi}\int_0^{\pi}\left|Y_{\ell\mathrm{m}}(\theta,\phi)\right|^2 sin\theta\, d\theta\, d\phi=1 \quad . \tag{9.4 - 24b}$$

The spherical harmonics $Y_{\ell\mathrm{m}}(\theta,\phi)$ as well as the surface harmonics $S_{\ell|\mathrm{m}|}(\theta,\phi)$ are eigenfunctions of the Laplace operator $\Delta_{\theta\phi}$ (9.4 - 15b)

$$\Delta_{\theta\phi}Y_{\ell\mathrm{m}}(\theta,\phi)=-\ell(\ell+1)\ Y_{\ell\mathrm{m}}(\theta,\phi) \quad . \tag{9.4 - 24c}$$

This relation is important in *quantum mechanics* since it shows that the spherical harmonics are eigenfunctions of the *angular-momentum* operator [Baym 1969 B, Blochinzew 1966 B, Kneubühl 1994 B, McGervey 1971 B, Messiah 1960 B, 1969 B, 1990 B, Landau & Lifschitz 1979 B].

9.5 Sturm-Liouville Wave Patterns

Many wave patterns $U(z)$ of harmonically oscillating one-dimensional linear standing waves

$$u(z,t) = sin(\omega t - \alpha)U(z) \quad \text{, or} \tag{9.5 - 1a}$$

$$u(z,t) = exp(-i\omega t)U(z) \tag{9.5 - 1b}$$

fulfill the Sturm-Liouville equation [Birkhoff & Rota 1989 B, Courant & Hilbert 1968 B, Hairer et al. 1987 B, Madelung 1943 B, Zwillinger 1989 B]

$$\frac{d}{dz}\left[p(z)\frac{d}{dz}U'(z)\right] + [\Lambda\,\rho(z) - q(z)]U(z) = 0 \quad . \tag{9.5 - 2}$$

This equation is called *regular* in the closed interval $a \leq z \leq b$, if $p(z)$, $\rho(z)$ and $q(z)$ are limited and $p(z)$ and $\rho(z)$ are positive in this interval.

The eigenvalue Λ is determined by the boundary conditions, e.g. (9.1 - 2a-d), (9.1 - 3) and (9.1 - 4a&b).

The Sturm-Liouville equation (9.5 - 2) can be transformed into the simple *Liouville normal form* that has the same eigenvalues Λ. This transformation is defined by [Courant & Hilbert 1960 B, Birkhoff & Rota 1989 B]

$$w = u[\rho(z)p(z)]^{1/4} \quad , \tag{9.5 - 3a}$$

$$y = \int^{z} [\rho(z)/p(z)]^{1/2}\,dz \quad . \tag{9.5 - 3b}$$

It transforms (9.5 - 2) into

$$\frac{d^2}{dy^2}w(y) + [\Lambda - r(y)]w(y) = 0 \tag{9.5 - 4a}$$

$$\text{with} \quad r(y) = (q/\rho) + (\rho p)^{-1/4}\frac{d^2}{dy^2}\left[(\rho p)^{1/4}\right] \quad , \tag{9.5 - 4b}$$

$$\text{where} \quad d/dy = (p/\rho)^{1/2}\,d/dz \quad .$$

The Sturm-Liouville equation (9.5 - 2) is *self-adjoint*. It can be replaced by the following *self-adjoint system of differential equations* (4.9 - 18b)

$$\begin{aligned} U' &= -R(z)V \\ V' &= +Q(z)U \quad , \end{aligned} \tag{9.5 - 5a}$$

$$\text{with} \quad R(z) = 1/p(z) \quad \text{and} \quad Q(z) = \Lambda\,\rho(z) + q(z) \quad . \tag{9.5 - 5b}$$

The functions $U(z)$ and $V(z)$ of (9.5 - 5a) fulfill the differential equations (4.9 - 18c&d)

$$\frac{d}{dz}\left[\frac{U'}{R(z)}\right] + Q(z)U = 0 \quad , \quad \text{and} \tag{9.5 - 6a}$$

$$\frac{d}{dz}\left[\frac{V'}{Q(z)}\right] + R(z)V = 0 \quad , \tag{9.5 - 6b}$$

where (9.5 - 6a) corresponds to (9.5 - 2).

9.5.1 Sturm-Liouville Systems

The combination of a Sturm-Liouville equation (9.5 - 2) with its boundary conditions is called a *Sturm-Liouville system* [Birkhoff-Rota 1989 B].

A Sturm-Liouville system is *regular* if a regular Sturm-Liouville equation (9.5 - 2) is subjected to the boundary conditions

$$A\,U(a) + A'U'(a) = 0 \ \text{ and } \ B\,U(b) + B'U'(b) = 0 \quad . \tag{9.5 - 7}$$

A Sturm-Liouville system is *singular* if a least one endpoint $z = a$, b of the interval $a \le z \le b$ has to be excluded [Birkhoff & Rota 1989 B]. This can occur for several reasons: i) Either one or both of the endpoints may be in infinity, $a = -\infty$ and/or $b = +\infty$. ii) Among the functions $p(z)$, $q(z)$ and $\rho(z)$ one or more may be singular in and endpoint, e.g. $z = a$. iii) The functions $p(z)$ and $\rho(z)$ vanish simultaneously in one or both endpoints, e.g. $z = a$

$$\lim_{z \to a} p(z) = 0 \quad \text{and} \quad \lim_{z \to a} \rho(z) = 0 \quad . \tag{9.5 - 8}$$

There exists a large variety of Sturm-Liouville systems: orthogonal polynomials, harmonic functions, Bessel functions, etc. [Abramowitz & Stegun 1965 B, Birkhoff & Rota 1989 B, Courant & Hilbert 1968 B, Gradshteyn & Ryzhik 1965 B, Hairer et al. 1989 B, Madelung 1943 B, Sommerfeld 1947 B, Zwillinger 1989 B]. Their comprehensive description would go far beyond the limits of this book.

Standing-wave patterns $U(z)$ determined by Sturm-Liouville systems are of various origins. This is demonstrated by the following examples.

a) Damped Klein-Gordon waves

A Sturm-Liouville system determines the harmonically oscillating standing Klein-Gordon waves with damping. They are governed by the equation

$$u_{tt} = \upsilon^2(z)u_{zz} + \delta(z)u_z - \omega_C^2(z)u \quad . \tag{9.5 - 9}$$

The Klein-Gordon term is characterized by $\omega_C(z)$, while the damping is represented by $\delta(z)$.

Harmonically oscillating standing waves correspond to the ansatz

$$u(z,t) = g(t)U(z) = sin(\omega t - \alpha)U(z) \quad , \tag{9.5 - 10a}$$

which is equivalent to the equation

$$u_{tt} = -\omega^2 u = -\Lambda u \quad . \tag{9.5 - 10b}$$

The combination of (9.5 - 9) and (9.5 - 10b) yields the differential equation for the pattern $U(z)$ of the standing wave

$$\upsilon^2(z)U'' + \delta(z)U' + \left[\Lambda - \omega_C^2(z)\right]U = 0 \quad . \tag{9.5 - 11}$$

The transformation

$$p = p_0 \, exp \int_0^z \frac{\delta}{\upsilon^2} dz \quad , \quad \rho = p / \upsilon^2 \quad , \quad q = p\omega_C^2 / \upsilon^2 \tag{9.5 - 12}$$

of its coefficients yields the Sturm-Liouville equation (9.5 - 2).

b) Multidimensional waves of high symmetry

Patterns of standing waves with high symmetry in multidimensional media often represent Sturm-Liouville systems. A simple example is the pattern $U(r)$ of a standing wave with circular symmetry on a thin circular membrane. The corresponding Helmholtz equation that can be derived from (9.2 - 12) is a Sturm-Liouville equation

$$\frac{d}{dr}\left[r\frac{dU}{dr}\right] + k^2 r U = 0 \quad \text{with} \quad k = \omega / \upsilon = \Lambda^{1/2} / \upsilon \quad . \tag{9.5 - 13}$$

c) Wave mechanics

In non-relativistic wave mechanics the time-independent Schrödinger equation that describes the dynamics of a particle with mass m in a time-independent potential $V(z)$

together with its boundary conditions forms a Sturm-Liouville system. The basic wave equation is the time-dependent Schrödinger equation [Baym 1969 B, Blochinzew 1966 B, Fick 1968 B, Flügge 1990 B, Landau & Lifschitz 1979 B, Messiah 1960 B, 1969 B, 1990 B, Schubert & Weber 1980 B]

$$i\hbar\,\psi_{\mathrm{t}} = -\left(\hbar^2/2m\right)\psi_{\mathrm{zz}} + V(z)\psi \quad \text{with} \quad \psi = \psi(z,t) \quad . \tag{9.5 - 14}$$

Because the potential $V(z)$ does not depend on time t, there exist stationary states with well-defined energies E. On this condition (9.5 - 14) can be solved by postulating a standing wave in the form

$$\psi(z,t) = g(t)\,\Psi(z) = exp(-i\,\omega\,t)\,\Psi(z) = exp(-i\,E\,t/\hbar)\,\Psi(z) \quad , \tag{9.5 - 15a}$$

where ω and E are connected by Planck's relation (7.2 - 16a). If this ansatz is applied to (9.5 - 14), then one finds that the wave function $\Psi(z)$ has to obey the *time-independent Schrödinger equation*

$$\Psi''(z) + \left(2m/\hbar^2\right)\left[E - V(z)\right]\Psi(z) = 0 \tag{9.5 - 15b}$$

that represents a Sturm-Liouville equation (9.5 - 2) with $E = \Lambda$. The energy eigenvalues $E = \Lambda$ of this equation are co-determined by the boundary conditions.

9.5.2 Eigenvalues and Eigenfunctions

Eigenvalues and eigenfunctions of Sturm-Liouville systems obey a multitude of *general laws* [Courant & Hilbert 1968 B, Birkhoff & Rota 1989 B]:

a) Regular Sturm-Liouville systems
Each regular Sturm-Liouville system possesses an infinite series of *non-degenerate positive eigenvalues*

$$\Lambda_0 < \Lambda_1 < \Lambda_2 <<< \Lambda_{\mathrm{n}} << \quad \text{with} \quad \lim_{\mathrm{n}\to\infty} \Lambda_{\mathrm{n}} = \infty \quad . \tag{9.5 - 16}$$

The *eigenfunction* $U_{\mathrm{n}}(z)$ that corresponds to the eigenvalue Λ_{n} has exactly *n zeros* in the internal $a \le z \le b$. Furthermore, it is determined uniquely except for a constant factor. The eigenfunctions $U_{\mathrm{n}}(z)$ with n = 0, 1, 2, ... form a *complete orthogonal system* with respect to a real weight function $\rho(z)$. This system can be *normalized* in a way that [Madelung 1943 B]

$$\int_{\mathrm{a}}^{\mathrm{b}} U_{\mathrm{m}}(z)\,U_{\mathrm{n}}(z)\,\rho(z)\,dz = \delta_{\mathrm{nm}} = \begin{cases} 1 \text{ for } n = m \\ 0 \text{ for } n \ne m \end{cases} \quad . \tag{9.5 - 17}$$

b) Singular Sturm-Liouville systems

Among the singular Sturm-Liouville systems those involving *square integrable functions* are of importance because of wave mechanics. A function $U(z)$ is square integrable with respect to a real weight function $\rho(z)$ if

$$\int_{-\infty}^{+\infty} |U(z)|^2 \rho(z)\, dz = C < +\infty \quad . \tag{9.5 - 18}$$

Square integrable eigenfunctions $U_n(z)$ of a singular Sturm-Liouville system, which correspond to different eigenvalues Λ_n, are orthogonal with respect to the weight function $\rho(z)$

$$\int_{-\infty}^{+\infty} U_m^*(z) U_n(z)\, \rho(z)\, dz = 0 \quad , \tag{9.5 - 19}$$

$$\text{if} \quad m \neq n \quad , \quad U_k(\pm\infty) = 0 \quad \text{and} \quad \frac{d}{dz} U_k(\pm\infty) = 0 \quad \text{for} \quad k = m, n \quad .$$

The asterisk * indicates the complex conjugate. Also these eigenfunctions can be *normalized* in a way that

$$\int_{-\infty}^{+\infty} U_m^*(z) U_n(z) \rho(z)\, dz = \delta_{nm} \quad . \tag{9.5 - 20}$$

c) Wave mechanics of the harmonic oscillator

In wave mechanics the harmonic oscillator is described by a singular *Sturm-Liouville system*. This oscillator is characterized by its mass m and the potential

$$V(z) = \frac{f}{2} z^2 \quad . \tag{9.5 - 21}$$

The corresponding time-independent Schrödinger equation (9.5 - 15b) is

$$\Psi''(z) + 2m / \hbar \left[E - \frac{f}{2} z^2 \right] \Psi(z) = 0 \quad . \tag{9.5 - 22}$$

The wave functions $\Psi(z)$ are required to be square integrable with respect to $\rho(z) = 1$ and to vanish at infinity, i.e. $\Psi(\pm\infty) = 0$. These boundary conditions together with (9.5 - 22) represent a singular Sturm-Liouville system.

The energy eigenvalues $E_n = \Lambda_n$ of the harmonic oscillator defined by (9.5 - 22) are

$$E_n = \hbar\,\omega_0\left(n+\frac{1}{2}\right) \quad \text{with} \quad \omega_0 = (f/m)^{1/2} \quad \text{and} \quad n = 1,2,\ldots \quad . \tag{9.5 - 23}$$

The corresponding square integrable eigenfunctions have the form

$$\begin{gathered}\Psi_n(z) = N_n\; H_n(z/z_0) exp(-z^2/2z_0^2) \\ \text{with} \quad z_0 = (\hbar/m\omega_0)^{1/2} \quad \text{and} \quad n = 0,1,2,\ldots\end{gathered} \tag{9.5 - 24}$$

In this equation the $H_n(z)$ are Hermite polynomials [Abramowitz & Stegun 1965 B], while the N_n represent normalization factors.

Fig. 9.5 - 1 illustrates the eigenfunctions $\Psi_n(z)$ of the three states n = 0, 1, 2 with the lowest energy eigenvalues E_n.

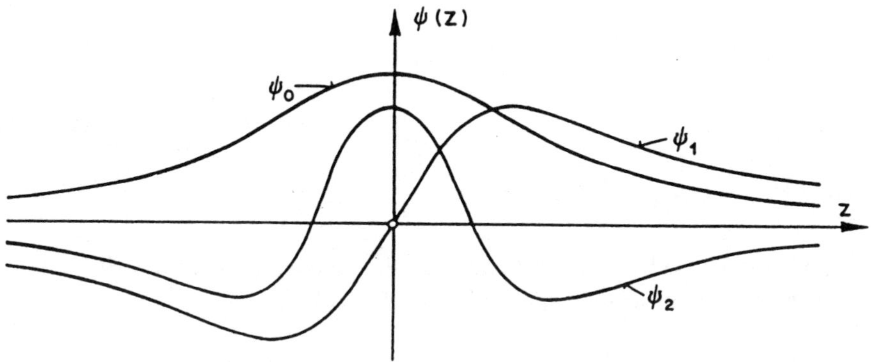

Fig. 9.5 - 1. Wave functions $\Psi_n(z)$ of the harmonic oscillator with n = 0, 1, 2

9.6 Free Nonlinear Standing Waves

An example of a free standing wave governed by a nonlinear wave equation is the solitary oscillating pulse called *breather* that is governed by the Sine-Gordon equation. It is defined by (8.5 - 13) and illustrated in Fig. 8.5 - 4. Two further examples of free nonlinear standing waves are discussed in the following.

9.6.1 Gradient Dominated Hertz Waves

The modified nonlinear Hertz equation

$$u_{tt} - \frac{c^2}{n^2(z, u_z/u)} u_{zz} = 0 \tag{9.6 - 1}$$

describes Hertz waves influenced by the *relative field gradient*

$$f = f(z,t) = u_z(z,t) / u(z,t) \quad . \tag{9.6 - 2}$$

This influence is manifested in a refractive index n that depends on f.

For a harmonically oscillating standing wave

$$u(z,t) = sin(\omega t - \alpha)U(z) \tag{9.5 - 10a}$$

$$\text{with} \quad u_{zz} = -\omega^2 u \tag{9.5 - 10b}$$

the Hertz equation (9.6 - 1) can be replaced by the Helmholtz equation

$$U_{zz} + (\omega / c)^2 n^2(z, U_z / U)U = 0 \quad . \tag{9.6 - 3}$$

It can be reduced to

$$f_z + f^2 + (\omega / c)^2 n^2(z, f) = 0 \tag{9.6 - 4}$$

$$\text{with} \quad f = f(z) = U_z(z) / U(z) \quad .$$

The solution of (9.6 - 4) by taking into account the boundary conditions (9.1 - 3a-d) or (9.1 - 4) and by the subsequent application of the relation

$$U(z) = U_0 exp \int_{-\infty}^{z} f(z) dz \tag{9.6 - 5}$$

yields the eigenfrequencies ω_m and the corresponding mode patterns.

An *example* of this type of wave is characterized by the index of refraction

$$n = n(f) = n_0\left(1 + \eta f^2\right) \quad , \tag{9.6 - 6}$$

which depends uniquely on the relative field gradient f defined by (9.6 - 2).

The homogeneous fixed-end conditions

$$u(0,t) = U(0) = 0 \quad \text{and} \quad U(L,t) = U(L) = 0 \tag{9.2 - 3a}$$

of the type (9.1 - 3a) yield the following resonant modes

$$u_m(z,t) = U_m sin(\omega_m t - \alpha)[sin\beta_m z]^{\gamma_m} \tag{9.6 - 7a}$$

with $\beta_m = m\pi / L$, $m = 1,2,3,\ldots$, (9.6 - 7b)

$$\gamma_m^{-1} = 1 + \gamma_m \beta_m^2 \eta \quad , \quad \gamma_m > 0 \quad , \tag{9.6 - 7c}$$

and $\omega_m = (c / n_0) \gamma_m^{1/2} \beta_m$. (9.6 - 7d)

The corresponding relative field gradient is

$$f_m(z) = \gamma_m \beta_m cos \beta_m z \quad . \tag{9.6 - 8}$$

9.6.2 Euler-Painlevé Waves

A second type of free nonlinear standing waves is governed by the wave equation

$$u\{u_{zz} + F(z)u_z - G(z)u\} + (a-1)u_z^2 = 0 \quad . \tag{9.6 - 9}$$

The wave patterns $U(z)$ of the corresponding standing waves

$$u(z,t) = sin(\omega t - \alpha)U(z) \tag{9.5 - 10a}$$

with $u_{tt} = -\omega^2 u = -\Lambda u$ (9.5 - 10b)

obey the *Euler-Painlevé equation* [Kamke 1957 B, eq. 6.129]

$$U\{U_{zz} + F(z)U_z + \omega^2 G(z)U\} + (a-1)U_z^2 = 0 \quad . \tag{9.6 - 10}$$

For a ≠ 0 it can be transformed into a linear differential equation by the ansatz

$$V = U^a \quad . \tag{9.6 - 11a}$$

The result is

$$V_{zz} + F(z)V_z + \omega^2 a G(z)V = 0 \quad . \tag{9.6 - 11b}$$

The additional transformation of the coefficients $F(z)$ and $G(z)$

$$p(z) = p(0)\int_0^z F(z)\,dz \quad \text{and} \quad \rho(z) = a\,G(z)p(z) \tag{9.6 - 12a}$$

yields the *Sturm-Liouville equation*

$$\frac{d}{dz}\left[p(z)V_z\right]+\omega^2\rho(z)V=0 \quad . \tag{9.6 - 12b}$$

The boundary conditions (9.1 - 3a-d) or (9.1 - 4) determine the resonant modes as eigenfunctions and their eigenfrequencies ω_m via the eigenvalues $\Lambda_m = \omega_m^2$ of this equation.

9.7 Forced Standing Waves

This section is dedicated to forced standing waves characterized by *inhomogeneous* linear and nonlinear partial differential equations.

9.7.1 Forced Waves on a String

The free standing waves on a string stretched between two clamps separated by the fixed distance L have been described in Section 9.2. This string fulfills the homogeneous fixed-end boundary conditions

$$u(0,t)=0 \quad \text{and} \quad u(L,t)=0 \quad . \tag{9.2 - 3a}$$

The free standing waves are governed by the homogeneous Hertz equation (7.4 - 4). Their resonant modes have the form

$$\begin{gathered} u_n(z,t)=U_n sin(n\omega_1 t-\alpha_n) sin\, n\beta_1 z \\ \text{with} \quad \beta_1=\pi / L \quad , \quad \omega_1=\pi \upsilon / L \quad , \quad n=1,2,3,\ldots \quad . \end{gathered} \tag{9.2 - 5a}$$

The forced standing waves on this string fulfill the inhomogeneous Hertz equation [Courant & Hilbert 1968 B]

$$u_{tt}-\upsilon^2 u_{zz}=F(z,t) \quad , \tag{9.7 - 1}$$

where υ indicates the constant wave velocity (7.1 - 5c) of the string.

The force $F(z, t)$ on the string acts only in the range $0 \leq z \leq L$. However, it can be extended periodically because of the term $sin\, n\beta_1 z$ in the representation (9.2 - 5a) of the resonant modes. Thus, it becomes a periodic function that can be written as a spatial Fourier series

$$\begin{gathered} F(z,t)=\sum_{n=1}^{\infty} F_n(t)\, sin\, n\beta_1 z=F(z+2L,t) \\ \text{with} \quad F_n(t)=\frac{2}{L}\int_0^L F(z,t) sin\, n\beta_1 z\, dz \quad \text{and} \quad \beta_1=\pi / L \quad . \end{gathered} \tag{9.7 - 2}$$

The same procedure is applied to the unknown forced standing wave

$$u(z,t) = \sum_{n=1}^{\infty} w_n(t) \, sin \, n\beta_1 z$$

$$\text{with} \quad w_n(t) = \frac{2}{L}\int_0^L u(z,t) sin \, n\beta_1 z \, dz \quad \text{and} \quad \beta_1 = \pi / L \quad . \tag{9.7 - 3}$$

The introduction of (9.7 - 2) and (9.7 - 3) in (9.7 - 1) yields the oscillation equations of all resonant modes

$$\ddot{w}_n(t) + n^2\omega_1^2 \, w_n(t) = F_n(t) \quad . \tag{9.7 - 4}$$

The solutions of these equations are [Courant & Hilbert 1968 B]

$$w_n(t) = \frac{1}{n\omega_1}\int_0^t F_n(\tau) \, sin \, n\omega_1(t-\tau) d\tau + U_n \, sin(n\omega_1 \, t - \alpha_n) \quad . \tag{9.7 - 5}$$

Consequently, the complete solution of (9.7 - 1) has the form

$$u(z,t) = \sum_{n=1}^{\infty} sin \, n\beta_1 z \left[\frac{1}{n\omega_1}\int_0^t F_n(\tau) sin \, n\omega_1(t-\tau) d\tau + U_n sin(n\omega_1 t - \alpha_n)\right] \tag{9.7 - 6}$$

where the arbitrary parameters U_n and α_n are determined by the initial conditions.

The method used to derive the general solution (9.7 - 6) of the inhomogeneous Hertz equation (9.7 - 1) can be applied to forced standing waves governed by other inhomogeneous linear wave equations, if their free standing waves are known. This is demonstrated in the next section by considering standing waves on a telegraph line.

9.7.2 Forced Waves on a Telegraph Line

Forced waves on a telegraph line are governed by the *inhomogeneous telegraph equation*

$$u_{tt} + (2/\tau)u_t + \Omega^2 u - \upsilon_0^2 u_{zz} = F(z,t) \quad . \tag{9.7 - 7}$$

For the homogeneous fixed-end or *short-circuit conditions*

$$u(0,t) = 0 \quad \text{and} \quad u(L,t=0) \tag{9.2 - 3a}$$

it is of advantage to make the ansatz (9.7 - 3) for $u(z, t)$ and to write the force $F(z, t)$ as Fourier series (9.7 - 2). In combination with (9.7 - 7) these expressions yield the equations

$$\ddot{u} + (2/\tau)\dot{w}_n + \left(\Omega^2 + n^2\beta_1^2\right)w_n = F_n(t) \tag{9.7 - 8}$$

$$\text{with} \quad \beta_1 = \pi / L \quad , \quad n = 1,2,3,\dots \quad .$$

In absence of the force $F(z, t)$ the solutions $w_n(t)$ of the homogenous equation (9.7 - 8) with $F_n(t) = 0$ represent the *transients* in the form of the damped oscillations described in Section 2.2.4 with reference to (2.2 - 28).

For an $F_n(t)$ different from zero the particular solution $w_n(t)$ of the corresponding inhomogeneous equation (9.7 - 8) constitutes a *stationary standing wave* and an oscillation with a resonance discussed in Section 3.2.2 on the basis of the inhomogeneous equation (3.2 - 4).

9.7.3 Forced Gradient Dominated Hertz Waves

Free gradient-dominated Hertz waves have been described in Section 9.6.1. Forced waves of this type are determined by the inhomogeneous nonlinear wave equation

$$u_{tt} - \frac{c^2}{n^2(z, u_z / u)} u_{zz} = F(z,t) \quad . \tag{9.7 - 9}$$

The following discussion is restricted to the wave described in Section 9.6.1. Its refractive index

$$n = n(f) = n_0\left(1 + \eta f^2\right)^{1/2} \tag{9.6 - 6}$$

varies with the relative field gradient f defined by (9.6 - 2). In addition, this wave is subjected to the homogeneous fixed-end boundary conditions

$$u(0,t) = 0 \quad \text{and} \quad u(L,t) = 0 \quad . \tag{9.2 - 3a}$$

Under these conditions (9.7 - 9) can be solved with the ansatz

$$F_m(z,t) = F_m sin\,\omega t\left[sin\beta_m z\right]^{\gamma_m} \tag{9.7 - 10a}$$

$$u_m(z,t) = U_m(\omega) sin\,\omega t\left[sin\beta_m z\right]^{\gamma_m} \quad , \tag{9.7 - 10b}$$

$$\text{with} \quad \beta_m = m\pi / L \quad , \quad m = 1,2,3,\dots \quad . \tag{9.6 - 7b}$$

The application of this ansatz to (9.7 - 9) and (9.6 - 6) yields the relations (9.6 - 7c) for γ_m and (9.6 - 8) for f_m. They are also valid for the free standing waves described in Section 9.6.1. In addition, it yields the ratio between the amplitude $U_m(\omega)$ and that of the applied force F_m

$$U_m(\omega)/F_m = \left[(c/n_0)^2 \gamma_m \beta_m^2 - \omega^2\right]^{-1} \quad . \tag{9.7 - 10c}$$

This equation demonstrates that a mode $u_m(z, t)$ of a free standing wave (9.6 - 7a) with the resonant frequency ω_m (9.6 - 7d) does not require an external force to oscillate with a finite amplitude $U_m = U_m(\omega_m)$.

9.7.4 Forced Bioche Waves

Forced standing waves governed by the nonlinear wave equation

$$u\left\{u_{tt} + (2/\tau)u_t + \Omega^2 u - \upsilon_0^2 u_{zz}\right\} + u_t^2 = F(z,t) \tag{9.7 - 11}$$

under the homogeneous fixed-end boundary conditions

$$u(0,t) = 0 \quad \text{and} \quad u(L,t) = 0 \tag{9.2 - 3a}$$

can be evaluated with the ansatz

$$\begin{aligned} &u_m(z,t) = w_m(t) sin\beta_m z \quad \text{and} \quad F_m(z,t) = f_m(t) sin^2\beta_m z \\ &\text{with} \quad \beta_m = m\pi / L \quad , \quad m = 1,2,3,\ldots \quad . \end{aligned} \tag{9.7 - 12}$$

Application of this ansatz to (9.7 - 11) yields the equations [Brioche 1910 J, Kamke 1956 B, eq. 6.128]

$$w_m\left\{\ddot{w}_m + (2/\tau)\dot{w}_m + \left(\Omega^2 + \beta_m^2 \upsilon_0^2\right) w_m\right\} + \dot{w}_m^2 = f_m(t) \quad . \tag{9.7 - 13}$$

They are transformed into linear differential equations by the introduction of the functions $s_m(t)$ defined by

$$s_m(t) = w_m^2(t) \geq 0 \quad . \tag{9.7 - 14a}$$

The result is

$$\ddot{s}_m + (2/\tau)\dot{s}_m + 2\left(\Omega^2 + \beta_m^2 \upsilon_0^2\right) s_m = 2 f_m(t) \quad . \tag{9.7 - 14b}$$

For $F(z, t) = 0$ and $f_m(t) = 0$ the solutions of these equations represent the *transients* in the form of damped oscillations discussed in Section 2.2.4. These oscillations need to be *critically or supercritically damped* because the functions $s_m(t)$ are positive according to (9.7 - 14a).

For a non-vanishing periodic $f_m(t)$ the particular solution $s_m(t)$ of the corresponding equation (9.7 - 14b) in combination with (9.7 - 12) describes a stationary standing wave whenever $s_m(t)$ remains positive.

Appendix

A.1 Fourier Series

[Benedetto 1996 B, Bracewell 1986 B, Carslaw 1930 B]

A.1.1 General Rules

period: $x(t) = x(t+T)$ $T = 2\pi / \omega = 2\pi / \omega_1$	basic circular frequency: $\omega = \omega_1 = 2\pi / T$
complex Fourier series: $x(t) = x(t+T) = \sum_{m=-\infty}^{m=\infty} F_m exp(-im\omega t)$	complex coefficients: $F_m = \frac{1}{T}\int_0^T x(t)\, exp(+im\omega t)dt$
real periodic functions: $x(t) = x(t+T) = x^*(t)$	coefficients: $F_0 = F_0{}^* = a_0 = A_0$ $F_m = F^*_{-m} = \frac{1}{2}(a_m + ib_m)$ $= \frac{1}{2} A_m exp(i\alpha_m)$
real Fourier series I: $x(t) = x(t+T) = x^*(t)$ $= a_0 + \sum_{m=1}^{\infty} a_m cos(m\omega t) + \sum_{m=1}^{\infty} b_m sin(m\omega t)$	real coefficients: $a_0 = \frac{1}{T}\int_0^T x(t)\, dt$ $a_m = \frac{2}{T}\int_0^T x(t)\, cos(m\omega t)dt$ $b_m = \frac{2}{T}\int_0^T x(t)\, sin(m\omega t)dt$
real Fourier series II: $x(t) = x(t+T) = x^*(t)$ $= A_0 + \sum_{m=1}^{\infty} A_m cos(m\omega t - \alpha_m)$	real parameters: $A_0 = a_0$ $A_m = \left[a_m^2 + b_m^2\right]^{1/2} \geq 0$ $\alpha_m = arc\ tan(b_m / a_m)$

real even periodic functions: $x(t) = x(t+T) = x^*(t) = x(-t)$	coefficients: $F_0 = F_0{}^* = a_0 = A_0$ $F_m = F_{-m} = F_m{}^* = \frac{1}{2} a_m = \frac{1}{2} A_m$ $\alpha_m = 0\,; b_m = 0$
real odd periodic functions: $x(t) = x(t+T) = x^*(t) = -x(-t)$	coefficients: $F_0 = a_0 = A_0 = 0$ $F_m = -F_{-m} = -F_m{}^* = \frac{1}{2} i b_m$ $\alpha_m = \pm\pi/2; a_m = 0$ $b_m = A_m\, sign(\alpha_m)$

A.1.2 Real Periodic Functions

Normalized parameters: $\omega = \omega_1 = 1$ and $T = 2\pi$

a) Square wave:

$$x(t) = x(t+2\pi) = sign(sin\,t) = \frac{4}{\pi}\left[sin\,t + \frac{1}{3} sin\,3t + \frac{1}{5} sin\,5t + \ldots\right]$$

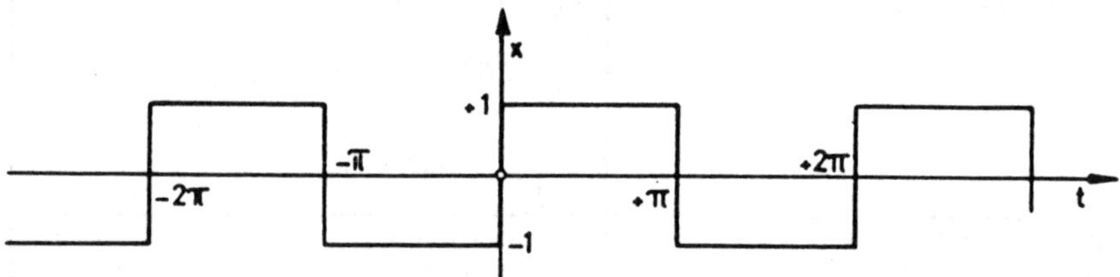

b) Asymmetric sawtooth:

$$x(t) = x(t+2\pi) = x/\pi \quad \text{for} \quad -\pi \le x + \pi = \frac{2}{\pi}\left(\frac{sin\,t}{1} - \frac{sin\,2t}{2} + \frac{sin\,3t}{3} - + \ldots\right)$$

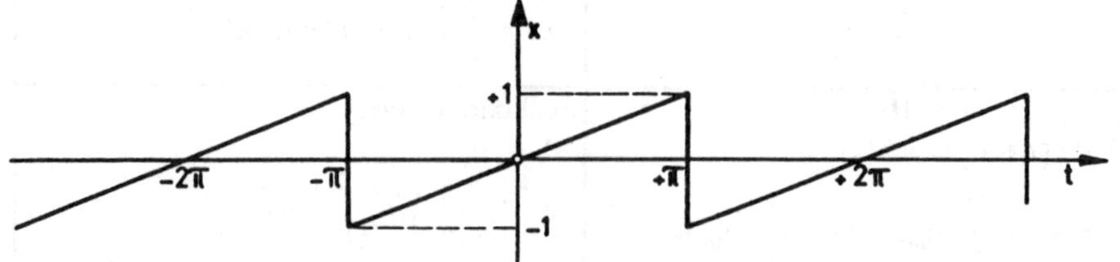

c) Symmetric sawtooth:

$$x(t) = x(t+2\pi) = \begin{Bmatrix} 2x/\pi & \text{for } -\pi/2 \le x \le +\pi/2 \\ 2-(2x/\pi) & \text{for } +\pi/2 \le x \le +3\pi/2 \end{Bmatrix}$$

$$= \frac{8}{\pi^2}\left(\sin t - \frac{1}{3^2}\sin 3t + \frac{1}{5^2}\sin 5t - + \ldots \right)$$

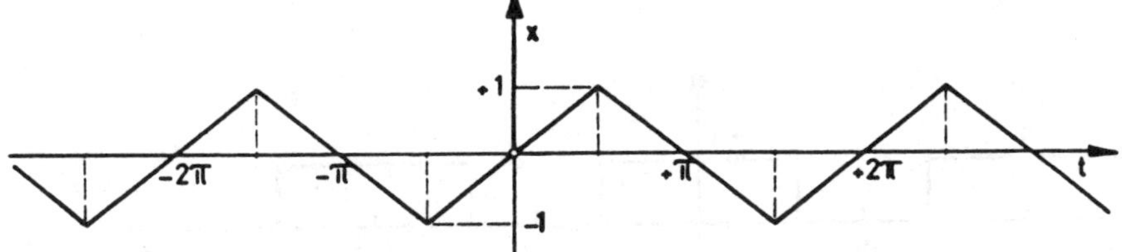

d) Modulus of sine:

$$x(t) = x(t+2\pi) = |\sin t| = \frac{2}{\pi} - \frac{4}{\pi}\left(\frac{\cos 2t}{1\cdot 3} + \frac{\cos 4t}{3\cdot 5} + \frac{\cos 6t}{5\cdot 7} + \ldots \right)$$

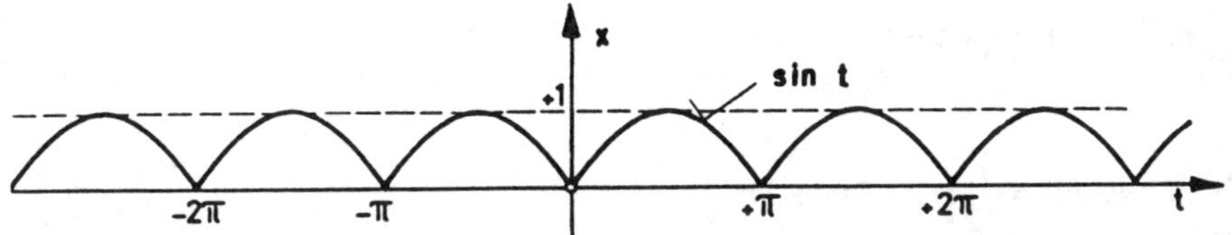

e) Rectified sine:

$$x(t) = x(t+2\pi) = H(\sin t)\cdot \sin t = \frac{1}{\pi} + \frac{1}{2}\sin t - \frac{2}{\pi}\left(\frac{\cos 2t}{1\cdot 3} + \frac{\cos 4t}{3\cdot 5} + \frac{\cos 6t}{5\cdot 7} + \ldots \right)$$

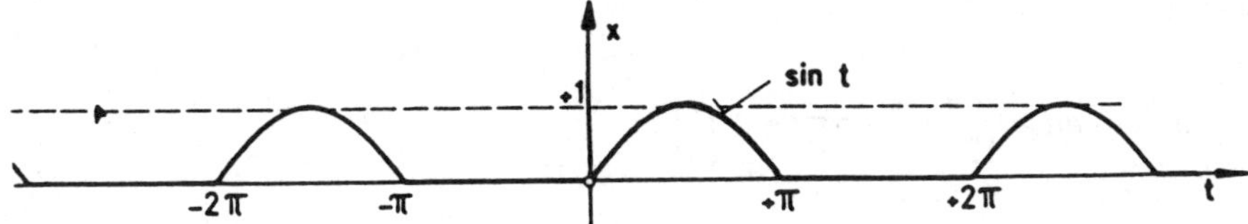

f) Rectangle:

$$x(t) = x(t+T) = \Pi(t/\tau) = \begin{Bmatrix} 1 & \text{for } |t| < \tau/2 \\ 1/2 & \text{for } |t| = \tau/2 \\ 0 & \text{for } |t| > \tau/2 \end{Bmatrix}$$

$$= \frac{\tau}{2\pi} + \frac{2}{\pi}\sum_{m=1}^{\infty} m^{-1} sin(m\tau/2) cos\, mt \quad \text{with} \quad \tau < 2\pi$$

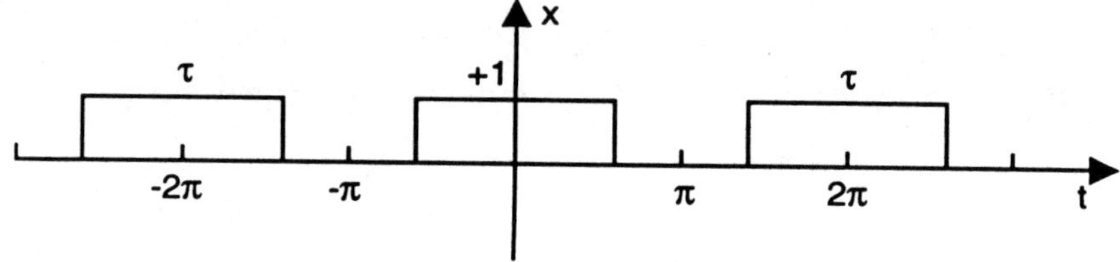

g) Simple Fourier series

α) Sums valid for $0 < t < 2\pi$:

$$\sum_{m=1}^{\infty} m^{-1} sin\, mt = \frac{1}{2}(\pi - t)$$

$$\sum_{m=1}^{\infty} m^{-3} sin\, mt = \frac{t}{12}\left[t^2 - 3\pi t + 2\pi^2\right]$$

$$\sum_{m=1}^{\infty} m^{-5} sin\, mt = -\frac{t}{720}\left[3t^4 - 15\pi t^3 + 20\pi^2 t^2 + 8\pi^4\right]$$

$$\sum_{m=1}^{\infty} m^{-1} cos\, mt = -\frac{1}{2}\,\ell n\left[2(1 - cos\, t)\right]$$

$$\sum_{m=1}^{\infty} m^{-2} cos\, mt = \frac{1}{12}\left[3t^2 - 6\pi t + 2\pi^2\right]$$

$$\sum_{m=1}^{\infty} m^{-4} cos\, mt = -\frac{1}{720}\left[15t^4 - 60\pi t^3 + 60\pi^2 t^2 - 8\pi^4\right]$$

β) Sums valid for $r^2 < 1$:

$$\sum_{m=1}^{\infty} r^{m} \sin mt = r \sin t \left[1 - 2r \cos t + r^{2}\right]^{-1}$$

$$\sum_{m=1}^{\infty} m^{-1} r^{m} \sin mt = \arctan\left\{r \sin t (1 - r \cos t)^{-1}\right\}$$

$$\sum_{m=1}^{\infty} r^{m} \cos mt = (1 - r \cos t)\left[1 - 2r \cos t + r^{2}\right]^{-1}$$

$$\sum_{m=1}^{\infty} m^{-1} r^{m} \cos mt = -\frac{1}{2} \ell n\left\{1 - 2r \cos t + r^{2}\right\}$$

A.2 Fourier Transformation

[Benedetto 1996 B, Bracewell 1986 B, Campbell & Foster 1948 B, Carslaw 1930 B, Champeney 1973 B, Erdelyi et al. 1954 B, Poularikas 1995 B, Zayed 1996 B]

A.2.1 General Rules

function: $x(t)=\frac{1}{2\pi}\int_{-\infty}^{+\infty} F(\omega)exp(-i\omega t)d\omega$	Fourier transform: $F(\omega)=\int_{-\infty}^{+\infty} x(t)exp(+i\omega t)dt$
real function: $x(t)=x^*(t)$	$F(-\omega)=F^*(\omega)$
even function: $x(t)=x(-t)=\frac{1}{\pi}\int_0^{\infty} F(\omega)\cos\omega t\cdot d\omega$	$F(\omega)=F(-\omega)=2\int_0^{\infty} x(t)\cos\omega t\cdot dt$
odd function: $x(t)=-x(-t)=-\frac{i}{\pi}\int_0^{\infty} F(\omega)\sin\omega t\cdot d\omega$	$F(\omega)=-F(-\omega)=2i\int_0^{\infty} x(t)\sin\omega t\cdot dt$
$x(-t)$	$F(-\omega)$
$x(t/\tau)$	$\tau F(\omega\tau)$
$x(t-\tau)$	$exp(i\omega\tau)F(\omega)$
$t^n x(t)$	$(-i)^n d^n F(\omega)/d\omega^n$
$t^{-1}x(t)$	$i\int_{-\infty}^{\omega} F(\omega')\,d\omega'$
$d^n x(t)/dt^n$	$(-i\omega)^n F(\omega)$
$\int_{-\infty}^{t} x(t')\,dt'$	$(i/\omega)F(\omega)$
product: $2\pi x_1(t)x_2(t)$	convolution: $\int_{-\infty}^{+\infty} F_1(\Omega)F_2(\omega-\Omega)\,d\Omega$
convolution: $\int_{-\infty}^{+\infty} x_1(\tau)x_2(t-\tau)d\tau$	product: $F_1(\omega)F_2(\omega)$
autocorrelation: $\int_{-\infty}^{+\infty} x(\tau)x(\tau+t)d\tau$	power spectrum: Wiener-Khintchine theorem $\lvert F(\omega)\rvert^2$
$x(t/\tau)\cos\Omega t$	$\frac{\tau}{2}\left[F(\tau(\omega+\Omega))+F(\tau(\omega-\Omega))\right]$
$x(t/\tau)\sin\Omega t$	$\frac{\tau}{2i}\left[F(\tau(\omega+\Omega))-F(\tau(\omega-\Omega))\right]$

A.2.2 Real Functions

function: $x(t)=\frac{1}{2\pi}\int_{-\infty}^{+\infty}F(\omega)\,exp(-i\omega t)\,d\omega$	Fourier transform: $F(\omega)=\int_{-\infty}^{+\infty}x(t)\,exp(+i\omega t)\,dt$
1	$2\pi\,\delta(\omega)$
$cos\,\Omega t$	$\pi[\delta(\omega-\Omega)+\delta(\omega+\Omega)]$
$sin\,\Omega\,\tau$	$i\pi[\delta(\omega-\Omega)-\delta(\omega+\Omega)]$
Dirac delta function δ: $\delta(t)$	1
$\delta(t-\tau)$	$exp(i\omega\tau)$
$d^{n}\delta(\tau)/d\,t^{n}$	$[-i\omega]^{n}$
sign function: $sign\,t$	$\frac{2i}{\omega}$
Heavyside unit step H: $H(t)=\frac{1}{2}[1+sign\,t]$	$\pi\,\delta(\omega)+\frac{i}{\omega}$
$H(t)\;exp(-at)$	$[a-i\omega]^{-1}$
rectangle function Π: $\Pi(t)=\begin{cases}1 & \text{for } \lvert x\rvert<1/2\\ 1/2 & \text{for } \lvert x\rvert=1/2\\ 0 & \text{for } \lvert x\rvert>1/2\end{cases}$	$sinc(\omega/2\pi)=\frac{2}{\omega}\,sin(\omega/2)$
triangle function Λ: $\Lambda(t)=\begin{cases}1-\lvert x\rvert & \text{for } \lvert x\rvert\le 1\\ 0 & \text{for } \lvert x\rvert\ge 1\end{cases}$	$sinc^{2}(\omega/2\pi)=\frac{4}{\omega^{2}}\,sin^{2}(\omega/2)$
$sinc\,t=\frac{sin\,\pi t}{\pi t}$	$\Pi(\omega/2\pi)$
$sinc^{2}t$	$\Lambda(\omega/2\pi)$
$\frac{1}{2}\Pi\left(t-\frac{1}{2}\right)-\Pi\left(t+\frac{1}{2}\right)$	$\frac{2i}{\omega}\,sin^{2}(\omega/2)$
$\Pi(t)\,cos\,\pi t$	$\frac{1}{2}\left[sinc\left(\frac{\omega}{2\pi}-\frac{1}{2}\right)+sinc\left(\frac{\omega}{2\pi}+\frac{1}{2}\right)\right]$
$\Pi(t)\,sin\,2\pi t$	$\frac{i}{2}\left[sinc\left(\frac{\omega}{2\pi}-1\right)-sinc\left(\frac{\omega}{2\pi}+1\right)\right]$
$exp\left\{-(t/\tau)^{2}\right\}$	$\pi^{1/2}\,\tau\,exp\left\{-(\omega\tau/2)^{2}\right\}$
$t\,exp\left(-\pi t^{2}\right)$	$i(2\pi)^{-1}\,\omega\,exp\left\{-\omega^{2}/4\pi\right\}$
$\lvert t\rvert^{-1/2}$	$\lvert 2\pi/\omega\rvert^{1/2}$

$\lvert t\rvert^{-1/2}\, sign\, t$	$i\lvert 2\pi / \omega\rvert^{1/2}\, sign\, \omega$
$exp(-\lvert t\rvert)$	$2\left[1+\omega^2\right]^{-1}$
$exp(-\lvert t\rvert)\, sign\, t$	$i\pi\omega\left[1+\omega^2\right]^{-1}$
$exp(-\lvert t\rvert)\, sinc(x / \pi)$	$arc\, tan\left(2 / \omega^2\right)$
$exp(-t)\, H(t)$	$[1-i\omega]^{-1}$
$t \cdot exp(-t)\, H(t)$	$[1-i\omega]^{-2}$
$sech\, \pi x = [cosh\, \pi x]^{-1}$	$sech(\omega / 2) = [cosh(\omega / 2)]^{-1}$
$sech^2 \pi x = [cosh\, \pi x]^{-2}$	$\omega[\pi\, sinh(\omega / 2)]^{-1}$
$tanh\, t$	$i[sinh(\omega / 2)]^{-1}$
$\Pi(t)\left[1-4t^2\right]$	$\frac{8}{\omega^2}\left[(\omega / 2)^{-1} sin(\omega / 2) - cos(\omega / 2)\right]$
$\Pi(t)\, cos^2 \pi t$	$\pi\, sinc\left(\frac{\omega}{2\pi}\right) + \left(\frac{\pi}{2}\right) sinc\left(\frac{\omega}{2\pi} - 1\right)$ $+ \frac{\pi}{2}\, sinc\left(\frac{\omega}{2\pi} + 1\right)$
$\ell n\left[1+(\tau / t)^2\right]$	$2\pi\, \omega^{-1}[1-exp(-\omega\tau)]$
$\ell n\left[\frac{\tau^2+t^2}{T^2+t^2}\right]$	$2\pi\, \omega^{-1}[exp(-\omega T) - exp(-\omega\tau)]$
$2\left[t^2+\tau^2\right]^{-1}$	$(\pi / \tau)\, exp[-\tau\lvert\omega\rvert]$
Bessel function J_0: $J_0(t / \tau)$	$2\Pi(\omega\tau / 2)\left[\tau^{-2} - \omega^2\right]^{-1/2}$
Bessel functions: J_{2n}: $J_{2n}(t / \tau)$	Chebyshev polynomials T_{2n}: $2\frac{\Pi(\omega\tau / 2) T_{2n}(\omega\tau)}{\left[\tau^{-2} - \omega^2\right]^{1/2}}$
Bessel functions J_m: $t^{-1/2} J_{n+\frac{1}{2}}(t)$	Legendre polynomials P_n: $i^n (2\pi)^{+1/2} \cdot \Pi(\omega / 2) \cdot P_n(\omega)$
Legendre polynomials P_n: $\Pi(t / 2)\, P_n(t)$	Bessel functions J_m: $i^n (2\pi / \omega)^{1/2} J_{n+\frac{1}{2}}(\omega)$
Chebyshev polynomials T_n: $\frac{\Pi(t / 2)\, T_n(t)}{\left[1-t^2\right]^{1/2}}$	Bessel functions J_n: $\pi i^n J_n(\omega)$

A.3 Laplace Transformation

[Abramowitz & Stegun 1965 B, Bracewell 1986 B, Doetsch 1970 B, Doetsch 1971/73 B, Erdelyi et al. 1954 B, Pöschl 1956 B, Poularikas 1995 B, Zayed 1996 B]

A.3.1 General Rules

For the definition of the convolution see (3.2 - 24a&b). If $F(t)$ and $\dot{F}(t)$ are continuous at $t = 0$, then $F(t = +0)$ and $\dot{F}(t = +0)$ can be replaced by $F(0)$ and $\dot{F}(0)$.

$$L\{F(t)\} = \int_0^\infty F(t)\,exp(-pt)\,dt$$

$$L\{aF_1(t) + bF_2(t)\} = aF_1(p) + bF_2(p)$$

$$L\{F(at)\} = F(p/a)$$

$$L\{H(t-t_0)\,F(t-t_0)\} = F(p)\,exp(-t_0 p),\ t_0 > 0$$

$$L\{(-t)^n F(t)\} = d^n F(p)/dp^n$$

$$L\{exp(+p_0 t)\,F(t)\} = F(p - p_0)$$

$$L\{\dot{F}(t)\} = pF(p) - F(t = +0)$$

$$L\{\ddot{F}(t)\} = p^2 F(p) - pF(t = +0) - \dot{F}(t = +0)$$

$$L\left\{\int_0^t F(t)dt\right\} = p^{-1}F(p)$$

$$L\{F(t) = F(t+T)\} = \left[1 - e^{-Tp}\right]^{-1} \int_0^T F(t)e^{-tp}dt, \quad F(t) \text{ periodic}$$

$$L\{F_1(t) * F_2(t)\} = F_1(p)\,F_2(p), \quad \text{convolution}$$

where $F_1(t<0) = F_2(t<0) = 0$

A.3.2 Heaviside and Dirac Functions

This list contains Laplace transforms of the Heaviside step function $H(t)$, the Dirac delta function $\delta(t)$ and its derivatives $\delta^{(n)}(t)$. In these equations $t = +0$ indicates a time just after $t = 0$.

$L\{\delta(t-t_0)\}$	$exp(-t_0\, p),\, t_0 > 0$
$L\left\{\frac{d^n}{dt^n}\delta(t-t_0)\right\} = L\{\delta^{(n)}(t-t_0)\}$	$= p^n\, exp(-t_0 p)$
$L\{\delta(t-(+0))\}$	$= 1$
$L\left\{\frac{d^n}{dt^n}\delta(t-(+0))\right\} = L\{\delta^{(n)}(t-(+0))\}$	$= p^n$
$L\{\delta(t)\}$	$= 1/2$
$L\{H(t-t_0)\}$	$= p^{-1} exp(-t_0 p), \quad t_0 > 0$
$L\{H(t)\}$	$= 1/p$
$L\left\{\sum_{n=0}^{\infty}(-1)^n H\left(t-\frac{1}{2}n\tau\right)\right\}$	$= p^{-1}\left[1+e^{-\tau p/2}\right]^{-1}$

A.3.3 Real Functions

$\boldsymbol{L}\left\{\frac{t^{n-1}}{(n-1)!}\right\}$	$=p^{-n}$
$\boldsymbol{L}\{exp(-at)\}$	$=(p+a)^{-1}$
$\boldsymbol{L}\{t\,exp(-at)\}$	$=(p+a)^{-2}$
$\boldsymbol{L}\left\{\frac{t^{n-1}}{(n-1)!}exp\,(-at)\right\}$	$=(p+a)^{-n}$
$\boldsymbol{L}\left\{\frac{exp(-at)-exp(-bt)}{b-a}\right\}$	$=[(p+a)(p+b)]^{-1}$
$\boldsymbol{L}\{sin\;at\}$	$=a\left[p^2+a^2\right]^{-1}$
$\boldsymbol{L}\{cos\;at\}$	$=p\left[p^2+a^2\right]^{-1}$
$\boldsymbol{L}\{sin\;at\,exp(-bt)\}$	$=a\left[(p+b)^2+a^2\right]^{-1}$
$\boldsymbol{L}\{cos\;at\,exp(-bt)\}$	$=(p+b)\left[(p+b)^2+a^2\right]^{-1}$
$\boldsymbol{L}\{sinh\;at\}$	$=a\left[p^2-a^2\right]^{-1}$
$\boldsymbol{L}\{cosh\;at\}$	$=p\left[p^2-a^2\right]^{-1}$
$\boldsymbol{L}\left\{(\pi t)^{-1/2}\right\}$	$=p^{-1/2}$
$\boldsymbol{L}\left\{\frac{4^n n!}{(2n)!\sqrt{\pi}}t^{(n-1/2)}\right\}$	$=p^{-\left(n+\frac{1}{2}\right)}$
$\boldsymbol{L}\{J_0(at)\}$	$=\left[p^2+a^2\right]^{-1/2}$
$\boldsymbol{L}\{t\;sin\;at\}$	$=2ap\left[p^2+a^2\right]^{-2}$
$\boldsymbol{L}\{t\;cos\;at\}$	$=\left(p^2-a^2\right)\left[p^2+a^2\right]^{-2}$
$\boldsymbol{L}\left\{t^{-1}sin\;at\right\}$	$=arc\;tan(a/p)$
$\boldsymbol{L}\left\{sin^2\left(\frac{at}{2}\right)\right\}$	$=\left(a^2/2p\right)\left[p^2+a^2\right]^{-1}$
$\boldsymbol{L}\{\lvert sin\,at\rvert\}$	$=a\left[p^2+a^2\right]^{-1}tanh(2a/\pi p)$
$\boldsymbol{L}\{+\ell n\,t\}$	$=-p^{-1}\left[\ell n\,p-C\right]$
$C=0.5772\ldots$	$=$ *Euler' s constant*

A.4 Convolution (Faltung)

[Bracewell 1986 B, Pöschl 1956 B. Schetzen 1989 B]

A.4.1 General Rules

Definition:

$$f(t)*g(t)=\int_{-\infty}^{+\infty}f(s)g(t-s)ds$$

Commutation:

$$f(t)*g(t)=g(t)*f(t)=\int_{-\infty}^{+\infty}g(s)f(t-s)ds$$

Association:

$$f(t)*[g(t)*h(t)]=[f(t)*g(t)]*h(t)=f(t)*g(t)*h(t)$$

Distribution over addition:

$$f(t)*[g(t)+h(t)]=f(t)*g(t)+f(t)*h(t)$$

Differentiation:

$$\frac{d}{dt}[f(t)*g(t)]=\frac{df(t)}{dt}*g(t)=f(t)*\frac{dg(t)}{dt}$$

Convolution with Dirac delta function:

$$\delta(t)*\delta(t)=\delta(t)$$

$$\delta(t)*f(t)=\left\{\begin{matrix}\frac{1}{2}[f(t-0)+f(t+0)]\\ f(t) \quad \text{if} \quad f(t)\ continuous\end{matrix}\right\}$$

Convolution with Heaviside unit step:

$$H(t)=\frac{1}{2}[1+sign(t)]$$

$$H(t)*\delta(t)=H(t)$$

$$H(t)*f(t)=\int_{-\infty}^{t}f(s)ds$$

$$H(t)t^{m}*f(t)=(m!)\iint\underset{\text{m fold}}{\cdots}\iint f(s)ds$$

$$H(t)exp(-at)*f(t)=exp(-at)\int_{-\infty}^{t}f(s)exp(as)ds$$

$$H(t)f(t)*g(t)=\int_{0}^{\infty}f(s)g(t-s)ds=\int_{-\infty}^{t}g(s)f(t-s)ds$$

$$H(t)f(t)*H(t)g(t)=\int_{0}^{t}f(s)g(t-s)ds=\int_{0}^{t}g(s)f(t-s)ds$$

Convolution with sampling function or shah function: III(t)

$$\mathrm{III}(t) = \sum_{m=-\infty}^{+\infty} \delta(t-m)$$

$$\mathrm{III}(t) f(t) = \sum_{m=-\infty}^{+\infty} f(m)\, \delta(t-m)$$

$$\mathrm{III}(t) * f(t) = \sum_{m=-\infty}^{+\infty} f(t-m)$$

A.4.2 Heaviside Unit Step

Definitions: $H(t)=\frac{1}{2}[1+sign\,t]=\begin{cases}0 & \text{for } t<0\\ 1/2 & \text{for } t=0\\ 1 & \text{for } t>0\end{cases}$ $[H(t)]^{*0}=\delta(t)$; $[H(t)]^{*1}=H(t)$, $[H(t)]^{*m}=\underbrace{H(t)*H(t)** \quad **H(t)*H(t)}_{\text{m fold}}$
Powers and products: $[H(t)]^{*m}=\frac{t^{(m-1)}}{(m-1)!}\cdot H(t)$ $[H(t)]^{*2}=H(t)*H(t)=t\cdot H(t)$ ramp function $[H(t)]^{*3}=H(t)*H(t)*H(t)=\frac{t^2}{2}\cdot H(t)$ $t^n H(t)=(n!)[H(t)]^{*(n+1)}$ $t^k H(t)*t^m H(t)=\frac{m!k!}{(m+k+1)!}t^{(k+m+1)}H(t)$
Exponentials: $exp(t)\,H(t)=\sum_{n=1}^{\infty}[H(t)]^{*n}$ $exp(t)\,H(t)*H(t)=[exp(t)-1]H(t)$ $exp(t)\,H(t)*(1-t)H(t)=t\,H(t)$ $[exp(t)\,H(t)]^{*2}=t\,exp(t)\,H(t)$ $[exp(t)\,H(t)]^{*3}=\frac{t^2}{2}exp(t)\,H(t)$ $[exp(t)\,H(t)]^{*m}=\frac{t^{(m-1)}}{(m-1)!}exp(t)\,H(t)$ $[exp(at)\,H(t)]^{*2}=t\,exp(at)\,H(t)$ $exp(at)\,H(t)*exp(-at)\,H(t)=sinh(at)\,H(t)$ $exp(at)\,H(t)*exp(bt)\,H(t)=-[a-b]^{-1}[exp(at)-exp(bt)]$ $exp(-at)\,H(t)*t\,H(t)=\frac{1}{a}\left\{t+\frac{1}{a}[exp(-at)-1]\right\}H(t)$

A.4.3 Special Real Functions

Rectangle function Π:

$$\Pi(t) = \begin{cases} 1 & \text{for } |x| < 1/2 \\ 1/2 & \text{for } |x| = 1/2 \\ 0 & \text{for } |x| > 1/2 \end{cases}$$

Triangle function Λ:

$$\Lambda(t) = \begin{cases} 1-|x| & \text{for } |x| \le 1 \\ 0 & \text{for } |x| \ge 1 \end{cases}$$

$$\Lambda(t) = \Pi(t) * \Pi(t)$$

$$\Lambda(t) * \sum_{m=-\infty}^{+\infty} a_m \delta(t-m) = \textit{polygon through } (m, a_m)$$

$$\Lambda(t) * \sum_{m=-\infty}^{+\infty} \delta(t-m) = 1$$

$$sinc(t) = \frac{sin \pi t}{\pi t}$$

$[sinc(t)]^{*m} = sinc(t)$; $m = 1, 2, 3, \ldots$

$sinc(t) * J_0(\pi t) = J_0(\pi t)$; Bessel function

A.4.4 Hilbert Transformation

[Bracewell 1989 B, Erdelyi et al. II 1954, Poularikas 1995 B, Zayed 1996 B]

Definition:

$$F_{\mathrm{Hi}}(t) = -\frac{1}{\pi t} * f(t) = \frac{1}{\pi} \int_{-\infty}^{+\infty} \frac{f(s)ds}{s-t}$$

Inversion:

$$-\frac{1}{\pi t} * F_{\mathrm{Hi}}(t) = -f(t)$$

Transforms:

$$-\frac{1}{\pi t} * 1 = 0$$

$$-\frac{1}{\pi t} * \delta(t) = -\frac{1}{\pi t}$$

$$-\frac{1}{\pi t} * \left(-\frac{1}{\pi t}\right) = \delta(t)$$

$$-\frac{1}{\pi t} * \sin t = \cos t$$

$$-\frac{1}{\pi t} * \frac{\sin t}{t} = \frac{\cos t - 1}{t}$$

$$-\frac{1}{\pi t} * \left[1+t^2\right]^{-1} = -t\left[1+t^2\right]^{-1}$$

$$-\frac{1}{\pi t} * \Pi(t) = \frac{1}{\pi} \ell n \left|\frac{t-1/2}{t+1/2}\right|$$

$$-\frac{1}{\pi t} * J_1(at) \sin at = J_1(at) \sin at \quad ; \quad 0 < a$$

$$-\frac{1}{\pi t} * J_{\mathrm{m}}(at) \sin bt = J_{\mathrm{m}}(at) \cos bt \quad ; \quad 0 < b < a, m = 0,1,2$$

References

B. Books

Abdullaev, F.K. (1994): Theory of Solitons in Inhomogeneous Media
Wiley, Chichester, UK

Ablowitz, M.J., Clarkson, P.A. (1991): Solitons, Nonlinear-Evolution Equations and Inverse Scattering
Cambridge Univ. Press, Cambridge, UK

Abramowitz, M., Stegun, I.A. (1965): Handbook of Mathematical Functions
Dover, New York

Agrawal, G.P. (1989): Nonlinear Fibre Optics
Academic Press, San Diego

Aimes, W.F. (1972): Nonlinear Partial Differential Equations in Engineering
Academic Press, New York

Allen, L., Eberly, J.H. (1975/1987): Optical Resonance and Two-Level Atoms
Wiley/Dover, New York

Alonso, M., Finn, E.J. (1967): Fundamental Universtiy Physics II, Fields and Waves
Addison-Wesley, Reading, MA

Alonso, M., Finn, E.J. (1969): Fundamental University Physics III, Quantum and Statistical Physics
Addison-Wesley, Reading, MA

Alonso, M., Finn, E.J. (1970): Physics
Addison-Wesley, Reading, MA

Alonso, M., Finn, E.J. (1988): Quantenphysik
2nd ed., Addison-Wesley, Reading, MA

Alonso, M., Finn, E.J. (1992): Physik
2nd ed., Addison-Wesley, Reading, MA

Anderson, J.D. jr. (1988): Fundamentals of Aerodynamics
McGraw Hill, New York

Andronov, A.A., Witt, A.A., Chaikin, S.E. (1966): Theory of Oscillations
Pergamon Press, Oxford, UK

Arbenz, K., Wohlhauser, A. (1986): Advanced Mathematics for Practicing Engineers
Artech House, Norwood, MA

Arnold, V.I. (1978): Mathematical Methods of Classical Mechanics
Springer, Berlin
Arnold, V.I. (1984): Catastrophe Theory
Springer, Berlin
Ashcroft, N.W., Mermin, N.D. (1976): Solid State Physics
Holt, Rinehart & Winston, New York
Atwater, H.A. (1962): Introduction to Microwave Theory
McGraw Hill, New York
Baden-Fuller, A.J. (1969): Microwaves
Pergamon, London
Bai-Lin, Hao (1984): Chaos
World Scientific, Singapore
Bainov, D.D., Mishev, D.P. (1991): Oscillation Theory
for Neutral Differential Equations with Delay
Institute of Physics Publishing, Bristol, UK
Baker, G.L., Gollub, J.R. (1990): Chaotic Dynamics
Cambridge University Press, Cambridge, UK
Balanis, C.A. (1982): Antenna Theory
Harper & Row, New York
Barnett, St. (1990): Matrices, Methods and Applications
Clarendon Press, Oxford, UK
Bateman, H. (1959): Partial Differential Equations of Mathematical Physics
Cambridge University Press, Cambridge, UK
Baym, G. (1969): Lectures on Quantum Mechanics
Benjamin Kummings, Reading, MA
Beletsky, V.V. (1995): Reguläre und chaotische Bewegung starrer Körper
Teubner, Stuttgart
Bellman, R. (1960): Introduction to Matrix Analysis
McGraw Hill, New York
Bellman, R. (1966): Perturbation Techniques in Mathematics, Physics,
and Engineering
Holt, Reinhart & Winston, New York
Beltrami, E. (1987): Mathematics for Dynamic Modeling
Academic Press, Boston/Orlando
Benedetto J.J. (1996): Harmonic Analysis and Applications
CRC Press, Bocaraton
Bergé, P., Pomeau, Y., Vidal, Ch. (1984): L'Ordre dans le Chaos
Hermann, Paris
Bethe, H., Sommerfeld, A. (1967): Elektronentheorie der Metalle
Springer, Berlin
Betz, A. (1964): Konforme Abbildung
Springer, Berlin
Bieberbach, L. (1986): Conformal Mapping
Chelsea, New York

Birkhoff, G., Rota, G.-C. (1989): Ordinary Differential Equations
4th ed., Wiley, New York
Bhatnagar, P.L. (1979): Nonlinear Waves in One-Dimensional Dispersive Systems
Clarendon Press, Oxford
Blakemore, J.S. (1974): Solid State Physics
2nd ed., Saunders, London
Blochinzew, D.I. (1966): Grundlagen der Quantenmechanik
5th ed., Harry Deutsch, Frankfurt / M & Zürich
Bloembergen, N. (1965): Nonlinear Optics
Benjamin, New York
Bogoljubow, N.N., Mitropolski (1965): Asymptotische Methoden
in der Theorie der nichtlinearen Schwingungen
Akademie Verlag, Berlin
Bohl, W. (1980): Technische Strömungslehre
4th ed., Vogel, Würzburg
Bracewell, R.N. (1986): The Fourier Transform and its Applications
2nd ed., McGraw Hill, New York
Brackbill, J.U., Cohen, B.L. eds. (1985): Multiple Time Scales
Academic Press, New York
Brillouin, L. (1946): Wave Propagation in Periodic Structures
McGraw Hill, New York
Brillouin, L. (1960): Wave Propagation and Group Velocity
Academic Press, New York
Bronson, R. (1993): Differential Equations, 2500 solved problems
in Schaum's Solved Problems Series
McGraw Hill, New York
Bronstein, I.N., Semendjajew, K.A., Musiol, G., Mühlig, H. (1993):
Taschenbuch der Mathematik
Harry Deutsch, Frankfurt / M & Zürich
Bullough, R.K., Caudrey, P.J., eds. (1980): Solitons
Springer, Berlin
Campell, G.A., Foster, R.M. (1948): Fourier Integrals for Practical Applications
van Nostrand, New York
Carslaw, H.S. (1930): Theory of Fourier's Series and Integrals
3rd ed., Dover, New York
Champeney, D.C. (1973): Fourier Transforms and Their Physical Applications
Academic Press, New York
Chow, S.N., Hale, J.R. (1982): Methods in Bifurcations Theory
Springer, New York
Chung, T.J. (1988): Continuum Mechanics
Prentice Hall, Englewood Cliffs
Churchill, P.V. (1948): Introduction to Complex Variables and Applications
McGraw Hill, New York
Coddington, E.A., Levinson, N. (1955): Theory of Ordinary Differential Equations
McGraw Hill, New York

Cohen-Tannoudji, C., Diu, B., Laloë, F. (1977): Quantum Mechanics I & II
Wiley, New York; Hermann, Paris
Collet, P., Eckmann, J.P. (1983): Iterated Maps on the Interval as Dynamical Systems
Birkhäuser, Basel
Convay, J.B. (1973): Functions of One Complex Variable
Springer, New York
Courant, R., Hilbert, D. (1968): Methoden der Mathematischen Physik I & II
2th/3rd ed., Springer, Berlin
Crawford, F.S. (1968): Waves, Berkeley Physics Course No. 3
McGraw Hill, New York
Critanovic, P. (1984): Universality in Chaos
Hilger, Bristol, UK
Darboux, G. (1889): Leçons sur la théorie générale des surfaces II
Gauthier-Villars, Paris
Davis, H.D. (1962): Introduction to Nonlinear Differential and Integral Equations
Dover, New York
De Santo, J.A. (1992): Scalar Wave Theory: Green's Functions and Applications
Springer, Berlin
Devaney, R.L. (1986): An Introduction to Chaotic Dynamical Systems
Benjamin/Cummings, Menlo Park
Dobrinski, P., Krakau, G., Vogel, A. (1993): Physik für Ingenieure
8th ed., Teubner, Stuttgart
Dodd, R.K., Eilbeck, J.C., Gibbon, J.D., Morris, H.C. (1982):
Solitons and Nonlinear Wave Equations
Academic Press, New York
Doetsch, G. (1970): Theorie und Anwendung der Laplace Transformation
Birkhäuser, Basel
Doetsch, G. (1971/73): Handbuch der Laplace Transformation
3 vol., Birkhäuser, Basel
Drazin, P.G. (1983): Solitons
Cambridge University Press, Cambridge, UK
Drazin, P.G. (1992): Nonlinear Systems
Cambridge University Press, Cambridge, UK
Drazin, P.G., Johnson, R.S. (1989): Solitons: an Introduction
Cambridge University Press, Cambridge, UK
Duffing, G. (1918): Erzwungene Schwingungen bei veränderlicher Eigenfrequenz
Vieweg, Braunschweig
Duley, W.W. (1976): CO_2 Lasers
Academic Press, New York
Erdelyi, A., Magnus, W., Oberhettinger, F., Triconi, F.G. (1952-1954):
Higher Transcendental Functions I-III
McGraw Hill, New York

Erdelyi, A, Magnus, W., Oberhettinger, F., Tricomi, F.G. (1954):
Tables of Integral Transforms I & II
McGraw Hill, New York
Farkas, M. (1994): Periodic Motions
Springer, New York
Feigenbaum, M.J. (1980): Universal Behavior of Nonlinear Systems
Los Alomos Science, Los Alomos
Fick, E. (1968): Einführung in die Grundlagen der Quantentheorie
Akademische Verlagsgesellschaft, Frankfurt / M
Flügge, S. (1990): Rechenmethoden der Quantenmechanik
4th ed., Springer, Berlin
Froyland, J. (1992): Introduction to Chaos and Coherence
Institute of Physics Publishing, Bristol, UK
Gallavotti, G., Zweifel, P.F. (1988): Nonlinear Evolution and Chaotic Phenomena
Plenum Press, New York
Gantmacher, F.R. (1959): Application of Matrices
Wiley, New York
Gaponov-Grekhov, A.V., Rabinovich, M.I. (1992): Nonlinearities in Action
Springer, Berlin
Gerard, A., Burch, J.M. (1975): Introduction to Matrix Methods in Optics
Wiley, London
Glass, L., Kaplan, D. (1995): Understanding Nonlinear Dynamics
Springer, Berlin
Gleick, J. (1990): Chaos
Droemer Knaur, München
Goldberg, H. (1978): Klassische Mechanik
Akademische Verlagsgesellschaft, Wiesbaden
Gradshteyn, I.S., Ryzhik, I.M. (1965): Table of Integrals, Series and Products
Academic Press, New York
Grimshaw, R. (1990): Nonlinear Ordinary Differential Equations
CRC Press, Bocaraton
Guckenheimer, J., Holmes, Ph. (1983): Nonlinear Oscillations Dynamical Systems
and Bifurcation of Vector Fields
Springer, New York
Hahn, W. (1959): Theorie und Anwendung der direkten Methode von Ljapunov
Springer, Berlin
Hairer, E., Norsett, S.P., Wanner, G. (1980):
Solving Ordinary Differential Equations I (Nonstiff Problems)
Springer, Berlin
Haken, H. (1983): Synergetics
3rd. ed., Springer, Berlin
Haken, H. (1984): Laser Theory
Springer, Berlin
Haken, H. (1985): Light, Vol. 2, Laser Light Dynamcis
North-Holland, Amsterdam

Hale, J.K. (1969): Ordinary Differential Equations
Wiley, New York
Halliday, D., Rosnick, R., Walker, J. (1993): Fundamentals of Physics
Wiley, New York
Hasegawa, A. (1989): Optical Solitons in Fibres
2nd ed., Springer, Berlin
Hazen, W.E., Pidd, R.W. (1965): Physics
Addision-Wesley, Reading, MA
Helmholtz, H. von (1896): Die Lehre von den Tonempfindungen
als physiologische Grundlage für die Theorie der Musik
5th ed., Vieweg, Braunschweig
Heuser, H. (1992): Gewöhnliche Differentialgleichungen
2nd ed., Teubner, Stuttgart
Holden, A. (1986): Chaos
Manchester University Press, Manchester
Holister, G.S. ed. (1979): Development in Stress Analysis
Applied Science Publishers, London
Huard, S. (1994): Polarisation de la Lumière
Masson, Paris
Hubbard, J.J., West, B.H. (1995): Differential Equations,
A Dynamical Systems Approach
Springer, Berlin
Hurwitz, A., Courant, R. (1929): Allg. Funktionentheorie
und elliptische Funktionen
Springer, Berlin
Ince, E.L. (1926): Ordinary Differential Equations
Longman-Greens, London
Infeld, E., Rowlands, G. (1990): Nonlinear Waves, Solitons and Chaos
Cambridge University Press, Cambridge, UK
Iooss, G., Joseph, D.D. (1980): Elementary Stability and Bifurcation Theory
Springer, New York
Ivanov, V.I., Trubetskov, M.K. (1995): Conformal Mapping
with Computer-Aided Visualization
CRC Press, Boca Raton
Jackson, J.D. (1975): Classical Electrodynamics
Wiley, New York
Jahnke, E., Emde, F. (1952): Tafeln höheren Funktionen
Teubner, Leipzig
Jakubovic, V.A. Starzinski, V.M. (1975): Linear Differential Equations
with Periodic Coefficients
Wiley, New York
Jeffrey, A. (1992): Complex Analysis and Applications
CRC Press, Boca Raton
Jeffrey, A. (1995): Handbook of Mathematical Formulas and Integrals
Academic Press, New York

John, F. (1982): Partial Differential Equations
4th ed., Springer, Berlin
Johnson, R.C. (1980): Designer Notes for Microwave Antennas
Artech House, Boston
Johnson, R.C., Jasik, H., ed. (1984): Antenna Engineering Handbook
2nd ed. McGraw Hill,, New York
Kachhava, C.M. (1990): Solid State Physics
Tata McGraw Hill, New York
Kamke, E. (1956): Differentialgleichungen, Lösungsmethoden und Lösungen:
I. gewöhnliche Differentialgleichungen
5th ed., Akademische Verlagsgesellschaft, Geest & Portig, Leipzig
Kaplan, W. (1962): Operational Methods for Linear Systems
Addison-Wesley, Reading, MA
Karpman, V.I. (1975): Nonlinear Waves in Dispersive Media
Pergamon Press, Oxford, UK
Kellogg, O.D. (1953): Foundations of Potential Theory
Dover, New York
Kittel, Ch. (1963): Quantum Theory of Solids
Wiley, New York
Kittel, Ch. (1971): Introduction to Solid State Physics
4th ed., Wiley, New York
Klages, G. (1956): Einführung in die Mikrowellenphysik
Steinkopff, Darmstadt
Kneubühl, F.K. (1993): Theories on Distributed Feedback Lasers
Harwood Acad. Publ., Chur
Kneubühl, F.K. (1994): Repetitorium der Physik
5th ed., Teubner, Stuttgart
Kneubühl, F.K. (1995): Lineare & nichtlineare Schwingungen und Wellen
Teubner, Stuttgart
Kneubühl, F.K., Sigrist, M.W. (1995): Laser
4th ed., Teubner, Stuttgart
Kober, H. (1957): Dictionary of Conformal Representations
2nd ed. Dover, London
Korsch, H.J., Jodl, H.J. (1994): Chaos
Springer, Berlin
Kravtsov, Yu.A., Orlov, Yu.I. (1993): Caustics, Catastrophes and Wave Fields
Springer, Berlin
Kreher, K. (1976): Festkörperphysik
Vieweg, Braunschweig
Kunik, A., Steeb, W.-H. (1986): Chaos in dynamischen Systemen
Bibliograph. Inst., Mannheim
Kuypers, F. (1982): Klassische Mechanik
Physik Verlag, Weinheim

Lamb, G.L. (1995): Introductory Applications of Partial Differential Equations
with Emphasis on Wave Propagation and Diffusion
Wiley, New York
Landau, L.D., Lifshitz, E.M. (1959): Fluid Mechanics
Pergamon, Oxford
Landau, L.D., Lifshitz, E.M. (1979): Quantenmechanik
= Lehrbuch der Theoretischen Physik, Bd III
Akademie Verlag, Berlin
LaSalle, J., Lefschetz, S. (1961): Stability by Ljapunov's Direct Method
with Applications
Academic Press, New York
LaSalle, J., Lefschetz, S. (1967): Die Stabilitätstheorie von Liapunow
Bibliographisches Institut, Mannheim
Levinson, N., Redheffer, R.M. (1970): Complex Variables
Holden Day, San Francisco
Lighthill, J. (1989): An Informal Introduction to Theoretical Fluid Mechanics
Clarendon Press, Oxford, UK
Lotka, A.J. (1925): Elements of Physical Biology
Williams & Wilkins, Baltimore
Lüst, R. (1978): Hydrodynamik
Bibliographisches Institut, Mannheim
Lyapunov, A.M. (1982 / 1992): The General Problem of the Stability of Motion
Taylor & Francis, Basingstoke
Madelung, E. (1943): Die mathematischen Hilfsmittel des Physikers
Dover, New York
Magnus, K., Popp, K. (1997): Schwingungen
5th ed., Teubner, Stuttgart
Magnus, W., Winkler, S. (1966): Hill's Equation
Interscience, New York
Mahnke, R. (1994): Nichtlineare Physik in Aufgaben
Teubner, Stuttgart
Mahnke, R., Schmelzer, J., Röpke, G. (1992): Nichtlineare Phänomene
und Selbstorganisation
Teubner, Stuttgart
Main, I.G. (1978): Vibrations and Waves in Physics
Cambridge University Press, Cambridge, UK
Mandelbrot, B.B. (1983): The Fractal Geometry of Nature
Freeman, New York
Marcuvitz, N. (1948): Waveguide Handbook, MIT Rad. Lab. Series 10
McGraw Hill, New York
Margenau, H., Murphy, G.M. (1956): The Mathematics of Physics and Chemistry
Van Nostrand, Princeton
McGervey, J.D. (1971): Introduction to Modern Physics
Academic Press, New York

McLachlan, N.W. (1950): Ordinary Non-Linear Differential Equations
Clarendon Press, Oxford, UK
Mehmeti, F.A. (1994): Nonlinear Waves in Networks
Akademie Verlag, Berlin
Messiah, A. (1960): Mecanique Quantique I et II
Dunod, Paris
Messiah, A. (1969): Quantum Mechanics I & II
North-Holland, Amsterdam
Messiah, A. (1990): Quantenmechanik I & II
Gruyter, Berlin
Meystre, P., Sargent III, M. (1990): Elements of Quantum Optics
Springer, Berlin
Mills, D.L. (1991): Nonlinear Optics
Springer, Berlin
Milne-Thomson, L.M. (1931): Die elliptischen Funktionen von Jacobi
Springer, Berlin
Minorsky, N. (1974) Nonlinear Oscillations
van Nostrand, Princeton
Mira, Ch. (1987): Chaotic Dynamics
World Scientific, Singapore
Möller, K.D. (1988): Optics
University Science Books, Mill Valley
Montgomery, C.G., Dicke, R.H., Purcell, E.M. (1948):
Principles of Microwave Circuits, MIT Rad. Lab. Series 8
McGraw Hill, New York
Moon, F.C. (1987): Chaotic Vibrations
Wiley, New York
Morse, P.M. (1936): Vibration and Sound
McGraw Hill, New York
Morse, P.M., Feshbach, H. (1953): Methods of Theoretical Physics
McGraw Hill, New York
Morse, P.M., Ingard, K.U. (1986): Theoretical Acoustics
Princeton University Press, Princeton
Moser, J. (1973): Stable and Random Motions in Dynamical Systems
Princeton University Press, Princeton
Myskis, D. (1955): Lineare Differential Gleichungen mit verzögertem Argument
Deutscher Verlag, Berlin
Naslin, P. (1968): Dynamik linearer und nichtlinearer Systeme
Oldenbourg Verlag, München & Wien
Nayfeh, A.H., Mook, D.T. (1979): Nonlinear Oscillations
Wiley, New York
Nehari, Z. (1952): Conformal Mapping
McGraw Hill, New York

Nemytskii, V.V., Stepanov, V.V. (1960): Qualitative Theory of Differential Equations
Princeton University Press, Princeton
Nettel, S. (1995): Wave Physics
2nd ed., Springer, Berlin
Ott, E. (1993): Chaos in Dynamical Systems
Cambridge University Press, Cambridge, UK
Oughstun, K.E., Sherman, G.C. (1994): Electromagnetic Pulse Propagation in Causal Dielectrics
Springer, Berlin
Pain, H.J. (1993): The Physics of Vibrations and Waves
4th ed., Wiley, Chichester, UK
Pauli, W. (1950): Die allgemeinen Prinzipien der Wellenmechanik
Handbuch der Physik, Band 24, 1. Teil
Edwards, Ann Arbor
Penney, E. (1959): Ordinary Difference - Differential Equations
University of California Press, Berkeley
Percival, I., Richards, D. (1982): Introduction to Dynamics
Cambridge University Press, Cambridge, UK
Pippard, A.B. (1978/1983): The Physics of Vibration
Cambridge University Press, Cambridge, UK
Plaschko, P., Brod, K. (1995): Nichtlineare Dynamik, Bifurkation und chaotische Systeme
Vieweg, Braunschweig
Poincaré, H. (1892a): Les Méthodes Nouvelles de la Méchanique Céleste
Gauthier-Villars, Paris
Poincaré, H. (1892b): Théorie Mathématique de la Lumière, Vol. 2
Gauthier-Villars, Paris
Polyanin, A.D., Zaitsev, V.F. (1995): Exact Solutions of Ordinary Differential Equations
CRC Press, Bocaraton
Pontrjagin, L.S. (1985): Gewöhnliche Differentialgleichungen
Deutscher Verlag der Wissenschaften, Berlin
Pöschl, K. (1956): Mathematische Methoden in der Hochfrequenztechnik
Springer, Berlin
Poston, T., Stewart, I. (1978): Catastrophe Theory and its Applications
Pitman, London
Poularikas, A.D. (1995): The Transforms and Applications Handbook
CRC Press, Bocaraton
Prandtl, L., Tietjens, O.G. (1957a): Fundamentals of Hydro- and Aeromechanics
Dover, New York
Prandtl, L., Tietjens, O.G. (1957b): Applied Hydro- and Aeromechanics
Dover, New York
Rajaraman, R. (1982): Solitons and Instantons
North Holland, New York

Ramo, S., Whinnery, J.R., van Duzen, Th. (1965): Fields and Waves in Communication Electronics
Wiley, New York
Rasband, S.N. (1990): Chaotic Dynamics of Nonlinear Systems
Wiley, New York
Rauch, J. (1991): Partial Differential Equations
Springer, Berlin
Reitman, V. (1996): Reguläre und chaotische Dynamik
Teubner, Stuttgart
Römer, H. (1994): Theoretische Optik
VCH, Weinheim
Rosencwaig, A. (1980): Photoacoustics and Photoacoustic Spectroscopy
Wiley, New York
Royal Society Mathematical Tables (1960): Bessel Functions, Vol. 7, Part. III
Cambridge University Press, Cambridge, UK
Ruelle, D. (1989a): Elements of Differentiable Dynamics and Bifurcation Theory
Academic Press, Boston
Ruelle, D. (1989b): Chaotic Evolution and Strange Attractors
Cambridge University Press, Cambridge, UK
Saathy, T.L. (1981): Modern Nonlinear Equations
2nd ed., Dover, New York
Saaty, T.L., Bram, J. (1964): Nonlinear Mathematics
McGraw Hill, New York
Schechter, M. (1977): Modern Methods in Partial Differential Equations
Mc Graw Hill, New York
Schetzen, M. (1989): Volterra-Wiener Theories of Nonlinear Systems
Wiley, New York
Schroeder, M.R. (1991): Fractals, Chaos, Power Laws
Freeman, New York
Schubert, M.; Weber, G. (1980): Quantentheorie
Deutscher Verlag der Wissenschaften, Berlin
Schubert, M.; Weber, G. (1993): Quantentheorie
Spektrum Akademischer Verlag, Heidelberg
Schuster, H.G. (1984): Deterministic Chaos
Physik-Verlag, Weinheim
Shewandrow, N.D. (1973): Polarisation des Lichts
Akademie Verlag, Berlin
Shurcliff, W.A. (1962): Polarized Light
Harvard University Press, Harvard
Slotine, J.-J.- E.; Li, W. (1991): Applied Nonlinear Control
Prentice-Hall, London
Smirnow, W.I. (1954): Lehrgang der höheren Mathematik
Deutscher Verlag der Wissenschaften, Berlin
Smoller, J. (1994): Shock, Waves and Reaction Diffusion Equations
2nd ed., Springer, Berlin

Sommerfeld, A. (1947): Partielle Differentialgleichungen der Physik
Dieterich, Wiesbaden
Sommerfeld, A. (1949): Elektrodynamik
Geest & Portig, Leipzig
Sommerfeld, A. (1959): Electrodynamics
Academic Press, New York
Sparrow, C.T. (1982): The Lorenz Equations
Springer, Berlin
Spurk, J.H. (1997): Fluid Mechanics
Springer, Berlin
Squire, Ch.F. (1971): Waves in Physical Systems
Prentice Hall, Englewood Cliffs
Steele, D. (1971): Theory of Vibrational Spectroscopy
Sanders, Philadelphia
Stephan, W., Postl, R. (1995): Schwingungen elastischer Kontinua
Teubner, Stuttgart
Stoker, J.J. (1957): Nonlinear Vibrations
Interscience, New York
Stratton, J.A. (1941): Electromagnetic Theory
McGraw Hill, New York
Strutt, M.J.O. (1932): Lamé'sche Matthieusche und verwandte Funktionen
Springer, Berlin
Svelto, O., Hanna, D.C. (1989): Principles of Lasers
3rd ed., Plenum Press, New York
Swindel, W. ed. (1995): Polarized Light
Dowden, Hutchison & Ross, Stroudsburg
Theocaris, P.S., Gdoutos, E.E. (1979): Matrix Theory of Photoelasticity
Springer, Berlin
Thom, R. (1972): Stabilité Structurelle et Morphogénèse
Benjamin, New York
Thom, R. (1975): Structural Stability and Morphogenesis
Benjamin, New York
Titchmarsh, E.C. (1932): The Theory of Functions
1st ed., Oxford University Press, Oxford, UK
Toda, M. (1981): Theory of Nonlinear Lattices
Springer, Berlin
Tu, P.N.V. (1992): Dynamical Systems
Springer, Berlin
van der Ziel, A. (1976): Noise in Measurements
Wiley, New York
Verhulst, F. (1990): Nonlinear Differential Equations and Dynamical Systems
Springer, Berlin
Volterra, V. (1931): Leçons sur la théorie mathematique de la lutte pour la vie
Gauthier-Villars, Paris

Volterra, V. (1959): Theory of Functionals and of Integral
and Integro-Differential Equations
Dover, New York
Wang, Sh. (1966): Solid-State Electronics
McGraw Hill, New York
Webster, A.G. (1927): Partial Differential Equations of Mathematical Physics
Teubner, Leipzig
Webster, A.G., Szegö, G. (1930): Partielle Differentialgleichungen
der mathematischen Physik
Teubner, Leipzig
Whitham, G.B. (1974): Linear and Nonlinear Waves
Wiley, New York
Whittaker, E.T., Watson, G.N. (1946): A Course of Modern Analysis
4th ed., Cambridge University Press, Cambridge, UK
Wiggins, S. (1990): Introduction to Applied Nonlinear Dynamical Systems
and Chaos
Springer, Berlin
Wilson, E.B. jr., Decius, J.C., Cross, P.C. (1955): Molecular Vibrations
McGraw Hill, New York
Wygodski, M.J. (1973): Höhere Mathematik, griffsbereit
Vieweg, Braunschweig
Wyld, H.W. (1976): Mathematical Methods of Physics
Benjamin, Reading, MA
Zauderer, E. (1983): Partial Differential Equations of Applied Mathematics
Wiley, New York
Zayed, A.I. (1996): Handbook of Function and Generalized Function Transforms
CRC Press, Bocaraton
Ziman, J.M. (1964): Principles of the Theory of Solids
Cambridge University Press, Cambridge, UK
Zwillinger, D. (1989): Handbook of Differential Equations
Academic Press, Boston

J. Publications in Journals

Acheson, D.J. (1993),
Proc. Roy. Soc. Lond. **A 443**, 239
Anderson, G.M., Geer, J.F. (1982),
SIAM, J. Appl. Math. **42**, 678
Arnold, V.I. (1963),
Russ. Math. Surveys, **18**, 5
Bäcklund, A.V. (1875),
Lund Universitets Arsskrift, **10**, 1
Barr, T.H. (1995),
J. Math. Anal. Appl. **195**, 261

Bioche, Ch. (1910),
Bulletin Soc. Math. France **38**, 160
Borgnis, F.E., Papas, Ch.H. (1958),
Encyclopedia of Physics, XVI, 423
Burgers, J.M. (1948),
Adv. Appl. Mech. **1**, 171
Bussinesq, J. (1871),
Comptes Rendus Acad. Sci. Paris **72**, 755
Cole, J.D. (1951),
Appl. Math. **9**, 225
Coullet, P., Tresser, J. (1978),
C.R. Hebd. Seances Acad. Sci. Series **A287**, 577;
J. Phys. (Paris) **C5**, 25
Delamotte, B. (1993),
Phys. Rev. Lett. **70**, 3361
Dicke, R.H. (1957),
Phys. Rev. **89**, 472
Duhamel (1833),
Journal de l'Ecole Polyt. **14**, 20
Feigenbaum, M.J. (1978/79),
J. Stat. Phys. **19**, 25; **21**, 669
Feynman, R.P., Vernon, F.L., Hellwarth, R.W. (1957),
J. Appl. Phys. **28**, 49
Froyland, J., Alfsen, K.H. (1984),
Phys. Rev. **A29**, 2928
Gnepf, S., Kneubühl, F.K. (1986),
in Infrared & Millimeter Waves **16**, Ch. 2, 35, Academic Press, New York
Goel, N.S., Maitra, S.C., Montroll, E.W. (1971),
Rev. Modern Phys. **43**, 231
Grossmann, S., Thomae, S. (1977),
Z. Naturforschung **32a**, 1353
Haken, H. (1975),
Phys. Lett. **53A**, 77
Harrison, R.G. und Biswas, D.J. (1985),
Progress in Quantum Electronics, **10**, 147
Helmholtz, H. von (1882),
Wissenschaftliche Abhandlungen (Barth, Leipzig) **1**, 303
Hess, P. (1983),
Topics in Current Chemistry **111**, 1
Hirota, R. (1971),
Phys. Rev. Lett. **27**, 1192
Hopf, E. (1942),
Math. Naturwiss. Klasse, Sächs. Akademie der Wiss., Leipzig, **94**, 1
Hopf, E. (1950),
Pure Appl. Math. **3**, 201

Ippen, E.P. (1994),
Appl. Phys. **B58**, 159

Jones, R.C. (1941),
J. Opt. Soc. Am. **31**, 488

Karbach, A., Hess, P. (1985),
J. Appl. Phys. **58**, 3851

Keller, U., t'Hooft, G.W., Knox, W.H., Cunningham, J.E. (1991),
Optics Lett. **16**, 1022

Kneubühl, F.K. (1989)
Infrared Physics **29**, 925

Kneubühl, F.K., Feng, J. (1991),
Chinese J. of Infrared and Millimeter Waves **10**, 386

Kogelnik, H., Shank, C.V. (1971),
Appl. Phys. Lett. **18**, 152

Kogelnik, H., Shank, C.V. (1972),
J. Appl. Phys. **43**, 2327

Kolmogorov, A.N. (1954),
Dokl. Akad. Nauk, USSR **98**, 525

Korteweg, D. J., de Vries, G. (1895),
Philos. Mag. Ser. 5, **39**, 422

Kramers, H.A. (1926),
Zeitschrift für Physik **39**, 828

Kramers, H.A., Kronig, R. de L. (1919),
Zeitschrift für Physik **30**, 521

Kronig, R. de L. (1926),
J. Opt. Soc. Am. **12**, 547

Kronig, R. de L., Penney, W.G. (1931),
Proc. Roy. Soc. London, Ser. A **130**, 499

Kurz, T., Lauterborn, W. (1988),
Phys. Rev. **A37**, 1029

Lamb, G.L. (1971),
Rev. Mod. Physics **43**, 99

Landau, L.D. (1944),
C.R. Dokl. Akad. Sci. USSR **44**, 311

Lauterborn, W., Meyer-Ilse, W. (1986),
Physik in unserer Zeit **17**, 177

Liénard, A. (1928),
Revue Générale de l'Electricité **22**, 1051; **23**, 901 & 946

Lindner, A., Freese, H. (1994)
J. Phys. A: Math. Gen. **27**, 5565

Lindstedt, A. (1883),
Mém. Acad. Sci. St. Petersbourg Vol. XXXI

Lorenz, E.N. (1963),
J. Atmos. Sci. **20**, 130

Lotka, A. (1920),
Proc. Natl. Acad. Sci. (Washington) **6**, 410
Lugiato, L.A., Narducci, L.M., eds. (1985),
Instabilities in Active Optical Media, J. Opt. Soc. **B2**, Heft 1
Magyari, E. (1990),
El. Math. **45**, 75
Maiman, T.H. (1960),
Nature **187**, 493
Mannewille, P., Pomeau, Y. (1979),
Phys. Lett. **75A**, 1
McCall, S.L., Hahn, E.L. (1967),
Phys. Rev. Lett. **18**, 908
McCall, S.L., Hahn, E.L. (1969),
Phys. Rev. **183**, 457
Meiman, N.N. (1977),
J. Math. Phys., **18**, 834
Meissner, E. (1918),
Schweizerische Bauzeitung
14. Sept., **72**, 95
Minden, M.L., Casperson, L.W. (1985),
J. Opt. Soc. Am. **B2**, 120
Morse, P.M. (1930),
Phys. Rev. **35**, 1310
Moser, J. (1967),
Math. Ann. **169**, 163
Newhouse, S., Ruelle, D., Takens, F. (1978),
Comm. Math. Phys. **64**, 35
Parlitz, U., Lauterborn, W. (1985),
Phys. Lett. **A107**, 351
Poincaré, H. (1885),
Acta Mathematica (16.9.1885), **7**, 259
Preiswerk, H.P., Lubanski, M., Kneubühl, F.K. (1984),
Appl. Phys. **B33**, 115
Rayleigh, Lord (1870),
"Scientific Papers of Lord Rayleigh", Reprint Vol. 1, Dover, New York
Rayleigh, Lord (1876),
Phil. Mag. (5), **1**, 257
Ruelle, D., Takens, F. (1971),
Comm. Math. Phys. **20**, 167, **23**, 343
Russell, J.S. (1844),
Report 14th Meeting of the British Assoc. for the Advancement of Science, London, John Murray
Salin, F., Squier, J., Piché, M. (1991),
Optics Lett. **16**, 1674

Scherrer, D.P. (1996),
private communication
Schötzau, H.J., Kneubühl, F.K. (1994),
European Trans. on Electrical Power Engineering **4**, 89
Shohat, J. (1994),
J. Appl. Phys. **15**, 568
Smith, R.A. (1961),
J. London Math. Soc. **36**, 33
Stokes, G.G. (1852),
Trans. Cambridge Phil. Soc. **9**, 399
Strutt, M.J.O. (1949),
Proc. Roy. Soc. Edinburgh **62**, 278
Taylor, G.I. (1910),
Proc. Roy. Soc. **A84**, 371
Toda, M. (1967),
J. Phys. Soc. Jpn, **22**, 431; **23**, 501
Ueda, Y. (1979),
J.Stat. Phys. **20**, 181
Ueda, Y. (1980),
New Approaches to Nonlinear Problems in Dynamics, 311, SIAM, Philadelphia
van der Pol, B. (1926)
Phil. Magazine **2**, 978
van der Pol, B. (1927),
Phil. Magazine **3**, 65
van der Pol, B. (1934),
Proc. Inst. of Radio Engineers, **22**, 1051
Verhulst, P.F. (1838),
Correspondance mathématique et physique **10**, 113
Volterra, V. (1937):
Acta Biotheoretica **3**, 1
Weiss, C.O., Godone, A., Olafson, A. (1983),
Phys. Rev. **28**, 892
Weiss, C.O., Klische, W., Ering, P.S., Cooper, M. (1985),
Opt. Comm. **52**, 405
Wentzel, G. (1926),
Zeitschrift für Physik **38**, 518
Whitney, C. (1971),
J. Opt. Soc. Am. **61**, 1207
Zabusky, N.J., Kruskal M.D. (1965),
Phys. Rev. Lett. **15**, 240
Zakharov, V.E., Shabat, A.B. (1972),
Trans. Soviet Phys., JETP **34**, 62

Index

Printing: Mercedesdruck, Berlin
Binding: Buchbinderei Lüderitz & Bauer, Berlin